E. Hairer
S. P. Nørsett
G. Wanner

Solving Ordinary Differential Equations I

Nonstiff Problems

Second Revised Edition
With 135 Figures

 Springer

Ernst Hairer
Gerhard Wanner

Université de Genève
Section de Mathématiques
2–4 rue du Lièvre
1211 Genève 4
Switzerland
Ernst.Hairer@math.unige.ch
Gerhard.Wanner@math.unige.ch

Syvert P. Nørsett
Norwegian University of Science
and Technology (NTNU)
Department of Mathematical Sciences
7491 Trondheim
Norway
norsett@math.ntnu.no

Corrected 3rd printing 2008

ISBN 978-3-642-08158-3   e-ISBN 978-3-540-78862-1

DOI 10.1007/978-3-540-78862-1

Springer Series in Computational Mathematics ISSN 0179-3632

Mathematics Subject Classification (2000): 65Lxx, 34A50

Cover design: WMX Design GmbH, Heidelberg

Printed on acid-free paper

9 8 7 6 5 4 3 2 1

springer.com

This edition is dedicated to

Professor John Butcher

on the occasion of his 60th birthday

His unforgettable lectures on Runge-Kutta methods, given in June 1970 at the University of Innsbruck, introduced us to this subject which, since then, we have never ceased to love and to develop with all our humble abilities.

From the Preface to the First Edition

> So far as I remember, I have never seen an Author's Preface which had any purpose but one — to furnish reasons for the publication of the Book. (Mark Twain)

> Gauss' dictum, "when a building is completed no one should be able to see any trace of the scaffolding," is often used by mathematicians as an excuse for neglecting the motivation behind their own work and the history of their field. Fortunately, the opposite sentiment is gaining strength, and numerous asides in this Essay show to which side go my sympathies. (B.B. Mandelbrot 1982)

> This gives us a good occasion to work out most of the book until the next year. (the Authors in a letter, dated Oct. 29, 1980, to Springer-Verlag)

There are two volumes, one on non-stiff equations, ..., the second on stiff equations, ... The first volume has three chapters, one on classical mathematical theory, one on Runge-Kutta and extrapolation methods, and one on multistep methods. There is an Appendix containing some Fortran codes which we have written for our numerical examples.

Each chapter is divided into sections. Numbers of formulas, theorems, tables and figures are consecutive in each section and indicate, in addition, the section number, but not the chapter number. Cross references to other chapters are rare and are stated explicitly. ... References to the Bibliography are by "Author" plus "year" in parentheses. The Bibliography makes no attempt at being complete; we have listed mainly the papers which are discussed in the text.

Finally, we want to thank all those who have helped and encouraged us to prepare this book. The marvellous "Minisymposium" which G. Dahlquist organized in Stockholm in 1979 gave us the first impulse for writing this book. J. Steinig and Chr. Lubich have read the whole manuscript very carefully and have made extremely valuable mathematical and linguistical suggestions. We also thank J.P. Eckmann for his troff software with the help of which the whole manuscript has been printed. For preliminary versions we had used textprocessing programs written by R. Menk. Thanks also to the staff of the Geneva computing center for their help. All computer plots have been done on their beautiful HP plotter. Last but not least, we would like to acknowledge the agreable collaboration with the planning and production group of Springer-Verlag.

October 29, 1986   The Authors

Preface to the Second Edition

The preparation of the second edition has presented a welcome opportunity to improve the first edition by rewriting many sections and by eliminating errors and misprints. In particular we have included new material on

– Hamiltonian systems (I.14) and symplectic Runge-Kutta methods (II.16);

– dense output for Runge-Kutta (II.6) and extrapolation methods (II.9);

– a new Dormand & Prince method of order 8 with dense output (II.5);

– parallel Runge-Kutta methods (II.11);

– numerical tests for first- and second order systems (II.10 and III.7).

Our sincere thanks go to many persons who have helped us with our work:

– all readers who kindly drew our attention to several errors and misprints in the first edition;

– those who read preliminary versions of the new parts of this edition for their invaluable suggestions: D.J. Higham, L. Jay, P. Kaps, Chr. Lubich, B. Moesli, A. Ostermann, D. Pfenniger, P.J. Prince, and J.M. Sanz-Serna.

– our colleague J. Steinig, who read the entire manuscript, for his numerous mathematical suggestions and corrections of English (and Latin!) grammar;

– our colleague J.P. Eckmann for his great skill in manipulating Apollo workstations, font tables, and the like;

– the staff of the Geneva computing center and of the mathematics library for their constant help;

– the planning and production group of Springer-Verlag for numerous suggestions on presentation and style.

This second edition now also benefits, as did Volume II, from the marvels of TEXnology. All figures have been recomputed and printed, together with the text, in Postscript. Nearly all computations and text processings were done on the Apollo DN4000 workstation of the Mathematics Department of the University of Geneva; for some long-time and high-precision runs we used a VAX 8700 computer and a Sun IPX workstation.

November 29, 1992   The Authors

Contents

Chapter I. Classical Mathematical Theory

Chapter II. Runge-Kutta and Extrapolation Methods

Chapter III. Multistep Methods and General Linear Methods

Chapter I. Classical Mathematical Theory

> ... halte ich es immer für besser, nicht mit dem Anfang anzufangen, der immer das Schwerste ist.
>
> (B. Riemann copied this from F. Schiller into his notebook)

This first chapter contains the classical theory of differential equations, which we judge useful and important for a profound understanding of numerical processes and phenomena. It will also be the occasion of presenting interesting examples of differential equations and their properties.

We first retrace in Sections I.2-I.6 the historical development of classical integration methods by series expansions, quadrature and elementary functions, from the beginning (Newton and Leibniz) to the era of Euler, Lagrange and Hamilton. The next part (Sections I.7-I.14) deals with theoretical properties of the solutions (existence, uniqueness, stability and differentiability with respect to initial values and parameters) and the corresponding flow (increase of volume, preservation of symplectic structure). This theory was initiated by Cauchy in 1824 and then brought to perfection mainly during the next 100 years. We close with a brief account of boundary value problems, periodic solutions, limit cycles and strange attractors (Sections I.15 and I.16).

I.1 Terminology

A *differential equation of first order* is an equation of the form

$$y' = f(x, y) \tag{1.1}$$

with a given function $f(x, y)$. A function $y(x)$ is called a *solution* of this equation if for all x,

$$y'(x) = f(x, y(x)). \tag{1.2}$$

It was observed very early by Newton, Leibniz and Euler that the solution usually contains a free parameter, so that it is uniquely determined only when an *initial value*

$$y(x_0) = y_0 \tag{1.3}$$

is prescribed. Cauchy's existence and uniqueness proof of this fact will be discussed in Section I.7. Differential equations arise in many applications. We shall see the first examples of such equations in Section I.2, and in Section I.3 how some of them can be solved explicitly.

A *differential equation of second order* for y is of the form

$$y'' = f(x, y, y'). \tag{1.4}$$

Here, the solution usually contains *two* parameters and is only uniquely determined by *two* initial values

$$y(x_0) = y_0, \qquad y'(x_0) = y_0'. \tag{1.5}$$

Equations of second order can rarely be solved explicitly (see I.3). For their numerical solution, as well as for theoretical investigations, one usually sets $y_1(x) := y(x)$, $y_2(x) := y'(x)$, so that equation (1.4) becomes

$$
\begin{aligned}
y_1' &= y_2 & y_1(x_0) &= y_0 \\
y_2' &= f(x, y_1, y_2) & y_2(x_0) &= y_0'.
\end{aligned} \tag{1.4'}
$$

This is an example of a *first order system of differential equations*, of dimension n (see Sections I.6 and I.9),

$$
\begin{aligned}
y_1' &= f_1(x, y_1, \ldots, y_n) & y_1(x_0) &= y_{10} \\
&\cdots & &\cdots \\
y_n' &= f_n(x, y_1, \ldots, y_n) & y_n(x_0) &= y_{n0}.
\end{aligned} \tag{1.6}
$$

Most of the theory of this book is devoted to the solution of the initial value problem for the system (1.6). At the end of the 19th century (Peano 1890) it became customary to introduce the vector notation

$$y = (y_1, \ldots, y_n)^T, \qquad f = (f_1, \ldots, f_n)^T$$

so that (1.6) becomes $y' = f(x, y)$, which is again the same as (1.1), but now with y and f interpreted as vectors.

Another possibility for the second order equation (1.4), instead of transforming it into a system (1.4'), is to develop *methods specially adapted to second order equations (Nyström methods)*. This will be done in special sections of this book (Sections II.13 and III.10). Nothing prevents us, of course, from considering (1.4) as a second order system of dimension n.

If, however, the initial conditions (1.5) are replaced by something like $y(x_0) = a$, $y(x_1) = b$, i.e., if the conditions determining the particular solution are not all specified at the same point x_0, we speak of a *boundary value problem*. The theory of the existence of a solution and of its numerical computation is here much more complicated. We give some examples in Section I.15.

Finally, a problem of the type

$$\frac{\partial u}{\partial t} = f\left(t, u, \frac{\partial u}{\partial x}, \frac{\partial^2 u}{\partial x^2}\right) \tag{1.7}$$

for an unknown function $u(t, x)$ of *two independent variables* will be called a *partial differential equation*. We can also deal with partial differential equations of higher order, with problems in three or four independent variables, or with systems of partial differential equations. Very often, initial value problems for partial differential equations can conveniently be transformed into a system of ordinary differential equations, for example with finite difference or finite element approximations in the variable x. In this way, the equation

$$\frac{\partial u}{\partial t} = a^2 \frac{\partial^2 u}{\partial x^2}$$

would become

$$\frac{du_i}{dt} = \frac{a^2}{\Delta x^2}\left(u_{i+1} - 2u_i + u_{i-1}\right),$$

where $u_i(t) \approx u(t, x_i)$. This procedure is called the "method of lines" or "method of discretization in space" (Berezin & Zhidkov 1965). We shall see in Section I.6 that this connection, the other way round, was historically the origin of partial differential equations (d'Alembert, Lagrange, Fourier). A similar idea is the "method of discretization in time" (Rothe 1930).

I.2 The Oldest Differential Equations

> ... So zum Beispiel die Aufgabe der umgekehrten Tangentenme-
> thode, von welcher auch Descartes eingestand, dass er sie nicht in
> seiner Gewalt habe. (Leibniz, 27. Aug 1676)
>
> ... et on sait que les seconds Inventeurs n'ont pas de droit à l'In-
> vention. (Newton, 29 mai 1716)
>
> Il ne paroist point que M. Newton ait eu avant moy la characteris-
> tique & l'algorithme infinitesimal ... (Leibniz)
>
> And by these words he acknowledged that he had not yet found the
> reduction of problems to differential equations. (Newton)

Newton

Differential equations are as old as differential calculus. Newton considered them
in his treatise on differential calculus (Newton 1671) and discussed their solution
by series expansion. One of the first examples of a first order equation treated by
Newton (see Newton (1671), Problema II, Solutio Casus II, Ex. I) was

$$y' = 1 - 3x + y + x^2 + xy. \tag{2.1}$$

For each value x and y, such an equation prescribes the derivative y' of the solutions. We thus obtain a *vector field,* which, for this particular equation, is sketched
in Fig. 2.1a. (So, contrary to the belief of many people, vector fields existed long
before Van Gogh). The solutions are the curves which respect these prescribed
directions everywhere (Fig. 2.1b).

Newton discusses the solution of this equation by means of infinite series,
whose terms he obtains recursively ("... & ils se jettent sur les series, où M. Newton m'a precedé sans difficulté; mais ...", Leibniz). The first term

$$y = 0 + \ldots$$

is the initial value for $x = 0$. Inserting this into the differential equation (2.1) he
obtains

$$y' = 1 + \ldots$$

which, integrated, gives

$$y = x + \ldots$$

Again, from (2.1), we now have

$$y' = 1 - 3x + x + \ldots = 1 - 2x + \ldots$$

and by integration

$$y = x - x^2 + \ldots.$$

E. x ɛ m ғ ʟ. I.

Sit Æquatio $\dfrac{\dot{y}}{x} = $ 1 — 3x + y + xx + xy ; cujus Terminos:

1 — 3x + xx non affectos *Relatâ* Quantitate difpofitos vides in la--
teralem Seriem primo loco , & reliquos y & xy in finiftrâ Columnâ..

	+ 1 — 3x + xx
+ y	$* + x - xx + \dfrac{1}{3}x^3 - \dfrac{1}{6}x^4 + \dfrac{1}{30}x^5$; &c.
+ xy	$* \quad x + xx - x^3 + \dfrac{1}{3}x^4 - \dfrac{1}{6}x^5 + \dfrac{1}{30}x^6$; &c.
Aggreg.	$+ 1 - 2x + xx - \dfrac{2}{3}x^3 + \dfrac{1}{6}x^4 - \dfrac{4}{30}x^5$; &c.
y =	$+ x - xx + \dfrac{1}{3}x^3 - \dfrac{1}{6}x^4 + \dfrac{1}{30}x^5 - \dfrac{1}{45}x^6$; &c.

Nunc:

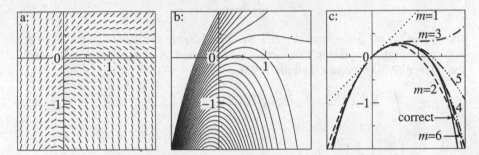

Fig. 2.1. a) vector field, b) various solution curves of equation (2.1),
c) Correct solution vs. approximate solution

The next round gives

$$y' = 1 - 2x + x^2 + \dots, \qquad y = x - x^2 + \frac{x^3}{3} + \dots.$$

Continuing this process, he finally arrives at

$$y = x - xx + \frac{1}{3}x^3 - \frac{1}{6}x^4 + \frac{1}{30}x^5 - \frac{1}{45}x^6; \&c. \tag{2.2}$$

These approximations, term after term, are plotted in Fig. 2.1c together with the
correct solution. It can be seen that these approximations are closer and closer
to the true solution for small values of x. For more examples see Exercises 1-3.
Convergence will be discussed in Section I.8.

Leibniz and the Bernoulli Brothers

A second access to differential equations is the consideration of geometrical problems such as *inverse tangent problems* (Debeaune 1638 in a letter to Descartes). A particular example describes the path of a silver pocket watch ("horologio portabili suae thecae argentae") and was proposed around 1674 by "Claudius Perraltus Medicus Parisinus" to Leibniz: a curve $y(x)$ is required whose tangent AB is given, say everywhere of constant length a (Fig. 2.2). This leads to

$$y' = -\frac{y}{\sqrt{a^2 - y^2}}, \tag{2.3}$$

a first order differential equation. Despite the efforts of the "plus célèbres mathématiciens de Paris et de Toulouse" (from a letter of Descartes 1645, "Toulouse" means "Fermat") the solution of these problems had to wait until Leibniz (1684) and above all until the famous paper of Jacob Bernoulli (1690). Bernoulli's idea applied to equation (2.3) is as follows: let the curve BM in Fig. 2.3 be such that LM is equal to $\sqrt{a^2 - y^2}/y$. Then (2.3), written as

$$dx = -\frac{\sqrt{a^2 - y^2}}{y}\, dy, \tag{2.3'}$$

shows that for *all* y the areas S_1 and S_2 (Fig. 2.3) are the same. Thus ("Ergo & horum integralia aequantur") the areas $BMLB$ and $A_1 A_2 C_2 C_1$ must be equal too. Hence (2.3') becomes (Leibniz 1693)

$$x = \int_y^a \frac{\sqrt{a^2 - y^2}}{y}\, dy = -\sqrt{a^2 - y^2} - a \cdot \log \frac{a - \sqrt{a^2 - y^2}}{y}. \tag{2.3''}$$

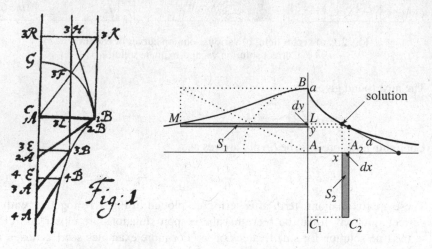

Fig. 2.2. Illustration from
Leibniz (1693)

Fig. 2.3. Jac. Bernoulli's
Solution of (2.3)

Variational Calculus

In 1696 Johann Bernoulli invited the brightest mathematicians of the world ("Profundioris in primis Mathesos cultori, Salutem!") to solve the *brachystochrone* (shortest time) problem, mainly in order to fault his brother Jacob, from whom he expected a wrong solution. The problem is to find a curve $y(x)$ connecting two points P_0, P_1, such that a point gliding on this curve under gravitation reaches P_1 in the shortest time possible. In order to solve his problem, Joh. Bernoulli (1697b) imagined thin layers of homogeneous media and knew from optics (Fermat's principle) that a light ray with speed v obeying the law of Snellius

$$\sin \alpha = Kv$$

passes through in the shortest time. Since the speed is known to be proportional to the square root of the fallen height, he obtains, by passing to thinner and thinner layers,

$$\sin \alpha = \frac{1}{\sqrt{1 + y'^2}} = K\sqrt{2g(y - h)}, \qquad (2.4)$$

a differential equation of the first order.

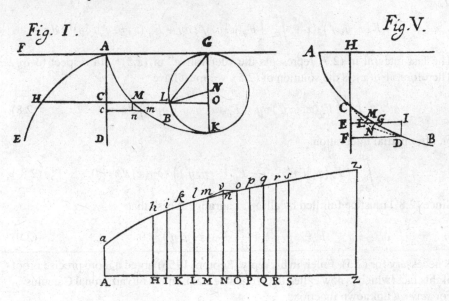

Fig. 2.4. Solutions of the variational problem (Joh. Bernoulli, Jac. Bernoulli, Euler)

The solutions of (2.4) can be shown to be cycloids (see Exercise 6 of Section I.3). Jacob, in his reply, also furnished a solution, much less elegant but unfortunately correct. Jacob's method (see Fig. 2.4) was something like today's (inverse) "finite element" method and more general than Johann's and led to the famous work of Euler (1744), which gives the general solution of the problem

$$\int_{x_0}^{x_1} F(x, y, y') \, dx = \min \tag{2.5}$$

with the help of the differential equation of the second order

$$F_y(x, y, y') - \frac{d}{dx}\left(F_{y'}(x, y, y')\right) = F_y - F_{y'y'}y'' - F_{y'y}y' - F_{y'x} = 0, \tag{2.6}$$

and treated 100 variational problems in detail. Equation (2.6), in the special case where F does not depend on x, can be integrated to give

$$F - F_{y'}y' = K. \tag{2.6'}$$

Euler's original proof used polygons in order to establish equation (2.6). Only the ideas of Lagrange, in 1755 at the age of 19, led to the proof which is today the usual one (letter of Aug. 12, 1755; Oeuvres vol. 14, p. 138): add an arbitrary "variation" $\delta y(x)$ to $y(x)$ and linearize (2.5).

$$\int_{x_0}^{x_1} F\left(x, y + \delta y, y' + (\delta y)'\right) dx \tag{2.7}$$

$$= \int_{x_0}^{x_1} F(x, y, y') \, dx + \int_{x_0}^{x_1} \left(F_y(x, y, y') \, \delta y + F_{y'}(x, y, y')(\delta y)'\right) dx + \dots$$

The last integral in (2.7) represents the "derivative" of (2.5) with respect to δy. Therefore, if $y(x)$ is the solution of (2.5), we must have

$$\int_{x_0}^{x_1} \left(F_y(x, y, y') \, \delta y + F_{y'}(x, y, y')(\delta y)'\right) dx = 0 \tag{2.8}$$

or, after partial integration,

$$\int_{x_0}^{x_1} \left(F_y(x, y, y') - \frac{d}{dx} F_{y'}(x, y, y')\right) \cdot \delta y(x) \, dx = 0. \tag{2.8'}$$

Since (2.8') must be fulfilled by all δy, Lagrange "sees" that

$$F_y(x, y, y') - \frac{d}{dx} F_{y'}(x, y, y') = 0 \tag{2.9}$$

is necessary for (2.5). Euler, in his reply (Sept. 6, 1755) urged a more precise proof of this fact (which is now called the "fundamental Lemma of variational calculus"). For *several* unknown functions

$$\int F(x, y_1, y_1', \dots, y_n, y_n') \, dx = \min \tag{2.10}$$

the same proof leads to the equations

$$F_{y_i}(x, y_1, y_1', \dots, y_n, y_n') - \frac{d}{dx} F_{y_i'}(x, y_1, y_1', \dots, y_n, y_n') = 0 \tag{2.11}$$

for $i = 1, \ldots, n$. Euler (1756) then gave, in honour of Lagrange, the name "Variational calculus" to the whole subject ("... tamen gloria primae inventionis acutissimo Geometrae Taurinensi La Grange erat reservata").

Clairaut

A class of equations with interesting properties was found by Clairaut (see Clairaut (1734), Problème III). He was motivated by the movement of a rectangular wedge (see Fig. 2.5), which led him to differential equations of the form

$$y - xy' + f(y') = 0. \tag{2.12}$$

This was the first *implicit* differential equation and possesses the particularity that not only the lines $y = Cx - f(C)$ are solutions, but also their enveloping curves (see Exercise 5). An example is shown in Fig. 2.6 with $f(C) = 5(C^3 - C)/2$.

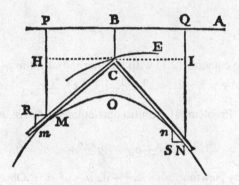

Fig. 2.5. Illustration from Clairaut (1734)

Since the equation is of the third degree in y', a given initial value may allow up to three different solution lines. Furthermore, where a line touches an enveloping curve, the solution may be continued either along the line or along the curve. There is thus a huge variety of different possible solution curves. This phenomenon attracted much interest in the classical literature (see e.g., Exercises 4 and 6). Today we explain this curiosity by the fact that at these points no Lipschitz condition is satisfied (see also Ince (1944), p. 538–539).

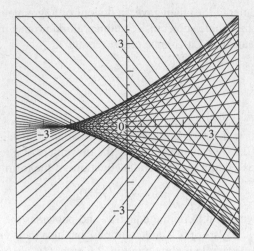

Fig. 2.6. Solutions of a Clairaut differential equation

Exercises

1. (Newton). Solve equation (2.1) with another initial value $y(0) = 1$.
   Newton's result: $y = 1 + 2x + x^3 + \frac{1}{4}x^4 + \frac{1}{4}x^5$, &c.

2. (Newton 1671, "Problema II, Solutio particulare"). Solve the total differential equation

$$3x^2 - 2ax + ay - 3y^2 y' + axy' = 0.$$

Solution given by Newton: $x^3 - ax^2 + axy - y^3 = 0$. Observe that he missed the arbitrary integration constant C.

3. (Newton 1671). Solve the equations

   a) $y' = 1 + \dfrac{y}{a} + \dfrac{xy}{a^2} + \dfrac{x^2 y}{a^3} + \dfrac{x^3 y}{a^4}$, &c.

   b) $y' = -3x + 3xy + y^2 - xy^2 + y^3 - xy^3 + y^4 - xy^4 + 6x^2 y$
   $\qquad - 6x^2 + 8x^3 y - 8x^3 + 10x^4 y - 10x^4$, &c.

Results given by Newton:

   a) $y = x + \dfrac{x^2}{2a} + \dfrac{x^3}{2a^2} + \dfrac{x^4}{2a^3} + \dfrac{x^5}{2a^4} + \dfrac{x^6}{2a^5}$, &c.

   b) $y = -\dfrac{3}{2}x^2 - 2x^3 - \dfrac{25}{8}x^4 - \dfrac{91}{20}x^5 - \dfrac{111}{16}x^6 - \dfrac{367}{35}x^7$, &c.

4. Show that the differential equation

$$x + yy' = y'\sqrt{x^2 + y^2 - 1}$$

possesses the solutions $2ay = a^2 + 1 - x^2$ for all a. Sketch these curves and find yet another solution of the equation (from Lagrange (1774), p. 7, which was written to explain the "Clairaut phenomenon").

5. Verify that the envelope of the solutions $y = Cx - f(C)$ of the Clairaut equation (2.12) is given in parametric representation by

$$x(p) = f'(p)$$
$$y(p) = pf'(p) - f(p).$$

Show that this envelope is also a solution of (2.12) and calculate it for $f(C) = 5(C^3 - C)/2$ (cf. Fig. 2.6).

6. (Cauchy 1824). Show that the family $y = C(x + C)^2$ satisfies the differential equation $(y')^3 = 8y^2 - 4xyy'$. Find yet another solution which is not included in this family (see Fig. 2.7).

Answer: $y = -\frac{4}{27}x^3$.

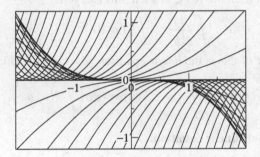

Fig. 2.7. Solution family of Cauchy's example in Exercise 6

I.3 Elementary Integration Methods

We now discuss some of the simplest types of equations, which can be solved by the computation of integrals.

First Order Equations

The equation with separable variables.

$$y' = f(x)g(y). \tag{3.1}$$

Extending the idea of Jacob Bernoulli (see (2.3')), we divide by $g(y)$, integrate and obtain the solution (Leibniz 1691, in a letter to Huygens)

$$\int \frac{dy}{g(y)} = \int f(x)\,dx + C.$$

A special example of this is the *linear equation* $y' = f(x)y$, which possesses the solution

$$y(x) = CR(x), \qquad R(x) = \exp\left(\int f(x)\,dx\right).$$

The inhomogeneous linear equation.

$$y' = f(x)y + g(x). \tag{3.2}$$

Here, the substitution $y(x) = c(x)R(x)$ leads to $c'(x) = g(x)/R(x)$ (Joh. Bernoulli 1697). One thus obtains the solution

$$y(x) = R(x)\left(\int_{x_0}^{x} \frac{g(s)}{R(s)}\,ds + C\right). \tag{3.3}$$

Total differential equations. An equation of the form

$$P(x,y) + Q(x,y)y' = 0 \tag{3.4}$$

is found to be immediately solvable if

$$\frac{\partial P}{\partial y} = \frac{\partial Q}{\partial x}. \tag{3.5}$$

One can then find by integration a potential function $U(x, y)$ such that

$$\frac{\partial U}{\partial x} = P, \qquad \frac{\partial U}{\partial y} = Q.$$

Therefore (3.4) becomes $\frac{d}{dx} U(x, y(x)) = 0$, so that the solutions can be expressed by $U(x, y(x)) = C$. For the case when (3.5) is not satisfied, Clairaut and Euler investigated the possibility of multiplying (3.4) by a suitable factor $M(x, y)$, which sometimes allows the equation $MP + MQy' = 0$ to satisfy (3.5).

Second Order Equations

Even more than for first order equations, the solution of *second* order equations by integration is very seldom possible. Besides linear equations with constant coefficients, whose solutions for the second order case were already known to Newton, several tricks of reduction are possible, as for example the following:

For a *linear equation*

$$y'' = a(x)y' + b(x)y$$

we make the substitution (Riccati 1723, Euler 1728)

$$y = \exp\left(\int p(x)\, dx\right). \tag{3.6}$$

The derivatives of this function contain only derivatives of p of lower order

$$y' = p \cdot \exp\left(\int p(x)\, dx\right), \qquad y'' = (p^2 + p') \cdot \exp\left(\int p(x)\, dx\right)$$

so that inserting this into the differential equation, after division by y, leads to a *lower order* equation

$$p^2 + p' = a(x)p + b(x) \tag{3.7}$$

which, however, is nonlinear.

If the equation is *independent* of y, $y'' = f(x, y')$, it is natural to put $y' = v$ which gives $v' = f(x, v)$.

An important case is that of *equations independent of x:*

$$y'' = f(y, y').$$

Here we consider y' as function of y: $y' = p(y)$. Then the chain rule gives $y'' = p'p = f(y, p)$, which is a first order equation. When the function $p(y)$ has been found, it remains to integrate $y' = p(y)$, which is an equation of type (3.1) (Riccati (1712): "Per liberare la premessa formula dalle seconde differenze, ..., chiamo p la sunnormale BF ... ", see also Euler (1769), Problema 96, p. 33).

The investigation of all possible differential equations which can be integrated by analytical methods was begun by Euler. His results have been collected, in more than 800 pages, in Volumes XXII and XXIII of Euler's Opera Omnia. For a more recent discussion see Ince (1944), p. 16-61. An irreplaceable document on this subject is the book of Kamke (1942). It contains, besides a description of the solution methods and general properties of the solutions, a systematically ordered list of more than 1500 differential equations with their solutions and references to the literature.

The computations, even for very simple looking equations, soon become very complicated and one quickly began to understand that elementary solutions would not always be possible. It was Liouville (1841) who gave the first *proof* of the fact that certain equations, such as $y' = x^2 + y^2$, cannot be solved in terms of elementary functions. Therefore, in the 19th century mathematicians became more and more interested in general existence theorems and in numerical methods for the computation of the solutions.

Exercises

1. Solve Newton's equation (2.1) by quadrature.

2. Solve Leibniz' equation (2.3) in terms of elementary functions.

   Hint. The integral for y might cause trouble. Use the substitution $a^2 - y^2 = u^2$, $-ydy = udu$.

3. Solve and draw the solutions of $y' = f(y)$ where $f(y) = \sqrt{|y|}$.

4. Solve the master-and-dog problem: a dog runs with speed w in the direction of his master, who walks with speed v along the y-axis. This leads to the differential equation

$$(xy')' = -\frac{v}{w}\sqrt{1 + (y')^2}.$$

5. Solve the equation $my'' = -k/y^2$, which describes a body falling according to Newton's law of gravitation.

6. Verify that the cycloid

$$x - x_0 = R(\tau - \sin\tau), \qquad y - h = R(1 - \cos\tau), \qquad R = \frac{1}{4gK^2}$$

   satisfies the differential equation (2.4) for the brachystochrone problem. Solving (2.4) in a forward manner, one arrives after some simplifications at the integral

$$\int \sqrt{\frac{y}{1-y}}\, dy,$$

   which is computed by the substitution $y = (\sin t)^2$.

7. Reduce the "Bernoulli equation" (Jac. Bernoulli 1695)

$$y' + f(x)y = g(x)y^n$$

with the help of the coordinate transformation $z(x) = (y(x))^q$ and a suitable choice of q, to a linear equation (Leibniz, Acta Erud. 1696, p. 145, Joh. Bernoulli, Acta Erud. 1697, p. 113).

8. Compute the "Linea Catenaria" of the hanging rope. The solution was given by Joh. Bernoulli (1691) and Leibniz (1691) (see Fig. 3.2) without any hint.

Hint. (Joh. Bernoulli, "Lectiones ... in usum Ill. Marchionis Hospitalii" 1691/92). Let H resp. V be the horizontal resp. vertical component of the tension in the rope (Fig. 3.1). Then $H = a$ is a constant and $V = q \cdot s$ is proportional to the arc length. This leads to $Cp = s$ or $Cdp = ds$ i.e., $Cdp = \sqrt{1 + p^2}dx$, where $p = y'$, a differential equation.

Result. $y = K + C \cosh\left(\dfrac{x - x_0}{C}\right)$.

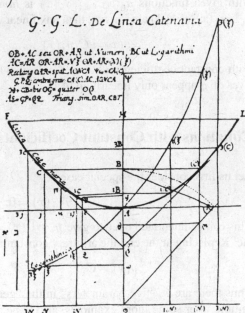

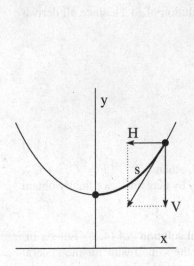

Fig. 3.1. Solution of the
Catenary problem

Fig. 3.2. "Linea Catenaria"
drawn by Leibniz (1691)

I.4 Linear Differential Equations

Following in the footsteps of Euler (1743), we want to understand the general solution of nth order linear differential equations. We say that the equation

$$\mathcal{L}(y) := a_n(x)y^{(n)} + a_{n-1}(x)y^{(n-1)} + \ldots + a_0(x)y = 0 \tag{4.1}$$

with given functions $a_0(x), \ldots, a_n(x)$ is *homogeneous*. If n solutions $u_1(x)$,
$\ldots, u_n(x)$ of (4.1) are known, then any linear combination

$$y(x) = C_1 u_1(x) + \ldots + C_n u_n(x) \tag{4.2}$$

with constant coefficients $C_1, \ldots, C_n$ is also a solution of (4.1), since all derivatives of y appear only linearly in (4.1).

Equations with Constant Coefficients

Let us first consider the special case

$$y^{(n)}(x) = 0. \tag{4.3}$$

This can be integrated once to give $y^{(n-1)}(x) = C_1$, then $y^{(n-2)}(x) = C_1 x + C_2$,
etc. Replacing at the end the arbitrary constants C_i by new ones, we finally obtain

$$y(x) = C_1 x^{n-1} + C_2 x^{n-2} + \ldots + C_n.$$

Thus there are n "free parameters" in the "general solution" of (4.3). Euler's in-
tuition, after some more examples, also expected the same result for the general
equation (4.1). This fact, however, only became completely clear many years later.

We now treat the general equation with constant coefficients,

$$y^{(n)} + A_{n-1}y^{(n-1)} + \ldots + A_0 y = 0. \tag{4.4}$$

Our problem is to find a basis of n linearly independent solutions $u_1(x), \ldots,$
$u_n(x)$. To this end, Euler's inspiration was guided by the transformation (3.6),
(3.7) above: if $a(x)$ and $b(x)$ are constants, we assume p constant in (3.7) so that
p' vanishes, and we obtain the quadratic equation $p^2 = ap + b$. For any root of this

equation, (3.6) then becomes $y = e^{px}$. In the general case we thus assume $y = e^{px}$ with an unknown constant p, so that (4.4) leads to the *characteristic equation*

$$p^n + A_{n-1}p^{n-1} + \ldots + A_0 = 0. \tag{4.5}$$

If the roots $p_1, \ldots, p_n$ of equation (4.5) are distinct, all solutions of (4.4) are given by

$$y(x) = C_1 e^{p_1 x} + \ldots + C_n e^{p_n x}. \tag{4.6}$$

It is curious to see that the "brightest mathematicians of the world" struggled for many decades to find this solution, which appears so trivial to today's students.

A difficulty arises with the solution (4.6) when (4.5) does not possess n distinct roots. Consider, with Euler, the example

$$y'' - 2qy' + q^2 y = 0. \tag{4.7}$$

Here $p = q$ is a double root of the corresponding characteristic equation. If we set

$$y = e^{qx} u, \tag{4.8}$$

(4.7) becomes $u'' = 0$, which brings us back to (4.3). So the general solution of (4.7) is given by $y(x) = e^{qx}(C_1 x + C_2)$ (see also Exercise 5 below). After some more examples of this type, one sees that the transformation (4.8) effects a *shift* of the characteristic polynomial, so that if q is a root of multiplicity k, we obtain for u an equation ending with $\ldots + Bu^{(k+1)} + Cu^{(k)} = 0$. Therefore

$$e^{qx}(C_1 x^{k-1} + \ldots + C_k)$$

gives us k independent solutions.

Finally, for a pair of complex roots $p = \alpha \pm i\beta$ the solutions $e^{(\alpha + i\beta)x}$, $e^{(\alpha - i\beta)x}$ can be replaced by the real functions

$$e^{\alpha x}(C_1 \cos \beta x + C_2 \sin \beta x).$$

The study of the *inhomogeneous* equation

$$\mathcal{L}(y) = f(x) \tag{4.9}$$

was begun in Euler (1750), p. 13. We mention from this work the case where $f(x)$ is a polynomial, say for example the equation

$$Ay'' + By' + Cy = ax^2 + bx + c. \tag{4.10}$$

Here Euler puts $y(x) = Ex^2 + Fx + G + v(x)$. Inserting this into (4.10) and eliminating all possible powers of x, one obtains

$$CE = a, \qquad CF + 2BE = b, \qquad CG + BF + 2AE = c,$$

$$Av'' + Bv' + Cv = 0.$$

This allows us, when C is different from zero, to compute E, F and G and we observe that *the general solution of the inhomogeneous equation is the sum of a particular solution of it and of the general solution of the corresponding homogeneous equation.* This is also true in the general case and can be verified by trivial linear algebra.

The above method of searching for a particular solution with the help of unknown coefficients works similarly if $f(x)$ is composed of exponential, sine, or cosine functions and is often called the "fast method". We see with pleasure that it was historically the first method to be discovered.

Variation of Constants

The *general treatment of the inhomogeneous equation*

$$a_n(x)y^{(n)} + \ldots + a_0(x)y = f(x) \tag{4.11}$$

is due to Lagrange (1775) ("... par une nouvelle méthode aussi simple qu'on puisse le désirer", see also Lagrange (1788), seconde partie, Sec. V.) We assume known n independent solutions $u_1(x), \ldots, u_n(x)$ of the *homogeneous* equation. We then set, in extension of the method employed for (3.2), instead of (4.2)

$$y(x) = c_1(x)u_1(x) + \ldots + c_n(x)u_n(x) \tag{4.12}$$

with unknown functions $c_i(x)$ ("method of variation of constants"). We have to insert (4.12) into (4.11) and thus compute the first derivative

$$y' = \sum_{i=1}^{n} c_i' u_i + \sum_{i=1}^{n} c_i u_i'.$$

If we continue blindly to differentiate in this way, we soon obtain complicated and useless formulas. Therefore Lagrange astutely requires the first term to vanish and puts

$$\sum_{i=1}^{n} c_i' u_i^{(j)} = 0 \qquad j = 0, \quad \text{then also for } j = 1, \ldots, n-2. \tag{4.13}$$

Then repeated differentiation of y, with continued elimination of the undesired terms (4.13), gives

$$y' = \sum_{i=1}^{n} c_i u_i', \qquad \ldots \qquad y^{(n-1)} = \sum_{i=1}^{n} c_i u_i^{(n-1)},$$

$$y^{(n)} = \sum_{i=1}^{n} c_i' u_i^{(n-1)} + \sum_{i=1}^{n} c_i u_i^{(n)}.$$

If we insert this into (4.11), we observe wonderful cancellations due to the fact that the $u_i(x)$ satisfy the homogeneous equation, and finally obtain, together with (4.13),

$$\begin{pmatrix} u_1 & \cdots & u_n \\ u_1' & \cdots & u_n' \\ \vdots & & \vdots \\ u_1^{(n-1)} & \cdots & u_n^{(n-1)} \end{pmatrix} \begin{pmatrix} c_1' \\ c_2' \\ \vdots \\ c_n' \end{pmatrix} = \begin{pmatrix} 0 \\ \vdots \\ 0 \\ f(x)/a_n(x) \end{pmatrix}. \tag{4.14}$$

This is a linear system, whose determinant is called the "Wronskian" and whose solution yields $c_1'(x), \ldots, c_n'(x)$ and after integration $c_1(x), \ldots, c_n(x)$.

Much more insight into this formula will be possible in Section I.11.

Exercises

1. Find the solution "huius aequationis differentialis quarti gradus" $a^4 y^{(4)} + y = 0$, $a^4 y^{(4)} - y = 0$; solve the equation "septimi gradus" $y^{(7)} + y^{(5)} + y^{(4)} + y^{(3)} + y^{(2)} + y = 0$. (Euler 1743, Ex. $4, 5, 6$).

2. Solve by Euler's technique $y'' - 3y' - 4y = \cos x$ and $y'' + y = \cos x$.

   Hint. In the first case the particular solution can be searched for in the form $E \cos x + F \sin x$. In the second case (which corresponds to a resonance in the equation) one puts $Ex \cos x + Fx \sin x$ just as in the solution of (4.7).

3. Find the solution of

   $$y'' - 3y' - 4y = g(x), \qquad g(x) = \begin{cases} \cos(x) & 0 \le x \le \pi/2 \\ 0 & \pi/2 \le x \end{cases}$$

   such that $y(0) = y'(0) = 0$,

   a) by using the solution of Exercise 2,

   b) by the method of Lagrange (variation of constants).

4. (Reduction of the order if one solution is known). Suppose that a nonzero solution $u_1(x)$ of $y'' + a_1(x)y' + a_0(x)y = 0$ is known. Show that a second independent solution can be found by putting $u_2(x) = c(x)u_1(x)$.

5. Treat the case of multiple characteristic values (4.7) by considering them as a limiting case $p_2 \to p_1$ and using the solutions

   $$u_1(x) = e^{p_1 x}, \qquad u_2(x) = \lim_{p_2 \to p_1} \frac{e^{p_2 x} - e^{p_1 x}}{p_2 - p_1} = \frac{\partial e^{p_1 x}}{\partial p_1}, \quad \text{etc.}$$

   (d'Alembert (1748), p. 284: "Enfin, si les valeurs de p & de p' sont égales, au lieu de les supposer telles, on supposera $p = a + \alpha$, $p' = a - \alpha$, α étant quantité infiniment petite ...").

I.5 Equations with Weak Singularities

> Der Mathematiker weiss sich ohnedies beim Auftreten von singu-
> lären Stellen gegebenenfalls leicht zu helfen.   (K. Heun 1900)

Many equations occurring in applications possess *singularities,* i.e., points at which
the function $f(x, y)$ of the differential equation becomes infinite. We study in some
detail the classical treatment of such equations, since numerical methods, which
will be discussed later in this book, often fail at the singular point, at least if they
are not applied carefully.

Linear Equations

As a first example, consider the equation

$$y' = \frac{q + bx}{x} y, \qquad q \neq 0 \tag{5.1}$$

with a singularity at $x = 0$. Its solution, using the method of separation of variables
(3.1), is

$$y(x) = Cx^q e^{bx} = C(x^q + bx^{q+1} + \ldots). \tag{5.2}$$

These solutions are plotted in Fig. 5.1 for different values of q and show the fundamental difference in the behaviour of the solutions in dependence of q.

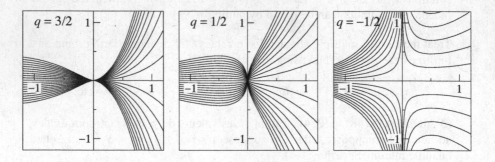

Fig. 5.1. Solutions of (5.1) for $b = 1$

Euler started a systematic study of equations with singularities. He asked which type of equation of the second order can conveniently be solved by a series as in (5.2) (Euler 1769, Problema 122, p. 177, "... quas commode per series resolvere licet..."). He found the equation

$$\mathcal{L}y := x^2(a+bx)y'' + x(c+ex)y' + (f+gx)y = 0. \tag{5.3}$$

Let us put $y = x^q(A_0 + A_1 x + A_2 x^2 + \ldots)$ with $A_0 \neq 0$ and insert this into (5.3). We observe that the powers x^2 and x which are multiplied by y'' and y', respectively, just re-establish what has been lost by the differentiations and obtain by comparing equal powers of x

$$\Big(q(q-1)a + qc + f\Big)A_0 = 0 \tag{5.4a}$$

$$\Big((q+i)(q+i-1)a + (q+i)c + f\Big)A_i \tag{5.4b}$$

$$= -\Big((q+i-1)(q+i-2)b + (q+i-1)e + g\Big)A_{i-1}$$

for $i = 1, 2, 3, \ldots$. In order to get $A_0 \neq 0$, q has to be a root of the *index equation*

$$\chi(q) := q(q-1)a + qc + f = 0. \tag{5.5}$$

For $a \neq 0$ there are two characteristic roots q_1 and q_2 of (5.5). Since the left-hand side of (5.4b) is of the form $\chi(q+i)A_i = \ldots$, this relation allows us to compute $A_1, A_2, A_3, \ldots$ at least for q_1 (if the roots are ordered such that $\operatorname{Re} q_1 \geq \operatorname{Re} q_2$). Thus we have obtained a first non-zero solution of (5.3). A second linearly independent solution for $q = q_2$ is obtained in the same way if $q_1 - q_2$ is not an integer.

Case of double roots. Euler found a second solution in this case with the inspiration of some acrobatic heuristics (Euler 1769, p. 150: "... quod $\frac{x^0}{0}$ aequivaleat ipsi $\ell x\, x \ldots$"). Fuchs (1866, 1868) then wrote a monumental paper on the form of all solutions for the general equation of order n, based on complicated calculations. A very elegant idea was then found by Frobenius (1873): fix A_0, say as $A_0(q) = 1$, completely ignore the index equation, choose q arbitrarily and consider the coefficients of the recursion (5.4b) as functions of q to obtain the series

$$y(x, q) = x^q \sum_{i=0}^{\infty} A_i(q)x^i, \tag{5.6}$$

whose convergence is discussed in Exercise 8 below. Since all conditions (5.4b) are satisfied, with the exception of (5.4a), we have

$$\mathcal{L}y(x, q) = \chi(q)x^q. \tag{5.7}$$

A second independent solution is now found simply by differentiating (5.7) with respect to q:

$$\mathcal{L}\Big(\frac{\partial y}{\partial q}(x, q)\Big) = \chi(q) \cdot \log x \cdot x^q + \chi'(q) \cdot x^q. \tag{5.8}$$

If we set $q = q_1$

$$\frac{\partial y}{\partial q}(x, q_1) = \log x \cdot y(x, q_1) + x^{q_1} \sum_{i=0}^{\infty} A_i'(q_1) x^i, \tag{5.9}$$

we obtain the desired second solution since $\chi(q_1) = \chi'(q_1) = 0$ (remember that q_1 is a double root of χ).

The case $q_1 - q_2 = m \in \mathbb{Z}$, $m \geq 1$. In this case we define a function $z(x)$ by satisfying $A_0(q) = 1$ and the recursion (5.4b) for all i with the exception of $i = m$. Then

$$\mathcal{L}z = \chi(q)x^q + Cx^{q+m} \tag{5.10}$$

where C is some constant. For $q = q_2$ the first term in (5.10) vanishes and a comparison with (5.8) shows that

$$\chi'(q_1)z(x) - C\frac{\partial y}{\partial q}(x, q_1) \tag{5.11}$$

is the required second solution of (5.3).

Euler (1778) later remarked that the formulas obtained become particularly elegant, if one starts from the differential equation

$$x(1 - x)y'' + (c - (a + b + 1)x)y' - aby = 0 \tag{5.12}$$

instead of from (5.3). Here, the above method leads to

$$q(q - 1) + cq = 0, \qquad q_1 = 0, \qquad q_2 = 1 - c, \tag{5.13}$$

$$A_{i+1} = \frac{(a + i)(b + i)}{(c + i)(1 + i)} A_i \qquad \text{for } q_1 = 0. \tag{5.14}$$

The resulting solutions, later named *hypergeometric functions,* became particularly famous throughout the 19th century with the work of Gauss (1812).

More generally, the above method works in the case of a differential equation

$$x^2 y'' + xa(x)y' + b(x)y = 0 \tag{5.15}$$

where $a(x)$ and $b(x)$ are regular analytic functions. One then says that 0 is a *regular singular point.* Similarly, we say that the equation $(x - x_0)^2 y'' + (x - x_0)a(x)y' + b(x)y = 0$ possesses the regular singular point x_0. In this case solutions can be obtained by the use of algebraic singularities $(x - x_0)^q$.

Finally, we also want to study the behaviour at *infinity* for an equation of the form

$$a(x)y'' + b(x)y' + c(x)y = 0. \tag{5.16a}$$

For this, we use the coordinate transformation $t = 1/x$, $z(t) = y(x)$ which yields

$$t^4 a\left(\frac{1}{t}\right)z'' + \left(2t^3 a\left(\frac{1}{t}\right) - t^2 b\left(\frac{1}{t}\right)\right)z' + c\left(\frac{1}{t}\right)z = 0. \tag{5.16b}$$

∞ is called a regular singular point of (5.16a) if 0 is a regular singular point of (5.16b). For examples see Exercise 9.

Nonlinear Equations

For nonlinear equations also, the above method sometimes allows one to obtain, if not the complete series of the solution, at least a couple of terms.

EXEMPLUM. Let us see what happens if we try to solve the classical brachystochrone problem (2.4) by a series. We suppose $h = 0$ and the initial value $y(0) = 0$. We write the equation as

$$(y')^2 = \frac{L}{y} - 1 \qquad \text{or} \qquad y(y')^2 + y = L. \tag{5.17}$$

At the initial point $y(0) = 0$, y' becomes infinite and most numerical methods would fail. We search for a solution of the form $y = A_0 x^q$. This gives in (5.17) $q^2 A_0^3 x^{3q-2} + A_0 x^q = L$. Due to the initial value we have that $y(x)$ becomes negligible for small values of x. We thus set the first term equal to L and obtain $3q - 2 = 0$ and $q^2 A_0^3 = L$. So

$$u(x) = \left(\frac{9Lx^2}{4}\right)^{1/3} \tag{5.18}$$

is a first approximate solution. The idea is now to use (5.18) just to escape from the initial point with a small x, and then to continue the solution with any numerical step-by-step procedure from the later chapters.

A more refined approximation could be tried in the form $y = A_0 x^q + A_1 x^{q+r}$. This gives with (5.17)

$$q^2 A_0^3 x^{3q-2} + q(3q + 2r)A_0^2 A_1 x^{3q+r-2} + A_0 x^q + \ldots = L.$$

We use the second term to neutralize the third one, which gives $3q + r - 2 = q$ or $r = q = 2/3$ and $5q^2 A_0 A_1 = -1$. Therefore

$$v(x) = \left(\frac{9Lx^2}{4}\right)^{1/3} - \left(\frac{9^2 x^4}{4^2 L5^3}\right)^{1/3} \tag{5.19}$$

is a better approximation. The following numerical results illustrate the utility of the approximations (5.18) and (5.19) compared with the correct solution $y(x)$ from I.3, Exercise 6, with $L = 2$:

$x = 0.10$	$y(x) = 0.342839$	$u(x) = 0.355689$	$v(x) = 0.343038$
$x = 0.01$	$y(x) = 0.076042$	$u(x) = 0.076631$	$v(x) = 0.076044.$

Exercises

1. Compute the general solution of the equation $x^2y'' + xy' + gx^n y = 0$ with g constant (Euler 1769, Problema 123, Exemplum 1).

2. Apply the technique of Euler to the *Bessel equation*

$$x^2 y'' + xy' + (x^2 - g^2)y = 0.$$

Sketch the solutions obtained for $g = 2/3$ and $g = 10/3$.

3. Compute the solutions of the equations

$$x^2 y'' - 2xy' + y = 0 \quad \text{and} \quad x^2 y'' - 3xy' + 4y = 0.$$

Equations of this type are often called Euler's or even Cauchy's equation. Its solution, however, was already known to Joh. Bernoulli.

4. (Euler 1769, Probl. 123, Exempl. 2). Let

$$y(x) = \int_0^{2\pi} \sqrt{\sin^2 s + x^2 \cos^2 s}\, ds$$

be the perimeter of the ellipse with axes 1 and $x < 1$.

a) Verify that $y(x)$ satisfies the differential equation

$$x(1 - x^2)y'' - (1 + x^2)y' + xy = 0. \tag{5.20}$$

b) Compute the solutions of this equation.

c) Show that the coordinate change $x^2 = t$, $y(x) = z(t)$ transforms (5.20) to a hypergeometric equation (5.12).

Hint. The computations for a) lead to the integral

$$\int_0^{2\pi} \frac{1 - 2\cos^2 s + q^2 \cos^4 s}{(1 - q^2 \cos^2 s)^{3/2}}\, ds, \qquad q^2 = 1 - x^2$$

which must be shown to be zero. Develop this into a power series in q^2.

5. Try to solve the equation

$$x^2 y'' + (3x - 1)y' + y = 0$$

with the help of a series (5.6) and study its convergence.

6. Find a series of the type

$$y = A_0 x^q + A_1 x^{q+s} + A_2 x^{q+2s} + \dots$$

which solves the nonlinear "Emden-Fowler equation" of astrophysics $(x^2 y')' + y^2 x^{-1/2} = 0$ in the neighbourhood of $x = 0$.

7. Approximate the solution of Leibniz's equation (2.3) in the neighbourhood of the singular initial value $y(0) = a$ by a function of the type $y(x) = a - Cx^q$. Compare the result with the correct solution of Exercise 2 of I.3.

8. Show that the radius of convergence of series (5.6) is given by

$$\text{i)} \quad r = |a/b| \qquad\qquad \text{ii)} \quad r = 1$$

   for the coefficients given by (5.4) and (5.14), respectively.

9. Show that the point ∞ is a regular singular point for the hypergeometric equation (5.12), but not for the Bessel equation of Exercise 2.

10. Consider the initial value problem

$$y' = \frac{\lambda}{x} y + g(x), \qquad y(0) = 0. \tag{5.21}$$

   a) Prove that if $\lambda \le 0$, the problem (5.21) possesses a unique solution for $x \ge 0$;

   b) If $g(x)$ is k-times differentiable and $\lambda \le 0$, then the solution $y(x)$ is $(k+1)$-times differentiable for $x \ge 0$ and we have

$$y^{(j)}(0) = \left(1 - \frac{\lambda}{j}\right)^{-1} g^{(j-1)}(0), \qquad j = 1, 2, \ldots.$$

I.6 Systems of Equations

Newton (1687) distilled from the known solutions of planetary motion (the Kepler laws) his "Lex secunda" together with the universal law of gravitation. It was mainly the "Dynamique" of d'Alembert (1743) which introduced, the·other way round, second order differential equations as a general tool for computing mechanical motion. Thus, Euler (1747) studied the movement of planets via the equations in 3-space

$$m\frac{d^2x}{dt^2} = X, \qquad m\frac{d^2y}{dt^2} = Y, \qquad m\frac{d^2z}{dt^2} = Z, \qquad (6.1)$$

where X, Y, Z are the forces in the three directions. ("... & par ce moyen j'evite quantité de recherches penibles").

The Vibrating String and Propagation of Sound

Suppose a string is represented by a sequence of identical and equidistant mass points and denote by $y_1(t)$, $y_2(t), \ldots$ the deviation of these mass points from the equilibrium position (Fig. 6.1a). If the deviations are supposed small ("fort petites"), the repelling force for the i-th mass point is proportional to $-y_{i-1} + 2y_i - y_{i+1}$ (Brook Taylor 1715, Johann Bernoulli 1727). Therefore equations (6.1) become

$$y_1'' = K^2(-2y_1 + y_2)$$
$$y_2'' = K^2(y_1 - 2y_2 + y_3)$$
$$\ldots$$
$$y_n'' = K^2(y_{n-1} - 2y_n). \qquad (6.2)$$

This is a system of n linear differential equations. Since the finite differences $y_{i-1} - 2y_i + y_{i+1} \approx c^2 \frac{\partial^2 y}{\partial x^2}$, equation (6.2) becomes, by the "inverse" method of lines, the famous partial differential equation (d'Alembert 1747)

$$\frac{\partial^2 u}{\partial t^2} = a^2 \frac{\partial^2 u}{\partial x^2}$$

for the vibrating string.

The *propagation of sound* is modelled similarly (Lagrange 1759): we suppose the medium to be a sequence of mass points and denote by $y_1(t)$, $y_2(t)$, ... their longitudinal displacements from the equilibrium position (see Fig. 6.1b). Then by Hooke's law of elasticity the repelling forces are proportional to the differences of displacements $(y_{i-1} - y_i) - (y_i - y_{i+1})$. This leads to equations (6.2) again ("En examinant les équations,... je me suis bientôt aperçu qu'elles ne différaient nullement de celles qui appartiennent au problème *de chordis vibrantibus*... ").

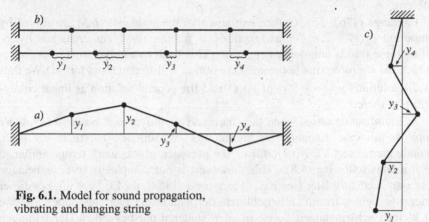

Fig. 6.1. Model for sound propagation, vibrating and hanging string

Another example, treated by Daniel Bernoulli (1732) and by Lagrange (1762, Nr. 36), is that of mass points attached to a *hanging* string (Fig. 6.1c). Here the tension in the string becomes greater in the upper part of the string and we have the following equations of movement

$$y_1'' = K^2(-y_1 + y_2)$$
$$y_2'' = K^2(y_1 - 3y_2 + 2y_3)$$
$$y_3'' = K^2(2y_2 - 5y_3 + 3y_4) \tag{6.3}$$
$$\cdots$$
$$y_n'' = K^2\big((n-1)y_{n-1} - (2n-1)y_n\big).$$

In all these examples, of course, the deviations y_i are supposed to be "infinitely" small, so that linear models are realistic.

Using a notation which came into use only a century later, we write these equations in the form

$$y_i'' = \sum_{j=1}^{n} a_{ij} y_j, \qquad i = 1, \ldots, n, \tag{6.4}$$

which is a *system of 2nd order linear equations with constant coefficients*. Lagrange solves system (6.4) by putting $y_i = c_i e^{pt}$, which leads to

$$p^2 c_i = \sum_{j=1}^{n} a_{ij} c_j, \qquad i = 1, \dots, n \qquad (6.5)$$

so that p^2 must be an *eigenvalue* of the matrix $A = (a_{ij})$ and $c = (c_1, \dots, c_n)^T$ a corresponding *eigenvector*. We see here the first appearance of an eigenvalue problem.

Lagrange (1762, Nr. 30) then explains that the equations (6.5) are solved by computing $c_2/c_1, \dots, c_n/c_1$ as functions of p from $n-1$ equations and by inserting these results into the last equation. This leads to a polynomial of degree n (in fact, the *characteristic polynomial*) to obtain n different roots for p^2. We thus get $2n$ solutions $y_i^{(j)} = c_i^{(j)} \exp(\pm p_j t)$ and the general solution as linear combinations of these.

A complication arises when the characteristic polynomial possesses *multiple roots*. In this case, Lagrange (in his famous "Mécanique Analytique" of 1788, seconde partie, sect.VI, No.7) affirms the presence of "secular" terms similar to the formulas following (4.8). This, however, is not completely true, as became clear only a century later (see e.g., Weierstrass (1858), p.243: "... um bei dieser Gelegenheit einen Irrtum zu berichtigen, der sich in der Lagrange'schen Theorie der kleinen Schwingungen, sowie in allen späteren mir bekannten Darstellungen derselben, findet."). We therefore postpone this subject to Section I.12.

We solve equations (6.2) in detail, since the results obtained are of particular importance (Lagrange 1759). The corresponding eigenvalue problem (6.5) becomes in this case $p^2 c_1 = K^2(-2c_1 + c_2)$, $p^2 c_i = K^2(c_{i-1} - 2c_i + c_{i+1})$ for $i = 2, \dots, n-1$ and $p^2 c_n = K^2(c_{n-1} - 2c_n)$. We introduce $p^2/K^2 + 2 = q$, so that

$$c_{j+1} - q c_j + c_{j-1} = 0, \qquad c_0 = 0, \qquad c_{n+1} = 0. \qquad (6.6)$$

This means that the c_i are the solutions of a *difference equation* and therefore $c_j = A a^j + B b^j$ where a and b are the roots of the corresponding characteristic equation $z^2 - qz + 1 = 0$, hence

$$a + b = q, \qquad ab = 1.$$

The condition $c_0 = 0$ of (6.6), which means that $A + B = 0$, shows that $c_j = A(a^j - b^j)$ with $A \neq 0$. The second condition $c_{n+1} = 0$, or equivalently $(a/b)^{n+1} = 1$, implies together with $ab = 1$ that

$$a = \exp\left(\frac{k\pi i}{n+1}\right), \quad b = \exp\left(\frac{-k\pi i}{n+1}\right)$$

for some $k = 1, \dots, n$. Thus we obtain

$$q_k = 2 \cos \frac{\pi k}{n+1}, \qquad k = 1, \dots, n, \qquad (6.7a)$$

$$p_k^2 = 2K^2\Big(\cos\frac{\pi k}{n+1} - 1\Big) = -4K^2\Big(\sin\frac{\pi k}{2n+2}\Big)^2. \tag{6.7b}$$

Finally, Euler's formula from 1740, $e^{ix} - e^{-ix} = 2i\sin x$ ("... si familière au-jourd'hui aux Géomètres") gives for the eigenvectors (with $A = -i/2$)

$$c_j^{(k)} = \sin\frac{jk\pi}{n+1}, \qquad j,k = 1,\ldots,n. \tag{6.8}$$

Since the p_k are purely imaginary, we also use for $\exp(\pm p_k t)$ the "familière" formula and obtain the general solution

$$y_j(t) = \sum_{k=1}^{n} \sin\frac{jk\pi}{n+1}(a_k\cos r_k t + b_k\sin r_k t), \qquad r_k = 2K\sin\frac{\pi k}{2n+2}. \tag{6.9}$$

Lagrange then observed after some lengthy calculations, which are today seen by using the orthogonality relations

$$\sum_{\ell=1}^{n} \sin\frac{\ell j\pi}{n+1}\sin\frac{\ell k\pi}{n+1} = \begin{cases} 0 & j \neq k \\ \frac{n+1}{2} & j = k \end{cases} \qquad j,k = 1,\ldots,n$$

that

$$a_k = \frac{2}{n+1}\sum_{j=1}^{n} \sin\frac{kj\pi}{n+1}\, y_j(0), \qquad b_k = \frac{1}{r_k}\frac{2}{n+1}\sum_{j=1}^{n} \sin\frac{kj\pi}{n+1}\, y_j'(0)$$

are determined by the initial positions and velocities of the mass points. He also studied the case where n, the number of mass points, tends to infinity (so that, in the formula for r_k, $\sin x$ can be replaced by x) and stood, 50 years before Fourier, at the portal of Fourier series theory. "Mit welcher Gewandtheit, mit welchem Aufwande analytischer Kunstgriffe er auch den ersten Theil dieser Untersuchung durchführte, so liess der Uebergang vom Endlichen zum Unendlichen doch viel zu wünschen übrig..." (Riemann 1854).

Fourier

> J'ajouterai que le livre de Fourier a une importance capitale dans
> l'histoire des mathématiques.   (H. Poincaré 1893)

The first *first order systems* were motivated by the problem of heat conduction (Biot 1804, Fourier 1807). Fourier imagined a rod to be a sequence of molecules, whose temperatures we denote by y_i, and deduced from a law of Newton that the energy which a particle passes to its neighbours is proportional to the difference of their temperatures, i.e., $y_{i-1} - y_i$ to the left and $y_{i+1} - y_i$ to the right ("Lorsque deux molécules d'un même solide sont extrêmement voisines et ont des températures inégales, la molécule plus échauffée communique à celle qui l'est moins une quantité de chaleur exactement exprimée par le produit formé de la durée de l'instant, de la différence extrêmement petite des températures, et d'une certaine fonction de la distance des molécules"). This long sentence means, in formulas, that the total gain of energy of the ith molecule is expressed by

$$y'_i = K^2(y_{i-1} - 2y_i + y_{i+1}), \tag{6.10}$$

or, in general by

$$y'_i = \sum_{j=1}^{n} a_{ij}y_i, \qquad i = 1, \ldots, n, \tag{6.11}$$

a first order system with constant coefficients.

By putting $y_i = c_i e^{pt}$, we now obtain the eigenvalue problem

$$pc_i = \sum_{j=1}^{n} a_{ij}c_j, \qquad i = 1, \ldots, n. \tag{6.12}$$

If we suppose the rod cooled to zero at both ends ($y_0 = y_{n+1} = 0$), we can use Lagrange eigenvectors from above and obtain the solution

$$y_j(t) = \sum_{k=1}^{n} a_k \sin \frac{jk\pi}{n+1} \exp(-w_k t), \qquad w_k = 4K^2 \left(\sin \frac{\pi k}{2n+2} \right)^2. \tag{6.13}$$

By taking n larger and larger, Fourier arrived from (6.10) (again the inverse "method of lines") at his famous heat equation

$$\frac{\partial u}{\partial t} = a^2 \frac{\partial^2 u}{\partial x^2} \tag{6.14}$$

which was the origin of Fourier series theory.

Lagrangian Mechanics

> Dies ist der kühne Weg, den *Lagrange* ..., freilich ohne ihn gehörig zu rechtfertigen, eingeschlagen hat.
>
> (Jacobi 1842/43, Vorl. Dynamik, p. 13)

This combines d'Alembert's dynamics, the "principle of least action" of Leibniz–Maupertuis and the variational calculus; published in the monumental treatise "Mécanique Analytique" (1788). It furnishes an excellent means for obtaining the differential equations of motion for complicated mechanical systems (arbitrary coordinate systems, constraints, etc.).

If we define (with Poisson 1809) the "Lagrange function"

$$\mathcal{L} = T - U \tag{6.15}$$

where

$$T = m \frac{\dot{x}^2 + \dot{y}^2 + \dot{z}^2}{2} \qquad \text{(kinetic energy)} \tag{6.16}$$

and U is the "potential energy" satisfying

$$\frac{\partial U}{\partial x} = -X, \qquad \frac{\partial U}{\partial y} = -Y, \qquad \frac{\partial U}{\partial z} = -Z \qquad (6.17)$$

then the equations of motion (6.1) are *identical* to Euler's equations (2.11) for the variational problem

$$\int_{t_0}^{t_1} \mathcal{L} \, dt = \min \qquad (6.18)$$

(this, mainly through a misunderstanding of Jacobi, is often called "Hamilton's principle"). The important idea is now to forget (6.16) and (6.17) and to apply (6.15) and (6.18) to *arbitrary mass points* and *arbitrary coordinate systems*.

Example. The *spherical pendulum* (Lagrange 1788, Seconde partie, Section VIII, Chap. II, §I). Let $\ell = 1$ and

$$x = \sin \theta \cos \varphi$$
$$y = \sin \theta \sin \varphi$$
$$z = -\cos \theta.$$

We set $m = g = 1$ and have

$$T = \frac{1}{2}(\dot{x}^2 + \dot{y}^2 + \dot{z}^2) = \frac{1}{2}(\dot{\theta}^2 + \sin^2 \theta \cdot \dot{\varphi}^2)$$
$$U = z = -\cos \theta \qquad (6.19)$$

so that (2.11) becomes

$$\mathcal{L}_\theta - \frac{d}{dt}(\mathcal{L}_{\dot{\theta}}) = -\sin\theta + \sin\theta\cos\theta \cdot \dot{\varphi}^2 - \ddot{\theta} = 0$$
$$\mathcal{L}_\varphi - \frac{d}{dt}(\mathcal{L}_{\dot{\varphi}}) = -\sin^2\theta \cdot \ddot{\varphi} - 2\sin\theta\cos\theta \cdot \dot{\varphi} \cdot \dot{\theta} = 0. \qquad (6.20)$$

We have thus obtained, by simple calculus, the equations of motion for the problem. These equations cannot be solved analytically. A solution, computed numerically by a Runge-Kutta method (see Chapter II) is shown in Fig. 6.2.

In general, suppose that the mechanical system in question is described by n coordinates $q_1, q_2, \ldots, q_n$ and that $\mathcal{L} = T - U$ depends on $q_1, q_2, \ldots, q_n$, $\dot{q}_1, \dot{q}_2, \ldots, \dot{q}_n$. Then the equations of motion are

$$\frac{d}{dt}\mathcal{L}_{\dot{q}_i} = \sum_{k=1}^{n} \mathcal{L}_{\dot{q}_i \dot{q}_k} \ddot{q}_k + \sum_{k=1}^{n} \mathcal{L}_{\dot{q}_i q_k} \dot{q}_k = \mathcal{L}_{q_i}, \qquad i = 1, \ldots, n. \qquad (6.21)$$

These equations allow several generalizations to time-dependent systems and non-conventional forces.

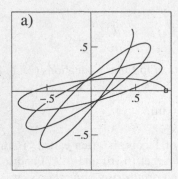 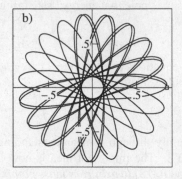

Fig. 6.2. Solution of the spherical pendulum, a) $0 \leq x \leq 20$, b) $0 \leq x \leq 100$
$(\varphi_0 = .0, \quad \dot{\varphi}_0 = 0.17, \quad \theta_0 = 1, \quad \dot{\theta}_0 = 0)$

Hamiltonian Mechanics

> Nach dem Erscheinen der ersten Ausgabe der Mécanique analy-
> tique wurde der wichtigste Fortschritt in der Umformung der Dif-
> ferentialgleichungen der Bewegung von *Poisson* ... gemacht ...
> im 15ten Hefte des polytechnischen Journals ... Hier führt *Pois-*
> *son* die Grössen $p = \partial T / \partial q'$ ... ein.
>
> (Jacobi 1842/43, Vorl. Dynamik, p. 67)

Hamilton, having worked for many years with variational principles (Fermat's principle) in his researches on optics, discovered at once that his ideas, after introducing a "principal function", allowed very elegant solutions for Kepler's motion of
a planet (Hamilton 1833). He then undertook in several papers (Hamilton 1834,
1835) to revolutionize mechanics. After many pages of computation he thereby discovered that it was "more convenient in many respects" (Hamilton 1834, Math. Papers II, p. 161) to work with the momentum coordinates (idea of Poisson)

$$p_i = \frac{\partial \mathcal{L}}{\partial \dot{q}_i} \tag{6.22}$$

instead of $\dot{q}_i$, and with the function

$$H = \sum_{k=1}^{n} \dot{q}_k p_k - \mathcal{L} \tag{6.23}$$

considered as function of $q_1, \ldots, q_n, p_1, \ldots, p_n$. This idea, to let derivatives
$\partial \mathcal{L} / \partial \dot{q}_i$ and independent variables p_i interchange their parts in order to simplify
differential equations, is due to Legendre (1787). Differentiating (6.23) by the

chain rule, we obtain

$$\frac{\partial H}{\partial p_i} = \sum_{k=1}^{n} \frac{\partial \dot{q}_k}{\partial p_i} \cdot p_k + \dot{q}_i - \sum_{k=1}^{n} \frac{\partial \mathcal{L}}{\partial \dot{q}_k} \frac{\partial \dot{q}_k}{\partial p_i}$$

and

$$\frac{\partial H}{\partial q_i} = \sum_{k=1}^{n} \frac{\partial \dot{q}_k}{\partial q_i} \cdot p_k - \frac{\partial \mathcal{L}}{\partial q_i} - \sum_{k=1}^{n} \frac{\partial \mathcal{L}}{\partial \dot{q}_k} \frac{\partial \dot{q}_k}{\partial q_i}.$$

By (6.22) and (6.21) both formulas simplify to

$$\dot{q}_i = \frac{\partial H}{\partial p_i}, \qquad \dot{p}_i = -\frac{\partial H}{\partial q_i}, \qquad i = 1, \dots, n. \tag{6.24}$$

These equations are marvellously symmetric "... and to integrate these differential equations of motion... is the chief and perhaps ultimately the only problem of mathematical dynamics" (Hamilton 1835). Jacobi (1843) called them *canonical differential equations*.

Remark. If the kinetic energy T is a quadratic function of the velocities $\dot{q}_i$, Euler's identity (Euler 1755, Caput VII, § 224, "... si V fuerit functio homogenea...") states that

$$2T = \sum_{k=1}^{n} \dot{q}_k \frac{\partial T}{\partial \dot{q}_k}. \tag{6.25}$$

If we further assume that the potential energy U is independent of $\dot{q}_i$, we obtain

$$H = \sum_{k=1}^{n} \dot{q}_k p_k - \mathcal{L} = \sum_{k=1}^{n} \dot{q}_k \frac{\partial T}{\partial \dot{q}_k} - \mathcal{L} = 2T - \mathcal{L} = T + U. \tag{6.26}$$

This is the *total* energy of the system.

Example. The spherical pendulum again. From (6.19) we have

$$p_\theta = \frac{\partial T}{\partial \dot{\theta}} = \dot{\theta}, \qquad p_\varphi = \frac{\partial T}{\partial \dot{\varphi}} = \sin^2 \theta \cdot \dot{\varphi} \tag{6.27}$$

and, by eliminating the undesired variables $\dot{\theta}$ and $\dot{\varphi}$,

$$H = T + U = \frac{1}{2} \left(p_\theta^2 + \frac{p_\varphi^2}{\sin^2 \theta} \right) - \cos \theta. \tag{6.28}$$

Therefore (6.26) becomes

$$\dot{p}_\theta = p_\varphi^2 \cdot \frac{\cos \theta}{\sin^3 \theta} - \sin \theta \qquad \dot{p}_\varphi = 0$$
$$\dot{\theta} = p_\theta \qquad\qquad\qquad \dot{\varphi} = \frac{p_\varphi}{\sin^2 \theta}. \tag{6.29}$$

These equations appear to be a little simpler than Lagrange's formulas (6.20). For example, we immediately see that $p_\varphi = Const$ (Kepler's second law).

Exercises

1. Verify that, if $u(x)$ is sufficiently differentiable,
$$\frac{u(x-\delta) - 2u(x) + u(x+\delta)}{\delta^2} = u''(x) + \frac{\delta^2}{12} u^{(4)}(x) + \mathcal{O}(\delta^4).$$

   Hint. Use Taylor series expansions for $u(x+\delta)$ and $u(x-\delta)$. This relation establishes the connection between (6.10) and (6.14) as well as between (6.2) and the wave equation.

2. Solve equation (6.3) for $n = 2$ and $n = 3$ by using the device of Lagrange described above (1762) and discover naturally the characteristic polynomial of the matrix.

3. Solve the first order system (6.11) with initial values $y_i(0) = (-1)^i$, where the matrix A is the same as in Exercise 2, and draw the solutions. Physically, this equation would represent a string with weights hanging, say, in honey.

4. Find the first terms of the development at the singular point $x = 0$ of the solutions of the following system of nonlinear equations
$$x^2 y'' + 2xy' = 2yz^2 + \lambda x^2 y(y^2 - 1), \qquad y(0) = 0$$
$$x^2 z'' = z(z^2 - 1) + x^2 y^2 z, \qquad z(0) = 1 \tag{6.30}$$

   where λ is a constant parameter. Equations (6.30) are the Euler equations for the variational problem
$$I = \int_0^\infty \left((z')^2 + \frac{x^2(y')^2}{2} + \frac{(z^2-1)^2}{2x^2} + y^2 z^2 + \frac{\lambda}{4} x^2 (y^2 - 1)^2 \right) dx,$$
$$y(\infty) = 1, \qquad z(\infty) = 0$$

   which gives the mass of a "monopole" in nuclear physics (see 't Hooft 1974).

5. Prove that the Hamiltonian function $H(q_1, \ldots, q_n, p_1, \ldots, p_n)$ is a first integral for the system (6.24), i.e., every solution satisfies
$$H\big(q_1(t), \ldots, q_n(t), p_1(t), \ldots, p_n(t)\big) = Const.$$

I.7 A General Existence Theorem

M. Cauchy annonce, que, pour se conformer au voeu du Conseil, il ne s'attachera plus à donner, comme il a fait jusqu'à présent, des démonstrations parfaitement rigoureuses.

(Conseil d'instruction de l'Ecole polytechnique, 24 nov. 1825)

You have all professional deformation of your minds; *convergence* does not matter here ...

(P. Henrici 1985)

We now enter a new era for our subject, more theoretical than the preceding one. It was inaugurated by the work of Cauchy, who was not as fascinated by long numerical calculations as was, say, Euler, but merely a fanatic for perfect mathematical rigor and exactness. He criticized in the work of his predecessors the use of infinite series and other infinite processes without taking much account of error estimates or convergence results. He therefore established around 1820 a convergence theorem for the polygon method of Euler and, some 15 years later, for the power series method of Newton (see Section I.8). Beyond the estimation of errors, these results also allow the statement of *general existence theorems* for the solutions of arbitrary differential equations ("d'une équation différentielle quelconque"), whose solutions were only known before in a very few cases. A second important consequence is to provide results about the *uniqueness* of the solution, which allow one to conclude that the computed solution (numerically or analytically) is the only one with the same initial value and that there are no others. Only then we are allowed to speak of *the* solution of the problem.

His very first proof has recently been discovered on fragmentary notes (Cauchy 1824), which were never published in Cauchy's lifetime (did his notes not satisfy the Minister of education?: "... mais que le second professeur, M. Cauchy, n'a présenté que des feuilles qui n'ont pu satisfaire la commission, et qu'il a été jusqu'à présent impossible de l'amener à se rendre au voeu du Conseil et à exécuter la décision du Ministre").

Convergence of Euler's Method

Let us now, with bared head and trembling knees, follow the ideas of this historical proof. We formulate it in a way which generalizes directly to higher dimensional systems.

Starting with the one-dimensional differential equation

$$y' = f(x, y), \qquad y(x_0) = y_0, \qquad y(X) = ? \tag{7.1}$$

we make use of the method explained by Euler (1768) in the last section of his "Institutiones Calculi Integralis I" (Caput VII, p. 424), i.e., we consider a subdivision of the interval of integration

$$x_0, x_1, \ldots, x_{n-1}, x_n = X \tag{7.2}$$

and replace in each subinterval the solution by the first term of its Taylor series

$$y_1 - y_0 = (x_1 - x_0)f(x_0, y_0)$$
$$y_2 - y_1 = (x_2 - x_1)f(x_1, y_1)$$
$$\cdots \tag{7.3}$$
$$y_n - y_{n-1} = (x_n - x_{n-1})f(x_{n-1}, y_{n-1}).$$

For the subdivision above we also use the notation

$$h = (h_0, h_1, \ldots, h_{n-1})$$

where $h_i = x_{i+1} - x_i$. If we connect y_0 and y_1, y_1 and y_2, ... etc by straight lines we obtain the *Euler polygon*

$$y_h(x) = y_i + (x - x_i)f(x_i, y_i) \qquad \text{for} \quad x_i \leq x \leq x_{i+1}. \tag{7.3a}$$

Lemma 7.1. *Assume that* $|f|$ *is bounded by* A *on*

$$D = \Big\{(x, y) \mid x_0 \leq x \leq X, \ |y - y_0| \leq b\Big\}.$$

If $X - x_0 \leq b/A$ *then the numerical solution* (x_i, y_i) *given by (7.3), remains in* D *for every subdivision (7.2) and we have*

$$|y_h(x) - y_0| \leq A \cdot |x - x_0|, \tag{7.4}$$

$$\Big|y_h(x) - \Big(y_0 + (x - x_0)f(x_0, y_0)\Big)\Big| \leq \varepsilon \cdot |x - x_0| \tag{7.5}$$

if $|f(x, y) - f(x_0, y_0)| \leq \varepsilon$ *on* D.

Proof. Both inequalities are obtained by adding up the lines of (7.3) and using the triangle inequality. Formula (7.4) then shows immediately that for $A(x - x_0) \leq b$ the polygon remains in D. $\qquad\square$

Our next problem is to obtain an estimate for the change of $y_h(x)$, when the initial value y_0 is changed: let z_0 be another initial value and compute

$$z_1 - z_0 = (x_1 - x_0)f(x_0, z_0). \tag{7.6}$$

We need an estimate for $|z_1 - y_1|$. Subtracting (7.6) from the first line of (7.3) we obtain

$$z_1 - y_1 = z_0 - y_0 + (x_1 - x_0)\Big(f(x_0, z_0) - f(x_0, y_0)\Big).$$

This shows that we need an estimate for $f(x_0, z_0) - f(x_0, y_0)$. If we suppose

$$|f(x, z) - f(x, y)| \leq L|z - y| \tag{7.7}$$

we obtain

$$|z_1 - y_1| \le \left(1 + (x_1 - x_0)L\right)|z_0 - y_0|. \tag{7.8}$$

Lemma 7.2. *For a fixed subdivision h let $y_h(x)$ and $z_h(x)$ be the Euler polygons corresponding to the initial values y_0 and z_0, respectively. If*

$$\left|\frac{\partial f}{\partial y}(x, y)\right| \le L \tag{7.9}$$

in a convex region which contains $(x, y_h(x))$ and $(x, z_h(x))$ for all $x_0 \le x \le X$, then

$$|z_h(x) - y_h(x)| \le e^{L(x - x_0)}|z_0 - y_0|. \tag{7.10}$$

Proof. (7.9) implies (7.7), (7.7) implies (7.8), (7.8) implies

$$|z_1 - y_1| \le e^{L(x_1 - x_0)}|z_0 - y_0|.$$

If we repeat the same argument for $z_2 - y_2$, $z_3 - y_3$, and so on, we finally obtain (7.10). $\qquad\qquad\square$

Remark. Condition (7.7) is called a "Lipschitz condition". It was Lipschitz (1876) who rediscovered the theory (footnote in the paper of Lipschitz: "L'auteur ne connaît pas évidemment les travaux de Cauchy ...") and advocated the use of (7.7) instead of the more stringent hypothesis (7.9). Lipschitz's proof is also explained in the classical work of Picard (1891-96), Vol. II, Chap. XI, Sec. I.

If the subdivision (7.2) is refined more and more, so that

$$|h| := \max_{i=0,\dots,n-1} h_i \to 0,$$

we expect that the Euler polygons converge to a solution of (7.1). Indeed, we have

Theorem 7.3. *Let $f(x, y)$ be continuous, and $|f|$ be bounded by A and satisfy the Lipschitz condition (7.7) on*

$$D = \left\{(x, y) \mid x_0 \le x \le X, \; |y - y_0| \le b\right\}.$$

If $X - x_0 \le b/A$, then we have:

a) *For $|h| \to 0$ the Euler polygons $y_h(x)$ converge uniformly to a continuous function $\varphi(x)$.*

b) *$\varphi(x)$ is continuously differentiable and solution of (7.1) on $x_0 \le x \le X$.*

c) *There exists no other solution of (7.1) on $x_0 \le x \le X$.*

Proof. a) Take an $\varepsilon > 0$. Since f is uniformly continuous on the compact set D, there exists a $\delta > 0$ such that

$$|u_1 - u_2| \le \delta \qquad \text{and} \qquad |v_1 - v_2| \le A \cdot \delta$$

imply

$$|f(u_1, v_1) - f(u_2, v_2)| \le \varepsilon. \tag{7.11}$$

Suppose now that the subdivision (7.2) satisfies

$$|x_{i+1} - x_i| \le \delta, \qquad \text{i.e.,} \quad |h| \le \delta. \tag{7.12}$$

We first study the effect of adding new mesh-points. In a first step, we consider a subdivision $h(1)$, which is obtained by adding new points only to the *first* subinterval (see Fig. 7.1). It follows from (7.5) (applied to this first subinterval) that for the new refined solution $y_{h(1)}(x_1)$ we have the estimate $|y_{h(1)}(x_1) - y_h(x_1)| \le \varepsilon |x_1 - x_0|$. Since the subdivisions h and $h(1)$ are identical on $x_1 \le x \le X$ we can apply Lemma 7.2 to obtain

$$|y_{h(1)}(x) - y_h(x)| \le e^{L(x-x_1)}(x_1 - x_0)\varepsilon \qquad \text{for} \quad x_1 \le x \le X.$$

We next add further points to the subinterval (x_1, x_2) and denote the new subdivision by $h(2)$. In the same way as above this leads to $|y_{h(2)}(x_2) - y_{h(1)}(x_2)| \le \varepsilon |x_2 - x_1|$ and

$$|y_{h(2)}(x) - y_{h(1)}(x)| \le e^{L(x-x_2)}(x_2 - x_1)\varepsilon \qquad \text{for} \quad x_2 \le x \le X.$$

The entire situation is sketched in Fig. 7.1. If we denote by $\widehat{h}$ the final refinement, we obtain for $x_i < x \le x_{i+1}$

$$|y_{\widehat{h}}(x) - y_h(x)| \tag{7.13}$$

$$\le \varepsilon \Big(e^{L(x-x_1)}(x_1 - x_0) + \ldots + e^{L(x-x_i)}(x_i - x_{i-1}) \Big) + \varepsilon(x - x_i)$$

$$\le \varepsilon \int_{x_0}^{x} e^{L(x-s)} \, ds = \frac{\varepsilon}{L}\Big(e^{L(x-x_0)} - 1 \Big).$$

If we now have two different subdivisions h and $\widetilde{h}$, which both satisfy (7.12), we introduce a *third* subdivision $\widehat{h}$ which is a refinement of both subdivisions (just as is usually done in proving the existence of Riemann's integral), and apply (7.13) twice. We then obtain from (7.13) by the triangle inequality

$$|y_h(x) - y_{\widetilde{h}}(x)| \le 2\frac{\varepsilon}{L}\Big(e^{L(x-x_0)} - 1 \Big).$$

For $\varepsilon > 0$ small enough, this becomes arbitrarily small and shows the uniform convergence of the Euler polygons to a continuous function $\varphi(x)$.

   b) Let

$$\varepsilon(\delta) := \sup\Big\{ |f(u_1, v_1) - f(u_2, v_2)| \, ; \, |u_1 - u_2| \le \delta, \, |v_1 - v_2| \le A\delta, \, (u_i, v_i) \in D \Big\}.$$

be the modulus of continuity. If x belongs to the subdivision h then we obtain from (7.5) (replace (x_0, y_0) by $(x, y_h(x))$ and x by $x + \delta$)

$$|y_h(x + \delta) - y_h(x) - \delta f(x, y_h(x))| \le \varepsilon(\delta)\delta. \tag{7.14}$$

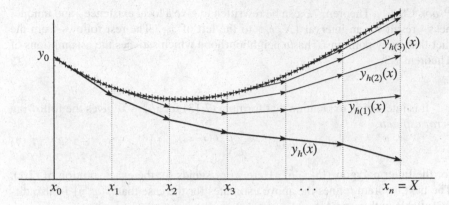

Fig. 7.1. Lady Windermere's Fan (O. Wilde 1892)

Taking the limit $|h| \to 0$ we get

$$|\varphi(x+\delta) - \varphi(x) - \delta f(x, \varphi(x))| \le \varepsilon(\delta)\delta. \tag{7.15}$$

Since $\varepsilon(\delta) \to 0$ for $\delta \to 0$, this proves the differentiability of $\varphi(x)$ and $\varphi'(x) = f(x, \varphi(x))$.

c) Let $\psi(x)$ be a second solution of (7.1) and suppose that the subdivision h satisfies (7.12). We then denote by $y_h^{(i)}(x)$ the Euler polygon to the initial value $(x_i, \psi(x_i))$ (it is defined for $x_i \le x \le X$). It follows from

$$\psi(x) = \psi(x_i) + \int_{x_i}^{x} f(s, \psi(s))\, ds$$

and (7.11) that

$$|\psi(x) - y_h^{(i)}(x)| \le \varepsilon |x - x_i| \qquad \text{for} \quad x_i \le x \le x_{i+1}.$$

Using Lemma 7.2 we deduce in the same way as in part a) that

$$|\psi(x) - y_h(x)| \le \frac{\varepsilon}{L}\left(e^{L(x-x_0)} - 1\right). \tag{7.16}$$

Taking the limits $|h| \to 0$ and $\varepsilon \to 0$ we obtain $|\psi(x) - \varphi(x)| \le 0$, proving uniqueness. $\qquad \square$

Theorem 7.3 is a *local* existence - and uniqueness - result. However, if we interpret the endpoint of the solution as a new initial value, we can apply Theorem 7.3 again and continue the solution. Repeating this procedure we obtain

Theorem 7.4. *Assume U to be an open set in $\mathbb{R}^2$ and let f and $\partial f/\partial y$ be continuous on U. Then, for every $(x_0, y_0) \in U$, there exists a unique solution of (7.1), which can be continued up to the boundary of U (in both directions).*

Proof. Clearly, Theorem 7.3 can be rewritten to give a local existence - and unique-ness - result for an interval (X, x_0) to the left of x_0. The rest follows from the fact that every point in U has a neighbourhood which satisfies the assumptions of Theorem 7.3.   □

It is interesting to mention that formula (7.13) for $|\widehat{h}| \to 0$ gives the following *error estimate*

$$|y(x) - y_h(x)| \leq \frac{\varepsilon}{L}\left(e^{L(x-x_0)} - 1\right) \tag{7.17}$$

for the Euler polygon $(|h| \leq \delta)$. Here $y(x)$ stands for the exact solution of (7.1). The next theorem refines the above estimates for the case that $f(x, y)$ is also dif-ferentiable with respect to x.

Theorem 7.5. *Suppose that in a neighbourhood of the solution*

$$|f| \leq A, \qquad \left|\frac{\partial f}{\partial y}\right| \leq L, \qquad \left|\frac{\partial f}{\partial x}\right| \leq M.$$

We then have the following error estimate for the Euler polygons:

$$|y(x) - y_h(x)| \leq \frac{M + AL}{L}\left(e^{L(x-x_0)} - 1\right) \cdot |h|, \tag{7.18}$$

provided that $|h|$ is sufficiently small.

Proof. For $|u_1 - u_2| \leq |h|$ and $|v_1 - v_2| \leq A|h|$ we obtain, due to the differentia-bility of f, the estimate

$$|f(u_1, v_1) - f(u_2, v_2)| \leq (M + AL)|h|$$

instead of (7.11). When we insert this amount for ε into (7.16), we obtain the stated result.   □

The estimate (7.18) shows that the global error of Euler's method is propor-tional to the maximal step size $|h|$. Thus, for an accuracy of, say, three decimal digits, we would need about a thousand steps; a precision of six digits will normally require a million steps etc. We see thus that the present method is not recommended for computations of high precision. In fact, the main subject of Chapter II will be to find methods which converge faster.

Existence Theorem of Peano

> Si a est un complexe d'ordre n, et b un nombre réel, alors on peut déterminer b' et f, où b' est une quantité plus grande que b, et f est un signe de fonction qui à chaque nombre de l'intervalle de b à b' fait correspondre un complexe (en d'autres mots, ft est un complexe fonction de la variable réelle t, définie pour toutes les valeurs de l'intervalle (b, b')); la valeur de ft pour $t = b$ est a; et dans tout l'intervalle (b, b') cette fonction ft satisfait à l'équation différentielle donnée.   (Original version of Peano's Theorem)

The Lipschitz condition (7.7) is a crucial tool in the proof of (7.10) and finally of the Convergence Theorem. If we completely abandon condition (7.7) and only require that $f(x, y)$ be continuous, the convergence of the Euler polygons is no longer guaranteed.

An example, plotted in Fig. 7.2, is given by the equation

$$y' = 4\left(\text{sign}\,(y)\sqrt{|y|} + \max\left(0, x - \frac{|y|}{x}\right)\cdot\cos\left(\frac{\pi\log x}{\log 2}\right)\right) \qquad (7.19)$$

with $y(0) = 0$. It has been constructed such that

$$f(h, 0) = 4(-1)^i h \qquad \text{for } h = 2^{-i},$$
$$f(x, y) = 4\,\text{sign}(y)\cdot\sqrt{|y|} \qquad \text{for } |y| \geq x^2.$$

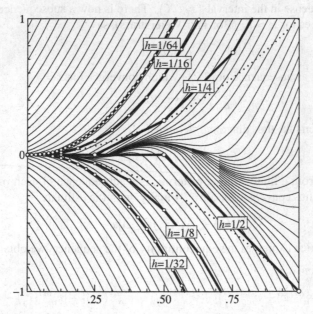

Fig. 7.2. Solution curves and Euler polygons for equation (7.19)

There is an infinity of solutions for this initial value, some of which are plotted in Fig. 7.2. The Euler polygons converge for $h = 2^{-i}$ and even i to the maximal solution $y = 4x^2$, and for odd i to $y = -4x^2$. For other sequences of h all intermediate solutions can be obtained as well.

Theorem 7.6 (Peano 1890). *Let $f(x, y)$ be continuous and $|f|$ be bounded by A on*

$$D = \Big\{ (x, y) \mid x_0 \leq x \leq X, \ |y - y_0| \leq b \Big\}.$$

If $X - x_0 \leq b/A$, then there is a subsequence of the sequence of the Euler polygons which converges to a solution of the differential equation.

The original proof of Peano is, in its crucial part on the convergence result, very brief and not clear to unexperienced readers such as us. Arzelà (1895), who took up the subject again, explains his ideas in more detail and emphasizes the need for an *equicontinuity* of the sequence. The proof usually given nowadays (for what has become the theorem of Arzelà-Ascoli), was only introduced later (see e.g. Perron (1918), Hahn (1921), p. 303) and is sketched as follows:

Proof. Let

$$v_1(x), v_2(x), v_3(x), \ldots \tag{7.20}$$

be a sequence of Euler polygons for decreasing step sizes. It follows from (7.4) that for fixed x this sequence is bounded. We choose a sequence of numbers $r_1, r_2, r_3, \ldots$ dense in the interval (x_0, X). There is now a subsequence of (7.20) which converges for $x = r_1$ (Bolzano-Weierstrass), say

$$v_1^{(1)}(x), v_2^{(1)}(x), v_3^{(1)}(x), \ldots \tag{7.21}$$

We next select a subsequence of (7.21) which converges for $x = r_2$

$$v_1^{(2)}(x), v_2^{(2)}(x), v_3^{(2)}(x), \ldots \tag{7.22}$$

and so on. Then take the "diagonal" sequence

$$v_1^{(1)}(x), v_2^{(2)}(x), v_3^{(3)}(x), \ldots \tag{7.23}$$

which, apart from a finite number of terms, is a subsequence of each of these sequences, and thus converges for all r_i. Finally, with the estimate

$$|v_n^{(n)}(x) - v_n^{(n)}(r_j)| \leq A|x - r_j|$$

(see (7.4)), which expresses the equicontinuity of the sequence, we obtain

$$|v_n^{(n)}(x) - v_m^{(m)}(x)|$$
$$\leq |v_n^{(n)}(x) - v_n^{(n)}(r_j)| + |v_n^{(n)}(r_j) - v_m^{(m)}(r_j)| + |v_m^{(m)}(r_j) - v_m^{(m)}(x)|$$
$$\leq 2A|x - r_j| + |v_n^{(n)}(r_j) - v_m^{(m)}(r_j)|.$$

For fixed $\varepsilon > 0$ we then choose a finite subset R of $\{r_1, r_2, \ldots\}$ satisfying

$$\min\{|x - r_j|\,;\; r_j \in R,\; x_0 \le x \le X\} \le \varepsilon/A$$

and secondly we choose N such that

$$|v_n^{(n)}(r_j) - v_m^{(m)}(r_j)| \le \varepsilon \qquad \text{for} \quad n, m \ge N \quad \text{and} \quad r_j \in R.$$

This shows the uniform convergence of (7.23). In the same way as in part b) of the proof of Theorem 7.3 it follows that the limit function is a solution of (7.1). One only has to add an $\mathcal{O}(|h|)$-term in (7.14), if x is not a subdivision point. $\square$

Exercises

1. Apply Euler's method with constant step size $x_{i+1} - x_i = 1/n$ to the differential equation $y' = ky$, $y(0) = 1$ and obtain a classical approximation for the solution $y(1) = e^k$. Give an estimate of the error.

2. Apply Euler's method with constant step size to
   a) $y' = y^2$, $y(0) = 1$, $y(1/2) = ?$
   b) $y' = x^2 + y^2$, $y(0) = 0$, $y(1/2) = ?$

   Make rigorous error estimates using Theorem 7.4 and compare these estimates with the actual errors. The main difficulty is to find a suitable region in which the estimates of Theorem 7.4 hold, without making the constants A, L, M too large and, at the same time, ensuring that the solution curves remain inside this region (see also I.8, Exercise 3).

3. Prove the result: if the differential equation $y' = f(x, y)$, $y(x_0) = y_0$ with f continuous, possesses a unique solution, then the Euler polygons converge to this solution.

4. "There is an elementary proof of Peano's existence theorem" (Walter 1971). Suppose that A is a bound for $|f|$. Then the sequence

   $$y_{i+1} = y_i + h \cdot \max\{f(x, y)|x_i \le x \le x_{i+1}, y_i - 3Ah \le y \le y_i + Ah\}$$

   converges for all continuous f to a (the maximal) solution. Try to prove this. Unfortunately, this proof does not extend to systems of equations, unless they are "quasimonotone" (see Section I.10, Exercise 3).

I.8 Existence Theory using Iteration Methods and Taylor Series

A second approach to existence theory is possible with the help of an iterative refinement of approximate solutions. The first appearances of the idea are very old. For instance many examples of this type can be found in the work of Lagrange, above all in his astronomical calculations. Let us consider here the following illustrative example of a Riccati equation

$$y' = x^2 + y + 0.1y^2, \qquad y(0) = 0. \tag{8.1}$$

Because of the quadratic term, there is no elementary solution. A very natural idea is therefore to neglect this term, which is in fact very small at the beginning, and to solve for the moment

$$y_1' = x^2 + y_1, \qquad y_1(0) = 0. \tag{8.2}$$

This gives, with formula (3.3), a first approximation

$$y_1(x) = 2e^x - (x^2 + 2x + 2). \tag{8.3}$$

With the help of this solution, we now know more about the initially neglected term $0.1y^2$; it will be close to $0.1y_1^2$. So the idea lies at hand to reintroduce this solution into (8.1) and solve now the differential equation

$$y_2' = x^2 + y_2 + 0.1 \cdot \left(y_1(x)\right)^2, \qquad y_2(0) = 0. \tag{8.4}$$

We can use formula (3.3) again and obtain after some calculations

$$y_2(x) = y_1(x) + \frac{2}{5}e^{2x} - \frac{2}{15}e^x(x^3 + 3x^2 + 6x - 54)$$
$$- \frac{1}{10}(x^4 + 8x^3 + 32x^2 + 72x + 76).$$

This is already much closer to the correct solution, as can be seen from the following comparison of the errors $e_1 = y(x) - y_1(x)$ and $e_2 = y(x) - y_2(x)$:

$$x = 0.2 \qquad e_1 = 0.228 \times 10^{-07} \qquad e_2 = 0.233 \times 10^{-12}$$
$$x = 0.4 \qquad e_1 = 0.327 \times 10^{-05} \qquad e_2 = 0.566 \times 10^{-09}$$
$$x = 0.8 \qquad e_1 = 0.534 \times 10^{-03} \qquad e_2 = 0.165 \times 10^{-05}.$$

It looks promising to continue this process, but the computations soon become very tedious.

Picard-Lindelöf Iteration

The general formulation of the method is the following: we try, if possible, to split up the function $f(x, y)$ of the differential equation

$$y' = f(x, y) = f_1(x, y) + f_2(x, y), \qquad y(x_0) = y_0 \qquad (8.5)$$

so that any differential equation of the form $y' = f_1(x, y) + g(x)$ can be solved analytically and so that $f_2(x, y)$ is small. Then we start with a first approximation $y_0(x)$ and compute successively $y_1(x)$, $y_2(x), \ldots$ by solving

$$y'_{i+1} = f_1(x, y_{i+1}) + f_2(x, y_i(x)), \qquad y_{i+1}(x_0) = y_0. \qquad (8.6)$$

The most primitive form of this process is obtained by choosing $f_1 = 0$, $f_2 = f$, in which case (8.6) is immediately integrated and becomes

$$y_{i+1}(x) = y_0 + \int_{x_0}^x f\big(s, y_i(s)\big)\, ds. \qquad (8.7)$$

This is called the *Picard-Lindelöf iteration method*. It appeared several times in the literature, e.g., in Liouville (1838), Cauchy, Peano (1888), Lindelöf (1894), Bendixson (1893). Picard (1890) considered it merely as a by-product of a similar idea for partial differential equations and analyzed it thoroughly in his famous treatise Picard (1891-96), Vol. II, Chap. XI, Sect. III.

The fast *convergence* of the method, for $|x - x_0|$ small, is readily seen: if we subtract formula (8.7) from the same with i replaced by $i - 1$, we have

$$y_{i+1}(x) - y_i(x) = \int_{x_0}^x \Big(f\big(s, y_i(s)\big) - f\big(s, y_{i-1}(s)\big) \Big)\, ds. \qquad (8.8)$$

We now apply the Lipschitz condition (7.7) and the triangle inequality to obtain

$$|y_{i+1}(x) - y_i(x)| \leq L \int_{x_0}^x |y_i(s) - y_{i-1}(s)|\, ds. \qquad (8.9)$$

When we assume $y_0(x) \equiv y_0$, the triangle inequality applied to (8.7) with $i = 0$ yields the estimate

$$|y_1(x) - y_0(x)| \leq A|x - x_0|$$

where A is a bound for $|f|$ as in Section I.7. We next insert this into the right hand side of (8.9) repeatedly to obtain finally the estimate (Lindelöf 1894)

$$|y_i(x) - y_{i-1}(x)| \leq AL^{i-1} \frac{|x - x_0|^i}{i!}. \qquad (8.10)$$

The right-hand side is a term of the Taylor series for $e^{L|x-x_0|}$, which converges for all x; we therefore conclude that $|y_{i+k} - y_i|$ becomes arbitrarily small when i is large. The error is bounded by the remainder of the above exponential series. So the sequence $y_i(x)$ converges uniformly to the solution $y(x)$. For example, if $L|x - x_0| \leq 1/10$ and the constant A is moderate, 10 iterations would provide a numerical solution with about 17 correct digits.

The main practical drawback of the method is the need for repeated computation of integrals, which is usually not very convenient, if at all analytically possible, and soon becomes very tedious. However, its fast convergence and new machine architectures (parallelism) coupled with numerical evaluations of the integrals have made the approach interesting for large problems (see Nevanlinna 1989).

Taylor Series

> Après avoir montré l'insuffisance des méthodes d'intégration fon-
> dées sur le développement en séries, il me reste à dire en peu de
> mots ce qu'on peut leur substituer.   (Cauchy)

A third existence proof can be based on a study of the convergence of the Taylor series of the solutions. This was mentioned in a footnote of Liouville (1836, p. 255), and brought to perfection by Cauchy (1839-42).

We have already seen the recursive computation of the Taylor coefficients in the work of Newton (see Section I.2). Euler (1768) then formulated the general procedure for the higher derivatives of the solution of

$$y' = f(x, y), \quad y(x_0) = y_0 \tag{8.11}$$

which, by successive differentiation, are obtained as

$$\begin{aligned}
y'' &= f_x + f_y y' = f_x + f_y f \\
y''' &= f_{xx} + 2f_{xy}f + f_{yy}f^2 + f_y(f_x + f_y f)
\end{aligned} \tag{8.12}$$

etc. Then the solution is

$$y(x_0 + h) = y(x_0) + y'(x_0)h + y''(x_0)\frac{h^2}{2!} + \ldots. \tag{8.13}$$

The formulas (8.12) for higher derivatives soon become very complicated. Euler therefore proposed to use only a few terms of this series with h sufficiently small and to repeat the computations from the point $x_1 = x_0 + h$ ("analytic continuation").

We shall now outline the main ideas of Cauchy's *convergence proof* for the series (8.13). We suppose that $f(x, y)$ is *analytic* in the neighbourhood of the initial value x_0, y_0, which for simplicity of notation we assume located at the origin $x_0 = y_0 = 0$:

$$f(x, y) = \sum_{i,j \geq 0} a_{ij} x^i y^j, \tag{8.14}$$

where the a_{ij} are multiples of the partial derivatives occurring in (8.12). If the series (8.14) is assumed to converge for $|x| \leq r$, $|y| \leq r$, then the Cauchy inequalities from classical complex analysis give

$$|a_{ij}| \leq \frac{M}{r^{i+j}}, \qquad \text{where} \qquad M = \max_{|x| \leq r, |y| \leq r} |f(x,y)|. \qquad (8.15)$$

The idea is now the following: since all signs in (8.12) are positive, we obtain the worst possible result if we replace in (8.14) all a_{ij} by the largest possible values (8.15) ("method of majorants"):

$$f(x,y) \to \sum_{i,j \geq 0} M \frac{x^i y^j}{r^{i+j}} = \frac{M}{(1-x/r)(1-y/r)}.$$

However, the majorizing differential equation

$$y' = \frac{M}{(1-x/r)(1-y/r)}, \qquad y(0) = 0$$

is readily integrated by separation of variables (see Section I.3) and has the solution

$$y = r \left(1 - \sqrt{1 + 2M \log\left(1 - \frac{x}{r}\right)} \right). \qquad (8.16)$$

This solution has a power series expansion which converges for all x such that $|2M \log(1 - x/r)| < 1$. Therefore, the series (8.13) also converges at least for all $|h| < r(1 - \exp(-1/2M))$. $\qquad \square$

Recursive Computation of Taylor Coefficients

> ... dieses Verfahren praktisch nicht in Frage kommen kann.
> (Runge & König 1924)

> The exact opposite is true, if we use the right approach ...
> (R.E. Moore 1979)

The "right approach" is, in fact, an extension of Newton's approach and has been rediscovered several times (e.g.,. Steffensen 1956) and implemented into computer programs by Gibbons (1960) and Moore (1966). For a more extensive bibliography see the references in Wanner (1969), p. 10-20.

The idea is the following: let

$$Y_i = \frac{1}{i!} y^{(i)}(x_0), \qquad F_i = \frac{1}{i!} \left(f\big(x, y(x)\big) \right)^{(i)} \Big|_{x=x_0} \qquad (8.17)$$

be the Taylor coefficients of $y(x)$ and of $f\big(x, y(x)\big)$, so that (8.13) becomes

$$y(x_0 + h) = \sum_{i=0}^{\infty} h^i Y_i.$$

Then, from (8.11),

$$Y_{i+1} = \frac{1}{i+1} F_i.$$ (8.18)

Now suppose that $f(x, y)$ is the composition of a sequence of algebraic operations and elementary functions. This leads to a sequence of items,

$$x, y, p, q, r, \ldots, \text{ and finally } f.$$ (8.19)

For each of these items we find formulas for generating the ith Taylor coefficient from the preceding ones as follows:

a) $r = p \pm q$:

$$R_i = P_i \pm Q_i, \qquad i = 0, 1, \ldots$$ (8.20a)

b) $r = pq$: the Cauchy product yields

$$R_i = \sum_{j=0}^{i} P_j Q_{i-j}, \qquad i = 0, 1, \ldots$$ (8.20b)

c) $r = p/q$: write $p = rq$, use formula b) and solve for R_i:

$$R_i = \frac{1}{Q_0} \left(P_i - \sum_{j=0}^{i-1} R_j Q_{i-j} \right), \qquad i = 0, 1, \ldots$$ (8.20c)

There also exist formulas for many elementary functions (in fact, because these functions are themselves solutions of rational differential equations).

d) $r = \exp(p)$: use $r' = p' \cdot r$ and apply (8.20b). This gives for $i = 1, 2, \ldots$

$$R_0 = \exp(P_0), \qquad R_i = \frac{1}{i} \sum_{j=0}^{i-1} (i - j) R_j P_{i-j}.$$ (8.20d)

e) $r = \log(p)$: use $p = \exp(r)$ and rearrange formula d). This gives

$$R_0 = \log(P_0), \qquad R_i = \frac{1}{P_0} \left(P_i - \frac{1}{i} \sum_{j=1}^{i-1} (i - j) P_j R_{i-j} \right).$$ (8.20e)

f) $r = p^c, c \neq 1$ constant. Use $pr' = crp'$ and apply (8.20b):

$$R_0 = P_0^c, \qquad R_i = \frac{1}{iP_0} \left(\sum_{j=0}^{i-1} (ci - (c+1)j) R_j P_{i-j} \right).$$ (8.20f)

g) $r = \cos(p)$, $s = \sin(p)$: as in d) we have

$$R_0 = \cos P_0, \qquad R_i = -\frac{1}{i}\sum_{j=0}^{i-1}(i-j)S_j P_{i-j},$$

$$(8.20g)$$

$$S_0 = \sin P_0, \qquad S_i = \frac{1}{i}\sum_{j=0}^{i-1}(i-j)R_j P_{i-j}.$$

The alternating use of (8.20) and (8.18) then allows us to compute the Taylor coefficients for (8.17) to any wanted order in a very economical way. It is not difficult to write subroutines for the above formulas, which have to be called in the same order as the differential equation (8.11) is composed of elementary operations. There also exist computer programs which "compile" Fortran statements for $f(x,y)$ into this list of subroutine calls. One has been written by T. Szymanski and J.H. Gray (see Knapp & Wanner 1969).

Example. The differential equation $y' = x^2 + y^2$ leads to the recursion

$$Y_0 = y(0), \qquad Y_{i+1} = \frac{1}{i+1}\Big(P_i + \sum_{j=0}^{i}Y_j Y_{i-j}\Big), \qquad i = 0, 1, \dots$$

where $P_i = 1$ for $i = 2$ and $P_i = 0$ for $i \neq 2$ are the coefficients for x^2. One can imagine how much easier this is than formulas (8.12).

An important property of this approach is that it can be executed in *interval analysis* and thus allows us to obtain *reliable error bounds* by the use of Lagrange's error formula for Taylor series. We refer to the books by R.E. Moore (1966) and (1979) for more details.

Exercises

1. Obtain from (8.10) the estimate

$$|y_i(x) - y_0| \le \frac{A}{L}\Big(e^{L(x-x_0)} - 1\Big)$$

and explain the similarity of this result with (7.16).

2. Apply the method of Picard to the problem $y' = Ky$, $y(0) = 1$.

3. Compute three Picard iterations for the problem $y' = x^2 + y^2$, $y(0) = 0$, $y(1/2) = ?$ and make a rigorous error estimate. Compare the result with the correct solution $y(1/2) = 0.04179114615468186322076880 6849179$.

4. Compute with an iteration method the solution of

$$y' = \sqrt{x} + \sqrt{y}, \qquad y(0) = 0$$

and observe that the method can work well for equations which pose serious problems with other methods. An even greater difference occurs for the equations

$$y' = \sqrt{x} + y^2, \quad y(0) = 0 \qquad \text{and} \qquad y' = \frac{1}{\sqrt{x}} + y^2, \quad y(0) = 0.$$

5. Define $f(x, y)$ by

$$f(x, y) = \begin{cases} 0 & \text{for } x \le 0 \\ 2x & \text{for } x > 0,\ y < 0 \\ 2x - \dfrac{4y}{x} & \text{for } 0 \le y \le x^2 \\ -2x & \text{for } x > 0,\ x^2 < y. \end{cases}$$

a) Show that $f(x, y)$ is continuous, but not Lipschitz.

b) Show that for the problem $y' = f(x, y)$, $y(0) = 0$ the Picard iteration method does not converge.

c) Show that there is a unique solution and that the Euler polygons converge.

6. Use the method of Picard iteration to prove: if $f(x, y)$ is continuous and satisfies a Lipschitz condition (7.7) on the infinite strip $D = \{(x, y)\,;\ x_0 \le x \le X\}$, then the initial value problem $y' = f(x, y)$, $y(x_0) = y_0$ possesses a unique solution on $x_0 \le x \le X$.

Compare this global result with Theorem 7.3.

7. Define a function $y(x)$ (the "inverse error function") by the relation

$$x = \frac{2}{\sqrt{\pi}} \int_0^y e^{-t^2}\, dt$$

and show that it satisfies the differential equation

$$y' = \frac{\sqrt{\pi}}{2} e^{y^2}, \qquad y(0) = 0.$$

Obtain recursion formulas for its Taylor coefficients.

I.9 Existence Theory for Systems of Equations

The first treatment of an existence theory for simultaneous systems of differential equations was undertaken in the last existing pages (p. 123-136) of Cauchy (1824). We write the equations as

$$y_1' = f_1(x, y_1, \ldots, y_n), \qquad y_1(x_0) = y_{10}, \qquad y_1(X) = ?$$
$$\ldots \qquad\qquad \ldots \qquad\qquad \ldots \tag{9.1}$$
$$y_n' = f_n(x, y_1, \ldots, y_n), \qquad y_n(x_0) = y_{n0}, \qquad y_n(X) = ?$$

and ask for the existence of the n solutions $y_1(x), \ldots, y_n(x)$. It is again natural to consider, in analogy to (7.3), the method of Euler

$$y_{k,i+1} = y_{ki} + (x_{i+1} - x_i) \cdot f_k(x_i, y_{1i}, \ldots, y_{ni}) \tag{9.2}$$

(for $k = 1, \ldots, n$ and $i = 0, 1, 2, \ldots$). Here y_{ki} is intended to approximate $y_k(x_i)$, where $x_0 < x_1 < x_2 \ldots$ is a subdivision of the interval of integration as in (7.2).

We now try to carry over everything we have done in Section I.7 to the new situation. Although we have no problem in extending (7.4) to the estimate

$$|y_{ki} - y_{k0}| \le A_k |x_i - x_0| \qquad \text{if} \qquad |f_k(x, y_1, \ldots, y_n)| \le A_k, \tag{9.3}$$

things become a little more complicated for (7.7): we have to estimate

$$f_k(x, z_1, \ldots, z_n) - f_k(x, y_1, \ldots, y_n) = \frac{\partial f_k}{\partial y_1} \cdot (z_1 - y_1) + \ldots + \frac{\partial f_k}{\partial y_n} \cdot (z_n - y_n), \tag{9.4}$$

where the derivatives $\partial f_k / \partial y_i$ are taken at suitable intermediate points. Here Cauchy uses the inequality now called the "Cauchy-Schwarz inequality" ("Enfin, il résulte de la formule (13) de la 11e leçon du calcul différentiel …") to obtain

$$|f_k(x, z_1, \ldots, z_n) - f_k(x, y_1, \ldots, y_n)| \tag{9.5}$$
$$\le \sqrt{\left(\frac{\partial f_k}{\partial y_1}\right)^2 + \ldots + \left(\frac{\partial f_k}{\partial y_n}\right)^2} \cdot \sqrt{(z_1 - y_1)^2 + \ldots + (z_n - y_n)^2}.$$

At this stage, we begin to feel that further development is advisable only after the introduction of vector notation.

Vector Notation

This was promoted in our subject by the papers of Peano, (1888) and (1890), who was influenced, as he says, by the famous "Ausdehnungslehre" of Grassmann and the work of Hamilton, Cayley, and Sylvester. We introduce the vectors (Peano called them "complexes")

$$y = (y_1, \ldots, y_n)^T, \quad y_i = (y_{1i}, \ldots, y_{ni})^T, \quad z = (z_1, \ldots, z_n)^T \quad \text{etc},$$

and hope that the reader will not confuse the components y_i of a vector y with vectors with indices. We consider the "vector function"

$$f(x, y) = \big(f_1(x, y), \ldots, f_n(x, y)\big)^T,$$

so that equations (9.1) become

$$y' = f(x, y), \qquad y(x_0) = y_0, \qquad y(X) =?, \tag{9.1'}$$

Euler's method (9.2) is

$$y_{i+1} = y_i + (x_{i+1} - x_i) f(x_i, y_i), \qquad i = 0, 1, 2, \ldots \tag{9.2'}$$

and the Euler polygon is given by

$$y_h(x) = y_i + (x - x_i) f(x_i, y_i) \qquad \text{for} \qquad x_i \leq x \leq x_{i+1}.$$

There is no longer any difference in notation with the one-dimensional cases (7.1), (7.3) and (7.3a).

In view of estimate (9.5), we introduce for a vector $y = (y_1, \ldots, y_n)^T$ the *norm* (originally "modulus")

$$\|y\| = \sqrt{y_1^2 + \ldots + y_n^2} \tag{9.6}$$

which satisfies all the usual properties of a norm, for example the triangle inequality

$$\|y + z\| \leq \|y\| + \|z\|, \qquad \left\|\sum_{i=1}^{n} y_i\right\| \leq \sum_{i=1}^{n} \|y_i\|. \tag{9.7}$$

The Euclidean norm (9.6) is not the only one possible, we also use ("on pourrait aussi définir par mx la plus grande des valeurs absolues des élements de x; alors les propriétes des modules sont presqu'évidentes.", Peano)

$$\|y\| = \max(|y_1|, \ldots, |y_n|), \tag{9.6'}$$

$$\|y\| = |y_1| + \ldots + |y_n|. \tag{9.6''}$$

We are now able to formulate estimate (9.3) as follows, in perfect analogy with (7.4): if for some norm $\|f(x, y)\| \leq A$ on $D = \{(x, y) \mid x_0 \leq x \leq X, \|y - y_0\| \leq b\}$ and if $X - x_0 \leq b/A$ then the numerical solution (x_i, y_i), given by (9.2'), remains in D and we have

$$\|y_h(x) - y_0\| \leq A \cdot |x - x_0|. \tag{9.8}$$

The analogue of estimate (7.5) can be obtained similarly.

In order to prove the implication "(7.9) $\Rightarrow$ (7.7)" for vector-valued functions it is convenient to work with norms of matrices.

Subordinate Matrix Norms

The relation (9.4) shows that the difference $f(x, z) - f(x, y)$ can be written as the product of a matrix with the vector $z - y$. It is therefore of interest to estimate $\|Qv\|$ and to find the best possible estimate of the form $\|Qv\| \leq \beta\|v\|$.

Definition 9.1. Let Q be a matrix (n columns, m rows) and $\|\ldots\|$ be one of the norms defined in (9.6), (9.6') or (9.6"). The *subordinate matrix norm* of Q is then defined by

$$\|Q\| = \sup_{v \neq 0} \frac{\|Qv\|}{\|v\|} = \sup_{\|u\|=1} \|Qu\|. \tag{9.9}$$

By definition, $\|Q\|$ is the smallest number such that

$$\|Qv\| \leq \|Q\| \cdot \|v\| \qquad \text{for all } v \tag{9.10}$$

holds. The following theorem gives explicit formulas for the computation of (9.9).

Theorem 9.2. *The norm of a matrix Q is given by the following formulas: for the Euclidean norm (9.6),*

$$\|Q\| = \sqrt{\text{largest eigenvalue of } Q^T Q}; \tag{9.11}$$

for the max-norm (9.6'),

$$\|Q\| = \max_{k=1,\ldots,m} \left(\sum_{i=1}^{n} |q_{ki}|\right); \tag{9.11'}$$

for the norm (9.6"),

$$\|Q\| = \max_{i=1,\ldots,n} \left(\sum_{k=1}^{m} |q_{ki}|\right). \tag{9.11"}$$

Proof. Formula (9.11) can be seen from $\|Qv\|^2 = v^T Q^T Q v$ with the help of an orthogonal transformation of $Q^T Q$ to diagonal form.

Formula (9.11') is obtained as follows (we denote (9.6') by $\|\ldots\|_\infty$):

$$\|Qv\|_\infty = \max_{k=1,\ldots,m} \left|\sum_{i=1}^{n} q_{ki} v_i\right| \leq \left(\max_{k=1,\ldots,m} \sum_{i=1}^{n} |q_{ki}|\right) \cdot \|v\|_\infty \tag{9.12}$$

shows that $\|Q\| \leq \max_k \sum_i |q_{ki}|$. The equality in (9.11') is then seen by choosing a vector of the form $v = (\pm 1, \pm 1, \ldots, \pm 1)^T$ for which equality holds in (9.12). The formula (9.11") is proved along the same lines. $\qquad \square$

All these formulas remain valid for *complex matrices.* Q^T has only to be replaced by Q^* (transposed and complex conjugate). See e.g., Wilkinson (1965), p. 55-61, Bakhvalov (1976), Chap. VI, Par. 3. With these preparations it is possible to formulate the desired estimate.

Theorem 9.3. *If $f(x,y)$ is differentiable with respect to y in an open convex region U and if*

$$\left\| \frac{\partial f}{\partial y}(x,y) \right\| \le L \qquad \text{for} \qquad (x,y) \in U \tag{9.13}$$

then

$$\| f(x,z) - f(x,y) \| \le L \, \| z - y \| \qquad \text{for} \qquad (x,y), (x,z) \in U. \tag{9.14}$$

(Obviously, the matrix norm in (9.13) is subordinate to the norm used in (9.14).)

Proof. This is the "mean value theorem" and its proof can be found in every textbook on calculus. In the case where $\partial f / \partial y$ is continuous, the following simple proof is possible. We consider $\varphi(t) = f(x, y + t(z-y))$ and integrate its derivative (componentwise) from 0 to 1

$$
\begin{aligned}
f(x,z) - f(x,y) = \varphi(1) - \varphi(0) &= \int_0^1 \varphi'(t) \, dt \\
&= \int_0^1 \frac{\partial f}{\partial y}(x, y + t(z-y)) \cdot (z-y) \, dt.
\end{aligned}
\tag{9.15}
$$

Taking the norm of (9.15), using

$$\left\| \int_0^1 g(t) \, dt \right\| \le \int_0^1 \| g(t) \| \, dt, \tag{9.16}$$

and applying (9.10) and (9.13) yields the estimate (9.14). The relation (9.16) is proved by applying the triangle inequality (9.7) to the finite Riemann sums which define the two integrals. □

We thus have obtained the analogue of (7.7). All that remains to do is, *Da capo al fine,* to read Sections I.7 and I.8 again: *Lemma 7.2, Theorems 7.3, 7.4, 7.5, and 7.6 together with their proofs and the estimates (7.10), (7.13), (7.15), (7.16), (7.17), and (7.18) carry over to the more general case with the only changes that some absolute values are to be replaced by norms.*

The *Picard-Lindelöf iteration* also carries over to systems of equations when in (8.7) we interpret $y_{i+1}(x), y_0$ and $f(s, y_i(s))$ as vectors, integrated componentwise. The convergence result with the estimate (8.10) also remains the same; for its proof we have to use, between (8.8) and (8.9), the inequality (9.16).

The Taylor series method, its convergence proof, and the recursive generation of the Taylor coefficients also generalize in a straightforward manner to systems of equations.

Exercises

1. Solve the system
$$y_1' = -y_2, \qquad y_1(0) = 1$$
$$y_2' = +y_1, \qquad y_2(0) = 0$$
by the methods of Euler and Picard, establish rigorous error estimates for all three norms mentioned. Verify the results using the correct solution $y_1(x) = \cos x$, $y_2(x) = \sin x$.

2. Consider the differential equations
$$y_1' = -100y_1 + y_2, \qquad y_1(0) = 1, \qquad y_1(1) = ?$$
$$y_2' = y_1 - 100y_2, \qquad y_2(0) = 0, \qquad y_2(1) = ?$$

   a) Compute the exact solution $y(x)$ by the method explained in Section I.6.
   b) Compute the error bound for $\|z(x) - y(x)\|$, where $z(x) = 0$, obtained from (7.10).
   c) Apply the method of Euler to this equation with $h = 1/10$.
   d) Apply Picard's iteration method.

3. Compute the Taylor series solution of the system with constant coefficients $y' = Ay$, $y(0) = y_0$. Prove that this series converges for all x. Apply this series to the equation of Exercise 1.

   Result.
   $$y(x) = \sum_{i=0}^{\infty} \frac{x^i}{i!} A^i y_0 =: e^{Ax} y_0.$$

I.10 Differential Inequalities

Differential inequalities are an elegant instrument for gaining a better understanding of equations (7.10), (7.17) and much new insight. This subject was inaugurated in the paper, once again, Peano (1890) and further developed by Perron (1915), Müller (1926), Kamke (1930). A classical treatise on the subject is the book of Walter (1970).

Introduction

The basic idea is the following: let $v(x)$ denote the Euler polygon defined in (7.3) or (9.2), so that

$$v'(x) = f(x_i, y_i) \qquad \text{for} \quad x_i < x < x_{i+1}. \tag{10.1}$$

For any chosen norm, we investigate the *error*

$$m(x) = \|v(x) - y(x)\| \tag{10.2}$$

as a function of x and we naturally try to estimate its growth.

Unfortunately, $m(x)$ is not necessarily differentiable, due firstly to the corners of the Euler polygons and secondly, to corners originating from the norms, especially the norms (9.6') and (9.6"). Therefore we consider the so-called *Dini derivatives* defined by

$$D^+ m(x) = \limsup_{h \to 0, h > 0} \frac{m(x+h) - m(x)}{h},$$

$$D_+ m(x) = \liminf_{h \to 0, h > 0} \frac{m(x+h) - m(x)}{h},$$

(see e.g., Scheeffer (1884), Hobson (1921), Chap. V, §260, §280). The property

$$\|w(x+h)\| - \|w(x)\| \le \|w(x+h) - w(x)\| \tag{10.3}$$

is a simple consequence of the triangle inequality (9.7). If we divide (10.3) by $h > 0$, we obtain the estimates

$$D_+ \|w(x)\| \le \|w'(x+0)\|, \qquad D^+ \|w(x)\| \le \|w'(x+0)\|, \tag{10.4}$$

where $w'(x+0)$ is the right derivative of the vector function $w(x)$. If we apply this to $m(x)$ of (10.2), we obtain

$$D_+m(x) \leq \|v'(x+0) - y'(x)\|$$
$$= \|v'(x+0) - f(x, v(x)) + f(x, v(x)) - f(x, y(x))\|$$

and, using the triangle inequality and the Lipschitz condition (9.14),

$$D_+m(x) \leq \delta(x) + L \cdot m(x). \tag{10.5}$$

Here, we have introduced

$$\delta(x) = \|v'(x+0) - f(x, v(x))\| \tag{10.6}$$

which is called the *defect* of the approximate solution $v(x)$. This fundamental quantity measures the extent to which the function $v(x)$ does *not* satisfy the imposed differential equation. (7.11) together with (10.1) tell us that $\delta(x) \leq \varepsilon$, so that (10.5) can be further estimated to become

$$D_+m(x) \leq L \cdot m(x) + \varepsilon, \qquad m(x_0) = 0. \tag{10.7}$$

Formula (10.7) (or (10.5)) is what one calls a *differential inequality*. The question is: are we allowed to replace "$\leq$" by "$=$", i.e., to solve instead of (10.7) the equation

$$u' = Lu + \varepsilon, \qquad u(x_0) = 0 \tag{10.8}$$

and to conclude that $m(x) \leq u(x)$? This would mean, by the formulas of Section I.3 or I.5, that

$$m(x) \leq \frac{\varepsilon}{L}\left(e^{L(x-x_0)} - 1\right). \tag{10.9}$$

We would thus have obtained (7.17) in a natural way and have furthermore discovered an elegant and powerful tool for many kinds of new estimates.

The Fundamental Theorems

A general theorem of the type

$$\left.\begin{array}{l} D_+m(x) \leq g(x, m(x)) \\ D_+u(x) \geq g(x, u(x)) \\ m(x_0) \leq u(x_0) \end{array}\right\} \implies m(x) \leq u(x) \quad \text{for} \quad x_0 \leq x \tag{10.10}$$

cannot be true. Counter-examples are provided by any differential equation with non-unique solutions, such as

$$g(x, y) = \sqrt{y}, \qquad m(x) = \frac{x^2}{4}, \qquad u(x) = 0. \tag{10.11}$$

The important observation, due to Peano and Perron, which allows us to overcome this difficulty, is that *one* of the first two inequalities must be replaced by a *strict* inequality (see Peano (1890), §3, Lemme 1):

Theorem 10.1. *Suppose that the functions $m(x)$ and $u(x)$ are continuous and satisfy for $x_0 \leq x < X$*

$$a) \quad D_+ m(x) \leq g(x, m(x))$$
$$b) \quad D_+ u(x) > g(x, u(x)) \tag{10.12}$$
$$c) \quad m(x_0) \leq u(x_0).$$

Then

$$m(x) \leq u(x) \quad for \quad x_0 \leq x \leq X. \tag{10.13}$$

The same conclusion is true if both D_+ are replaced by D^+.

Proof. In order to be able to compare the derivatives $D_+ m$ and $D_+ u$ in (10.12), we consider points at which $m(x) = u(x)$. This is the main idea.

If (10.13) were not true, we could choose a point x_2 with $m(x_2) > u(x_2)$ and look for the first point x_1 to the left of x_2 with $m(x_1) = u(x_1)$. Then for small $h > 0$ we would have

$$\frac{m(x_1 + h) - m(x_1)}{h} > \frac{u(x_1 + h) - u(x_1)}{h}$$

and, by taking limits, $D_+ m(x_1) \geq D_+ u(x_1)$. This, however, contradicts (a) and (b), which give

$$D_+ m(x_1) \leq g(x_1, m(x_1)) = g(x_1, u(x_1)) < D_+ u(x_1). \qquad \square$$

Many variant forms of this theorem are possible, for example by using left Dini derivates (Walter 1970, Chap. II, §8, Theorem V).

Theorem 10.2 (The "fundamental lemma"). *Suppose that $y(x)$ is a solution of the system of differential equations $y' = f(x, y)$, $y(x_0) = y_0$, and that $v(x)$ is an approximate solution. If*

$$a) \quad \|v(x_0) - y(x_0)\| \leq \varrho$$
$$b) \quad \|v'(x+0) - f(x, v(x))\| \leq \varepsilon$$
$$c) \quad \|f(x, v) - f(x, y)\| \leq L\|v - y\|,$$

then, for $x \geq x_0$, we have the error estimate

$$\|y(x) - v(x)\| \leq \varrho e^{L(x - x_0)} + \frac{\varepsilon}{L}\left(e^{L(x - x_0)} - 1\right). \tag{10.14}$$

Remark. The two terms in (10.14) express, respectively, the influence of the error ϱ in the initial values and the influence of the defect ε to the error of the approximate solution. It implies that the error depends continuously on both, and that for $\varrho = \varepsilon = 0$ we have $y(x) = v(x)$, i.e., uniqueness of the solution.

Proof. We put $m(x) = \|y(x) - v(x)\|$ and obtain, as in (10.7),

$$D_+ m(x) \leq L \cdot m(x) + \varepsilon, \qquad m(x_0) \leq \varrho.$$

We shall try to compare this with the differential equation

$$u' = Lu + \varepsilon, \qquad u(x_0) = \varrho. \tag{10.15}$$

Theorem 10.1 is not directly applicable. We therefore replace in (10.15) ε by $\varepsilon + \eta, \eta > 0$ and solve instead

$$u' = Lu + \varepsilon + \eta > Lu + \varepsilon, \qquad u(x_0) = \varrho.$$

Now Theorem 10.1 gives the estimate (10.14) with ε replaced by $\varepsilon + \eta$. Since this estimate is true for *all* $\eta > 0$, it is also true for $\eta = 0$. □

Variant form of Theorem 10.2. *The conditions*

$$a) \quad \|v(x_0) - y(x_0)\| \leq \varrho$$
$$b) \quad \|v'(x+0) - f(x, v(x))\| \leq \delta(x)$$
$$c) \quad \|f(x, v) - f(x, y)\| \leq \ell(x)\|v - y\|$$

imply for $x \geq x_0$

$$\|y(x) - v(x)\| \leq e^{L(x)}\left(\varrho + \int_{x_0}^{x} e^{-L(s)} \delta(s) \, ds\right), \qquad L(x) = \int_{x_0}^{x} \ell(s) \, ds.$$

Proof. This is simply formula (3.3). □

Theorem 10.3. *If the function* $g(x, y)$ *is continuous and satisfies a Lipschitz condition, then the implication (10.10) is true for continuous functions* $m(x)$ *and* $u(x)$.

Proof. Define functions $w_n(x), v_n(x)$ by

$$w_n'(x) = g(x, w_n(x)) + 1/n, \qquad w_n(x_0) = m(x_0),$$
$$v_n'(x) = g(x, v_n(x)) - 1/n, \qquad v_n(x_0) = u(x_0),$$

so that from Theorem 10.1

$$m(x) \leq w_n(x), \qquad v_n(x) \leq u(x) \qquad \text{for} \qquad x_0 \leq x \leq X. \tag{10.16}$$

It follows from Theorem 10.2 that the functions $w_n(x)$ and $v_n(x)$ converge for $n \to \infty$ to the solutions of

$$w'(x) = g(x, w(x)), \qquad w(x_0) = m(x_0),$$
$$v'(x) = g(x, v(x)), \qquad v(x_0) = u(x_0),$$

since the defect is $\pm 1/n$. Finally, because of $m(x_0) \le u(x_0)$ and uniqueness we have $w(x) \le v(x)$. Taking the limit $n \to \infty$ in (10.16) thus gives $m(x) \le u(x)$.

$\square$

A further generalization of Theorem 10.2 is possible if the Lipschitz condition (c) is replaced by something nonlinear such as

$$\|f(x,v) - f(x,y)\| \le \omega(x, \|v - y\|).$$

Then the differential inequality for the error $m(x)$ is to be compared with the solution of

$$u' = \omega(x, u) + \delta(x) + \eta, \qquad u(x_0) = \varrho, \qquad \eta > 0.$$

See Walter (1970), Chap. II, §11 for more details.

Estimates Using One-Sided Lipschitz Conditions

As we already observed in Exercise 2 of I.9, and as has been known for a long time, much information about the errors can be lost by the use of positive Lipschitz constants L (e.g (9.11), (9.11'), or (9.11")) in the estimates (7.16), (7.17), or (7.18). The estimates all grow exponentially with x, even if the solutions and errors decay. Therefore many efforts have been made to obtain better error estimates, as for example the papers Eltermann (1955), Uhlmann (1957), Dahlquist (1959), and the references therein. We follow with great pleasure the particularly clear presentation of Dahlquist.

Let us estimate the derivative of $m(x) = \|v(x) - y(x)\|$ with more care than we did in (10.5): for $h > 0$ we have

$$
\begin{aligned}
m(x + h) &= \|v(x + h) - y(x + h)\| \\
&= \|v(x) - y(x) + h(v'(x + 0) - y'(x))\| + \mathcal{O}(h^2) \qquad (10.17) \\
&\le \left\| v(x) - y(x) + h\Big(f(x, v(x)) - f(x, y(x))\Big) \right\| + h\delta(x) + \mathcal{O}(h^2)
\end{aligned}
$$

by the use of (10.6) and (9.7). Here, we apply the mean value theorem to the function $y + hf(x, y)$ and obtain

$$m(x + h) \le \left(\max_{\eta \in [y(x), v(x)]} \left\| I + h\frac{\partial f}{\partial y}(x, \eta) \right\| \right) \cdot m(x) + h\delta(x) + \mathcal{O}(h^2)$$

and finally for $h > 0$,

$$\frac{m(x + h) - m(x)}{h} \le \max_{\eta \in [y(x), v(x)]} \frac{\|I + h\frac{\partial f}{\partial y}(x, \eta)\| - 1}{h} m(x) + \delta(x) + \mathcal{O}(h).$$

$$(10.18)$$

The expression on the right hand side of (10.18) leads us to the following definition:

Definition 10.4. Let Q be a square matrix, then we call

$$\mu(Q) = \lim_{h \to 0, h > 0} \frac{\|I + hQ\| - 1}{h} \tag{10.19}$$

the *logarithmic norm* of Q.

Here are formulas for its computation (Dahlquist (1959), p. 11, Eltermann (1955), p. 498, 499):

Theorem 10.5. *The logarithmic norm (10.19) is obtained by the following formulas: for the Euclidean norm (9.6),*

$$\mu(Q) = \lambda_{\max} = \text{largest eigenvalue of } \frac{1}{2}(Q^T + Q); \tag{10.20}$$

for the max-norm (9.6'),

$$\mu(Q) = \max_{k=1,\dots,n} \left(q_{kk} + \sum_{i \neq k} |q_{ki}| \right); \tag{10.20'}$$

for the norm (9.6"),

$$\mu(Q) = \max_{i=1,\dots,n} \left(q_{ii} + \sum_{k \neq i} |q_{ki}| \right). \tag{10.20"}$$

Proofs. Formulas (10.20') and (10.20") follow quite trivially from (9.11') and (9.11") and the definition (10.19). The point is that the presence of I suppresses, for h sufficiently small, the absolute values for the diagonal elements. (10.20) is seen from the fact that the eigenvalues of

$$(I + hQ)^T (I + hQ) = I + h(Q^T + Q) + h^2 Q^T Q,$$

for $h \to 0$, converge to $1 + h\lambda_i$, where λ_i are the eigenvalues of $Q^T + Q$. $\quad\square$

Remark. For complex-valued matrices the above formulas remain valid if one replaces Q by Q^* and q_{kk}, q_{ii} by $\text{Re}\,q_{kk}$, $\text{Re}\,q_{ii}$.

We now obtain from (10.18) the following improvement of Theorem 10.3.

Theorem 10.6. *Suppose that we have the estimates*

$$\mu\left(\frac{\partial f}{\partial y}(x, \eta) \right) \leq \ell(x) \qquad for \qquad \eta \in [y(x), v(x)] \qquad and \tag{10.21}$$

$$\|v'(x+0) - f(x, v(x))\| \leq \delta(x), \qquad \|v(x_0) - y(x_0)\| \leq \varrho.$$

Then for $x > x_0$ we have

$$\|y(x) - v(x)\| \leq e^{L(x)} \left(\varrho + \int_{x_0}^{x} e^{-L(s)} \delta(s)\, ds \right), \tag{10.22}$$

with $L(x) = \int_{x_0}^{x} \ell(s)\, ds$.

Proof. Since, for a fixed x, the segment $[v(x), y(x)]$ is compact,

$$K = \max_i \max_{[v(x), y(x)]} \left| \frac{\partial f_i}{\partial y_i} \right|$$

is finite. Then (see the proof of Theorem 10.5)

$$\frac{\|I + h\frac{\partial f}{\partial y}(x, \eta)\| - 1}{h} = \mu\left(\frac{\partial f}{\partial y}(x, \eta)\right) + \mathcal{O}(h)$$

where the $\mathcal{O}(h)$-term is uniformly bounded in η. (For the norms (9.6') and (9.6")) this term is in fact zero for $h < 1/K$). Thus the condition (10.21) inserted into (10.18) gives

$$D_+ m(x) \leq \ell(x)m(x) + \delta(x).$$

Now the estimate (10.22) follows in the same way as that of Theorem 10.3.

$\square$

Exercises

1. Apply Theorem 10.6 to the example of Exercise 2 of I.9. Observe the substantial improvement of the estimates.

2. Prove the following (a variant form of the famous "Gronwall lemma", Gronwall 1919): suppose that a positive function $m(x)$ satisfies

$$m(x) \leq \varrho + \varepsilon(x - x_0) + L\int_{x_0}^x m(s)\, ds =: w(x) \qquad (10.23)$$

   then

$$m(x) \leq \varrho e^{L(x - x_0)} + \frac{\varepsilon}{L}\left(e^{L(x - x_0)} - 1\right); \qquad (10.24)$$

   a) directly, by subtracting from (10.23)

$$u(x) = \varrho + \varepsilon(x - x_0) + L\int_{x_0}^x u(s)\, ds;$$

   b) by differentiating $w(x)$ in (10.23) and using Theorem 10.1.

   c) Prove Theorem 10.2 with the help of the above lemma of Gronwall. The same interrelations are, of course, also valid in more general situations.

3. Consider the problem $y' = \lambda y$, $y(0) = 1$ with $\lambda \geq 0$ and apply Euler's method with constant step size $h = 1/n$. Prove that

$$\frac{\lambda}{1 + \lambda/n} y_h(x) \leq D_+ y_h(x) \leq \lambda y_h(x).$$

and derive the estimate

$$\left(1 + \frac{\lambda}{n}\right)^n \le e^\lambda \le \left(1 + \frac{\lambda}{n}\right)^{n+\lambda} \qquad \text{for} \quad \lambda \ge 0.$$

4. Prove the following properties of the logarithmic norm:

   a) $\mu(\alpha Q) = \alpha \mu(Q)$ for $\alpha \ge 0$

   b) $-\|Q\| \le \mu(Q) \le \|Q\|$

   c) $\mu(Q + P) \le \mu(Q) + \mu(P),$ $\mu\left(\int Q(t)\,dt\right) \le \int \mu(Q(t))\,dt$

   d) $|\mu(Q) - \mu(P)| \le \|Q - P\|.$

5. For the Euclidean norm (10.20), $\mu(Q)$ is the smallest number satisfying

$$\langle v, Qv \rangle \le \mu(Q) \|v\|^2.$$

This property is valid for all norms associated with a scalar product. Prove this.

6. Show that for the Euclidean norm the condition (10.21) is equivalent to

$$\langle y - z, f(x, y) - f(x, z) \rangle \le \ell(x) \|y - z\|^2.$$

7. Observe, using an example of the form

$$y_1' = y_2, \qquad y_2' = -y_1,$$

that a generalization of Theorem 10.1 to *systems* of first order differential equations, with inequalities interpreted component-wise, is not true in general (Müller 1926).

However, it is possible to prove such a generalization of Theorem 10.1 under the additional hypothesis that the functions $g_i(x, y_1, \ldots, y_n)$ are *quasimonotone*, i.e., that

$$g_i(x, y_1, \ldots, y_j, \ldots, y_n) \le g_i(x, y_1, \ldots, z_j, \ldots, y_n)$$

$$\text{if} \quad y_j < z_j \quad \text{for all} \quad j \ne i.$$

Try to prove this.

An important fact is that many systems from parabolic differential equations, such as equation (6.10), *are* quasimonotone. This allows many interesting applications of the ideas of this section (see Walter (1970), Chap. IV).

I.11 Systems of Linear Differential Equations

With more knowledge about existence and uniqueness, and with more skill in linear algebra, we shall now, as did the mathematicians of the 19th century, better understand many points which had been left somewhat obscure in Sections I.4 and I.6 about linear differential equations of higher order.

Equation (4.9) divided by $a_n(x)$ (which is $\neq 0$ away from singular points) becomes

$$y^{(n)} + b_{n-1}(x)y^{(n-1)} + \ldots + b_0(x)y = g(x), \qquad b_i(x) = a_i(x)/a_n(x). \quad (11.1)$$

with $g(x) = f(x)/a_n(x)$. Introducing $y = y_1$, $y' = y_2, \ldots, y^{(n-1)} = y_n$ we arrive at

$$\begin{pmatrix} y_1' \\ y_2' \\ \vdots \\ y_n' \end{pmatrix} = \begin{pmatrix} 0 & 1 & & \\ 0 & 0 & \ddots & \\ \vdots & \vdots & & 1 \\ -b_0(x) & -b_1(x) & \ldots & -b_{n-1}(x) \end{pmatrix} \begin{pmatrix} y_1 \\ y_2 \\ \vdots \\ y_n \end{pmatrix} + \begin{pmatrix} 0 \\ \vdots \\ 0 \\ g(x) \end{pmatrix}. \quad (11.1')$$

We again denote by y the vector $(y_1, \ldots, y_n)^T$ and by $f(x)$ the inhomogeneity, so that (11.1') becomes a special case of the following *system of linear differential equations*

$$y' = A(x)y + f(x), \tag{11.2}$$

$$A(x) = \big(a_{ij}(x)\big), \qquad f(x) = \big(f_i(x)\big), \qquad i,j = 1, \ldots, n.$$

Here, the theorems of Section I.9 and I.10 apply without difficulty. Since the partial derivatives of the right hand side of (11.2) with respect to y_i are given by $a_{ki}(x)$, we have the Lipschitz estimate (see condition (c) of the variant form of Theorem 10.2), where $\ell(x) = \|A(x)\|$ in any subordinate matrix norm (9.11, 11', 11''). We apply Theorem 7.4, and the variant form of Theorem 10.2 with $v(x) = 0$ as "approximate solution". We may also take $\ell(x) = \mu(A(x))$ (see (10.20, 20', 20'')) and apply Theorem 10.6.

Theorem 11.1. *Suppose that $A(x)$ is continuous on an interval $[x_0, X]$. Then for any initial values $y_0 = (y_{10}, \ldots, y_{n0})^T$ there exists for all $x_0 \leq x \leq X$ a unique* *solution of (11.2) satisfying*

$$\|y(x)\| \le e^{L(x)}\left(\|y_0\| + \int_{x_0}^{x} e^{-L(s)}\|f(s)\|\,ds\right) \tag{11.3}$$

$$L(x) = \int_{x_0}^{x} \ell(s)\,ds, \qquad \ell(x) = \|A(x)\| \quad or \quad \ell(x) = \mu\big(A(x)\big).$$

For $f(x) \equiv 0$, $y(x)$ depends linearly on the initial values, i.e., there is a matrix $R(x, x_0)$ (the "resolvent"), such that

$$y(x) = R(x, x_0)\, y_0. \tag{11.4}$$

Proof. Since $\ell(x)$ is continuous and therefore bounded on any compact interval $[x_0, X]$, the estimate (11.3) shows that the solutions can be continued until the end. The linear dependence follows from the fact that, for $f \equiv 0$, linear combinations of solutions are again solutions, and from uniqueness. $\qquad\square$

Resolvent and Wronskian

From uniqueness we have that the solutions with initial values y_0 at x_0 and $y_1 = R(x_1, x_0)\, y_0$ at x_1 (see (11.4)) must be the same. Hence we have

$$R(x_2, x_0) = R(x_2, x_1)R(x_1, x_0) \tag{11.5}$$

for $x_0 \le x_1 \le x_2$. Finally by integrating backward from x_1, y_1, i.e., by the co-ordinate transformation $x = x_1 - t$, $0 \le t \le x_1 - x_0$, we must arrive, again by uniqueness, at the starting values. Hence

$$R(x_0, x_1) = \Big(R(x_1, x_0)\Big)^{-1} \tag{11.6}$$

and (11.5) is true without any restriction on x_0, x_1, x_2.

Let $y_i(x) = (y_{1i}(x), \ldots, y_{ni}(x))^T$ (for $i = 1, \ldots, n$) be a set of n solutions of the homogeneous differential equation

$$y' = A(x)\, y \tag{11.7}$$

which are *linearly independent* at $x = x_0$ (i.e., they form a *fundamental system*). We form the *Wronskian matrix* (Wronski 1810)

$$W(x) = \begin{pmatrix} y_{11}(x) & \cdots & y_{1n}(x) \\ \vdots & & \vdots \\ y_{n1}(x) & \cdots & y_{nn}(x) \end{pmatrix},$$

so that

$$W'(x) = A(x)W(x)$$

and all solutions can be written as

$$c_1 y_1(x) + \ldots + c_n y_n(x) = W(x)\,c \qquad \text{where} \qquad c = (c_1, \ldots, c_n)^T. \quad (11.8)$$

If this solution must satisfy the initial conditions $y(x_0) = y_0$, we obtain $c = W^{-1}(x_0)y_0$ and we have the formula

$$R(x, x_0) = W(x)W^{-1}(x_0). \tag{11.9}$$

Therefore all solutions are known if one has found n linearly independent solutions.

Inhomogeneous Linear Equations

Extending the idea of Joh. Bernoulli for (3.2) and Lagrange for (4.9), we now compute the solutions of the *inhomogeneous* equation (11.2) by letting c be "variable" in the "general solution" (11.8): $y(x) = W(x)c(x)$ (Liouville 1838). Exactly as in Section I.3 for (3.2) we obtain from (11.2) and (11.7) by differentiation

$$y' = W'c + Wc' = AWc + Wc' = AWc + f.$$

Hence $c' = W^{-1}f$. If we integrate this with integration constants c, we obtain

$$y(x) = W(x) \int_{x_0}^{x} W^{-1}(s)f(s)\,ds + W(x)\,c.$$

The initial conditions $y(x_0) = y_0$ imply $c = W^{-1}(x_0)y_0$ and we obtain:

Theorem 11.2 ("Variation of constants formula"). *Let $A(x)$ and $f(x)$ be continuous. Then the solution of the inhomogeneous equation $y' = A(x)y + f(x)$ satisfying the initial conditions $y(x_0) = y_0$ is given by*

$$y(x) = W(x)\left(W^{-1}(x_0)\,y_0 + \int_{x_0}^{x} W^{-1}(s)f(s)\,ds\right)$$
$$= R(x, x_0)\,y_0 + \int_{x_0}^{x} R(x, s)f(s)\,ds. \tag{11.10}$$

The Abel-Liouville-Jacobi-Ostrogradskii Identity

We already know from (11.6) that $W(x)$ remains regular for all x. We now show that the *determinant* of $W(x)$ can be given explicitly as follows (Abel 1827, Liouville 1838, Jacobi 1845, §17):

$$\det\big(W(x)\big) = \det\big(W(x_0)\big) \cdot \exp\Big(\int_{x_0}^{x} \operatorname{tr}\big(A(s)\big)\, ds\Big), \qquad (11.11)$$

$$\operatorname{tr}\big(A(x)\big) = a_{11}(x) + a_{22}(x) + \ldots + a_{nn}(x)$$

which connects the determinant of $W(x)$ to the *trace* of $A(x)$.

For the *proof* of (11.11) (see also Exercise 2) we compute the derivative $\frac{d}{dx}\det\big(W(x)\big)$. Since $\det\big(W(x)\big)$ is multilinear, this derivative (by the Leibniz rule) is a sum of n terms, whose first is

$$T_1 = \det \begin{pmatrix} y'_{11} & y'_{12} & \cdots & y'_{1n} \\ y_{21} & y_{22} & \cdots & y_{2n} \\ \vdots & \vdots & & \vdots \\ y_{n1} & y_{n2} & \cdots & y_{nn} \end{pmatrix}.$$

We insert $y'_{1i} = a_{11}(x)y_{1i} + \ldots + a_{1n}(x)y_{ni}$ from (11.7). All terms $a_{12}(x)y_{2i}$, $\ldots, a_{1n}(x)y_{ni}$ disappear by subtracting multiples of lines 2 to n, so that $T_1 = a_{11}(x)\det\big(W(x)\big)$. Summing all these terms we obtain finally

$$\frac{d}{dx}\det\big(W(x)\big) = \big(a_{11}(x) + \ldots + a_{nn}(x)\big) \cdot \det\big(W(x)\big) \qquad (11.12)$$

and (11.11) follows by integration. $\qquad\square$

Exercises

1. Compute the resolvent matrix $R(x, x_0)$ for the two systems

$$\begin{aligned} y'_1 &= y_1 & y'_1 &= y_2 \\ y'_2 &= 3y_2 & y'_2 &= -y_1 \end{aligned}$$

and check the validity of (11.5), (11.6) as well as (11.11).

2. Reconstruct Abel's original proof for (11.11), which was for the case

$$y''_1 + py'_1 + qy_1 = 0, \qquad y''_2 + py'_2 + qy_2 = 0.$$

Multiply the equations by y_2 and y_1 respectively and subtract to eliminate q. Then integrate.

Use the result to obtain an identity for the two integrals

$$y_1(a) = \int_0^\infty e^{ax - x^2} x^{\alpha - 1}\, dx, \qquad y_2(a) = \int_0^\infty e^{-ax - x^2} x^{\alpha - 1}\, dx,$$

which both satisfy

$$\frac{d^2 y_i}{da^2} - \frac{a}{2} \cdot \frac{dy_i}{da} - \frac{\alpha}{2} y_i = 0. \qquad (11.13)$$

Hint. To verify (11.13), integrate from 0 to infinity the expression for $\frac{d}{dx}(\exp(ax - x^2)x^\alpha)$ (Abel 1827, case IV).

3. (Kummer 1839). Show that the general solution of the equation

$$y^{(n)}(x) = x^m y(x) \tag{11.14}$$

can be obtained by quadrature.

Hint. Differentiate (11.14) to obtain

$$y^{(n+1)} = x^m y' + m x^{m-1} y. \tag{11.15}$$

Suppose by recursion that the general solution of

$$\psi^{(n+1)} = x^{m-1}\psi, \quad \text{i.e.,} \quad \frac{d^{n+1}}{dx^{n+1}}\psi(xu) = x^{m-1}u^{m+n}\psi(xu) \tag{11.16}$$

is already known. Show that then

$$y(x) = \int_0^\infty u^{m-1} \exp\left(-\frac{u^{m+n}}{m+n}\right)\psi(xu)\,dx$$

is the general solution of (11.15), and, under some conditions on the parameters, also of (11.14). To simplify the computations, consider the function

$$g(u) = u^m \exp\left(-\frac{u^{m+n}}{m+n}\right)\psi(xu),$$

compute its derivative with respect to u, multiply by x^{m-1}, and integrate from 0 to infinity.

4. (Weak singularities for systems). Show that the linear system

$$y' = \frac{1}{x}\left(A_0 + A_1 x + A_2 x^2 + \dots\right)y \tag{11.17}$$

possesses solutions of the form

$$y(x) = x^q\left(v_0 + v_1 x + v_2 x^2 + \dots\right) \tag{11.18}$$

where $v_0, v_1, \dots$ are vectors. Determine first q and v_0, then recursively v_1, v_2, etc. Observe that there exist n independent solutions of the form (11.18) if the eigenvalues of A_0 satisfy $\lambda_i \neq \lambda_j \bmod (\mathbb{Z})$ (Fuchs 1866).

5. Find the general solution of the weakly singular systems

$$y' = \frac{1}{x}\begin{pmatrix} \frac{3}{4} & 1 \\ \frac{1}{4} & -\frac{1}{4} \end{pmatrix} y \quad \text{and} \quad y' = \frac{1}{x}\begin{pmatrix} \frac{3}{4} & 1 \\ -\frac{1}{4} & -\frac{1}{4} \end{pmatrix} y. \tag{11.19}$$

Hint. While the first is easy from Exercise 4, the second needs an additional idea (see formula (5.9)). A second possibility is to use the transformation $x = e^t$, $y(x) = z(t)$, and apply the methods of Section I.12.

I.12 Systems with Constant Coefficients

> Die Technik der Integration der linearen Differentialgleichungen
> mit constanten Coeffizienten wird hier auf das Höchste entwickelt.
> (F. Klein in Routh 1898)

Linearization

Systems of linear differential equations with constant coefficients form a class of
equations for which the resolvent $R(x, x_0)$ can be computed explicitly. They generally occur by *linearization* of time-independent (i.e., *autonomous* or *permanent*)
nonlinear differential equations

$$y'_i = f_i(y_1, \ldots, y_n) \qquad \text{or} \qquad y''_i = f_i(y_1, \ldots, y_n) \tag{12.1}$$

in the neighbourhood of a stationary point (Lagrange (1788), see also Routh (1860),
Chap. IX, Thomson & Tait 1879). We choose the coordinates so that the stationary
point under consideration is the origin, i.e., $f_i(0, \ldots, 0) = 0$. We then expand f_i
in its Taylor series and neglect all nonlinear terms:

$$y'_i = \sum_{k=1}^{n} \frac{\partial f_i}{\partial y_k}(0) y_k \qquad \text{or} \qquad y''_i = \sum_{k=1}^{n} \frac{\partial f_i}{\partial y_k}(0) y_k. \tag{12.1'}$$

This is a system of equations with constant coefficients, as introduced in Section
I.6 (see (6.4), (6.11)),

$$y' = Ay \qquad \text{or} \qquad y'' = Ay. \tag{12.1''}$$

Autonomous systems are invariant under a *shift* $x \to x + C$. We may therefore
always assume that $x_0 = 0$. For arbitrary x_0 the resolvent is given by

$$R(x, x_0) = R(x - x_0, 0). \tag{12.2}$$

Diagonalization

We have seen in Section I.6 that the assumption $y(x) = v \cdot e^{\lambda x}$ leads to

$$Av = \lambda v \qquad \text{or} \qquad Av = \lambda^2 v, \tag{12.3}$$

hence $v \neq 0$ must be an *eigenvector* of A and λ the corresponding *eigenvalue* (in
the first case; a square root of the eigenvalue in the second case, which we do not

consider any longer). From (12.3) we obtain by subtraction that there exists such a $v \neq 0$ if and only if the determinant

$$\chi_A(\lambda) := \det(\lambda I - A) = (\lambda - \lambda_1)(\lambda - \lambda_2)\ldots(\lambda - \lambda_n) = 0. \tag{12.4}$$

This determinant is called the *characteristic polynomial of A*.

Suppose now that for the n eigenvalues λ_i the n eigenvectors v_i can be chosen linearly independent. We then have from (12.3)

$$A\Big(v_1, v_2, \ldots, v_n\Big) = \Big(v_1, v_2, \ldots, v_n\Big) \operatorname{diag}\Big(\lambda_1, \lambda_2, \ldots, \lambda_n\Big),$$

or, if T is the matrix *whose columns are the eigenvectors of A*,

$$T^{-1}AT = \operatorname{diag}\Big(\lambda_1, \lambda_2, \ldots, \lambda_n\Big). \tag{12.5}$$

On comparing (12.5) with (12.1"), we see that the differential equation simplifies considerably if we use the coordinate transformation

$$y(x) = Tz(x), \qquad y'(x) = Tz'(x) \tag{12.6}$$

which leads to

$$z'(x) = \operatorname{diag}\Big(\lambda_1, \lambda_2, \ldots, \lambda_n\Big)z(x). \tag{12.7}$$

Thus the original system of differential equations decomposes into n single equations which are readily integrated to give

$$z(x) = \operatorname{diag}\Big(\exp(\lambda_1 x), \exp(\lambda_2 x), \ldots, \exp(\lambda_n x)\Big)z_0,$$

from which (12.6), yields

$$y(x) = T \operatorname{diag}\Big(\exp(\lambda_1 x), \exp(\lambda_2 x), \ldots, \exp(\lambda_n x)\Big)T^{-1}y_0. \tag{12.8}$$

The Schur Decomposition

> Der Beweis ist leicht zu erbringen.   (Schur 1909)

The foregoing theory, beautiful as it may appear, has several drawbacks:

a) Not all $n \times n$ matrices have a set of n linearly independent eigenvectors;

b) Even if it is invertible, the matrix T can behave very badly (see Exercise 1). However, for *symmetric* matrices a classical theory tells that A can always be diagonalized by orthogonal transformations. Let us therefore, with Schur (1909), extend this classical theory to non-symmetric matrices. A real matrix Q is called *orthogonal* if its column vectors are mutually orthogonal and of norm 1, i.e., if $Q^TQ = I$ or $Q^T = Q^{-1}$. A complex matrix Q is called *unitary* if $Q^*Q = I$ or $Q^* = Q^{-1}$, where Q^* is the *adjoint* matrix of Q, i.e., transposed and complex conjugate.

Theorem 12.1. a) (Schur 1909). *For each complex matrix A there exists a unitary matrix Q such that*

$$Q^*AQ = \begin{pmatrix} \lambda_1 & \times & \times & \cdots & \times \\ & \lambda_2 & \times & \cdots & \times \\ & & \ddots & & \vdots \\ & & & & \lambda_n \end{pmatrix}; \tag{12.9}$$

b) (Wintner & Murnaghan 1931). *For a real matrix A the matrix Q can be chosen real and orthogonal, if for each pair of conjugate eigenvalues $\lambda, \overline{\lambda} = \alpha \pm i\beta$ one allows the block*

$$\begin{pmatrix} \lambda & \times \\ & \overline{\lambda} \end{pmatrix} \qquad \text{to be replaced by} \qquad \begin{pmatrix} \times & \times \\ \times & \times \end{pmatrix}.$$

Proof. a) The matrix A has at least one eigenvector with eigenvalue λ_1. We use this (normalized) vector as the first column of a matrix Q_1. Its other columns are then chosen by arbitrarily completing the first one to an orthonormal basis. Then

$$AQ_1 = Q_1 \left(\begin{array}{c|ccc} \lambda_1 & \times & \cdots & \times \\ \hline 0 & & A_2 & \end{array} \right). \tag{12.10}$$

We then apply the same argument to the $(n-1)$-dimensional matrix A_2. This leads to

$$A_2 \widetilde{Q}_2 = \widetilde{Q}_2 \left(\begin{array}{c|ccc} \lambda_2 & \times & \cdots & \times \\ \hline 0 & & A_3 & \end{array} \right).$$

With the unitary matrix

$$Q_2 = \left(\begin{array}{c|c} 1 & 0 \\ \hline 0 & \widetilde{Q}_2 \end{array} \right)$$

we obtain

$$Q_1^* A Q_1 Q_2 = Q_2 \left(\begin{array}{ccccc} \lambda_1 & \times & \times & \cdots & \times \\ & \lambda_2 & \times & \cdots & \times \\ \hline & & 0 & & A_3 \end{array} \right).$$

A continuation of this process leads finally to a triangular matrix as in (12.9) with $Q = Q_1 Q_2 \cdots Q_{n-1}$.

b) Suppose A to be a real matrix. If λ_1 is real, Q_1 can be chosen real and orthogonal. Now let $\lambda_1 = \alpha + i\beta$ ($\beta \neq 0$) be a *non-real* eigenvalue with a corresponding eigenvector $u + iv$, i.e.,

$$A(u \pm iv) = (\alpha \pm i\beta)(u \pm iv) \tag{12.11}$$

or

$$Au = \alpha u - \beta v, \qquad Av = \beta u + \alpha v. \tag{12.11'}$$

Since $\beta \neq 0$, u and v are linearly independent. We choose an orthogonal basis $\widehat{u}$, $\widehat{v}$ of the subspace spanned by u and v and take $\widehat{u}$, $\widehat{v}$ as the first two columns of the orthogonal matrix Q_1. We then have from (12.11')

$$AQ_1 = Q_1 \left(\begin{array}{cc|ccc} \times & \times & \times & \cdots & \times \\ \times & \times & \times & \cdots & \times \\ \hline & 0 & & A_3 & \end{array} \right).$$

$\square$

Schur himself was not very proud of "his" decomposition, he just derived it as a tool for proving interesting properties of eigenvalues (see e.g., Exercise 2).

Clearly, if A is real and symmetric, $Q^T A Q$ will also be symmetric, and therefore diagonal (see also Exercise 3).

Numerical Computations

The above theoretical proof is still not of much practical use. It requires that one know the eigenvalues, but the computation of eigenvalues from the characteristic polynomial is one of the best-known stupidities of numerical analysis. Good numerical analysis turns it the other way round: the real matrix A is directly reduced, first to Hessenberg form, and then by a sequence of orthogonal transformations to the real Schur form of Wintner & Murnaghan ("QR-algorithm" of Francis, coded by Martin, Peters & Wilkinson, contribution II/14 in Wilkinson & Reinsch 1970). The eigenvalues then drop out. However, the produced code, called "HQR2", does *not* give the Schur form of A, since it continues for the eigenvectors of A. Some manipulations must therefore be done to interrupt the code at the right moment (in the FORTRAN translation HQR2 of Eispack (1974), for example, the "340" of statement labelled "60" has to be replaced by "1001"). Happy "Matlab"-users just call "SCHUR".

Whenever the Schur form has been obtained, the transformation $y(x) = Qz(x)$, $y'(x) = Qz'(x)$ (see (12.6)) leads to

$$\begin{pmatrix} z_1' \\ \vdots \\ z_{n-1}' \\ z_n' \end{pmatrix} = \begin{pmatrix} \lambda_1 & b_{12} & \cdots & b_{1,n-1} & b_{1n} \\ & \ddots & & \vdots & \vdots \\ & & & \lambda_{n-1} & b_{n-1,n} \\ & & & & \lambda_n \end{pmatrix} \begin{pmatrix} z_1 \\ \vdots \\ z_{n-1} \\ z_n \end{pmatrix}. \tag{12.12}$$

The last equation of this system is $z_n' = \lambda_n z_n$, and it can be integrated to give $z_n = \exp(\lambda_n x) z_{n0}$. Next, the equation for z_{n-1} is

$$z_{n-1}' = \lambda_{n-1} z_{n-1} + b_{n-1,n} z_n \tag{12.12'}$$

with z_n known. This is a linear equation (inhomogeneous, if $b_{n-1,n} \neq 0$) which can be solved by Euler's technique (Section I.4). Two different cases arise:

a) If $\lambda_{n-1} \neq \lambda_n$ we put $z_{n-1} = E \exp(\lambda_{n-1}x) + F \exp(\lambda_n x)$, insert into (12.12') and compare coefficients. This gives $F = b_{n-1,n}z_{n0}/(\lambda_n - \lambda_{n-1})$ and $E = z_{n-1,0} - F$.

b) If $\lambda_{n-1} = \lambda_n$ we set $z_{n-1} = (E + Fx)\exp(\lambda_n x)$ and obtain $F = b_{n-1,n}z_{n0}$ and $E = z_{n-1,0}$.

The next stage, following the same ideas, gives z_{n-2}, etc. Simple recursive formulas for the elements of the resolvent, which work in the case $\lambda_i \neq \lambda_j$, are obtained as follows (Parlett 1976): we assume

$$z_i(x) = \sum_{j=i}^{n} E_{ij} \exp(\lambda_j x) \tag{12.13}$$

and insert this into (12.12). After comparing coefficients, we obtain for $i = n$, $n-1$, $n-2$, etc.

$$E_{ik} = \frac{1}{\lambda_k - \lambda_i}\Big(\sum_{j=i+1}^{k} b_{ij}E_{jk} \Big), \qquad k = i+1, i+2, \dots$$

$$E_{ii} = z_{i0} - \sum_{j=i+1}^{n} E_{ij}. \tag{12.13'}$$

The Jordan Canonical Form

Simpler Than You Thought
(Amer. Math. Monthly 87 (1980) Nr. 9)

Whenever one is not afraid of badly conditioned matrices (see Exercise 1), and many mathematicians are not, the Schur form obtained above can be further transformed into the famous *Jordan canonical form*:

Theorem 12.2 (Jordan 1870, Livre deuxième, §5 and 6). *For every matrix A there exists a non-singular matrix T such that*

$$T^{-1}AT = \mathrm{diag}\left\{ \begin{pmatrix} \lambda_1 & 1 & \\ & \ddots & 1 \\ & & \lambda_1 \end{pmatrix}, \begin{pmatrix} \lambda_2 & 1 & \\ & \ddots & 1 \\ & & \lambda_2 \end{pmatrix}, \dots \right\}. \tag{12.14}$$

(The dimensions (≥ 1) of the blocks may vary and the λ_i are not necessarily distinct).

Proof. We may suppose that the matrix is already in the Schur form. This is of course possible in such a way that identical eigenvalues are grouped together on the principal diagonal.

The next step (see Fletcher & Sorensen 1983) is to remove all nonzero elements outside the upper-triangular blocks containing identical eigenvalues. We let

$$A = \begin{pmatrix} B & C \\ 0 & D \end{pmatrix}$$

where B and D are upper-triangular. The diagonal elements of B are all equal to λ_1, whereas those of D are $\lambda_2, \lambda_3, \ldots$ and all different from λ_1. We search for a matrix S such that

$$\begin{pmatrix} B & C \\ 0 & D \end{pmatrix} \begin{pmatrix} I & S \\ 0 & I \end{pmatrix} = \begin{pmatrix} I & S \\ 0 & I \end{pmatrix} \begin{pmatrix} B & 0 \\ 0 & D \end{pmatrix}$$

or, equivalently,

$$BS + C = SD. \tag{12.15}$$

From this relation the matrix S can be computed column-wise as follows: the first column of (12.15) is $BS_1 + C_1 = \lambda_2 S_1$ (here S_j and C_j denote the jth column of S and C, respectively) which yields S_1 because λ_2 is not an eigenvalue of B. The second column of (12.15) yields $BS_2 + C_2 = \lambda_3 S_2 + d_{12} S_1$ and allows us to compute S_2, etc.

In the following steps we treat each of the remaining blocks separately: we thus assume that all diagonal elements are equal to λ and transform the block recursively to the form stated in the theorem. Since $(A - \lambda I)^n = 0$ (n is the dimension of the matrix A) there exists an integer k ($1 \leq k \leq n$) such that

$$(A - \lambda I)^k = 0, \qquad (A - \lambda I)^{k-1} \neq 0. \tag{12.16}$$

We fix a vector v such that $(A - \lambda I)^{k-1} v \neq 0$ and put

$$v_j = (A - \lambda I)^{k-j} v, \qquad j = 1, \ldots, k$$

so that

$$Av_1 = \lambda v_1, \qquad Av_j = \lambda v_j + v_{j-1} \quad \text{for} \quad j = 2, \ldots, k.$$

The vectors $v_1, \ldots, v_k$ are linearly independent, because a multiplication of the expression $\sum_{j=1}^{k} c_j v_j = 0$ with $(A - \lambda I)^{k-1}$ yields $c_k = 0$, then a multiplication with $(A - \lambda I)^{k-2}$ yields $c_{k-1} = 0$, etc. As in the proof of the Schur decomposition (Theorem 12.1) we complete $v_1, \ldots, v_k$ to a basis of $\mathbb{C}^n$ in such a way that (with $V = (v_1, \ldots, v_n)$)

$$AV = V \begin{pmatrix} J & C \\ 0 & D \end{pmatrix}, \qquad J = \left. \begin{pmatrix} \lambda & 1 & & \\ & \ddots & \ddots & \\ & & & 1 \\ & & & \lambda \end{pmatrix} \right\} k \tag{12.17}$$

where D is upper-triangular with λ on its diagonal.

Our next aim is to eliminate the nonzero elements of C in (12.17). In analogy to (12.15) it is natural to search for a matrix S such that $JS + C = SD$. Unfortunately, such an S does not always exist because the eigenvalues of J and of D are the same. However, it is possible to find S such that all elements of C are removed with the exception of its last line, i.e.,

$$\begin{pmatrix} J & C \\ 0 & D \end{pmatrix} \begin{pmatrix} I & S \\ 0 & I \end{pmatrix} = \begin{pmatrix} I & S \\ 0 & I \end{pmatrix} \begin{pmatrix} J & e_k c^T \\ 0 & D \end{pmatrix} \tag{12.18}$$

or equivalently

$$JS + C = e_k c^T + SD,$$

where $e_k = (0, \dots, 0, 1)^T$ and $c^T = (c_1, \dots, c_{n-k})$. This can be seen as follows: the first column of this relation becomes $(J - \lambda I)S_1 + C_1 = c_1 e_k$. Its last component yields c_1 and the other components determine the 2nd to kth elements of S_1. The first element of S_1 can arbitrarily be put equal to zero. Then we compute S_2 from $(J - \lambda I)S_2 + C_2 = c_2 e_k + d_{12}S_1$, etc. We thus obtain a matrix S (with vanishing first line) such that (12.18) holds.

We finally show that the assumption $(A - \lambda I)^k = 0$ implies $c = 0$ in (12.18). Indeed, a simple calculation yields

$$\begin{pmatrix} J - \lambda I & e_k c^T \\ 0 & D - \lambda I \end{pmatrix}^k = \begin{pmatrix} 0 & \widehat{C} \\ 0 & 0 \end{pmatrix}$$

where the first row of $\widehat{C}$ is equal to the row-vector c^T.

We have thus transformed A to block-diagonal form with blocks J of (12.17) and D. The procedure can now be repeated with the lower-dimensional matrix D. The product of all the occurring transformation matrices is then the matrix T in (12.14).   □

Corollary 12.3. *For every matrix A and for every number $\varepsilon \neq 0$ there exists a non-singular matrix T (depending on ε) such that*

$$T^{-1}AT = \text{diag} \left\{ \begin{pmatrix} \lambda_1 & \varepsilon & \\ & \ddots & \varepsilon \\ & & \lambda_1 \end{pmatrix}, \begin{pmatrix} \lambda_2 & \varepsilon & \\ & \ddots & \varepsilon \\ & & \lambda_2 \end{pmatrix}, \dots \right\}. \tag{12.14'}$$

Proof. Multiply equation (12.14) from the right by $D = \text{diag}(1, \varepsilon, \varepsilon^2, \varepsilon^3, \dots)$ and from the left by D^{-1}.   □

Numerical difficulties in determining the Jordan canonical form are described in Golub & Wilkinson (1976). There exist also several computer programs, for example the one described in Kågström & Ruhe (1980).

When the matrix A has been transformed to Jordan canonical form (12.14), the solutions of the differential equation $y' = Ay$ can be calculated by the method explained in (12.12'), case b):

$$y(x) = TDT^{-1}y_0 \tag{12.19}$$

where D is a block-diagonal matrix with blocks of the form

$$\begin{pmatrix} e^{\lambda x} & xe^{\lambda x} & \cdots & \dfrac{x^k}{k!}e^{\lambda x} \\ & e^{\lambda x} & & \vdots \\ & & \ddots & xe^{\lambda x} \\ & & & e^{\lambda x} \end{pmatrix}$$

This is an extension of formula (12.8).

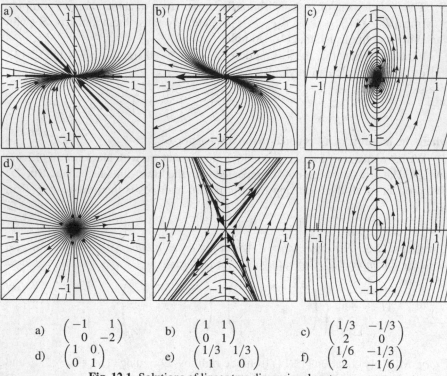

a) $\begin{pmatrix} -1 & 1 \\ 0 & -2 \end{pmatrix}$ b) $\begin{pmatrix} 1 & 1 \\ 0 & 1 \end{pmatrix}$ c) $\begin{pmatrix} 1/3 & -1/3 \\ 2 & 0 \end{pmatrix}$

d) $\begin{pmatrix} 1 & 0 \\ 0 & 1 \end{pmatrix}$ e) $\begin{pmatrix} 1/3 & 1/3 \\ 1 & 0 \end{pmatrix}$ f) $\begin{pmatrix} 1/6 & -1/3 \\ 2 & -1/6 \end{pmatrix}$

Fig. 12.1. Solutions of linear two dimensional systems

Geometric Representation

The geometric shapes of the solution curves of $y' = Ay$ are presented in Fig. 12.1 for dimension $n = 2$. They are plotted as paths in the phase-space (y_1, y_2). The cases a), b), c) and e) are the linearized equations of (12.20) at the four critical points (see Fig. 12.2).

Much of this structure remains valid also for *nonlinear* systems (12.1) in the *neighbourhood of equilibrium points*. Exceptions may be "structurally unstable" cases such as complex eigenvalues with $\alpha = \text{Re}(\lambda) = 0$. This has been the subject of many papers discussing "critical points" or "singularities" (see e.g., the famous treatise of Poincaré (1881, 82, 85)).

In Fig. 12.2 we show solutions of the quadratic system

$$y_1' = \frac{1}{3}(y_1 - y_2)(1 - y_1 - y_2)$$
$$y_2' = y_1(2 - y_2)$$

(12.20)

which possesses four critical points of all four possible structurally stable types (Exercise 4).

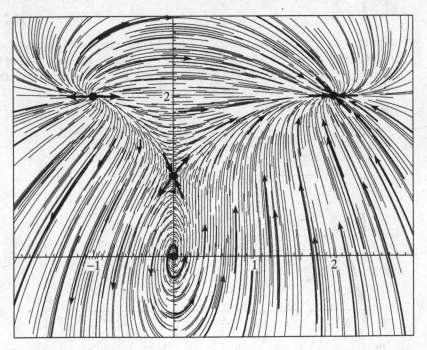

Fig. 12.2. Solution flow of System (12.20)

Exercises

1. a) Compute the eigenvectors of the matrix

$$A = \begin{pmatrix} -1 & 20 & & & & \\ & -2 & 20 & & & \\ & & -3 & 20 & & \\ & & & \ddots & \ddots & \\ & & & & -19 & 20 \\ & & & & & -20 \end{pmatrix} \tag{12.21}$$

by solving $(A - \lambda_i I)v_i = 0$.

Result. $v_1 = (1, 0, \ldots)^T$, $v_2 = (1, -1/20, 0, \ldots)^T$, $v_3 = (1, -2/20, 2/400, 0, \ldots)^T$, $v_4 = (1, -3/20, 6/400, -6/8000, 0, \ldots)^T$, etc.

b) Compute numerically the inverse of $T = (v_1, v_2, \ldots, v_n)$ and determine its largest element (answer: 4.5×10^{12}). The matrix T is thus very badly conditioned.

c) Compute numerically or analytically from (12.13) the solutions of

$$y' = Ay, \qquad y_i(0) = 1, \qquad i = 1, \ldots, 20. \tag{12.22}$$

Observe the "hump" (Moler & Van Loan 1978): although all eigenvalues of A are negative, the solutions first grow enormously before decaying to zero. This is typical of non-symmetric matrices and is connected with the bad condition of T (see Fig. 12.3).

Result.

$$y_1 = -\frac{20^{19}}{19!} e^{-20x} + \frac{(1+20)20^{18}}{18!} e^{-19x} - \frac{(1+20+20^2/2!)20^{17}}{17!} e^{-18x} \pm \ldots$$

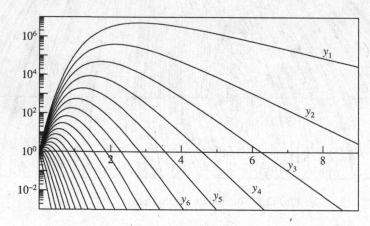

Fig. 12.3. Solutions of equation (12.22) with matrix (12.21)

2. (Schur). Prove that the eigenvalues of a matrix A satisfy the estimate

$$\sum_{i=1}^{n} |\lambda_i|^2 \le \sum_{i,j=1}^{n} |a_{ij}|^2$$

and that equality holds iff A is orthogonally diagonalizable (see also Exercise 3).

Hint. $\sum_{i,j} |a_{ij}|^2$ is the trace of A^*A and thus invariant under unitary transformations Q^*AQ.

3. Show that the Schur decomposition $S = Q^*AQ$ is diagonal iff $A^*A = AA^*$. Such matrices are called *normal*. Examples are symmetric and skew-symmetric matrices.

Hint. The condition is equivalent to $S^*S = SS^*$.

4. Compute the four critical points of System (12.20), and for each of these points the eigenvalues and eigenvectors of the matrix $\partial f/\partial y$. Compare the results with Figs. 12.2 and 12.1.

5. Compute a Schur decomposition and the Jordan canonical form of the matrix

$$A = \frac{1}{9} \begin{pmatrix} 14 & 4 & 2 \\ -2 & 20 & 1 \\ -4 & 4 & 20 \end{pmatrix}.$$

Result. The Jordan canonical form is

$$\begin{pmatrix} 2 & 1 & \\ & 2 & \\ & & 2 \end{pmatrix}.$$

6. Reduce the matrices

$$A = \begin{pmatrix} \lambda & 1 & b & c \\ & \lambda & 1 & d \\ & & \lambda & 1 \\ & & & \lambda \end{pmatrix}, \qquad A = \begin{pmatrix} \lambda & 1 & b & c \\ & \lambda & 0 & d \\ & & \lambda & 1 \\ & & & \lambda \end{pmatrix}$$

to Jordan canonical form. In the second case distinguish the possibilities $b + d = 0$ and $b + d \ne 0$.

I.13 Stability

The Examiners give notice that the following is the subject of the Prize to be adjudged in 1877: *The Criterion of Dynamical Stability.*
(S.G. Phear
(Vice-Chancellor), J. Challis, G.G. Stokes, J. Clerk Maxwell)

Introduction

"To illustrate the meaning of the question imagine a particle to slide down inside a smooth inclined cylinder along the lowest generating line, or to slide down outside along the highest generating line. In the former case a slight derangement of the motion would merely cause the particle to oscillate about the generating line, while in the latter case the particle would depart from the generating line altogether. The motion in the former case would be, in the sense of the question, stable, in the latter unstable ... what is desired is, a corresponding condition enabling us to decide when a dynamically possible motion of a system is such, that *if slightly deranged the motion shall continue to be only slightly departed from.*" ("The Examiners" in Routh 1877).

Whenever no analytical solution of a problem is known, numerical solutions can only be obtained for specified initial values. But often one needs information about the stability behaviour of the solutions for all initial values in the neighbourhood of a certain equilibrium point. We again transfer the equilibrium point to the origin and define:

Definition 13.1. Let

$$y_i' = f_i(y_1, \ldots, y_n), \qquad i = 1, \ldots, n \qquad (13.1)$$

be a system with $f_i(0, \ldots, 0) = 0$, $i = 1, \ldots, n$. Then the origin is called *stable in the sense of Liapunov* if for any $\varepsilon > 0$ there is a $\delta > 0$ such that for the solutions, $\|y(x_0)\| < \delta$ implies $\|y(x)\| < \varepsilon$ for all $x > x_0$.

The first step, taken by Routh in his famous Adams Prize essay (Routh 1877), was to study the *linearized equation*

$$y_i' = \sum_{j=1}^n a_{ij} y_j, \qquad a_{ij} = \frac{\partial f_i}{\partial y_j}(0). \qquad (13.2)$$

("The quantities $x, y, z, \ldots$ etc are said to be *small* when their squares can be neglected.") From the general solution of (13.2) obtained in Section I.12, we immediately have

Theorem 13.1. *The linearized equation (13.2) is stable (in the sense of Liapunov) iff all roots of the characteristic equation*

$$\det(\lambda I - A) = a_0 \lambda^n + a_1 \lambda^{n-1} + \ldots + a_{n-1}\lambda + a_n = 0 \qquad (13.3)$$

satisfy $\mathrm{Re}\,(\lambda) \leq 0$, *and the multiple roots, which give rise to Jordan chains, satisfy the strict inequality* $\mathrm{Re}\,(\lambda) < 0$.

Proof. See (12.12) and (12.19). For Jordan chains the "secular" term (e.g., $E + Fx$ in the solution of (12.12), case (b)) which tends to infinity for increasing x, must be "killed" by an exponential with strictly negative exponent.   □

The Routh-Hurwitz Criterion

The next task, which leads to the famous Routh-Hurwitz criterion, was the verification of the conditions $\mathrm{Re}\,(\lambda) < 0$ directly from the coefficients of (13.3), without computing the roots. To solve this problem, Routh combined two known ideas: the first was Cauchy's *argument principle,* saying that the number of roots of a polynomial $p(z) = u(z) + iv(z)$ inside a closed contour is equal to the number of (positive) rotations of the vector $(u(z), v(z))$, as z travels along the boundary in the positive sense (see e.g., Henrici (1974), p. 276). An example is presented in Fig. 13.1 for the polynomial

$$z^6 + 6z^5 + 16z^4 + 25z^3 + 24z^2 + 14z + 4$$
$$= (z+1)(z+2)(z^2 + z + 1)(z^2 + 2z + 2). \qquad (13.4)$$

On the half-circle $z = Re^{i\theta}$ ($\pi/2 \leq \theta \leq 3\pi/2$, R very large) the argument of $p(z)$, due to the dominant term z^n, makes $n/2$ positive rotations. In order to have all zeros of p in the negative half plane, we therefore need an additional $n/2$ positive rotations along the imaginary axis:

Lemma 13.2. *Let $p(z)$ be a polynomial of degree n and suppose that $p(iy) \neq 0$ for $y \in \mathbb{R}$. Then all roots of $p(z)$ are in the negative half-plane iff, along the imaginary axis, $\arg(p(iy))$ makes $n/2$ positive rotations for y from $-\infty$ to $+\infty$.*   □

The second idea was the use of Sturm's theorem (Sturm 1829) which had its origin in Euclid's algorithm for polynomials. Sturm made the discovery that in the division of the polynomial $p_{i-1}(y)$ by $p_i(y)$ it is better to take the remainder $p_{i+1}(y)$ with negative sign

$$p_{i-1}(y) = p_i(y)q_i(y) - p_{i+1}(y). \qquad (13.5)$$

Then, due to the "Sturm sequence property"

$$\mathrm{sign}\,(p_{i+1}(y)) \neq \mathrm{sign}\,(p_{i-1}(y)) \qquad \text{if} \qquad p_i(y) = 0, \qquad (13.6)$$

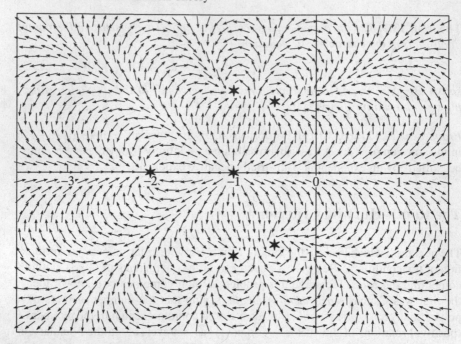

Fig. 13.1. Vector field of arg $(p(z))$ for the polynomial $p(z)$ of (13.4)

the number of *sign changes*

$$w(y) = \text{No. of sign changes of } \Big(p_0(y), p_1(y), \ldots, p_m(y)\Big) \tag{13.7}$$

does not vary at the zeros of $p_1(y), \ldots, p_{m-1}(y)$. A consequence is the following

Lemma 13.3. *Suppose that a sequence* $p_0(y), p_1(y), \ldots, p_m(y)$ *of real polynomials satisfies*

  i) $\deg(p_0) > \deg(p_1)$,

  ii) $p_0(y)$ *and* $p_1(y)$ *not simultaneously zero,*

  iii) $p_m(y) \neq 0$ *for all* $y \in \mathbb{R}$,

  iv) *and the Sturm sequence property (13.6).*

Then

$$\frac{w(\infty) - w(-\infty)}{2} \tag{13.8}$$

is equal to the number of rotations, measured in the positive direction, of the vector $(p_0(y), p_1(y))$ *as* y *tends from* $-\infty$ *to* $+\infty$.

Proof. Due to the Sturm sequence property, $w(y)$ does not change at zeros of $p_1(y), \ldots, p_{m-1}(y)$. By assumption (iii) also $p_m(y)$ has no influence. Therefore $w(y)$ can change only at zeros of $p_0(y)$. If $w(y)$ increases by one at $\widehat{y}$, either $p_0(y)$ changes from $+$ to $-$ and $p_1(\widehat{y}) > 0$ or it changes from $-$ to $+$ and $p_1(\widehat{y}) < 0$ ($p_1(\widehat{y}) = 0$ is impossible by (ii)). In both situations the vector $(p_0(y), p_1(y))$ crosses the imaginary axis in the positive direction (see Fig. 13.2). If $w(y)$ decreases by one, $(p_0(y), p_1(y))$ crosses the imaginary axis in the negative direction. The result now follows from (i), since the vector $(p_0(y), p_1(y))$ is horizontal for $y \to -\infty$ and for $y \to +\infty$. $\qquad\qquad\square$

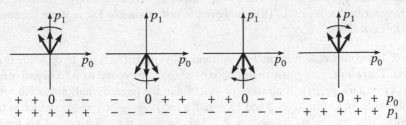

$$
\begin{array}{ccccc}
+ & + & 0 & - & - \\
+ & + & + & + & +
\end{array}
\quad
\begin{array}{ccccc}
- & - & 0 & + & + \\
- & - & - & - & -
\end{array}
\quad
\begin{array}{ccccc}
+ & + & 0 & - & - \\
- & - & - & - & -
\end{array}
\quad
\begin{array}{cccccc}
- & - & 0 & + & + & p_0 \\
+ & + & + & + & + & p_1
\end{array}
$$

Fig. 13.2. Rotations of $(p_0(y), p_1(y))$ compared to $w(y)$

The two preceding lemmas together give us the desired criterion for stability: let the characteristic polynomial (13.3)

$$p(z) = a_0 z^n + a_1 z^{n-1} + \ldots + a_n = 0, \qquad a_0 > 0$$

be given. We divide $p(iy)$ by i^n and separate real and imaginary parts,

$$
\begin{aligned}
p_0(y) &= \operatorname{Re} \frac{p(iy)}{i^n} = a_0 y^n - a_2 y^{n-2} + a_4 y^{n-4} \pm \ldots \\
p_1(y) &= -\operatorname{Im} \frac{p(iy)}{i^n} = a_1 y^{n-1} - a_3 y^{n-3} + a_5 y^{n-5} \pm \ldots.
\end{aligned}
\tag{13.9}
$$

Due to the special structure of these polynomials, the Euclidean algorithm (13.5) is here particularly simple: we write

$$p_i(y) = c_{i0} y^{n-i} + c_{i1} y^{n-i-2} + c_{i2} y^{n-i-4} + \ldots, \tag{13.10}$$

and have for the quotient in (13.5) $q_i(y) = (c_{i-1,0}/c_{i0})y$, provided that $c_{i0} \neq 0$. Now (13.10) inserted into (13.5) gives the following recursive formulas for the computation of the coefficients c_{ij}:

$$c_{i+1,j} = c_{i,j+1} \cdot \frac{c_{i-1,0}}{c_{i0}} - c_{i-1,j+1} = \frac{1}{c_{i0}} \det \begin{pmatrix} c_{i-1,0} & c_{i-1,j+1} \\ c_{i,0} & c_{i,j+1} \end{pmatrix}. \tag{13.11}$$

If $c_{i0} = 0$ for some i, the quotient $q_i(y)$ is a higher degree polynomial and the Euclidean algorithm stops at $p_m(y)$ with $m < n$.

The sequence $(p_i(y))$ obtained in this way obviously satisfies conditions (i) and (iv) of Lemma 13.3. Condition (ii) is equivalent to $p(iy) \neq 0$ for $y \in \mathbb{R}$, and (iii) is a consequence of (ii) since $p_m(y)$ is the *greatest common divisor* of $p_0(y)$ and $p_1(y)$.

Theorem 13.4 (Routh 1877). *All roots of the real polynomial (13.3) with $a_0 > 0$ lie in the negative half plane* $\operatorname{Re} \lambda < 0$ *if and only if*

$$c_{i0} > 0 \qquad for \qquad i = 0, 1, 2, \ldots, n. \tag{13.12}$$

Remark. Due to the condition $c_{i0} > 0$, the division by c_{i0} in formula (13.11) can be omitted (common positive factor of $p_{i+1}(y)$), which leads to the same theorem (Routh (1877), p. 27: "... so that by remembering this simple cross-multiplication we may write down ..."). This, however, is not advisable for n large because of possible overflow.

Proof. The coordinate systems (p_0, p_1) and $(\operatorname{Re}(p), \operatorname{Im}(p))$ are of *opposite* orientation. Therefore, $n/2$ positive rotations of $p(iy)$ correspond to $n/2$ negative rotations of $(p_0(y), p_1(y))$. If all roots of $p(\lambda)$ lie in the negative half plane $\operatorname{Re} \lambda < 0$, it follows from Lemmas 13.2 and 13.3 that $w(\infty) - w(-\infty) = -n$, which is only possible if $w(\infty) = 0$, $w(-\infty) = n$. This implies the positivity of all leading coefficients of $p_i(y)$.

On the other hand, if (13.12) is satisfied, we see that $p_n(y) \equiv c_{n0}$. Hence the polynomials $p_0(y)$ and $p_1(y)$ cannot have a common factor and $p(\lambda) \neq 0$ on the imaginary axis. We can now apply Lemmas 13.2 and 13.3 again to obtain the result. $\qquad \square$

Table 13.1.
Routh tableau for (13.4)

	$j=0$	$j=1$	$j=2$	$j=3$
$i=0$	1	-16	24	-4
$i=1$	6	-25	14	
$i=2$	11.83	-21.67	4	
$i=3$	14.01	-11.97		
$i=4$	11.56	-4		
$i=5$	7.12			
$i=6$	4			

Table 13.2.
Routh tableau for (13.13)

	$j=0$	$j=1$	$j=2$
$i=0$	1	$-q$	s
$i=1$	p	$-r$	
$i=2$	$pq - r$	$-ps$	
$i=3$	$(pq-r)r - p^2 s$		
$i=4$	$((pq-r)r - p^2 s)ps$		

Example 1. The Routh tableau (13.11) for equation (13.4) is given in Table 13.1. It clearly satisfies the conditions for stability.

Example 2 (Routh 1877, p. 27). Express the stability conditions for the biquadratic
$$z^4 + pz^3 + qz^2 + rz + s = 0. \tag{13.13}$$
The c_{ij} values (without division) are given in Table 13.2. We have stability iff
$$p > 0, \qquad pq - r > 0, \qquad (pq - r)r - p^2 s > 0, \qquad s > 0.$$

Computational Considerations

The actual computational use of Routh's criterion, in spite of its high historical importance and mathematical elegance, has two drawbacks for higher dimensions:

1) It is not easy to compute the characteristic polynomial for higher order matrices;

2) The use of the characteristic polynomial is very dangerous in the presence of rounding errors.

So, whenever one is not working with exact algebra or high precision, it is advisable to avoid the characteristic polynomial and use numerically stable algorithms for the eigenvalue problem (e.g., Eispack 1974).

Numerical experiments. 1. The $2n \times 2n$ dimensional matrix

$$A = \begin{pmatrix} -.05 & & & -1 & & \\ & \ddots & & & \ddots & \\ & & -.05 & & & -n \\ \hline 1 & & & -.05 & & \\ & \ddots & & & \ddots & \\ & & n & & & -.05 \end{pmatrix}$$

has the characteristic polynomial

$$p(z) = \prod_{j=1}^{n} (z^2 + 0.1z + j^2 + 0.0025).$$

We computed the coefficients of p using double precision, and then applied the Routh algorithm in single precision (machine precision $= 6 \times 10^{-8}$). The results indicated stability for $n \leq 15$, but not for $n \geq 16$, although the matrix always has its eigenvalues $-0.05 \pm ki$ in the negative half plane. On the other hand, a direct computation of the eigenvalues of A with the use of Eispack subroutines gave no problem for any n.

2. We also tested the Routh algorithm at the (scaled) *numerators of the diagonal Padé approximations to* $\exp(z)$

$$1 + \frac{n}{2n}(nz) + \frac{n(n-1)}{(2n)(2n-1)}\frac{(nz)^2}{2!} + \frac{n(n-1)(n-2)}{(2n)(2n-1)(2n-2)}\frac{(nz)^3}{3!} + \dots, \quad (13.14)$$

which are also known to possess all zeros in $\mathbb{C}^-$. Here, the results were correct only for $n \leq 21$, and wrong for larger n due to rounding errors.

Liapunov Functions

We now consider the question whether the stability of the nonlinear system (13.1) "can really be determined by examination of the terms of the first order only" (Routh 1877, Chapt. VII). This theory, initiated by Routh and Poincaré, was brought to perfection in the famous work of Liapunov (1892). As a general reference to the enormous theory that has developed in the meantime we mention Rouche, Habets & Laloy (1977) and W. Hahn (1967).

Liapunov's (and Routh's) main tools are the so-called *Liapunov functions* $V(y_1, \ldots, y_n)$, which should satisfy

$$V(y_1, \ldots, y_n) \geq 0,$$

$$V(y_1, \ldots, y_n) = 0 \qquad \text{iff} \qquad y_1 = \ldots = y_n = 0 \qquad (13.15)$$

and along the solutions of (13.1)

$$\frac{d}{dx} V(y_1(x), \ldots, y_n(x)) \leq 0. \qquad (13.16)$$

Usually $V(y)$ behaves quadratically for small y and condition (13.15) means that

$$c\|y\|^2 \leq V(y) \leq C\|y\|^2, \qquad C \geq c > 0. \qquad (13.17)$$

The existence of such a Liapunov function is then a sufficient condition for stability of the origin.

We start with the *construction of a Liapunov function* for the linear case

$$y' = Ay. \qquad (13.18)$$

This is best done in the basis which is naturally given by the *eigenvectors* (or Jordan chains) of A. We therefore introduce $y = Tz$, $z = T^{-1}y$, so that A is transformed to Jordan canonical form (12.14') $J = T^{-1}AT$ and (13.18) becomes

$$z' = Jz. \qquad (13.19)$$

If we put

$$V_0(z) = \|z\|^2 \qquad \text{and} \qquad V(y) = V_0(T^{-1}y) = V_0(z), \qquad (13.20)$$

the derivative of $V(y(x))$ becomes

$$\frac{d}{dx} V(y(x)) = \frac{d}{dx} V_0(z(x)) = 2\mathrm{Re}\,\langle z(x), z'(x) \rangle$$
$$= 2\mathrm{Re}\,\langle z(x), Jz(x) \rangle \leq 2\mu(J) V(y(x)). \qquad (13.21)$$

By (10.20) the logarithmic norm is given by

$$2\mu(J) = \text{largest eigenvalue of } (J + J^*).$$

The matrix $J + J^*$ is block-diagonal with tridiagonal blocks

$$\begin{pmatrix} 2\operatorname{Re}\lambda_i & \varepsilon & & \\ \varepsilon & 2\operatorname{Re}\lambda_i & \ddots & \\ & \ddots & \ddots & \varepsilon \\ & & \varepsilon & 2\operatorname{Re}\lambda_i \end{pmatrix}. \tag{13.22}$$

Subtracting the diagonal and using formula (6.7a), we see that the eigenvalues of the m-dimensional matrix (13.22) are given by

$$2\left(\operatorname{Re}\lambda_i + \varepsilon\cos\frac{\pi k}{m+1}\right), \qquad k = 1,\dots,m. \tag{13.23}$$

As a consequence of this formula or by the use of Exercise 4 we have:

Lemma 13.5. *If all eigenvalues of A satisfy $\operatorname{Re}\lambda_i < -\varrho < 0$, then there exists a (quadratic) Liapunov function for equation (13.18) which satisfies*

$$\frac{d}{dx}V\big(y(x)\big) \le -\varrho\, V\big(y(x)\big). \tag{13.24}$$

$\square$

This last differential inequality implies that (Theorem 10.1)

$$V\big(y(x)\big) \le V(y_0)\cdot\exp\big(-\varrho(x - x_0)\big)$$

and ensures that $\lim_{x\to\infty}\|y(x)\| = 0$, i.e., *asymptotic stability*.

Stability of Nonlinear Systems

It is now easy to extend the same ideas to *nonlinear* equations. The following theorem is an example of such a result.

Theorem 13.6. *Let the nonlinear system*

$$y' = Ay + g(x,y) \tag{13.25}$$

be given with $\operatorname{Re}\lambda_i < -\varrho < 0$ for all eigenvalues of A. Further suppose that for each $\varepsilon > 0$ there is a $\delta > 0$ such that

$$\|g(x,y)\| \le \varepsilon\|y\| \qquad \text{for} \qquad \|y\| < \delta,\; x \ge x_0. \tag{13.26}$$

Then the origin is (asymptotically) stable in the sense of Liapunov.

Proof. We use the Liapunov function $V(y)$ constructed for Lemma 13.5 and obtain from (13.25)

$$\frac{d}{dx}V\big(y(x)\big) \le -\varrho\, V\big(y(x)\big) + 2\operatorname{Re}\big\langle T^{-1}y(x),\, T^{-1}g\big(x,y(x)\big)\big\rangle. \tag{13.27}$$

Cauchy's inequality together with (13.26) yields

$$\frac{d}{dx}V\big(y(x)\big) \leq \big(-\varrho + \|T\| \cdot \|T^{-1}\|\varepsilon\big) \cdot V\big(y(x)\big). \tag{13.28}$$

For sufficiently small ε the right hand side is negative and we obtain asymptotic stability.   □

We see that, for nonlinear systems, stability *is only assured in a neighbourhood* of the origin. This can also be observed in Fig. 12.2. Another difference is that the *stability for eigenvalues on the imaginary axis can be destroyed.* An example for this (Routh 1877, pp. 95-96) is the system

$$y_1' = y_2, \qquad y_2' = -y_1 + y_2^3. \tag{13.29}$$

Here, with the Liapunov function $V = (y_1^2 + y_2^2)/2$, we obtain $V' = y_2^4$ which is > 0 for $y_2 \neq 0$. Therefore all solutions with initial value $\neq 0$ increase. A survey of this question ("the center problem") together with its connection to limit cycles is given in Wanner (1983).

Stability of Non-Autonomous Systems

When the coefficients are not constant,

$$y' = A(x)y, \tag{13.30}$$

it is *not* a sufficient test of stability that the eigenvalues of A satisfy the conditions of stability for each instantaneous value of x.

Examples. 1. (Routh 1877, p. 96).

$$y_1' = y_2, \qquad y_2' = -\frac{1}{4x^2}y_1 \tag{13.31}$$

which is satisfied by $y_1(x) = a\sqrt{x}$.

2. An example with eigenvalues strictly negative: we start with

$$B = \begin{pmatrix} -1 & 0 \\ 4 & -1 \end{pmatrix}, \qquad y' = By.$$

An inspection of the derivative of $V = (y_1^2 + y_2^2)/2$ shows that V *increases* in the sector $2 - \sqrt{3} < y_2/y_1 < 2 + \sqrt{3}$. The idea is to take the initial value in this region and, for x increasing, to rotate the coordinate system with the same speed as the solution rotates:

$$y' = T(x)BT(-x)y = A(x)y, \qquad T(x) = \begin{pmatrix} \cos ax & -\sin ax \\ \sin ax & \cos ax \end{pmatrix}. \tag{13.32}$$

For $y(0) = (1, 1)^T$, a good choice for a is $a = 2$ and (13.32) possesses the solution

$$y(x) = \left((\cos 2x - \sin 2x)e^x, (\cos 2x + \sin 2x)e^x \right)^T. \tag{13.33}$$

This solution is clearly unstable, while -1 remains for all x the double eigenvalue of $A(x)$. For more examples see Exercises 6 and 7 below.

We observe that stability theory for non-autonomous systems is more compli-cated. Among the cases in which stability can be shown are the following:
1) $a_{ii}(x) < 0$ and $A(x)$ is diagonally dominant; then $\mu(A(x)) \leq 0$ such that sta-bility follows from Theorem 10.6.
2) $A(x) = B + C(x)$, with B constant and satisfying Re $\lambda_i < -\varrho < 0$ for its eigen-values, and $\|C(x)\| < \varepsilon$ with ε so small that the proof of Theorem 13.6 can be applied.

Exercises

1. Express the stability conditions for the polynomials $z^2 + pz + q = 0$ and $z^3 + pz^2 + qz + r = 0$.

   Result. a) $p > 0$ and $q > 0$; b) $p > 0$, $r > 0$ and $pq - r > 0$.

2. (Hurwitz 1895). Verify that condition (13.12) is equivalent to the positivity of the principal minors of the matrix

$$H = \begin{pmatrix} a_1 & a_3 & a_5 & \cdots \\ a_0 & a_2 & a_4 & \cdots \\ & a_1 & a_3 & \cdots \\ & a_0 & a_2 & \cdots \\ & & \cdots & \cdots \end{pmatrix} = \left(a_{2j-i} \right)_{i,j=1}^n$$

   ($a_k = 0$ for $k < 0$ and $k > n$). Understand that Routh's algorithm (13.11) is identical to a sort of Gaussian elimination transforming H to triangular form.

3. The polynomial

$$\frac{5 \cdot 4 \cdot 3 \cdot 2 \cdot 1}{10 \cdot 9 \cdot 8 \cdot 7 \cdot 6} \frac{z^5}{5!} + \frac{5 \cdot 4 \cdot 3 \cdot 2}{10 \cdot 9 \cdot 8 \cdot 7} \frac{z^4}{4!} + \frac{5 \cdot 4 \cdot 3}{10 \cdot 9 \cdot 8} \frac{z^3}{3!} + \frac{5 \cdot 4}{10 \cdot 9} \frac{z^2}{2!} + \frac{5}{10} z + 1$$

   is the numerator of the $(5, 5)$-Padé approximation to $\exp(z)$. Verify that all its roots satisfy Re $z < 0$. Try to establish the result for general n (see e.g., Birkhoff & Varga (1965), Lemma 7).

4. (Gerschgorin). Prove that the eigenvalues of a matrix $A = (a_{ij})$ lie in the union of the discs

$$\left\{ z \,;\, |z - a_{ii}| \leq \sum_{j \neq i} |a_{ij}| \right\}.$$

Hint. Write the formula $Ax = \lambda x$ in coordinates $\sum_j a_{ij}x_j = \lambda x_i$, put the diagonal elements on the right hand side and choose i such that $|x_i|$ is maximal.

5. Determine the stability of the origin for the system

$$y_1' = -y_2 - y_1^2 - y_1 y_2 ,$$
$$y_2' = y_1 + 2y_1 y_2 .$$

Hint. Find a Liapunov function of degree 4 starting with $V = (y_1^2 + y_2^2)/2 + \ldots$ such that $V' = K(y_1^2 + y_2^2)^2 + \ldots$ and determine the sign of K.

6. (J. Lambert 1987). Consider the system

$$y' = A(x) \cdot y \qquad \text{where} \qquad A(x) = \begin{pmatrix} -1/4x & 1/x^2 \\ -1/4 & -1/4x \end{pmatrix}. \qquad (13.34)$$

a) Show that both eigenvalues of $A(x)$ satisfy $\operatorname{Re} \lambda < 0$ for all $x > 0$.

b) Compute $\mu(A)$ (from (10.20)) and show that

$$\mu(A) \leq 0 \qquad \text{iff} \qquad \sqrt{5} - 1 \leq x \leq \sqrt{5} + 1.$$

c) Compute the general solution of (13.34).

Hint. Introduce the new functions $z_2(x) = y_2(x)$, $z_1(x) = xy_1(x)$ which leads to the second equation of (11.19) (Exercise 5 of Section I.11). The solution is

$$y_1(x) = x^{-3/4}\left(a + b\log x\right), \qquad y_2(x) = x^{1/4}\left(-\frac{a}{2} + b\left(1 - \frac{1}{2}\log x\right)\right). \tag{13.35}$$

d) Determine a and b such that $\|y(x)\|_2^2$ is *increasing* for $0 < x < \sqrt{5} - 1$.

e) Determine a and b such that $\|y(x)\|_2^2$ is *increasing* for $\sqrt{5} + 1 < x < \infty$.

Results. $b = 1.8116035 \cdot a$ for (d) and $b = 0.2462015 \cdot a$ for (e).

7. Find a counter-example for Fatou's conjecture

    If $\ddot{y} + A(t)y = 0$ and $\forall t \quad 0 < C_1 \leq A(t) \leq C_2$ then $y(t)$ is stable

(C.R. 189 (1929), p.967-969; for a solution see Perron (1930)).

8. Help James Watt (see original drawing from 1788 in Fig. 13.3) to solve the stability problem for his steam engine governor: if ω is the rotation speed of the engine, its acceleration is influenced by the steam supply and exterior work as follows:

$$\omega' = k\cos(\varphi + \alpha) - F, \qquad k, F > 0.$$

Here α is a fixed angle and φ describes the motion of the governor. The acceleration of φ is determined by centrifugal force, weight, and friction as

$$\varphi'' = \omega^2 \sin\varphi \cos\varphi - g\sin\varphi - b\varphi', \qquad g, b > 0.$$

Compute the equilibrium point $\varphi'' = \varphi' = \omega' = 0$ and determine under which conditions it is stable (the solution is easier for $\alpha = 0$).

Correct solutions should be sent to: James Watt, famous inventor of the steam engine, Westminster Abbey, 6HQ 1FX London.

Remark. Hurwitz' paper (1895) was motivated by a similar practical problem, namely "... die Regulirung von Turbinen des Badeortes Davos".

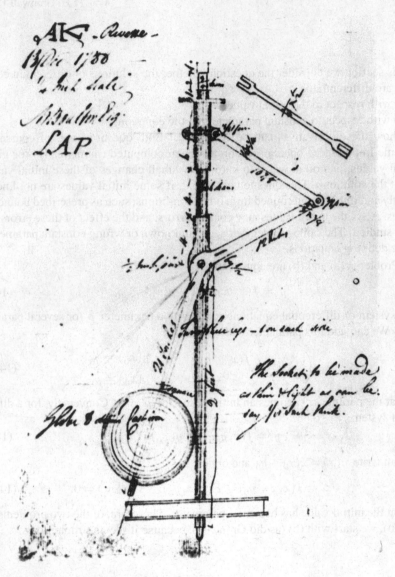

Fig. 13.3. James Watt's steam engine governor

I.14 Derivatives with Respect to Parameters and Initial Values

For a single equation, Dr. Ritt has solved the problem indicated in the title by a very simple and direct method ... Dr. Ritt's proof cannot be extended immediately to a system of equations.

(T.H. Gronwall 1919)

In this section we consider the question whether the solutions of differential equations are differentiable

a) with respect to the initial values;

b) with respect to constant parameters in the equation;

and how these derivatives can be computed. Both questions are, of course, of extreme importance: once a solution has been computed (numerically) for given initial values, one often wants to know how small changes of these initial values affect the solutions. This question arises e.g. if some initial values are not known exactly and must be determined from other conditions, such as prescribed boundary values. Also, the initial values may contain errors, and the effect of these errors has to be studied. The same problems arise for unknown or wrong constant parameters in the defining equations.

Problems (a) and (b) are equivalent: let

$$y' = f(x, y, p), \qquad y(x_0) = y_0 \tag{14.1}$$

be a system of differential equations containing a parameter p (or several parameters). We can add this parameter to the solutions

$$\begin{pmatrix} y' \\ p' \end{pmatrix} = \begin{pmatrix} f(x, y, p) \\ 0 \end{pmatrix}, \qquad \begin{aligned} y(x_0) &= y_0 \\ p(x_0) &= p, \end{aligned} \tag{14.1'}$$

so that the parameter becomes an initial value for $p' = 0$. Conversely, for a differential system

$$y' = f(x, y), \qquad y(x_0) = y_0 \tag{14.2}$$

we can write $y(x) = z(x) + y_0$ and obtain

$$z' = f(x, z + y_0) = F(x, z, y_0), \qquad z(x_0) = 0, \tag{14.2'}$$

so that the initial value has become a parameter. Therefore, of the two problems (a) and (b), we start with (b) (as did Gronwall), because it seems simpler to us.

The Derivative with Respect to a Parameter

Usually, a given problem contains *several* parameters. But since we are interested in partial derivatives, we can treat one parameter after another while keeping the remaining ones fixed. It is therefore sufficient in the following theory to suppose that $f(x, y, p)$ depends only on *one* scalar parameter p.

When we replace the parameter p in (14.1) by q we obtain another solution, which we denote by $z(x)$:

$$z' = f(x, z, q), \qquad z(x_0) = y_0. \tag{14.3}$$

It is then natural to subtract (14.1) from (14.3) and to linearize

$$z' - y' = f(x, z, q) - f(x, y, p) \tag{14.4}$$
$$= \frac{\partial f}{\partial y}(x, y, p)(z - y) + \frac{\partial f}{\partial p}(x, y, p)(q - p) + \varrho_1 \cdot (z - y) + \varrho_2 \cdot (q - p).$$

If we put $(z(x) - y(x))/(q - p) = \psi(x)$ and drop the error terms, we obtain

$$\psi' = \frac{\partial f}{\partial y}(x, y(x), p)\psi + \frac{\partial f}{\partial p}(x, y(x), p), \qquad \psi(x_0) = 0. \tag{14.5}$$

This equation is the key to the problem.

Theorem 14.1 (Gronwall 1919). *Suppose that for $x_0 \le x \le X$ the partial derivatives $\partial f/\partial y$ and $\partial f/\partial p$ exist and are continuous in the neighbourhood of the solution $y(x)$. Then the partial derivatives*

$$\frac{\partial y(x)}{\partial p} = \psi(x)$$

exist, are continuous, and satisfy the differential equation (14.5).

Proof. This theorem was the origin of the famous Gronwall lemma (see I.10, Exercise 2). We prove it here by the equivalent Theorem 10.2. Set

$$L = \max\left\|\frac{\partial f}{\partial y}\right\|, \qquad A = \max\left\|\frac{\partial f}{\partial p}\right\| \tag{14.6}$$

where the max is taken over the domain under consideration. When we consider $z(x)$ as an approximate solution for (14.1) we have for the defect

$$\|z'(x) - f(x, z(x), p)\| = \|f(x, z(x), q) - f(x, z(x), p)\| \le A|q - p|,$$

therefore from Theorem 10.2

$$\|z(x) - y(x)\| \le \frac{A}{L}|q - p|(e^{L(x - x_0)} - 1). \tag{14.7}$$

So for $|q - p|$ sufficiently small and $x_0 \le x \le X$, we can have $\|z(x) - y(x)\|$ arbitrarily small. By definition of differentiability and by (14.7), for each $\varepsilon > 0$

there is a δ such that the error terms in (14.4) satisfy

$$\|\varrho_1 \cdot (z-y) + \varrho_2 \cdot (q-p)\| \leq \varepsilon |q-p| \qquad \text{if} \qquad |q-p| \leq \delta. \qquad (14.8)$$

(The situation is, in fact, a little more complicated: the δ for the bounds $\|\varrho_1\| < \varepsilon$ and $\|\varrho_2\| < \varepsilon$ may depend on x. But due to compactness and continuity, it can then be replaced by a uniform bound. Another possibility to overcome this little obstacle would be a bound on the second derivatives. But why should we worry about this detail? Gronwall himself did not mention it).

We now consider $(z(x) - y(x))/(q-p)$ as an approximate solution for (14.5) and apply Theorem 10.2 a second time. Its defect is by (14.8) and (14.4) bounded by ε and the linear differential equation (14.5) *also* has L as a Lipschitz constant (see (11.2)). Therefore from (10.14) we obtain

$$\left\| \frac{z(x) - y(x)}{q-p} - \psi(x) \right\| \leq \frac{\varepsilon}{L} \left(e^{L(x-x_0)} - 1 \right)$$

which becomes arbitrarily small; this proves that $\psi(x)$ is the derivative of $y(x)$ with respect to p.

Continuity. The partial derivatives $\partial y / \partial p = \psi(x)$ are solutions of the differential equation (14.5), which we write as $\psi' = g(x, \psi, p)$, where by hypothesis g depends continuously on p. Therefore the continuous dependence of ψ on p follows again from Theorem 10.2. □

Theorem 14.2. *Let $y(x)$ be the solution of equation (14.1) and consider the Jacobian*

$$A(x) = \frac{\partial f}{\partial y}(x, y(x), p). \qquad (14.9)$$

Let $R(x, x_0)$ be the resolvent of the equation $y' = A(x)y$ (see (11.4)). Then the solution $z(x)$ of (14.3) with a slightly perturbed parameter q is given by

$$z(x) = y(x) + (q-p) \int_{x_0}^{x} R(x, s) \frac{\partial f}{\partial p}(s, y(s), p)\, ds + o(|q-p|) \qquad (14.10)$$

Proof. This is the variation of constants formula (11.10) applied to (14.5). □

It can be seen that the sensitivity of the solutions to changes of parameters is influenced firstly by the partial derivatives $\partial f / \partial p$ (which is natural), and secondly by the size of $R(x, s)$, i.e., by the stability of the differential equation with matrix (14.9).

Derivatives with Respect to Initial Values

Notation. We denote by $y(x, x_0, y_0)$ the solution $y(x)$ at the point x satisfying the initial values $y(x_0) = y_0$, and hope that no confusion arises from the use of the same letter y for two different functions.

The following identities are trivial by definition or follow from uniqueness arguments as for (11.6):

$$\frac{\partial y(x, x_0, y_0)}{\partial x} = f\big(x, y(x, x_0, y_0)\big) \tag{14.11}$$

$$y(x_0, x_0, y_0) = y_0 \tag{14.12}$$

$$y\big(x_2, x_1, y(x_1, x_0, y_0)\big) = y(x_2, x_0, y_0). \tag{14.13}$$

Theorem 14.3. *Suppose that the partial derivative of f with respect to y exists and is continuous. Then the solution $y(x, x_0, y_0)$ is differentiable with respect to y_0 and the derivative is given by the matrix*

$$\frac{\partial y(x, x_0, y_0)}{\partial y_0} = \Psi(x) \tag{14.14}$$

where $\Psi(x)$ is the resolvent of the so-called "variational equation"

$$\Psi'(x) = \frac{\partial f}{\partial y}\big(x, y(x, x_0, y_0)\big) \cdot \Psi(x),$$
$$\Psi(x_0) = I. \tag{14.15}$$

Proof. We know from (14.2) and (14.2') that $\partial F/\partial z$ and $\partial F/\partial y_0$ are both equal to $\partial f/\partial y$, so the derivatives are known to *exist* by Theorem 14.1. In order to obtain formula (14.15), we just have to differentiate (14.11) and (14.12) with respect to y_0. $\square$

We finally compute the derivative of $y(x, x_0, y_0)$ with respect to x_0.

Theorem 14.4. *Under the same hypothesis as in Theorem 14.3, the solutions are also differentiable with respect to x_0 and the derivative is given by*

$$\frac{\partial y(x, x_0, y_0)}{\partial x_0} = -\frac{\partial y(x, x_0, y_0)}{\partial y_0} \cdot f(x_0, y_0). \tag{14.16}$$

Proof. Differentiate the identity

$$y\big(x_1, x_0, y(x_0, x_1, y_1)\big) = y_1,$$

which follows from (14.13), with respect to x_0 and apply (14.11) (see Exercise 1). $\square$

The Nonlinear Variation-of-Constants Formula

The following theorem is an extension of Theorem 11.2 to systems of non-linear differential equations.

Theorem 14.5 (Alekseev 1961, Gröbner 1960). *Denote by y and z the solutions of*

$$y' = f(x, y), \qquad\qquad y(x_0) = y_0, \qquad\qquad (14.17a)$$

$$z' = f(x, z) + g(x, z), \qquad z(x_0) = y_0, \qquad\qquad (14.17b)$$

respectively and suppose that $\partial f / \partial y$ exists and is continuous. Then the solutions of (14.17a) and of the "perturbed" equation (14.17b) are connected by

$$z(x) = y(x) + \int_{x_0}^{x} \frac{\partial y}{\partial y_0} \big(x, s, z(s)\big) \cdot g\big(s, z(s)\big)\, ds. \qquad (14.18)$$

Proof. We choose a subdivision $x_0 = s_0 < s_1 < s_2 < \ldots < s_N = x$ (see Fig. 14.1). The descending curves represent the solutions of the unperturbed equation (14.17a) with initial values s_i, $z(s_i)$. The differences d_i are, due to the different slopes of $z(s)$ and $y(s)$ ((14.17b) minus (14.17a)), equal to $d_i = g(s_i, z(s_i)) \cdot \Delta s_i + o(\Delta s_i)$. This "error" at s_i is then "transported" to the final value x by the amount given in Theorem 14.3, to give

$$D_i = \frac{\partial y}{\partial y_0} \big(x, s_i, z(s_i)\big) \cdot g\big(s_i, z(s_i)\big) \cdot \Delta s_i + o(\Delta s_i). \qquad (14.19)$$

Since $z(x) - y(x) = \sum_{i=1}^{N} D_i$, we obtain the integral in (14.18) after insertion of (14.19) and passing to the limit $\Delta s_i \to 0$. $\qquad\qquad \square$

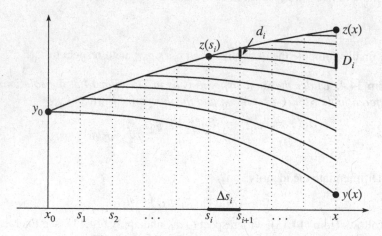

Fig. 14.1. Lady Windermere's fan, Act 2

If we also want to take into account a possible difference in the initial values, we may formulate:

Corollary 14.6. *Let $y(x)$ and $z(x)$ be the solutions of*

$$y' = f(x, y), \qquad\qquad y(x_0) = y_0,$$
$$z' = f(x, z) + g(x, z), \qquad z(x_0) = z_0,$$

then

$$z(x) = y(x) + \int_0^1 \frac{\partial y}{\partial y_0}\Big(x, x_0, y_0 + s(z_0 - y_0)\Big) \cdot (z_0 - y_0)\, ds$$
$$+ \int_{x_0}^x \frac{\partial y}{\partial y_0}\Big(x, s, z(s)\Big) \cdot g\big(s, z(s)\big)\, ds. \qquad\qquad\square$$

(14.20)

These two theorems allow many estimates of the stability of general nonlinear systems. For *linear* systems, $\partial y / \partial y_0(x, s, z)$ is independent of z, and formulas (14.20) and (14.18) become the variation-of-constants formula (11.10). Also, by majorizing the integrals in (14.20) in a trivial way, one obtains the fundamental lemma (10.14) and also the variant form of Theorem 10.2.

Flows and Volume-Preserving Flows

> Considérons des molécules fluides dont l'ensemble forme à l'origine des temps une certaine figure F_0; quand ces molécules se déplaceront, leur ensemble formera une nouvelle figure qui ira en se déformant d'une manière continue, et à l'instant t l'ensemble des molécules envisagées formera une nouvelle figure F.
>
> (H. Poincaré, Mécanique Céleste 1899, Tome III, p.2)

We now turn our attention to a new interpretation of the Abel-Liouville-Jacobi-Ostrogradskii formula (11.11). Liouville and above all Jacobi (in his "Dynamik" 1843) used this formula extensively to obtain "first integrals", i.e., relations between the solutions, so that the dimension of the system could be decreased and the analytic integration of the differential equations of mechanics becomes a little less hopeless. Poincaré then (see the quotation) introduced a much more geometric point of view: for an autonomous system of differential equations [1]

$$\frac{dy}{dt} = f(y) \qquad\qquad (14.21)$$

we define the *flow* $\varphi_t : \mathbb{R}^n \to \mathbb{R}^n$ to be the function which associates, for a given t, to the initial value $y^0 \in \mathbb{R}^n$ the corresponding solution value at time t

$$\varphi_t(y^0) := y(t, 0, y^0). \qquad\qquad (14.22)$$

---

[1] Due to the origin of these topics in Mechanics and Astronomy, we here use t for the independent variable.

For sets A of initial values we also study its behaviour under the action of the flow and write

$$\varphi_t(A) = \{y \mid y = y(t, 0, y^0),\ y^0 \in A\}\,. \tag{14.22'}$$

We can imagine, with Poincaré, sets of "molecules" moving (and being deformed) with the flow.

Example 14.7. Fig. 14.2 shows, for the two-dimensional system (12.20) (see Fig. 12.2), the transformations which three sets A, B, C [2] undergo when t passes from 0 to 0.2, 0.4 and (for C) 0.6. It can be observed that these sets quickly lose very much of their beauty.

Fig. 14.2. Transformation of three sets under a flow

Now divide A into "infinitely small" cubes I of sides $dy_1^0, \ldots, dy_n^0$. The image $\varphi_t(I)$ of such a cube is an infinitely small parallelepiped. It is created by the columns of $\partial y / \partial y^0(t, 0, y^0)$ scaled by dy_i^0, and its volume is $\det\big(\partial y / \partial y^0(t, 0, y^0)\big) \cdot dy_1^0 \ldots dy_n^0$. Adding up all these volumes (over A) or, more precisely, using the transformation formula for multiple integrals

---

[2] The resemblance of these sets with a certain feline animal is not entirely accidental; we chose it in honour of V.I. Arnol'd.

(Euler 1769b, Jacobi 1841), we obtain

$$\text{Vol}\left(\varphi_t(A)\right) = \int_{\varphi_t(A)} dy = \int_A \left| \det\left(\frac{\partial y}{\partial y^0}(t, 0, y^0)\right)\right| dy^0.$$

Next we use formula (11.11) together with (14.15)

$$\det\left(\frac{\partial y}{\partial y^0}(t, 0, y^0)\right) = \exp\left(\int_0^t \text{tr}\left(f'(y(s, 0, y^0))\right) ds\right) \tag{14.23}$$

and we obtain

Theorem 14.8. *Consider the system (14.21) with continuously differentiable function $f(y)$.*

a) For a set $A \subset \mathbb{R}^n$ the total volume of $\varphi_t(A)$ satisfies

$$\text{Vol}\left(\varphi_t(A)\right) = \int_A \exp\left(\int_0^t \text{tr}\left(f'(y(s, 0, y^0))\right) ds\right) dy^0. \tag{14.24}$$

b) If $\text{tr}\left(f'(y)\right) = 0$ along the solution, the flow is volume-preserving, i.e.,
 $\text{Vol}\left(\varphi_t(A)\right) = \text{Vol}(A)$. □

Example 14.9. For the system (12.20) we have

$$f'(y) = \begin{pmatrix} (1 - 2y_1)/3 & (2y_2 - 1)/3 \\ 2 - y_2 & -y_1 \end{pmatrix} \qquad \text{and} \qquad \text{tr}\left(f'(y)\right) = (1 - 5y_1)/3.$$

The trace of $f'(y)$ changes sign at the line $y_1 = 1/5$. To its left the volume increases, to the right we have decreasing volumes. This can clearly be seen in Fig. 14.2.

Example 14.10. For the mathematical pendulum (with y_1 the angle of deviation from the vertical)

$$\begin{aligned} \dot{y}_1 &= y_2 \\ \dot{y}_2 &= -\sin y_1 \end{aligned} \qquad f'(y) = \begin{pmatrix} 0 & 1 \\ -\cos y_1 & 0 \end{pmatrix} \tag{14.25}$$

we have $\text{tr}\left(f'(y)\right) = 0$. Therefore the flow, although treating the cats quite badly, at least preserves their areas (Fig. 14.3).

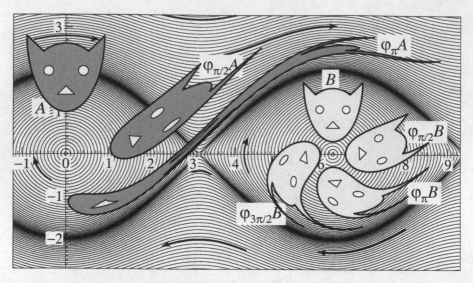

Fig. 14.3. Cats, beware of pendulums!

Canonical Equations and Symplectic Mappings

Let $H(p_1, \ldots, p_n, q_1, \ldots, q_n)$ be a twice continuously differentiable function of $2n$ variables and (see (6.26))

$$\dot{p}_i = -\frac{\partial H}{\partial q_i}(p, q), \qquad \dot{q}_i = \frac{\partial H}{\partial p_i}(p, q) \qquad (14.26)$$

the corresponding canonical system of differential equations. Small variations of the initial values lead to variations $\delta p_i(t), \delta q_i(t)$ of the solution of (14.26). By Theorem 14.3 (variational equation) these satisfy

$$
\begin{aligned}
\dot{\delta p}_i &= -\sum_{j=1}^{n} \frac{\partial^2 H}{\partial p_j \partial q_i}(p, q) \cdot \delta p_j - \sum_{j=1}^{n} \frac{\partial^2 H}{\partial q_j \partial q_i}(p, q) \cdot \delta q_j \\
\dot{\delta q}_i &= \sum_{j=1}^{n} \frac{\partial^2 H}{\partial p_j \partial p_i}(p, q) \cdot \delta p_j + \sum_{j=1}^{n} \frac{\partial^2 H}{\partial q_j \partial p_i}(p, q) \cdot \delta q_j.
\end{aligned}
\qquad (14.27)
$$

The upper left block of the Jacobian matrix is the negative transposed of the lower right block. As a consequence, the trace of the Jacobian of (14.27) is identically zero and *the corresponding flow is volume-preserving* ("Theorem of Liouville").

But there is much more than that (Poincaré 1899, vol. III, p. 43): consider a two-dimensional manifold A in the $2n$-dimensional flow. We represent it as a (differentiable) map of a compact set $K \subset \mathbb{R}^2$ into $\mathbb{R}^{2n}$ (Fig. 14.4)

$$
\begin{array}{cccc}
\Phi: & K & \longrightarrow & A \subset \mathbb{R}^{2n} \\
& (u, v) & \longmapsto & (p^0(u, v), q^0(u, v))
\end{array}
\qquad (14.28)
$$

We let $\pi_i(A)$ be the projection of A onto the (p_i, q_i)-coordinate plane and consider the *sum of the oriented areas of* $\pi_i(A)$. We shall see that this is also an invariant.

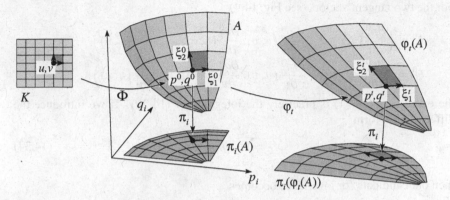

Fig. 14.4. Two-dimensional manifold in the flow

The oriented area of $\pi_i(A)$ is a surface integral over A which is defined, with the transformation formula in mind, as

$$\text{or.area}\big(\pi_i(A)\big) = \iint_K \det \begin{pmatrix} \dfrac{\partial p_i^0}{\partial u} & \dfrac{\partial p_i^0}{\partial v} \\ \dfrac{\partial q_i^0}{\partial u} & \dfrac{\partial q_i^0}{\partial v} \end{pmatrix} du\, dv\,. \tag{14.29}$$

For the computation of the area of $\pi_i\big(\varphi_t(A)\big)$, after the action of the flow, we use the composition $\varphi_t \circ \Phi$ as coordinate map (Fig. 14.4). This produces, with p_i^t, q_i^t being the ith respectively $(n+i)$th component of this map,

$$\text{or.area}\big(\pi_i(\varphi_t(A))\big) = \iint_K \det \begin{pmatrix} \dfrac{\partial p_i^t}{\partial u} & \dfrac{\partial p_i^t}{\partial v} \\ \dfrac{\partial q_i^t}{\partial u} & \dfrac{\partial q_i^t}{\partial v} \end{pmatrix} du\, dv\,. \tag{14.30}$$

There is no theoretical difficulty in differentiating this expression with respect to t and summing for $i = 1, \ldots, n$. This will give zero and the invariance is established.

The proof, however, becomes more elegant if we introduce *exterior differential forms* (E. Cartan 1899). These, originally "expressions purement symboliques", are today understood as *multilinear maps* on the *tangent space* (for more details see "Chapter 7" of Arnol'd 1974). In our case the one-forms dp_i, respectively dq_i, map a tangent vector ξ to its ith, respectively $(n+i)$th, component. The *exterior product* $dp_i \wedge dq_i$ is a bilinear map acting on a pair of vectors

$$\begin{aligned} (dp_i \wedge dq_i)(\xi_1, \xi_2) &= \det \begin{pmatrix} dp_i(\xi_1) & dp_i(\xi_2) \\ dq_i(\xi_1) & dq_i(\xi_2) \end{pmatrix} \\ &= dp_i(\xi_1)\,dq_i(\xi_2) - dp_i(\xi_2)\,dq_i(\xi_1) \end{aligned} \tag{14.31}$$

and satisfies Grassmann's rules for exterior multiplication

$$dp_i \wedge dp_j = -dp_j \wedge dp_i , \qquad dp_i \wedge dp_i = 0 . \tag{14.32}$$

For the two tangent vectors (see Fig. 14.4)

$$\xi_1^0 = \left(\frac{\partial p_1^0}{\partial u}(u,v), \ldots, \frac{\partial p_n^0}{\partial u}(u,v), \frac{\partial q_1^0}{\partial u}(u,v), \ldots, \frac{\partial q_n^0}{\partial u}(u,v) \right)^T$$
$$\xi_2^0 = \left(\frac{\partial p_1^0}{\partial v}(u,v), \ldots, \frac{\partial p_n^0}{\partial v}(u,v), \frac{\partial q_1^0}{\partial v}(u,v), \ldots, \frac{\partial q_n^0}{\partial v}(u,v) \right)^T \tag{14.33}$$

the expression (14.31) is precisely the integrand of (14.29). If we introduce the differential 2-form

$$\omega^2 = \sum_{i=1}^n dp_i \wedge dq_i \tag{14.34}$$

then our candidate for invariance becomes

$$\sum_{i=1}^n \text{or.area}\big(\pi_i(A)\big) = \iint_K \omega^2(\xi_1^0, \xi_2^0)\, du\, dv.$$

After the action of the flow we have the tangent vectors

$$\xi_1^t = \varphi_t'(p^0, q^0) \cdot \xi_1^0 , \qquad \xi_2^t = \varphi_t'(p^0, q^0) \cdot \xi_2^0$$

and

$$\sum_{i=1}^n \text{or.area}\big(\pi_i(\varphi_t(A))\big) = \iint_K \omega^2(\xi_1^t, \xi_2^t)\, du\, dv$$

(see (14.30)). We shall see that $\omega^2(\xi_1^t, \xi_2^t) = \omega^2(\xi_1^0, \xi_2^0)$.

Definition 14.11. For a differentiable function $g : \mathbb{R}^{2n} \to \mathbb{R}^{2n}$ we define the differential form $g^*\omega^2$ by

$$(g^*\omega^2)(\xi_1, \xi_2) := \omega^2\big(g'(p,q)\xi_1, g'(p,q)\xi_2\big) . \tag{14.35}$$

Such a function g is called *symplectic* (a name suggested by H. Weyl 1939, p. 165) if

$$g^*\omega^2 = \omega^2, \tag{14.36}$$

i.e., if the 2-form ω^2 is invariant under g.

Theorem 14.12. *The flow of a canonical system (14.26) is symplectic, i.e.,*

$$(\varphi_t)^*\omega^2 = \omega^2 \qquad \text{for all } t. \tag{14.37}$$

Proof. We compute the derivative of $\omega^2(\xi_1^t, \xi_2^t)$ (see (14.35)) with respect to t by the Leibniz rule. This gives

$$\frac{d}{dt}\left(\sum_{i=1}^{n}(dp_i \wedge dq_i)(\xi_1^t, \xi_2^t)\right) = \sum_{i=1}^{n}(dp_i \wedge dq_i)(\dot{\xi}_1^t, \xi_2^t) + \sum_{i=1}^{n}(dp_i \wedge dq_i)(\xi_1^t, \dot{\xi}_2^t).$$
(14.38)

Since the vectors ξ_1^t and ξ_2^t satisfy the variational equation (14.27), we have

$$\frac{d}{dt}\omega^2(\xi_1^t, \xi_2^t) = \sum_{i,j=1}^{n}\left(-\frac{\partial^2 H}{\partial p_j \partial q_i}\,dp_j \wedge dq_i - \frac{\partial^2 H}{\partial q_j \partial q_i}\,dq_j \wedge dq_i\right.$$
(14.39)

$$\left. + \frac{\partial^2 H}{\partial p_j \partial p_i}\,dp_i \wedge dp_j + \frac{\partial^2 H}{\partial q_j \partial p_i}\,dp_i \wedge dq_j\right)(\xi_1^t, \xi_2^t).$$

The first and last terms in this formula cancel by symmetry of the partial derivatives. Further, the properties (14.32) imply that

$$\sum_{i,j=1}^{n}\frac{\partial^2 H}{\partial p_i \partial p_j}(p,q)\,dp_i \wedge dp_j = \sum_{i<j}\left(\frac{\partial^2 H}{\partial p_i \partial p_j}(p,q) - \frac{\partial^2 H}{\partial p_j \partial p_i}(p,q)\right)dp_i \wedge dp_j$$

vanishes. Since the last remaining term cancels in the same way, the derivative (14.38) vanishes identically.   □

Example 14.13. We use the spherical pendulum in canonical form (6.28)

$$\dot{p}_1 = p_2^2\,\frac{\cos q_1}{\sin^3 q_1} - \sin q_1 \qquad \dot{p}_2 = 0$$
$$\dot{q}_1 = p_1 \qquad\qquad\qquad \dot{q}_2 = \frac{p_2}{\sin^2 q_1}$$
(14.40)

and for A the familiar two-dimensional cat placed in $\mathbb{R}^4$ such that its projection to (p_1, q_1) is a line; i.e., with zero area. It can be seen that with increasing t the area in (p_1, q_1) increases and the area in (p_2, q_2) decreases. Their sum remains constant. Observe that for larger t the left ear in (p_1, q_1) is twisted, i.e., surrounded in the negative sense, so that this part counts for negative area (Fig. 14.5). If time proceeded in the negative sense, *both* areas would increase, but the first area would be oriented negatively.

Between the two-dimensional invariant of Theorem 14.12 and the $2n$-dimensional of Liouville's theorem, there are many others; e.g., the differential 4-form

$$\omega^4 = \sum_{i<j}dp_i \wedge dp_j \wedge dq_i \wedge dq_j.$$
(14.41)

These invariants, however, are not really new, because (14.41) is proportional to the exterior square of ω^2, $\omega^2 \wedge \omega^2 = -2\omega^4$.

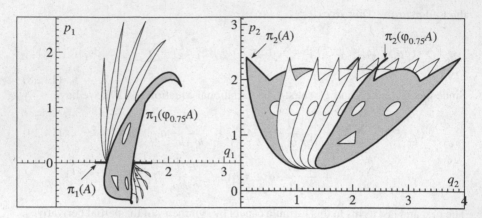

Fig. 14.5. Invariance of $\omega^2 = \sum_{i=1}^n dp_i \wedge dq_i$ for the spherical pendulum

Writing (14.31) in matrix notation

$$\omega^2(\xi_1, \xi_2) = \xi_1^T J \xi_2 \qquad \text{with} \qquad J = \begin{pmatrix} 0 & I \\ -I & 0 \end{pmatrix} \tag{14.42}$$

we obtain the following criterion:

Theorem 14.14. *A differentiable transformation* $g : \mathbb{R}^{2n} \to \mathbb{R}^{2n}$ *is symplectic if and only if its Jacobian* $R = g'(p, q)$ *satisfies*

$$R^T J R = J \tag{14.43}$$

with J given in (14.42).

Proof. This follows at once from (see (14.35))

$$(g^* \omega^2)(\xi_1, \xi_2) = (R\xi_1)^T J(R\xi_2) = \xi_1^T R^T J R \xi_2. \qquad \square$$

Exercises

1. Prove the following lemma from elementary calculus which is used in the proof
   of Theorem 14.4: if for a function $F(x, y)$, $\partial F / \partial y$ exists and $y(x)$ is differentiable and such that $F(x, y(x)) = Const$, then $\partial F / \partial x$ exists at $(x, y(x))$
   and is equal to

   $$\frac{\partial F}{\partial x}(x, y(x)) = -\frac{\partial F}{\partial y}(x, y(x)) \cdot y'(x).$$

   Hint. Use the identity

   $$F(x_1, y(x_1)) - F(x_0, y(x_1)) = F(x_0, y(x_0)) - F(x_0, y(x_1)).$$

I.15 Boundary Value and Eigenvalue Problems

Although our book is mainly concerned with initial value problems, we want to include in this first chapter some properties of boundary and eigenvalue problems.

Boundary Value Problems

They arise in systems of differential equations, say

$$y_1' = f_1(x, y_1, y_2),$$
$$y_2' = f_2(x, y_1, y_2), \qquad (15.1)$$

when there is *no* initial point x_0 at which $y_1(x_0)$ and $y_2(x_0)$ are known simultaneously. Questions of existence and uniqueness then become much more complicated.

Example 1. Consider the differential equation

$$y'' = \exp(y) \qquad \text{or} \qquad y_1' = y_2, \quad y_2' = \exp(y_1) \qquad (15.2a)$$

with the *boundary conditions*

$$y_1(0) = a, \qquad y_1(1) = b. \qquad (15.2b)$$

In order to apply our existence theorems or to do numerical computations (say by Euler's method (7.3)), we can proceed as follows: guess the missing initial value y_{20}. We can then compute the solution and check whether the computed value for $y_1(1)$ is equal to b or not. So our problem is, whether the function of the single variable y_{20}

$$F(y_{20}) := y_1(1) - b \qquad (15.3)$$

possesses a zero or not.

Equation (15.2a) is *quasimonotone,* which implies that $F(y_{20})$ depends monotonically on y_{20} (Fig. 15.1a, see Exercise 7 of I.10). Also, for y_{20} very small or very large, $y_1(1)$ is arbitrarily small or large, or even infinite. Therefore, (15.2) possesses for all a, b a unique solution (see Fig. 15.1b).

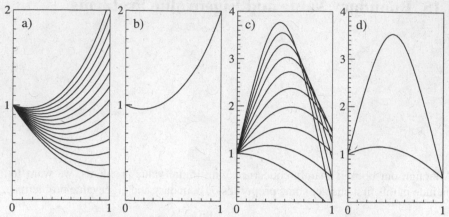

Fig. 15.1. a) Solutions of (15.2a) for different initial values $y_{20} = -1.7, \ldots, -0.4$
b) Unique solution of (15.2a) for $a = 1, b = 2, y_{20} = -0.476984656$
c) Solutions of (15.4a) for $y(0) = 1$ and $y_{20} = 0, 1, 2, \ldots, 9$
d) The two solutions of (15.4a), $y(0) = 1$, $y(1) = 0.5$, $y_{20} = 7.93719$, $y_{20} = 0.97084$

The root of $F(y_{20}) = 0$ can be computed by an iterative method, (bisection, regula falsi,...; if the derivative of $y_1(1)$ with respect to y_{20} is used from Theorem 14.3 or numerically from finite differences, also by Newton's method). The initial value problem is then computed several times. Small problems, such as the above example, can be done by a simple dialogue with the computer. Harder problems with more unknown initial values need more programming skills. This method is one of the most commonly used and is called *the shooting method.*

Example 2. For the differential equation

$$y'' = -\exp(y) \qquad \text{or} \qquad y_1' = y_2, \quad y_2' = -\exp(y_1) \tag{15.4a}$$

with the *boundary conditions*

$$y_1(0) = a, \qquad y_1(1) = b \tag{15.4b}$$

the monotonicity of $F(y_{20})$ is lost and things become more complicated: solutions for different initial values y_{20} are sketched for $a = 1$ in Fig. 15.1c. It can be seen that for b above a certain value (which is 1.499719998) there exists *no* solution of the problem at all, and for b below this value there exist *two* solutions (Fig. 15.1d).

Example 3.

$$y_1' = y_2, \qquad y_2' = y_1^3, \qquad y_1(0) = 1, \qquad y_1(100) = 2. \tag{15.5}$$

This equation is similar to (15.2) and the same statement of existence and uniqueness holds as above. However, if one tries to compute the solutions by the shooting method, one gets into trouble because of the length of the interval: *the solution nearly never exists on the whole interval;* in fact, the correct solution is

$y_{20} = -0.70710616655$. But already for $y_{20} = -0.7071061$, $y_1(x)$ tends to $+\infty$ for $x \to 98.2$. On the other side, for $y_{20} = -0.70711$, we have $y_1(94.1) = -\infty$. So the domain where $F(y_{20})$ of (15.3) *exists* is of length less than 4×10^{-6}.

In a case like this, one can use the *multiple shooting technique:* the interval is split up into several subintervals, on each of which the problem is solved with well-chosen initial values. At the endpoints of the subintervals, the solutions are then matched together. Equation (15.3) thereby becomes a system of higher dimension to be solved. Another possibility is to apply *global methods* (finite differences, collocation). Instead of integrating a sequence of initial value problems, a global representation of the approximate solution is sought. There exists an extensive literature on methods for boundary value problems. As a general reference we give Ascher, Mattheij & Russel (1988) and Deuflhard (1980).

Sturm-Liouville Eigenvalue Problems

This subject originated with a remarkable paper of Sturm (Sturm 1836) in Liouville's newly founded Journal. This paper was followed by a series of papers by Liouville and Sturm published in the following volumes. It is today considered as the starting point of the "geometric theory", where the main effort is not to try to integrate the equation, but merely to obtain geometric properties of the solution, such as its form, oscillations, sign changes, zeros, existence of maxima or minima and so on, *directly from the differential equation* ("Or on peut arriver à ce but par la seule considération des équations différentielles en elles-mêmes, sans qu'on ait besoin de leur intégration.")

The physical origin was, as in Section I.6, the study of heat and small oscillations of elastic media. Let us consider the heat equation with non-constant conductivity

$$\frac{\partial u}{\partial t} = \frac{\partial}{\partial x}\left(k(x)\frac{\partial u}{\partial x}\right) - \ell(x)u, \qquad k(x) > 0, \qquad (15.6)$$

which was studied extensively in Poisson's "Théorie de la chaleur". Poisson (1835) assumes $u(x,t) = y(x)e^{-\lambda t}$, so that (15.6) becomes

$$\frac{d}{dx}\left(k(x)\frac{dy}{dx}\right) - \ell(x)y = -\lambda y. \qquad (15.7)$$

We write (15.7) in the form

$$(k(x)y')' + G(x)y = 0 \qquad (15.8)$$

and state the following comparison theorem of Sturm:

Theorem 15.1. *Consider, with (15.8), the differential equation*

$$(\widehat{k}(x)\widehat{y}')' + \widehat{G}(x)\widehat{y} = 0, \tag{15.9}$$

and assume k, $\widehat{k}$ *differentiable,* G, $\widehat{G}$ *continuous,*

$$0 < \widehat{k}(x) \le k(x), \qquad \widehat{G}(x) \ge G(x) \tag{15.10}$$

for all x *and let* $y(x)$, $\widehat{y}(x)$ *be linearly independent solutions of (15.8) and (15.9), respectively. Then, between any two zeros of* $y(x)$ *there is at least one zero of* $\widehat{y}(x)$*, i.e., if* $y(x_1) = y(x_2) = 0$ *with* $x_1 < x_2$ *then there exists* x_3 *in the open interval* (x_1, x_2) *such that* $\widehat{y}(x_3) = 0$*.*

Proof. The original proof of Sturm is based on the quotient

$$q(x) = \frac{y(x)}{k(x)y'(x)}$$

which is the slope of the line connecting the origin with the solution point in the (ky', y)-plane and satisfies a first-order differential equation. In order to avoid the singularities caused by the zeros of $y'(x)$, we prefer the use of polar coordinates (Prüfer 1926)

$$k(x)y'(x) = \varrho(x)\cos\varphi(x), \qquad y(x) = \varrho(x)\sin\varphi(x). \tag{15.11}$$

Differentiation of (15.11) yields the following differential equations for φ and ϱ:

$$\varphi' = \frac{1}{k(x)}\cos^2\varphi + G(x)\sin^2\varphi \tag{15.12}$$

$$\varrho' = \left(\frac{1}{k(x)} - G(x)\right) \cdot \sin\varphi \cdot \cos\varphi \cdot \varrho. \tag{15.13}$$

In the same way we also introduce functions $\widehat{\varrho}(x)$ and $\widehat{\varphi}(x)$ for the second differential equation (15.9). They satisfy analogous relations with $k(x)$ and $G(x)$ replaced by $\widehat{k}(x)$ and $\widehat{G}(x)$.

Suppose now that x_1, x_2 are two consecutive zeros of $y(x)$. Then $\varphi(x_1)$ and $\varphi(x_2)$ must be multiples of π, since $\varrho(x)$ is always different from zero (uniqueness of the initial value problem). By (15.12) $\varphi'(x)$ is positive at x_1 and at x_2. Therefore we may assume that

$$\varphi(x_1) = 0, \qquad \varphi(x_2) = \pi, \qquad \widehat{\varphi}(x_1) \in [0, \pi). \tag{15.14}$$

The fact that equation (15.12) is first-order and the inequalities (15.10) allow the application of Theorem 10.3 to give

$$\widehat{\varphi}(x) \ge \varphi(x) \qquad \text{for} \qquad x_1 \le x \le x_2.$$

It is impossible that $\widehat{\varphi}(x) = \varphi(x)$ everywhere, since this would imply $\widehat{G}(x) = G(x)$, $\cos\widehat{\varphi}(x)/\widehat{k}(x) = \cos\varphi(x)/k(x)$ by (15.12) and (15.10). As a consequence of (15.13) we would have $\widehat{\varrho}(x) = C \cdot \varrho(x)$ and the solutions $y(x)$, $\widehat{y}(x)$ would be linearly dependent. Therefore, there exists $x_0 \in (x_1, x_2)$ such that $\widehat{\varphi}(x_0) > \varphi(x_0)$. In this situation $\widehat{\varphi}(x) > \varphi(x)$ for all $x \geq x_0$ and the existence of $x_3 \in (x_1, x_2)$ with $\widehat{\varphi}(x_3) = \pi$ is assured.   □

The next theorem shows that our eigenvalue problem possesses an *infinity* of solutions. We add to (15.7) the boundary conditions

$$y(x_0) = y(x_1) = 0. \tag{15.15}$$

Theorem 15.2. *The eigenvalue problem (15.7), (15.15) possesses an infinite sequence of eigenvalues* $\lambda_1 < \lambda_2 < \lambda_3 < \ldots$ *whose corresponding solutions* $y_i(x)$ *("eigenfunctions") possess respectively* $0, 1, 2, \ldots$ *zeros in the interval* (x_0, x_1). *The zeros of* $y_{j+1}(x)$ *separate those of* $y_j(x)$. *If* $0 < K_1 \leq k(x) \leq K_2$ *and* $L_1 \leq \ell(x) \leq L_2$, *then*

$$L_1 + K_1 \frac{j^2 \pi^2}{(x_1 - x_0)^2} \leq \lambda_j \leq L_2 + K_2 \frac{j^2 \pi^2}{(x_1 - x_0)^2}. \tag{15.16}$$

Proof. Let $y(x, \lambda)$ be the solution of (15.7) with initial values $y(x_0) = 0$, $y'(x_0) = 1$. Theorem 15.1 (with $\widehat{k}(x) = k(x)$, $\widehat{G}(x) = G(x) + \Delta\lambda$) implies that for increasing λ the zeros of $y(x, \lambda)$ move towards x_0, so that the number of zeros in (x_0, x_1) is a non-decreasing function of λ.

Comparing next (15.7) with the solution $(\lambda > L_1)$

$$\sin\left(\sqrt{(\lambda - L_1)/K_1} \cdot (x - x_0)\right)$$

of $K_1 y'' + (\lambda - L_1)y = 0$ we see that for $\lambda < L_1 + K_1 j^2 \pi^2/(x_1 - x_0)^2$, $y(x, \lambda)$ has at most $j - 1$ zeros in $(x_0, x_1]$. Similarly, a comparison with

$$\sin\left(\sqrt{(\lambda - L_2)/K_2} \cdot (x - x_0)\right)$$

which is a solution of $K_2 y'' + (\lambda - L_2)y = 0$, shows that $y(x, \lambda)$ possesses at least j zeros in (x_0, x_1), if $\lambda > L_2 + K_2 j^2 \pi^2/(x_1 - x_0)^2$. The statements of the theorem are now simple consequences of these three properties.   □

Example. Fig. 15.2 shows the first 5 solutions of the problem

$$((1 - 0.8\sin^2 x)y')' - (x - \lambda)y = 0, \qquad y(0) = y(\pi) = 0. \tag{15.17}$$

The first eigenvalues are 2.1224, 3.6078, 6.0016, 9.3773, 13.7298, 19.053, 25.347, 32.609, 40.841, 50.041, etc.

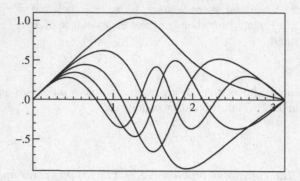

Fig. 15.2. Solutions of the Sturm-Liouville eigenvalue problem (15.17)

For more details about this theory, which is a very important page of history, we refer to the book of Reid (1980).

Exercises

1. Consider the equation

$$L(x)y'' + M(x)y' + N(x)y = 0.$$

Multiply it with a suitable function $\varphi(x)$, so that the ensuing equation is of the form (15.8) (Sturm 1836, p. 108).

2. Prove that two solutions of (15.7), (15.15) satisfy the orthogonality relations

$$\int_{x_0}^{x_1} y_j(x)y_k(x)dx = 0 \qquad \text{for} \qquad \lambda_j \neq \lambda_k.$$

Hint. Multiply this by λ_j, replace $\lambda_j y_j(x)$ from (15.7) and do partial integration (Liouville 1836, p. 257).

3. Solve the problem (15.5) by elementary functions. Explain why the given value for y_{20} is so close to $-\sqrt{2}/2$.

4. Show that the boundary value problem (see Collatz 1967)

$$y'' = -y^3, \qquad y(0) = 0, \qquad y(A) = B \qquad (15.18)$$

possesses infinitely many solutions for each pair (A, B) with $A \neq 0$.

Hint. Draw the solution $y(x)$ of (15.18) with $y(0) = 0$, $y'(0) = 1$. Show that for each constant a, $z(x) = ay(ax)$ is also a solution.

I.16 Periodic Solutions, Limit Cycles, Strange Attractors

2° Les demi-spirales que l'on suit sur un arc infini sans arriver à un nœud ou à un foyer et sans revenir au point de départ ; ...

(H. Poincaré 1882, Oeuvres vol. 1, p. 54)

The phenomenon of limit cycles was first described theoretically by Poincaré (1882) and Bendixson (1901), and has since then found many applications in Physics, Chemistry and Biology. In higher dimensions things can become much more chaotic and attractors may look fairly "strange".

Van der Pol's Equation

I have a theory that whenever you want to get in trouble with a method, look for the Van der Pol equation.

(P.E. Zadunaisky 1982)

The first practical examples were studied by Rayleigh (1883) and later by Van der Pol (1920-1926) in a series of papers on nonlinear oscillations: the solutions of

$$y'' + \alpha y' + y = 0$$

are *damped* for $\alpha > 0$, and *unstable* for $\alpha < 0$. The idea is to change α (with the help of a triode, for example) so that $\alpha < 0$ for small y and $\alpha > 0$ for large y. The simplest expression, which describes the physical situation in a somewhat idealized form, would be $\alpha = \varepsilon(y^2 - 1)$, $\varepsilon > 0$. Then the above equation becomes

$$y'' + \varepsilon(y^2 - 1)y' + y = 0, \tag{16.1}$$

or, written as a system,

$$\begin{aligned} y_1' &= y_2 \\ y_2' &= \varepsilon(1 - y_1^2)y_2 - y_1, \quad \varepsilon > 0. \end{aligned} \tag{16.2}$$

In this equation, small oscillations are amplified and large oscillations are damped. We therefore expect the existence of a stable periodic solution to which all other solutions converge. We call this a *limit cycle* (Poincaré 1882, "Chap. VI"). The original illustrations of the paper of Van der Pol are reproduced in Fig. 16.1.

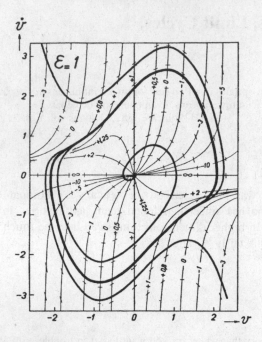

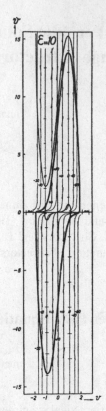

Fig. 16.1. Illustrations from
Van der Pol (1926)
(with permission)

Existence proof. The existence of limit cycles is studied by the method of *Poincaré sections* (Poincaré 1882, "Chap. V, Théorie des conséquents"). The idea is to cut the solutions transversally by a hyperplane Π and, for an initial value $y_0 \in \Pi$, to study the first point $\Phi(y_0)$ where the solution again crosses the plane Π in the same direction.

For our example (16.2), we choose for Π the half-line $y_2 = 0$, $y_1 > 0$. We then examine the signs of y_1' and y_2' in (16.2). The sign of y_2' changes at the curve

$$y_2 = \frac{y_1}{\varepsilon(1 - y_1^2)}, \tag{16.3}$$

which is drawn as a broken line in Fig. 16.2. It follows (see Fig. 16.2) that $\Phi(y_0)$ exists for all $y_0 \in \Pi$. Since two different solutions *cannot intersect* (due to uniqueness), the map Φ is *monotone*. Further, Φ is bounded (e.g., by every solution starting on the curve (16.3)), so $\Phi(y_0) < y_0$ for y_0 large. Finally, since the origin is unstable, $\Phi(y_0) > y_0$ for y_0 small. Hence there must be a fixed point of $\Phi(y_0)$, i.e., a limit cycle.   □

The limit cycle is, in fact, *unique*. The proof for this is more complicated and is indicated in Exercise 8 below (Liénard 1928).

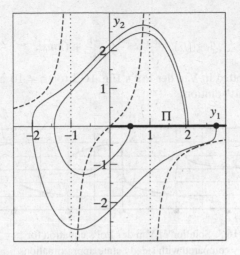

Fig. 16.2. The Poincaré map for Van der Pol's equation, $\varepsilon = 1$

With similar ideas one proves the following general result:

Theorem 16.1 (Poincaré 1882, Bendixson 1901). *Each bounded solution of a two-dimensional system*

$$y'_1 = f_1(y_1, y_2), \qquad y'_2 = f_2(y_1, y_2) \tag{16.4}$$

must

i) *tend to a critical point $f_1 = f_2 = 0$ for an infinity of points $x_i \to \infty$; or*

ii) *be periodic; or*

iii) *tend to a limit cycle.*   □

Remark. Exercise 1 below explains why the possibility (i) is written in a form somewhat more complicated than seems necessary.

Steady-state approximations for ε large. An important tool for simplifying complicated nonlinear systems is that of steady-state approximations. Consider (16.2) with ε very large. Then, in the neighbourhood of $f_2(y_1, y_2) = 0$ for $|y_1| > 1$, the derivative of $y'_2 = f_2$ with respect to y_2 is very large negative. Therefore the solution will very rapidly approach an equilibrium state in the neighbourhood of $y'_2 = f_2(y_1, y_2) = 0$, i.e., in our example, $y_2 = y_1/(\varepsilon(1 - y_1^2))$. This can be inserted into (16.2) and leads to

$$y'_1 = \frac{y_1}{\varepsilon(1 - y_1^2)}, \tag{16.5}$$

an equation of lower dimension. Using the formulas of Section I.3, (16.5) is easily solved to give

$$\log(y_1) - \frac{y_1^2}{2} = \frac{x - x_0}{\varepsilon} + Const.$$

These curves are dotted in Van der Pol's Fig. 16.3 for $\varepsilon = 10$ and show the good approximation of this solution.

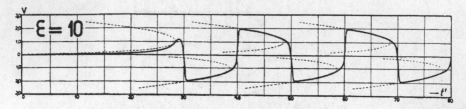

Fig. 16.3. Solution of Van der Pol's equation for $\varepsilon = 10$
compared with steady state approximations

Asymptotic solutions for ε small. The computation of periodic solutions for *small* parameters was initiated by astronomers such as Newcomb and Lindstedt and brought to perfection by Poincaré (1893). We demonstrate the method for the Van der Pol equation (16.1). The idea is to develop the solution as a series in powers of ε. Since the period will change too, we also introduce a coordinate change

$$t = x(1 + \gamma_1 \varepsilon + \gamma_2 \varepsilon^2 + \ldots) \tag{16.6}$$

and put

$$y(x) = z(t) = z_0(t) + \varepsilon z_1(t) + \varepsilon^2 z_2(t) + \ldots. \tag{16.7}$$

Inserting now $y'(x) = z'(t)(1 + \gamma_1 \varepsilon + \ldots)$, $y''(x) = z''(t)(1 + \gamma_1 \varepsilon + \ldots)^2$ into (16.1) we obtain

$$\begin{aligned}
& (z_0'' + \varepsilon z_1'' + \varepsilon^2 z_2'' + \ldots)(1 + 2\gamma_1 \varepsilon + (2\gamma_2 + \gamma_1^2)\varepsilon^2 + \ldots) \\
& + \varepsilon((z_0 + \varepsilon z_1 + \ldots)^2 - 1)(z_0' + \varepsilon z_1' + \ldots)(1 + \gamma_1 \varepsilon + \ldots) \\
& + (z_0 + \varepsilon z_1 + \varepsilon^2 z_2 + \ldots) = 0.
\end{aligned} \tag{16.8}$$

We first compare the coefficients of ε^0 and obtain

$$z_0'' + z_0 = 0. \tag{16.8;0}$$

We fix the initial value on the Poincaré section P, i.e., $z'(0) = 0$, so that $z_0 = A\cos t$ with A, for the moment, a free parameter. Next, the coefficients of ε yield

$$\begin{aligned}
z_1'' + z_1 &= -2\gamma_1 z_0'' - (z_0^2 - 1)z_0' \\
&= 2\gamma_1 A \cos t + \left(\frac{A^3}{4} - A\right)\sin t + \frac{A^3}{4}\sin 3t.
\end{aligned} \tag{16.8;1}$$

Here, the crucial idea is that we are looking for *periodic* solutions, hence the terms in $\cos t$ and $\sin t$ on the right-hand side of (16.8;1) must disappear, in order to avoid that $z_1(t)$ contain terms of the form $t \cdot \cos t$ and $t \cdot \sin t$ ("... et de faire disparaître ainsi les termes dits *séculaires* ..."). We thus obtain $\gamma_1 = 0$ and $A = 2$. Then (16.8;1) can be solved and gives, together with $z_1'(0) = 0$,

$$z_1 = B \cos t + \frac{3}{4} \sin t - \frac{1}{4} \sin 3t. \tag{16.9}$$

The continuation of this process is now clear: the terms in ε^2 in (16.8) lead to, after insertion of (16.9) and simplification,

$$z_2'' + z_2 = \left(4\gamma_2 + \frac{1}{4}\right) \cos t + 2B \sin t + 3B \sin 3t - \frac{3}{2} \cos 3t + \frac{5}{4} \cos 5t. \tag{16.8;2}$$

Secular terms are avoided if we set $B = 0$ and $\gamma_2 = -1/16$. Then

$$z_2 = C \cos t + \frac{3}{16} \cos 3t - \frac{5}{96} \cos 5t.$$

The next round will give $C = -1/8$ and $\gamma_3 = 0$, so that we have: *the periodic orbit of the Van der Pol equation (16.1) for ε small is given by*

$$y(x) = z(t), \qquad t = x(1 - \varepsilon^2/16 + \ldots),$$
$$z(t) = 2 \cos t + \varepsilon \left(\frac{3}{4} \sin t - \frac{1}{4} \sin 3t\right)$$
$$+ \varepsilon^2 \left(-\frac{1}{8} \cos t + \frac{3}{16} \cos 3t - \frac{5}{96} \cos 5t\right) + \ldots \tag{16.10}$$

and is of period $2\pi(1 + \varepsilon^2/16 + \ldots)$.

Chemical Reactions

The laws of chemical kinetics give rise to differential equations which, for multimolecular reactions, become nonlinear and have interesting properties. Some of them possess periodic solutions (e.g. the Zhabotinski-Belousov reaction) and have important applications to the interpretation of biological phenomena (e.g. Prigogine, Lefever).

Let us examine in detail the model of Lefever and Nicolis (1971), the so-called "Brusselator": suppose that six substances A, B, D, E, X, Y undergo the following reactions:

$$
\begin{aligned}
A &\xrightarrow{\ k_1\ } X \\
B + X &\xrightarrow{\ k_2\ } Y + D \quad \text{(bimolecular reaction)} \\
2X + Y &\xrightarrow{\ k_3\ } 3X \quad\ \ \text{(autocatalytic trimol. reaction)} \\
X &\xrightarrow{\ k_4\ } E
\end{aligned}
\tag{16.11}
$$

If we denote by $A(x), B(x), \ldots$ the *concentrations* of $A, B, \ldots$ as functions of the time x, the reactions (16.11) become by the mass action law the following differential equations

$$A' = -k_1 A$$
$$B' = -k_2 BX$$
$$D' = k_2 BX$$
$$E' = k_4 X$$
$$X' = k_1 A - k_2 BX + k_3 X^2 Y - k_4 X$$
$$Y' = k_2 BX - k_3 X^2 Y.$$

This system is now simplified as follows: the equations for D and E are left out, because they do not influence the others; A and B are supposed to be maintained constant (positive) and all reaction rates k_i are set equal to 1. We further set $y_1(x) := X(x)$, $y_2(x) := Y(x)$ and obtain

$$\begin{aligned} y_1' &= A + y_1^2 y_2 - (B+1)y_1 \\ y_2' &= By_1 - y_1^2 y_2. \end{aligned} \qquad (16.12)$$

The resulting system has one critical point $y_1' = y_2' = 0$ at $y_1 = A$, $y_2 = B/A$. The linearized equation in the neighbourhood of this point is unstable iff $B > A^2 + 1$. Further, a study of the domains where y_1', y_2', or $(y_1 + y_2)'$ is positive or negative leads to the result that all solutions remain bounded. Thus, for $B > A^2 + 1$ there must be a limit cycle which, by numerical calculations, is seen to be unique (Fig. 16.4).

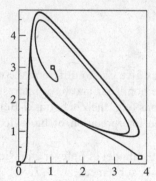

Fig. 16.4. Solutions of the Brusselator, $A = 1$, $B = 3$

An interesting phenomenon (Hopf bifurcation, see below) occurs, when B approaches $A^2 + 1$. Then the limit cycle becomes smaller and smaller and finally disappears in the critical point. Another example of this type is given in Exercise 2.

Limit Cycles in Higher Dimensions, Hopf Bifurcation

The Theorem of Poincaré-Bendixson is apparently true only in two dimensions. Higher dimensional counter-examples are given by nearly every mechanical movement without friction, as for example the spherical pendulum (6.20), see Fig. 6.2. Therefore, in higher dimensions limit cycles are usually found by numerical studies of the Poincaré section map Φ defined above.

There is, however, one situation where limit cycles occur quite naturally (Hopf 1942): namely when at a critical point of $y' = f(y, \alpha)$, $y, f \in \mathbb{R}^n$, all eigenvalues of $(\partial f/\partial y)(y_0, \alpha)$ have strictly negative real part with the exception of *one* pair which, by varying α, crosses the imaginary axis. The eigenspace of the stable eigenvalues then continues into an analytic two dimensional manifold, inside which a limit cycle appears. This phenomenon is called "Hopf bifurcation". The proof of this fact is similar to Poincaré's parameter expansion method (16.7) (see Exercises 6 and 7 below), so that Hopf even hesitated to publish it ("... ich glaube kaum, dass an dem obigen Satz etwas wesentlich Neues ist ...").

As an example, we consider the "full Brusselator" (16.11): we no longer suppose that B is kept constant, but that B is constantly added to the mixture with

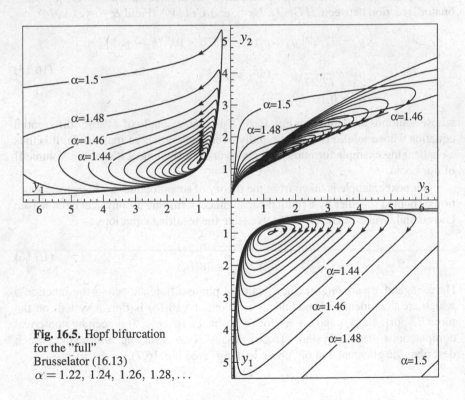

Fig. 16.5. Hopf bifurcation for the "full" Brusselator (16.13)
$\alpha = 1.22, 1.24, 1.26, 1.28, \ldots$

rate α. When we set $y_3(x) := B(x)$, we obtain instead of (16.12) (with $A = 1$)

$$
\begin{aligned}
y_1' &= 1 + y_1^2 y_2 - (y_3 + 1)y_1 \\
y_2' &= y_1 y_3 - y_1^2 y_2 \\
y_3' &= -y_1 y_3 + \alpha.
\end{aligned}
\tag{16.13}
$$

This system possesses a critical point at $y_1 = 1$, $y_2 = y_3 = \alpha$ with derivative

$$
\frac{\partial f}{\partial y} = \begin{pmatrix} \alpha - 1 & 1 & -1 \\ -\alpha & -1 & 1 \\ -\alpha & 0 & -1 \end{pmatrix}.
\tag{16.14}
$$

This matrix has $\lambda^3 + (3 - \alpha)\lambda^2 + (3 - 2\alpha)\lambda + 1$ as characteristic polynomial and satisfies the condition for stability iff $\alpha < (9 - \sqrt{17})/4 = 1.21922$ (see I.13, Exercise 1). Thus when α increases beyond this value, there arises a limit cycle which exists for all values of α up to approximately 1.5 (see Fig. 16.5). When α continues to grow, the limit cycle "explodes" and $y_1 \to 0$ while y_2 and $y_3 \to \infty$. So the system (16.13) has a behaviour completely different from the simplified model (16.12).

A famous chemical reaction with a limit cycle in three dimensions is the "Oregonator" reaction between $HBrO_2, Br^-$, and $Ce\,(IV)$ (Field & Noyes 1974)

$$
\begin{aligned}
y_1' &= 77.27\Big(y_2 + y_1(1 - 8.375 \times 10^{-6} y_1 - y_2)\Big) \\
y_2' &= \frac{1}{77.27}(y_3 - (1 + y_1)y_2) \\
y_3' &= 0.161(y_1 - y_3)
\end{aligned}
\tag{16.15}
$$

whose solutions are plotted in Fig. 16.6. This is an example of a "stiff" differential equation whose solutions change rapidly over many orders of magnitude. It is thus a challenging example for numerical codes and we shall meet it again in Volume II of our book.

Our next example is taken from the theory of superconducting Josephson junctions, coupled together by a mutual capacitance. Omitting all physical details, (see Giovannini, Weiss & Ulrich 1978), we state the resulting equations as

$$
\begin{aligned}
c(y_1'' - \alpha y_2'') &= i_1 - \sin(y_1) - y_1' \\
c(y_2'' - \alpha y_1'') &= i_2 - \sin(y_2) - y_2'.
\end{aligned}
\tag{16.16}
$$

Here, y_1 and y_2 are *angles* (the "quantum phase difference across the junction") which are thus identified modulo 2π. Equation (16.16) is thus a system on the torus T^2 for (y_1, y_2), and on $\mathbb{R}^2$ for the voltages (y_1', y_2'). It is seen by numerical computations that the system (16.16) possesses an attracting limit cycle, which describes the phenomenon of "phase locking" (see Fig. 16.7).

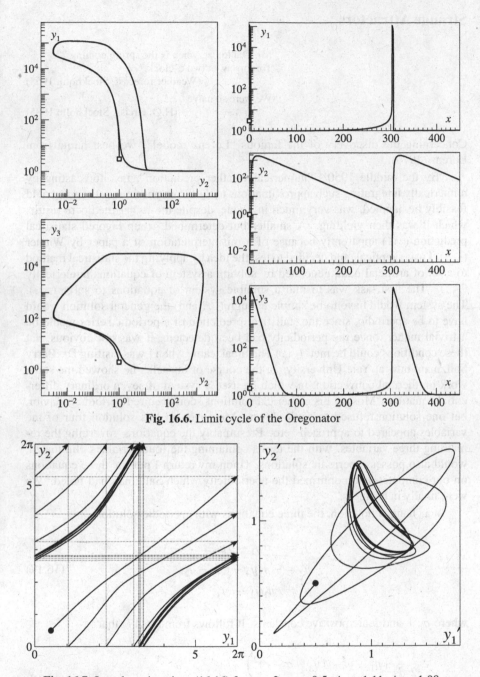

Fig. 16.6. Limit cycle of the Oregonator

Fig. 16.7. Josephson junctions (16.16) for $c = 2$, $\alpha = 0.5$, $i_1 = 1.11$, $i_2 = 1.08$

Strange Attractors

> "Mr. Dahlquist, when is the spring coming ?"
> "Tomorrow, at two o'clock."
>
> (Weather forecast, Stockholm 1955)
>
> "We were **so** naïve ..."
>
> (H.O. Kreiss, Stockholm 1985)

Concerning the discovery of the famous "Lorenz model", we best quote from Lorenz (1979):

"By the middle 1950's "numerical weather prediction", i.e., forecasting by numerically integrating such approximations to the atmospheric equations as could feasibly be handled, was very much in vogue, despite the rather mediocre results which it was then yielding. A smaller but determined group favored statistical prediction (...) apparently because of a misinterpretation of a paper by Wiener (...). I was skeptical, and decided to test the idea by applying the statistical method to a set of artificial data, generated by solving a system of equations numerically (...). The first task was to find a suitable system of equations to solve (...). The system would have to be simple enough (... and) the general solution would have to be aperiodic, since the statistical prediction of a periodic series would be a trivial matter, once the periodicity had been detected. It was not obvious that these conditions could be met. (...) The break came when I was visiting Dr. Barry Saltzman, now at Yale University. In the course of our talks he showed me some work on thermal convection, in which he used a system of seven ordinary differential equations. Most of his numerical solutions soon acquired periodic behavior, but one solution refused to settle down. Moreover, in this solution four of the variables appeared to approach zero. Presumably the equations governing the remaining three variables, with the terms containing the four variables eliminated, would also possess aperiodic solutions. Upon my return I put the three equations on our computer, and confirmed the aperiodicity which Saltzman had noted. We were finally in business."

In a changed notation, the three equations with aperiodic solutions are

$$
\begin{aligned}
y_1' &= -\sigma y_1 + \sigma y_2 \\
y_2' &= -y_1 y_3 + r y_1 - y_2 \\
y_3' &= y_1 y_2 - b y_3
\end{aligned}
\tag{16.17}
$$

where σ, r and b are positive constants. It follows from (16.17) that

$$
\begin{aligned}
\frac{1}{2} \frac{d}{dx} &\left(y_1^2 + y_2^2 + (y_3 - \sigma - r)^2 \right) \\
&= -\left(\sigma y_1^2 + y_2^2 + b(y_3 - \frac{\sigma}{2} - \frac{r}{2})^2 \right) + b\left(\frac{\sigma}{2} + \frac{r}{2} \right)^2.
\end{aligned}
\tag{16.18}
$$

Therefore the ball

$$R_0 = \left\{ (y_1, y_2, y_3) \mid y_1^2 + y_2^2 + (y_3 - \sigma - r)^2 \le c^2 \right\} \tag{16.19}$$

is mapped by the flow φ_1 (see (14.22)) into itself, provided that c is sufficiently large so that R_0 wholly contains the ellipsoid defined by equating the right side of (16.18) to zero. Hence, if x assumes the increasing values $1, 2, 3, \ldots$, R_0 is carried into regions $R_1 = \varphi_1(R_0)$, $R_2 = \varphi_2(R_0)$ etc., which satisfy $R_0 \supset R_1 \supset R_2 \supset R_3 \supset \ldots$ (applying φ_1 to the inclusion $R_0 \supset R_1$ gives $R_1 \supset R_2$ and so on).

Since the trace of $\partial f / \partial y$ for the system (16.17) is the negative constant $-(\sigma + b + 1)$, the *volumes* of R_k tend exponentially to zero (see Theorem 14.8). Every orbit is thus ultimately trapped in a set $R_\infty = R_0 \cap R_1 \cap R_2 \ldots$ of zero volume.

System (16.17) possesses an obvious critical point $y_1 = y_2 = y_3 = 0$; this becomes unstable when $r > 1$. In this case there are two additional critical points C and C' respectively given by

$$y_1 = y_2 = \pm \sqrt{b(r-1)}, \qquad y_3 = r - 1. \tag{16.20}$$

These become unstable (e.g. by the Routh criterion, Exercise 1 of Section I.13) when $\sigma > b + 1$ and

$$r \ge r_c = \frac{\sigma(\sigma + b + 3)}{\sigma - b - 1}. \tag{16.21}$$

In the first example we shall use Saltzman's values $b = 8/3$, $\sigma = 10$, and $r = 28$. ("Here we note another lucky break: Saltzman used $\sigma = 10$ as a crude approximation to the Prandtl number (about 6) for water. Had he chosen to study air, he would probably have let $\sigma = 1$, and the aperiodicity would not have been discovered", Lorenz 1979). In Fig. 16.8 we have plotted the solution curve of (16.17) with the initial value $y_1 = -8$, $y_2 = 8$, $y_3 = r - 1$, which, indeed, looks pretty chaotic.

For a clearer understanding of the phenomenon, we choose the plane $y_3 = r - 1$, especially the square region between the critical points C and C', as Poincaré section Π. The critical point $y_1 = y_2 = y_3 = 0$ possesses (since $r > 1$) one unstable eigenvalue $\lambda_1 = (-1 - \sigma + \sqrt{(1-\sigma)^2 + 4r\sigma})/2$ and two stable eigenvalues $\lambda_2 = -b$, $\lambda_3 = (-1 - \sigma - \sqrt{(1-\sigma)^2 + 4r\sigma})/2$. The eigenspace of the stable eigenvalues continues into a two-dimensional manifold of initial values, whose solutions tend to 0 for $x \to \infty$. This "stable manifold" cuts Π in a curve Σ (see Fig. 16.9). The one-dimensional *unstable* manifold (created by the unstable eigenvalue λ_1) cuts Π in the points D and D' (Fig. 16.9).

All solutions starting in Π_u *above* Σ (the dark cat) surround the above critical point C and are, at the first return, mapped to a narrow stripe S_u, while the solutions starting in Π_d *below* Σ surround C' and go to the left stripe S_d. At the *second* return, the two stripes are mapped into two very narrow stripes *inside* S_u and S_d. After the third return, we have 8 stripes closer and closer together, and so on. The intersection of all these stripes is a Cantor-like set and, continued

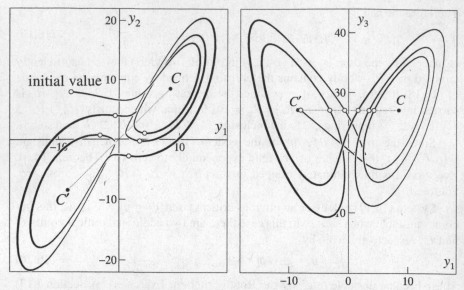

Fig. 16.8. Two views of a solution of (16.17)
(small circles indicate intersection of solution with plane $y_3 = r - 1$)

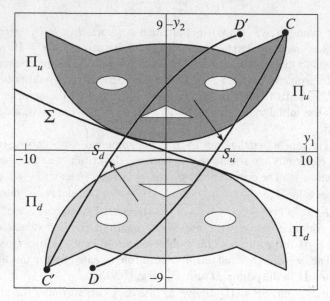

Fig. 16.9. Poincaré map for (16.17)

into 3-space by the flow, forms the *strange attráctor* ("An attractor of the type just described can therefore not be thrown away as non-generic pathology", Ruelle & Takens 1971).

The Ups and Downs of the Lorenz Model

> "Mr. Laurel and Mr. Hardy have many ups and downs — Mr. Hardy
> takes charge of the upping, and Mr. Laurel does most of the downing
> — "   (from "Another Fine Mess", Hal Roach 1930)

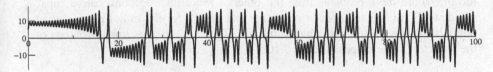

If one watches the solution $y_1(x)$ of the Lorenz equation being calculated, one wonders who decides for the solution to go up or down in an apparently unpredictable fashion. Fig. 16.9 shows that Σ cuts both stripes S_d and S_u. Therefore the *inverse image* of Σ (see Fig. 16.10) consists of *two* lines Σ_0 and Σ_1 which cut, together with Σ, the plane Π into *four* sets Π_{uu}, Π_{ud}, Π_{du}, Π_{dd}. If the initial value is in one of these, the corresponding solution goes up-up, up-down, down-up, down-down. Further, the inverse images of Σ_0 and Σ_1 lead to four lines Σ_{00}, Σ_{01}, Σ_{10}, Σ_{11}. The plane Π is then cut into 8 stripes and we now know the fate of the first three ups and downs. The more inverse images of these curves we compute, the finer the plane Π is cut into stripes and all the future ups and downs are coded in the position of the initial value with respect to these stripes (see Fig. 16.10). It appears that a *very small* change in the initial value gives rise, after a couple of rotations, to a *totally different* solution curve. This phenomenon, discovered merely by accident by Lorenz (see Lorenz 1979), is highly interesting

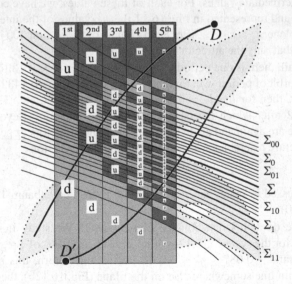

Fig. 16.10. Stripes deciding for the ups and downs

and explains why the theorem of uniqueness (Theorem 7.4), of whose philosophical consequences Laplace was so proud, has its practical limits.

Remark. It appears in Fig. 16.10 that not all stripes have the same width. The sequences of "u"'s and "d"'s which repeat u or d a couple of times (but not too often) are more probable than the others. More than 25 consecutive "ups" or "downs" are (for the chosen constants and except for the initial phase) never possible. This has to do with the position of D and D', the outermost frontiers of the attractor, in the stripes of Fig. 16.10.

Feigenbaum Cascades

However nicely the beginning of Lorenz' (1979) paper is written, the affirmations of his last section are only partly true. As Lorenz did, we now vary the parameter b in (16.17), letting at the same time $r = r_c$ (see (16.21)) and

$$\sigma = b + 1 + \sqrt{2(b+1)(b+2)}. \tag{16.22}$$

This is the value of σ for which r_c is minimized. Numerical integration shows that for b very small (say $b \leq 0.139$), the solutions of (16.17) evidently converge to a stable limit cycle, which cuts the Poincaré section $y_3 = r - 1$ twice at two different locations and surrounds both critical points C and C'. Further, for b large (for example $b = 8/3$) the coefficients are not far from those studied above and we have a strange attractor. But what happens in between? We have computed the solutions of the Lorenz model (16.17) for b varying from 0.1385 to 0.1475 with 1530 intermediate values. For each of these values, we have computed 1500 Poincaré cuts and represented in Fig. 16.11 the y_1-values of the intersections with the Poincaré plane $y_3 = r - 1$. After each change of b, the first 300 iterations were not drawn so that only the attractor becomes visible.

For b small, there is one periodic orbit; then, at $b = b_1 = 0.13972$, it suddenly splits into an orbit of period two, this then splits for $b = b_2 = 0.14327$ into an orbit of period four, then for $b = b_3 = 0.14400$ into period eight, etc. There is a point $b_\infty = 0.14422$ after which the movement becomes chaotic. Beyond this value, however, there are again and again intervals of stable attractors of periods 5, 3, etc. The whole picture resembles what is obtained by the recursion

$$x_{n+1} = a(x_n - x_n^2) \tag{16.23}$$

which is discussed in many papers (e.g. May 1976, Feigenbaum 1978, Collet & Eckmann 1980).

But where does this resemblance come from? We study in Fig. 16.12 the Poincaré map for the system (16.17) with b chosen as 0.146 of a region $-0.095 \leq y_1 \leq -0.078$ and $-0.087 \leq y_2 \leq -0.07$. After one return, this region is compressed to a thin line somewhere else on the plane (Fig. 16.12b), the second return bends this line to U-shape and maps it into the original region (Fig. 16.12c).

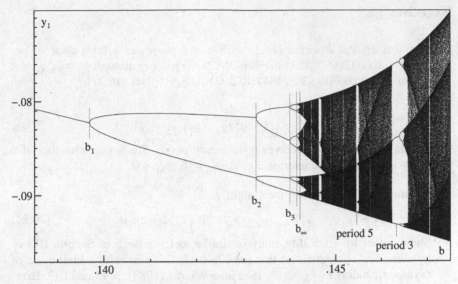

Fig. 16.11. Poincaré cuts y_1 for (16.17) as function of b

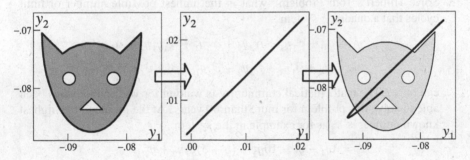

Fig. 16.12. Poincaré map for system (16.17) with $b = 0.146$

Therefore, the Poincaré map is essentially a map of the interval $[0, 1]$ to itself similar to (16.23). It is a great discovery of Feigenbaum that for *all* maps of a similar shape, the phenomena are always the same, in particular that

$$\lim_{i \to \infty} \frac{b_i - b_{i-1}}{b_{i+1} - b_i} = 4.6692016091029906715 \ldots$$

is a universal constant, the *Feigenbaum number*. The repeated doublings of the periods at $b_1, b_2, b_3, \ldots$ are called *Feigenbaum cascades*.

Exercises

1. The Van der Pol equation (16.2) with $\varepsilon = 1$ possesses a limit cycle of period $T = 6.6632868593231301896996820305$ passing through $y_2 = 0$, $y_1 = A$ where $A = 2.00861986087484313650940188$. Replace (16.2) by

$$y_1' = y_2(A - y_1)$$
$$y_2' = ((1 - y_1^2)y_2 - y_1)(A - y_1)$$

so that the limit cycle receives a stationary point. Study the behaviour of a solution starting in the interior, e.g. at $y_{10} = 1$, $y_{20} = 0$.

2. (Frommer 1934). Consider the system

$$y_1' = -y_2 + 2y_1y_2 - y_2^2, \qquad y_2' = y_1 + (1+\varepsilon)y_1^2 + 2y_1y_2 - y_2^2. \tag{16.24}$$

Show, either by a stability analysis similar to Exercise 5 of Section I.13 or by numerical computations, that for $\varepsilon > 0$ (16.24) possesses a limit cycle of asymptotic radius $r = \sqrt{6\varepsilon/7}$. (See also Wanner (1983), p. 15 and I.13, Exercise 5).

3. Solve Hilbert's 16th Problem: what is the highest possible number of limit cycles that a quadratic system

$$y_1' = \alpha_0 + \alpha_1 y_1 + \alpha_2 y_2 + \alpha_3 y_1^2 + \alpha_4 y_1 y_2 + \alpha_5 y_2^2$$
$$y_2' = \beta_0 + \beta_1 y_1 + \beta_2 y_2 + \beta_3 y_1^2 + \beta_4 y_1 y_2 + \beta_5 y_2^2$$

can have? The mathematical community is waiting for *you*: nobody has been able to solve this problem for more than 80 years. At the moment, the highest known number is 4, as for example in the system

$$y_1' = \lambda y_1 - y_2 - 10y_1^2 + (5 + \delta)y_1 y_2 + y_2^2$$
$$y_2' = y_1 + y_1^2 + (-25 + 8\varepsilon - 9\delta)y_1 y_2,$$
$$\delta = -10^{-13}, \qquad \varepsilon = -10^{-52}, \qquad \lambda = -10^{-200}$$

(see Shi Songling 1980, Wanner 1983, Perko 1984).

4. Find a change of coordinates such that the equation

$$my'' + (-A + B(y')^2)y' + ky = 0$$

becomes the Van der Pol equation (16.2) (see Kryloff & Bogoliuboff (1947), p. 5).

5. Treat the pendulum equation

$$y'' + \sin y = y'' + y - \frac{y^3}{6} + \frac{y^5}{120} \pm \ldots = 0, \qquad y(0) = \varepsilon, \quad y'(0) = 0,$$

by the method of asymptotic expansions (16.6) and (16.7) and study the period as a function of ε.

Result. The period is $2\pi(1 + \varepsilon^2/16 + \ldots)$.

6. Compute the limit cycle (Hopf bifurcation) for

$$y'' + y = \varepsilon^2 y' - (y')^3$$

for ε small by the method of Poincaré (16.6), (16.7) with $z'(0) = 0$.

7. Treat in a similar way as in Exercise 6 the Brusselator (16.12) with $A = 1$ and $B = 2 + \varepsilon^2$.

Hint. With the new variable $y = y_1 + y_2 - 3$ the differential equation (16.12) becomes equivalent to $y' = 1 - y_1$ and

$$y'' + y = -\varepsilon^2(y' - 1) - (y')^2(y + y') + 2yy'.$$

Result. $z(t) = \varepsilon(2/\sqrt{3})\cos t + \ldots$, $t = x(1 - \varepsilon^2/18 + \ldots)$, so that the period is asymptotically $2\pi(1 + \varepsilon^2/18 + \ldots)$.

8. (Liénard 1928). Prove that the limit cycle of the Van der Pol equation (16.1) *is unique* for every $\varepsilon > 0$.

Hint. The identity

$$y'' + \varepsilon(y^2 - 1)y' = \frac{d}{dx}\left(y' + \varepsilon\left(\frac{y^3}{3} - y\right)\right)$$

suggests the use of the coordinate system $y_1(x) = y(x)$, $y_2(x) = y' + \varepsilon(y^3/3 - y)$. Write the resulting first order system, study the signs of y_1', y_2' and the increase of the "energy" function $V(x) = (y_1^2 + y_2^2)/2$.

Also generalize the result to equations of the form $y'' + f(y)y' + g(y) = 0$. For more details see e.g. Simmons (1972), p. 349.

9. (Rayleigh 1883). Compute the periodic solution of

$$y'' + \kappa y' + \lambda(y')^3 + n^2 y = 0$$

for κ and λ small.

Result. $y = A\sin(nx) + (\lambda nA^3/32)\cos(3nx) + \ldots$ where A is given by $\kappa + (3/4)\lambda n^2 A^2 = 0$.

10. (Bendixson 1901). If in a certain region Ω of the plane the expression

$$\frac{\partial f_1}{\partial y_1} + \frac{\partial f_2}{\partial y_2}$$

is always negative or always positive, then the system (16.4) cannot have closed solutions in Ω.

Hint. Apply Green's formula

$$\int \int \left(\frac{\partial f_1}{\partial y_1} + \frac{\partial f_2}{\partial y_2} \right) dy_1 dy_2 = \int \left(f_1 \, dy_2 - f_2 \, dy_1 \right).$$

Chapter II. Runge-Kutta and Extrapolation Methods

Numerical methods for ordinary differential equations fall naturally into two classes: those which use *one* starting value at each step ("one-step methods") and those which are based on *several* values of the solution ("multistep methods" or "multi-value methods"). The present chapter is devoted to the study of one-step methods, while multistep methods are the subject of Chapter III. Both chapters can, to a large extent, be read independently of each other.

We start with the theory of Runge-Kutta methods: the derivation of order conditions with the help of labelled trees, error estimates, convergence proofs, implementation, methods of higher order, dense output. Section II.7 introduces implicit Runge-Kutta methods. More attention will be drawn to these methods in Volume II on stiff differential equations. Two sections then discuss the elegant idea of *extrapolation* (Richardson, Romberg, etc) and its use in obtaining high order codes. The methods presented are then tested and compared on a series of problems. The potential of parallelism is discussed in a separate section. We then turn our attention to an algebraic theory of the composition of methods. This will be the basis for the study of order properties for many general classes of methods in the following chapter. The chapter ends with special methods for second order differential equations $y'' = f(x, y)$, for Hamiltonian systems (symplectic methods) and for problems with delay.

We illustrate the methods of this chapter with an example from Astronomy, the restricted three body problem. One considers two bodies of masses $1 - \mu$ and μ in circular rotation in a plane and a third body of negligible mass moving around in the same plane. The equations are (see e.g., the classical textbook Szebehely 1967)

$$
\begin{aligned}
y_1'' &= y_1 + 2y_2' - \mu' \frac{y_1 + \mu}{D_1} - \mu \frac{y_1 - \mu'}{D_2}, \\
y_2'' &= y_2 - 2y_1' - \mu' \frac{y_2}{D_1} - \mu \frac{y_2}{D_2}, \\
D_1 &= ((y_1 + \mu)^2 + y_2^2)^{3/2}, \qquad D_2 = ((y_1 - \mu')^2 + y_2^2)^{3/2}, \\
\mu &= 0.012277471, \qquad \mu' = 1 - \mu.
\end{aligned}
\tag{0.1}
$$

There exist initial values, for example

$$y_1(0) = 0.994, \qquad y_1'(0) = 0, \qquad y_2(0) = 0,$$
$$y_2'(0) = -2.00158510637908252240537862224,$$
$$x_{\text{end}} = 17.0652165601579625588917206249,$$

(0.2)

such that the solution is periodic with period x_{end}. Such periodic solutions have fascinated astronomers and mathematicians for many decades (Poincaré; extensive numerical calculations are due to Sir George Darwin (1898)) and are now often called "Arenstorf orbits" (see Arenstorf (1963) who did numerical computations "on high speed electronic computers"). The problem is C^∞ with the exception of the two singular points $y_1 = -\mu$ and $y_1 = 1 - \mu$, $y_2 = 0$, therefore the Euler polygons of Section I.7 are known to converge to the solution. But are they really numerically useful here? We have chosen 24000 steps of step length $h = x_{\text{end}}/24000$ and plotted the result in Figure 0.1. The result is not very striking.

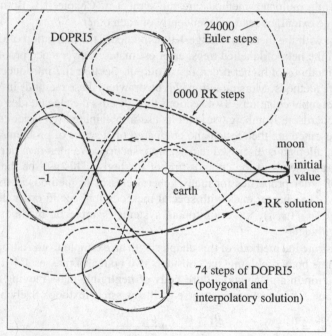

Fig. 0.1. An Arenstorf orbit computed by equidistant Euler, equidistant Runge-Kutta and variable step size Dormand & Prince

The performance of the Runge-Kutta method (left tableau of Table 1.2) is already much better and converges faster to the solution. We have used 6000 steps of step size $x_{\text{end}}/6000$, so that the numerical work becomes equivalent. Clearly, most accuracy is lost in those parts of the orbit which are close to a singularity. Therefore, codes with automatic step size selection, described in Section II.4, perform much better and the code DOPRI5 (Table 5.2) computes the orbit with a precision of 10^{-3} in 98 steps (74 accepted and 24 rejected). The step size becomes very large in some regions and the graphical representation as polygons connecting the solution points becomes unsatisfactory. The solid line is the interpolatory solution (Section II.6), which is also precise for all intermediate values and useful for many other questions such as delay differential equations, event location or discontinuities in the differential equation.

For still higher precision one needs methods of higher order. For example, the code DOP853 (Section II.5) computes the orbit faster than DOPRI5 for more stringent tolerances, say smaller than about 10^{-6}. The highest possible order is obtained by extrapolation methods (Section II.9) and the code ODEX (with $K_{max} = 15$) obtains the orbit with a precision of 10^{-30} with about 25000 function evaluations, precisely the same amount of work as for the above Euler solution.

II.1 The First Runge-Kutta Methods

Die numerische Berechnung irgend einer Lösung einer gegebenen Differentialgleichung, deren analytische Lösung man nicht kennt, hat, wie es scheint, die Aufmerksamkeit der Mathematiker bisher wenig in Anspruch genommen ...

(C. Runge 1895)

The Euler method for solving the initial value problem

$$y' = f(x, y), \qquad y(x_0) = y_0 \tag{1.1}$$

was described by Euler (1768) in his "Institutiones Calculi Integralis" (Sectio Secunda, Caput VII). The method is easy to understand and to implement. We have studied its convergence extensively in Section I.7 and have seen that the global error behaves like Ch, where C is a constant depending on the problem and h is the maximal step size. If one wants a precision of, say, 6 decimals, one would thus need about a million steps, which is not very satisfactory. On the other hand, one knows since the time of Newton that much more accurate methods can be found, if f in (1.1) is independent of y, i.e., if we have a quadrature problem

$$y' = f(x), \qquad y(x_0) = y_0 \tag{1.1'}$$

with solution

$$y(X) = y_0 + \int_{x_0}^{X} f(x)\, dx. \tag{1.2}$$

As an example consider the midpoint rule (or first Gauss formula)

$$y(x_0 + h_0) \approx y_1 = y_0 + h_0 f\left(x_0 + \frac{h_0}{2}\right)$$

$$y(x_1 + h_1) \approx y_2 = y_1 + h_1 f\left(x_1 + \frac{h_1}{2}\right) \tag{1.3'}$$

$$\cdots$$

$$y(X) \approx Y = y_{n-1} + h_{n-1} f\left(x_{n-1} + \frac{h_{n-1}}{2}\right),$$

where $h_i = x_{i+1} - x_i$ and $x_0, x_1, \ldots, x_{n-1}, x_n = X$ is a subdivision of the integration interval. Its global errror $y(X) - Y$ is known to be bounded by Ch^2. Thus for a desired precision of 6 decimals, a thousand steps will usually do, i.e., the method here is a thousand times faster. Therefore Runge (1895) asked whether it would also be possible to extend method (1.3') to problem (1.1). The first step with $h = h_0$ would read

$$y(x_0 + h) \approx y_0 + h f\left(x_0 + \frac{h}{2}, y\left(x_0 + \frac{h}{2}\right)\right), \tag{1.3}$$

but which value should we take for $y(x_0 + h/2)$? In the absence of something better, it is natural to use one small Euler step with step size $h/2$ and obtain from (1.3) [1]

$$k_1 = f(x_0, y_0)$$
$$k_2 = f\left(x_0 + \frac{h}{2}, y_0 + \frac{h}{2} k_1\right) \tag{1.4}$$
$$y_1 = y_0 + h k_2.$$

One might of course be surprised that we propose an Euler step for the computation of k_2, just half a page after preaching its inefficiency. The crucial point is, however, that k_2 is multiplied by h in the third expression and therefore its error becomes less important. To be more precise, we compute the Taylor expansion of y_1 in (1.4) as a function of h,

$$\begin{aligned}
y_1 &= y_0 + h f\left(x_0 + \frac{h}{2}, y_0 + \frac{h}{2} f_0\right) \\
&= y_0 + h f(x_0, y_0) + \frac{h^2}{2}\left(f_x + f_y f\right)(x_0, y_0) \\
&\quad + \frac{h^3}{8}\left(f_{xx} + 2 f_{xy} f + f_{yy} f^2\right)(x_0, y_0) + \cdots.
\end{aligned} \tag{1.5}$$

This can be compared with the Taylor series of the exact solution, which is obtained from (1.1) by repeated differentiation and replacing y' by f every time it appears (Euler (1768), Problema 86, §656, see also (8.12) of Chap. I)

$$\begin{aligned}
y(x_0 + h) &= y_0 + h f(x_0, y_0) + \frac{h^2}{2}\left(f_x + f_y f\right)(x_0, y_0) \\
&\quad + \frac{h^3}{6}\left(f_{xx} + 2 f_{xy} f + f_{yy} f^2 + f_y f_x + f_y^2 f\right)(x_0, y_0) + \cdots.
\end{aligned} \tag{1.6}$$

Subtracting these two equations, we obtain for the error of the first step

$$y(x_0 + h) - y_1 = \frac{h^3}{24}\left(f_{xx} + 2 f_{xy} f + f_{yy} f^2 + 4(f_y f_x + f_y^2 f)\right)(x_0, y_0) + \cdots. \tag{1.7}$$

When all second partial derivatives of f are bounded, we thus obtain $\|y(x_0 + h) - y_1\| \le K h^3$.

In order to obtain an approximation of the solution of (1.1) at the endpoint X, we apply formula (1.4) successively to the intervals (x_0, x_1), $(x_1, x_2), \ldots$, (x_{n-1}, X), very similarly to the application of Euler's method in Section I.7. Again similarly to the convergence proof of Section I.7, it will be shown in Section II.3 that, as in the case (1.1'), the error of the numerical solution is bounded by Ch^2 (h the maximal step size). Method (1.4) is thus an improvement on the Euler method. For high precision computations we need to find still better methods; this will be the main task of what follows.

---

[1] The analogous extension of the *trapezoidal rule* has been given in an early publication by Coriolis in 1837; see Chapter II.4.2 of the thesis of D. Tournès, Paris VII, 1996.

General Formulation of Runge-Kutta Methods

Runge (1895) and Heun (1900) constructed methods by including additional Euler steps in (1.4). It was Kutta (1901) who then formulated the general scheme of what is now called a Runge-Kutta method:

Definition 1.1. Let s be an integer (the "number of stages") and $a_{21}, a_{31}, a_{32}, \ldots,$ $a_{s1}, a_{s2}, \ldots, a_{s,s-1}, b_1, \ldots, b_s, c_2, \ldots, c_s$ be real coefficients. Then the method

$$
\begin{aligned}
k_1 &= f(x_0, y_0) \\
k_2 &= f(x_0 + c_2 h, y_0 + h a_{21} k_1) \\
k_3 &= f(x_0 + c_3 h, y_0 + h(a_{31} k_1 + a_{32} k_2)) \\
&\cdots \\
k_s &= f(x_0 + c_s h, y_0 + h(a_{s1} k_1 + \ldots + a_{s,s-1} k_{s-1})) \\
y_1 &= y_0 + h(b_1 k_1 + \ldots + b_s k_s)
\end{aligned}
\tag{1.8}
$$

is called an *s-stage explicit Runge-Kutta method* (ERK) for (1.1).

Usually, the c_i satisfy the conditions

$$
c_2 = a_{21}, \quad c_3 = a_{31} + a_{32}, \quad \ldots \quad c_s = a_{s1} + \ldots + a_{s,s-1},
\tag{1.9}
$$

or briefly,

$$
c_i = \sum_{j=1}^{i-1} a_{ij}.
\tag{1.9'}
$$

These conditions, already assumed by Kutta, express that all points where f is evaluated are first order approximations to the solution. They greatly simplify the derivation of order conditions for high order methods. For low orders, however, these assumptions are not necessary (see Exercise 6).

Definition 1.2. A Runge-Kutta method (1.8) has *order p* if for sufficiently smooth problems (1.1),

$$
\|y(x_0 + h) - y_1\| \le K h^{p+1},
\tag{1.10}
$$

i.e., if the Taylor series for the exact solution $y(x_0 + h)$ and for y_1 coincide up to (and including) the term h^p.

With the paper of Butcher (1964b) it became customary to symbolize method (1.8) by the tableau (1.8').

$$
\begin{array}{c|ccccc}
0 \\
c_2 & a_{21} \\
c_3 & a_{31} & a_{32} \\
\vdots & \vdots & \vdots & \ddots \\
c_s & a_{s1} & a_{s2} & \cdots & a_{s,s-1} \\
\hline
& b_1 & b_2 & \cdots & b_{s-1} & b_s
\end{array}
\tag{1.8'}
$$

Examples. The above method of Runge as well as methods of Runge and Heun of order 3 are given in Table 1.1.

Table 1.1. Low order Runge-Kutta methods

$$
\begin{array}{c|cc}
0 \\
1/2 & 1/2 \\
\hline
& 0 & 1
\end{array}
\qquad
\begin{array}{c|ccc}
0 \\
1/2 & 1/2 \\
1 & 0 & 1 \\
1 & 0 & 0 & 1 \\
\hline
& 1/6 & 2/3 & 0 & 1/6
\end{array}
\qquad
\begin{array}{c|ccc}
0 \\
1/3 & 1/3 \\
2/3 & 0 & 2/3 \\
\hline
& 1/4 & 0 & 3/4
\end{array}
$$

Runge, order 2   Runge, order 3   Heun, order 3

Discussion of Methods of Order 4

> Von den neueren Verfahren halte ich das folgende von Herrn Kutta
> angegebene für das beste.   (C. Runge 1905)

Our task is now to determine the coefficients of 4-stage Runge-Kutta methods (1.8) in order that they be of order 4. We have seen above what we must do: compute the derivatives of $y_1 = y_1(h)$ for $h = 0$ and compare them with those of the true solution for orders 1, 2, 3, and 4. In theory, with the known rules of differential calculus, this is a completely trivial task and, by the use of (1.9), results in the following conditions:

$$\sum_i b_i = b_1 + b_2 + b_3 + b_4 = 1 \tag{1.11a}$$

$$\sum_i b_i c_i = b_2 c_2 + b_3 c_3 + b_4 c_4 = 1/2 \tag{1.11b}$$

$$\sum_i b_i c_i^2 = b_2 c_2^2 + b_3 c_3^2 + b_4 c_4^2 = 1/3 \tag{1.11c}$$

$$\sum_{i,j} b_i a_{ij} c_j = b_3 a_{32} c_2 + b_4 (a_{42} c_2 + a_{43} c_3) = 1/6 \tag{1.11d}$$

$$\sum_i b_i c_i^3 = b_2 c_2^3 + b_3 c_3^3 + b_4 c_4^3 = 1/4 \tag{1.11e}$$

$$\sum_{i,j} b_i c_i a_{ij} c_j = b_3 c_3 a_{32} c_2 + b_4 c_4 (a_{42} c_2 + a_{43} c_3) = 1/8 \tag{1.11f}$$

$$\sum_{i,j} b_i a_{ij} c_j^2 = b_3 a_{32} c_2^2 + b_4 (a_{42} c_2^2 + a_{43} c_3^2) = 1/12 \qquad (1.11\text{g})$$

$$\sum_{i,j,k} b_i a_{ij} a_{jk} c_k = b_4 a_{43} a_{32} c_2 = 1/24. \qquad (1.11\text{h})$$

These computations, which are not reproduced in Kutta's paper (they are, however, in Heun 1900), are very tedious. And they grow enormously with higher orders. We shall see in Section II.2 that by using an appropriate notation, they can become very elegant.

Kutta gave the general solution of (1.11) without comment. A clear derivation of the solutions is given in Runge & König (1924), p. 291. We shall follow here the ideas of J.C. Butcher, which make clear the role of the so-called *simplifying assumptions*, and will also apply to higher order cases.

Lemma 1.3. *If*

$$\sum_{i=j+1}^{s} b_i a_{ij} = b_j (1 - c_j), \qquad j = 1, \dots, s, \qquad (1.12)$$

then the equations (d), (g), and (h) in (1.11) follow from the others.

Proof. We demonstrate this for (g):

$$\sum_{i,j} b_i a_{ij} c_j^2 = \sum_j b_j c_j^2 - \sum_j b_j c_j^3 = \frac{1}{3} - \frac{1}{4} = \frac{1}{12}$$

by (c) and (e). Equations (d) and (h) are derived similarly.   □

We shall now show that (1.12) is also *necessary* in our case:

Lemma 1.4. *For $s = 4$, the equations (1.11) and (1.9) imply (1.12).*

The proof of this lemma will be based on the following:

Lemma 1.5. *Let U and V be 3×3 matrices such that*

$$UV = \begin{pmatrix} a & b & 0 \\ c & d & 0 \\ 0 & 0 & 0 \end{pmatrix}, \qquad \det \begin{pmatrix} a & b \\ c & d \end{pmatrix} \neq 0. \qquad (1.13)$$

Then either $V e_3 = 0$ or $U^T e_3 = 0$ where $e_3 = (0, 0, 1)^T$.

Proof of Lemma 1.5. If $\det U \neq 0$, then $UV e_3 = 0$ implies $V e_3 = 0$. If $\det U = 0$, there exists $x = (x_1, x_2, x_3)^T \neq 0$ such that $U^T x = 0$, and therefore $V^T U^T x = 0$. But (1.13) implies that x must be a multiple of e_3.   □

Proof of Lemma 1.4. Define

$$d_j = \sum_i b_i a_{ij} - b_j(1 - c_j) \qquad \text{for} \qquad j = 1, \ldots, 4,$$

so that we have to prove $d_j = 0$. We now introduce the matrices

$$U = \begin{pmatrix} b_2 & b_3 & b_4 \\ b_2 c_2 & b_3 c_3 & b_4 c_4 \\ d_2 & d_3 & d_4 \end{pmatrix}, \qquad V = \begin{pmatrix} c_2 & c_2^2 & \sum_j a_{2j} c_j - c_2^2/2 \\ c_3 & c_3^2 & \sum_j a_{3j} c_j - c_3^2/2 \\ c_4 & c_4^2 & \sum_j a_{4j} c_j - c_4^2/2 \end{pmatrix}. \qquad (1.14)$$

Multiplication of these two matrices, using the conditions of (1.11), gives

$$UV = \begin{pmatrix} 1/2 & 1/3 & 0 \\ 1/3 & 1/4 & 0 \\ 0 & 0 & 0 \end{pmatrix} \qquad \text{with} \qquad \det \begin{pmatrix} 1/2 & 1/3 \\ 1/3 & 1/4 \end{pmatrix} \neq 0.$$

Now the last column of V cannot be zero, since $c_1 = 0$ implies

$$\sum_j a_{2j} c_j - c_2^2/2 = -c_2^2/2 \neq 0$$

by condition (h). Thus $d_2 = d_3 = d_4 = 0$ follows from Lemma 1.5. The last identity $d_1 = 0$ follows from $d_1 + d_2 + d_3 + d_4 = 0$, which is a consequence of (1.11a,b) and (1.9). $\square$

From Lemmas 1.3 and 1.4 we obtain

Theorem 1.6. *Under the assumption (1.9) the equations (1.11) are equivalent to*

$$b_1 + b_2 + b_3 + b_4 = 1 \tag{1.15a}$$
$$b_2 c_2 + b_3 c_3 + b_4 c_4 = 1/2 \tag{1.15b}$$
$$b_2 c_2^2 + b_3 c_3^2 + b_4 c_4^2 = 1/3 \tag{1.15c}$$
$$b_2 c_2^3 + b_3 c_3^3 + b_4 c_4^3 = 1/4 \tag{1.15e}$$
$$b_3 c_3 a_{32} c_2 + b_4 c_4 (a_{42} c_2 + a_{43} c_3) = 1/8 \tag{1.15f}$$
$$b_3 a_{32} + b_4 a_{42} = b_2 (1 - c_2) \tag{1.15i}$$
$$b_4 a_{43} = b_3 (1 - c_3) \tag{1.15j}$$
$$0 = b_4 (1 - c_4). \tag{1.15k}$$

$\square$

It follows from (1.15j) and (1.11h) that

$$b_3 b_4 c_2 (1 - c_3) \neq 0. \tag{1.16}$$

In particular this implies $c_4 = 1$ by (1.15k).

Solution of equations (1.15). Equations (a)-(e) and (k) just state that b_i and c_i are the coefficients of a fourth order quadrature formula with $c_1 = 0$ and $c_4 = 1$. We distinguish four cases for this:

1) $c_2 = u$, $c_3 = v$ and $0, u, v, 1$ are all distinct;   (1.17)

then (a)-(e) form a regular linear system for b_1, b_2, b_3, b_4. This system has the solution

$$b_1 = \frac{1 - 2(u+v) + 6uv}{12uv}, \qquad b_2 = \frac{2v - 1}{12u(1-u)(v-u)},$$

$$b_3 = \frac{1 - 2u}{12v(1-v)(v-u)}, \qquad b_4 = \frac{3 - 4(u+v) + 6uv}{12(1-u)(1-v)}.$$

Due to (1.16) we have to assume that u, v are such that $b_3 \neq 0$ and $b_4 \neq 0$. The three other cases with double nodes are built upon the Simpson rule:

2) $c_3 = 0$, $c_2 = 1/2$, $b_3 = w \neq 0$, $b_1 = 1/6 - w$, $b_2 = 4/6$, $b_4 = 1/6$;

3) $c_2 = c_3 = 1/2$, $b_1 = 1/6$, $b_3 = w \neq 0$, $b_2 = 4/6 - w$, $b_4 = 1/6$;

4) $c_2 = 1$, $c_3 = 1/2$, $b_4 = w \neq 0$, $b_2 = 1/6 - w$, $b_1 = 1/6$, $b_3 = 4/6$.

Once b_i and c_i are chosen, we obtain a_{43} from (j), and then (f) and (i) form a linear system of two equations for a_{32} and a_{42}. The determinant of this system is

$$\det \begin{pmatrix} b_3 & b_4 \\ b_3 c_3 c_2 & b_4 c_4 c_2 \end{pmatrix} = b_3 b_4 c_2 (c_4 - c_3)$$

which is $\neq 0$ by (1.16). Finally we obtain a_{21}, a_{31}, and a_{41} from (1.9).

Two particular choices of Kutta (1901) have become especially popular: case (3) with $w = 2/6$ and case (1) with $u = 1/3$, $v = 2/3$. They are given in Table 1.2. Both methods generalize classical quadrature rules in keeping the same order. The first is more popular, the second is more precise ("Wir werden diese Näherung als im allgemeinen beste betrachten ...", Kutta).

Table 1.2. Kutta's methods

0					0				
1/2	1/2				1/3	1/3			
1/2	0	1/2			2/3	−1/3	1		
1	0	0	1		1	1	−1	1	
	1/6	2/6	2/6	1/6		1/8	3/8	3/8	1/8

"The" Runge-Kutta method   3/8–Rule

"Optimal" Formulas

Much research has been undertaken, in order to choose the "best" possibilities from the variety of possible 4th order RK-formulas.

The first attempt in this direction was the very popular method of Gill (1951), with the aim of reducing the need for computer storage ("registers") as much as possible. The first computers in the fifties largely used this method which is therefore of historical interest. Gill observed that most computer storage is needed for the computation of k_3, where "registers are required to store in some form"

$$y_0 + a_{31}hk_1 + a_{32}hk_2, \quad y_0 + a_{41}hk_1 + a_{42}hk_2, \quad y_0 + b_1hk_1 + b_2hk_2, \quad hk_3.$$

"Clearly, three registers will suffice for the third stage if the quantities to be stored are linearly dependent, i.e., if"

$$\det \begin{pmatrix} 1 & a_{31} & a_{32} \\ 1 & a_{41} & a_{42} \\ 1 & b_1 & b_2 \end{pmatrix} = 0.$$

Gill observed that this condition is satisfied for the methods of type (3) if $w = (1+\sqrt{0.5})/3$. The resulting method can then be reformulated as follows ("As each quantity is calculated it is stored in the register formerly holding the corresponding quantity of the previous stage, which is no longer required"):

$$y := \text{initial value}, \quad k := hf(y), \quad y := y + 0.5k, \quad q := k,$$
$$k := hf(y), \quad y := y + (1 - \sqrt{0.5})(k - q),$$
$$q := (2 - \sqrt{2})k + (-2 + 3\sqrt{0.5})q,$$
$$k := hf(y), \quad y := y + (1 + \sqrt{0.5})(k - q), \tag{1.18}$$
$$q := (2 + \sqrt{2})k + (-2 - 3\sqrt{0.5})q,$$
$$k := hf(y), \quad y := y + \frac{k}{6} - \frac{q}{3}, \quad (\rightarrow \text{compute next step}).$$

Today, in large high-speed computers, this method is no longer used, but could still be of interest for very high dimensional equations.

Other attempts have been made to choose u and v in (1.17), case (1), such that the *error terms* (terms in h^5, see Section II.3) become as small as possible. We shall discuss this question in Section II.3.

Numerical Example

> Zu grosses Gewicht darf man natürlich solchen Beispielen nicht
> beilegen ...
> (W. Kutta 1901)

We compare five different choices of 4th order methods on the Van der Pol equation
(I.16.2) with $\varepsilon = 1$. As initial values we take $y_1(0) = A$, $y_2(0) = 0$ on the limit
cycle and we integrate over one period T (the values of A and T are given in
Exercise I.16.1). For a comparison of these methods with lower order ones we have
also included the explicit Euler method, Runge's method of order 2 and Heun's
method of order 3 (see Table 1.1).

We have applied the methods with several fixed step sizes. The errors of both
components and the number of function evaluations (fe) are displayed in logarithmic scales in Fig. 1.1. Whenever the error behaves like $C \cdot h^p = C_1 \cdot (fe)^{-p}$, the
curves appear as straight lines with slope $1/p$. We have chosen the scales such that
the theoretical slope of the 4th order methods appears to be 45°.

These tests clearly show up the importance of higher order methods. Among
the various 4th order methods there is usually no big difference. It is interesting to
note that in our example the method with the smallest error in y_1 has the biggest
error in y_2 and vice versa.

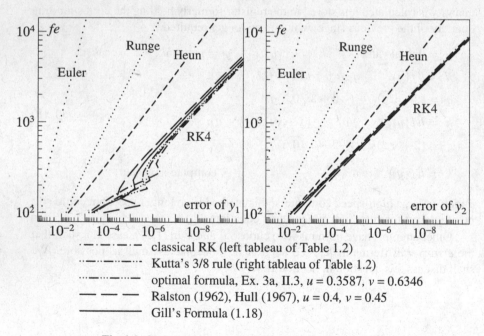

‒ · ‒ · ‒   classical RK (left tableau of Table 1.2)
···‒···‒···   Kutta's 3/8 rule (right tableau of Table 1.2)
·····‒·····‒   optimal formula, Ex. 3a, II.3, $u = 0.3587$, $v = 0.6346$
‒ ‒ ‒   Ralston (1962), Hull (1967), $u = 0.4$, $v = 0.45$
———   Gill's Formula (1.18)

Fig. 1.1. Global errors versus number of function evaluations

Exercises

1. Show that every s-stage explicit RK method of order s, when applied to the problem $y' = \lambda y$ (λ a complex constant), gives

$$y_1 = \left(\sum_{j=0}^{s} \frac{z^j}{j!} \right) y_0, \qquad z = h\lambda.$$

   Hint. Show first that y_1/y_0 must be a polynomial in z of degree s and then determine its coefficients by comparing the derivatives of y_1, with respect to h, to those of the true solution.

2. (Runge 1895, p. 175; see also the introduction to Adams methods in Chap. III.1). The theoretical form of drops of fluids is determined by the differential equation of Laplace (1805)

$$-z = \alpha^2 \frac{(K_1 + K_2)}{2} \tag{1.21}$$

   where α is a constant, $(K_1 + K_2)/2$ the mean curvature, and z the height (see Fig. 1.2). If we insert $1/K_1 = r/\sin\varphi$ and $K_2 = d\varphi/ds$, the curvature of the meridian curve, we obtain

$$-2z = \alpha^2 \left(\frac{\sin\varphi}{r} + \frac{d\varphi}{ds} \right), \tag{1.22}$$

   where we put $\alpha = 1$. Add

$$\frac{dr}{ds} = \cos\varphi, \qquad \frac{dz}{ds} = -\sin\varphi, \tag{1.22'}$$

   to obtain a system of three differential equations for $\varphi(s)$, $r(s)$, $z(s)$, s being the arc length. Compute and plot different solution curves by the method of Runge (1.4) with initial values $\varphi(0) = 0$, $r(0) = 0$ and $z(0) = z_0$ ($z_0 < 0$ for lying drops; compute also hanging drops with appropriate sign changes in (1.22)). Use different step sizes and compare the results.

   Hint. Be careful at the singularity in the beginning: from (1.22) and (1.22') we have for small s that $r = s$, $\varphi = \zeta s$ with $\zeta = -z_0$, hence $(\sin\varphi)/r \to -z_0$. A more precise analysis gives for small s the expansions ($\zeta = -z_0$)

$$\varphi = \zeta s + \frac{\zeta}{4}s^3 + \left(\frac{\zeta}{48} - \frac{\zeta^3}{120} \right)s^5 + \dots$$

$$r = s - \frac{\zeta^2}{6}s^3 + \left(-\frac{\zeta^2}{20} + \frac{\zeta^4}{120} \right)s^5 + \dots$$

$$z = -\zeta - \frac{\zeta}{2}s^2 + \left(-\frac{\zeta}{16} + \frac{\zeta^3}{24} \right)s^4 + \left(-\frac{\zeta}{288} + \frac{\zeta^3}{45} - \frac{\zeta^5}{720} \right)s^6 + \dots \,.$$

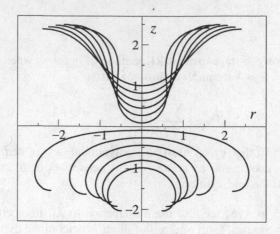

Fig. 1.2. Drops

3. Find the conditions for a 2-stage explicit RK-method to be of order two and determine all such methods ("... wozu eine weitere Erörterung nicht mehr nötig ist", Kutta).

4. Find all methods of order three with three stages (i.e., solve (1.11;a-d) with $b_4 = 0$).

   Result. $c_2 = u$, $c_3 = v$, $a_{32} = v(v-u)/(u(2-3u))$, $b_2 = (2-3v)/(6u(u-v))$, $b_3 = (2-3u)/(6v(v-u))$, $b_1 = 1 - b_2 - b_3$, $a_{31} = c_3 - a_{32}$, $a_{21} = c_2$ (Kutta 1901, p. 438).

5. Construct all methods of order 2 of the form

$$
\begin{array}{c|ccc}
0 & & & \\
c_2 & c_2 & & \\
c_3 & 0 & c_3 & \\
\hline
 & 0 & 0 & 1
\end{array}
$$

   Such methods "have the property that the corresponding Runge-Kutta process requires relatively less storage in a computer" (Van der Houwen (1977), §2.7.2). Apply them to $y' = \lambda y$ and compare with Exercise 1.

6. Determine the conditions for order two of the RK methods with two stages which do *not* satisfy the conditions (1.9):

$$
\begin{aligned}
k_1 &= f(x_0 + c_1 h, y_0) \\
k_2 &= f(x_0 + c_2 h, y_0 + a_{21} h k_1) \\
y_1 &= y_0 + h(b_1 k_1 + b_2 k_2).
\end{aligned}
$$

   Discuss the use of this extra freedom for c_1 and c_2 (Oliver 1975).

II.2 Order Conditions for Runge-Kutta Methods

∴.. I heard a lecture by Merson ...

(J. Butcher's first contact with RK methods)

In this section we shall derive the general structure of the order conditions (Merson 1957, Butcher 1963). The proof has evolved very much in the meantime, mainly under the influence of Butcher's later work, many personal discussions with him, the proof of "Theorem 6" in Hairer & Wanner (1974), and our teaching experience. We shall see in Section II.11 that exactly the same ideas of proof lead to a general theorem of composition of methods (= B-series), which gives access to order conditions for a much larger class of methods.

A big advantage is obtained by transforming (1.1) to *autonomous* form by appending x to the dependent variables as

$$\begin{pmatrix} x \\ y \end{pmatrix}' = \begin{pmatrix} 1 \\ f(x,y) \end{pmatrix}. \tag{2.1}$$

The main difficulty in the derivation of the order conditions is to understand the correspondence of the formulas to certain rooted labelled trees; this comes out most naturally if we use well-chosen indices and tensor notation (as in Gill (1951), Henrici (1962), p. 118, Gear (1971), p. 32). As is usual in tensor notation, we denote (in this section) the components of vectors by *superscript* indices which, in order to avoid confusion, we choose as *capitals*. Then (2.1) can be written as

$$(y^J)' = f^J(y^1, \ldots, y^n), \qquad J = 1, \ldots, n. \tag{2.2}$$

We next rewrite the method (1.8) for the autonomous differential equation (2.2). In order to get a better symmetry in all formulas of (1.8), we replace k_i by the argument g_i such that $k_i = f(g_i)$. Then (1.8) becomes

$$g_i^J = y_0^J + \sum_{j=1}^{i-1} a_{ij} h f^J(g_j^1, \ldots, g_j^n), \qquad i = 1, \ldots, s$$

$$y_1^J = y_0^J + \sum_{j=1}^{s} b_j h f^J(g_j^1, \ldots, g_j^n). \tag{2.3}$$

If the system (2.2) originates from (2.1), then, for $J = 1$,

$$g_i^1 = y_0^1 + \sum_{j=1}^{i-1} a_{ij} h = x_0 + c_i h$$

by (1.9). We see that (1.9) becomes a natural condition. If it is satisfied, then for the derivation of order conditions only the autonomous equation (2.2) has to be considered.

As indicated in Section II.1 we have to compare the Taylor series of y_1^J with that of the exact solution. Therefore we compute the derivatives of y_1^J and g_i^J with respect to h at $h = 0$. Due to the similarity of the two formulas, it is sufficient to do this for g_i^J. On the right hand side of (2.3) there appear expressions of the form $h\varphi(h)$, so we make use of Leibniz' formula

$$\left(h\varphi(h)\right)^{(q)}\big|_{h=0} = q \cdot \left(\varphi(h)\right)^{(q-1)}\big|_{h=0}. \tag{2.4}$$

The reader is now asked to take a deep breath, take five sheets of reversed computer paper, remember the basic rules of differential calculus, and begin the following computations:

$q = 0$: from (2.3)

$$(g_i^J)^{(0)}\big|_{h=0} = y_0^J. \tag{2.5;0}$$

$q = 1$: from (2.3) and (2.4)

$$(g_i^J)^{(1)}\big|_{h=0} = \sum_j a_{ij} f^J\big|_{y=y_0}. \tag{2.5;1}$$

$q = 2$: because of (2.4) we shall need the first derivative of $f^J(g_j)$

$$\left(f^J(g_j)\right)^{(1)} = \sum_K f_K^J(g_j) \cdot (g_j^K)^{(1)}, \tag{2.6;1}$$

where, as usual, f_K^J denotes $\partial f^J / \partial y^K$. Inserting formula (2.5;1) (with i, j, J replaced by j, k, K) into (2.6;1) we obtain with (2.4)

$$(g_i^J)^{(2)}\big|_{h=0} = 2\sum_{j,k} a_{ij} a_{jk} \sum_K f_K^J f^K\big|_{y=y_0}. \tag{2.5;2}$$

$q = 3$: we differentiate (2.6;1) to obtain

$$\left(f^J(g_j)\right)^{(2)} = \sum_{K,L} f_{KL}^J(g_j) \cdot (g_j^K)^{(1)}(g_j^L)^{(1)} + \sum_K f_K^J(g_j)(g_j^K)^{(2)}. \tag{2.6;2}$$

The derivatives $(g_j^K)^{(1)}$ and $(g_j^K)^{(2)}$ at $h = 0$ are already available in (2.5;1) and (2.5;2). So we have from (2.3) and (2.4)

$$\begin{aligned}
(g_i^J)^{(3)}\big|_{h=0} = {} & 3\sum_{j,k,l} a_{ij} a_{jk} a_{jl} \sum_{K,L} f_{KL}^J f^K f^L\big|_{y=y_0} \\
& + 3 \cdot 2 \sum_{j,k,l} a_{ij} a_{jk} a_{kl} \sum_{K,L} f_K^J f_L^K f^L\big|_{y=y_0}.
\end{aligned} \tag{2.5;3}$$

The same formula holds for $(y_1^J)^{(3)}\big|_{h=0}$ with a_{ij} replaced by b_j.

The Derivatives of the True Solution

The derivatives of the correct solution are obtained much more easily just by differentiating equation (2.2): first

$$(y^J)^{(1)} = f^J(y). \tag{2.7;1}$$

Differentiating (2.2) and inserting (2.2) again for the derivatives we get

$$(y^J)^{(2)} = \sum_K f_K^J(y) \cdot (y^K)^{(1)} = \sum_K f_K^J(y) f^K(y). \tag{2.7;2}$$

Differentiating (2.7;2) again we obtain

$$(y^J)^{(3)} = \sum_{K,L} f_{KL}^J(y) f^K(y) f^L(y) + \sum_{K,L} f_K^J(y) f_L^K(y) f^L(y). \tag{2.7;3}$$

Conditions for Order 3

For order 3, the derivatives (2.5;1-3), (with a_{ij} replaced by b_j) must be equal to the derivatives (2.7;1-3), and this for every differential equation. Thus, comparing the corresponding expressions, we obtain:

Theorem 2.1. *The RK method (2.3) (and thus (1.8)) is of order 3 iff*

$$\sum_j b_j = 1, \qquad\qquad 2 \sum_{j,k} b_j a_{jk} = 1,$$

$$3 \sum_{j,k,l} b_j a_{jk} a_{jl} = 1, \qquad 6 \sum_{j,k,l} b_j a_{jk} a_{kl} = 1. \tag{2.8}$$

□

Inserting $\sum_k a_{jk} = c_j$ from (1.9), we can simplify these expressions still further and obtain formulas (a)-(d) of (1.11).

Trees and Elementary Differentials

> But without a more convenient notation, it would be difficult to find the corresponding expressions ... This, however, can be at once effected by means of the analytical forms called trees ...
>
> (A. Cayley 1857)

The continuation of this process, although theoretically clear, soon leads to very complicated formulas. It is therefore advantageous to use a graphical representation: indeed, the indices j, k, l and J, K, L in the terms of (2.5;3) are linked together as pairs of indices in $a_{jk}, a_{jl}, \ldots$ in exactly the same way as upper and lower indices in the expressions f_{KL}^J, f_K^J, namely

$$
t_{31} = \underset{j}{\diagdown\diagup}\overset{l\quad k}{} \qquad \text{and} \qquad t_{32} = \overset{l}{\underset{j}{\diagdown}}\, k \tag{2.9}
$$

for the first and second term respectively. We call these objects *labelled trees,* because they are connected graphs (trees) whose vertices are labelled with summation indices. They can also be represented as *mappings,* e.g.,

$$
l \mapsto j, \quad k \mapsto j \qquad \text{and} \qquad l \mapsto k, \quad k \mapsto j \tag{2.9'}
$$

for the above trees. This mapping indicates to which lower letter the corresponding vertices are attached.

Definition 2.2. Let A be an ordered chain of indices $A = \{j < k < l < m < \ldots\}$ and denote by A_q the subset consisting of the first q indices. A *(rooted) labelled tree* of *order* q $(q \geq 1)$ is a mapping (the son-father mapping)

$$
t : A_q \setminus \{j\} \to A_q
$$

such that $t(z) < z$ for all $z \in A_q \setminus \{j\}$. The *set of all labelled trees of order* q is denoted by LT_q. We call "z" the *son* of "$t(z)$" and "$t(z)$" the *father* of "z". The vertex "j", the forefather of the whole dynasty, is called the *root* of t. The *order* q of a labelled tree is equal to the number of its vertices and is usually denoted by $q = \varrho(t)$.

Definition 2.3. For a labelled tree $t \in LT_q$ we call

$$
F^J(t)(y) = \sum_{K,L,\ldots} f_{K,\ldots}^J(y) f_{\ldots}^K(y) f_{\ldots}^L(y) \cdots
$$

the corresponding *elementary differential.* The summation is over $q-1$ indices $K, L, \ldots$ (which correspond to $A_q \setminus \{j\}$) and the summand is a product of q f's, where the upper index runs through all vertices of t and the lower indices are the corresponding sons. We denote by $F(t)(y)$ the vector $\big(F^1(t)(y), \ldots, F^n(t)(y)\big)$.

If the set A_q is written as

$$
A_q = \{j_1 < j_2 < \ldots < j_q\}, \tag{2.10}
$$

then we can write the definition of $F(t)$ as follows:

$$
F^{J_1}(t) = \sum_{J_2,\ldots,J_q} \prod_{i=1}^{q} f_{t^{-1}(J_i)}^{J_i}, \tag{2.11}
$$

since the sons of an index are its inverse images under the map t.

Examples of elementary differentials are

$$\sum_{K,L} f_{KL}^J f^K f^L \quad \text{and} \quad \sum_{K,L} f_K^J f_L^K f^L$$

for the labelled trees t_{31} and t_{32} above. These expressions appear in formulas (2.5;3) and (2.7;3).

The three labelled trees

(2.12)

all look topologically alike, moreover the corresponding elementary differentials

$$\sum_{K,L,M} f_{KM}^J f^M f_L^K f^L, \quad \sum_{K,L,M} f_{KL}^J f^L f_M^K f^M, \quad \sum_{K,L,M} f_{LK}^J f^K f_M^L f^M \quad (2.12')$$

are the same, because they just differ by an exchange of the summation indices. Thus we give

Definition 2.4. Two labelled trees t and u are *equivalent,* if they have the same order, say q, and if there exists a permutation $\sigma : A_q \to A_q$, such that $\sigma(j) = j$ and $t\sigma = \sigma u$ on $A_q \setminus \{j\}$.

This clearly defines an equivalence relation.

Definition 2.5. An equivalence class of qth order labelled trees is called a *(rooted) tree* of *order* q. The set of all trees of order q is denoted by T_q. The *order* of a tree is defined as the order of a representative and is again denoted by $\varrho(t)$. Furthermore we denote by $\alpha(t)$ (for $t \in T_q$) the number of elements in the equivalence class t; i.e., the number of possible different monotonic labellings of t.

Geometrically, a tree is distinguished from a labelled tree by omitting the labels. Often it is advantageous to include $\emptyset$, the empty tree, as the only tree of order 0. The only tree of order 1 is denoted by τ. The number of trees of orders $1, 2, \ldots, 10$ are given in Table 2.1. Representatives of all trees of order ≤ 5 are shown in Table 2.2.

Table 2.1. Number of trees up to order 10

q	1	2	3	4	5	6	7	8	9	10
card(T_q)	1	1	2	4	9	20	48	115	286	719

Table 2.2. Trees and elementary differentials up to order 5

q	t	graph	$\gamma(t)$	$\alpha(t)$	$F^J(t)(y)$	$\Phi_j(t)$
0	$\emptyset$	$\emptyset$	1	1	y^J	
1	τ	$\bullet j$	1	1	f^J	1
2	t_{21}		2	1	$\sum_K f_K^J f^K$	$\sum_k a_{jk}$
3	t_{31}		3	1	$\sum_{K,L} f_{KL}^J f^K f^L$	$\sum_{k,l} a_{jk} a_{jl}$
	t_{32}		6	1	$\sum_{K,L} f_K^J f_L^K f^L$	$\sum_{k,l} a_{jk} a_{kl}$
4	t_{41}		4	1	$\sum_{K,L,M} f_{KLM}^J f^K f^L f^M$	$\sum_{k,l,m} a_{jk} a_{jl} a_{jm}$
	t_{42}		8	3	$\sum_{K,L,M} f_{KM}^J f_L^K f^L f^M$	$\sum_{k,l,m} a_{jk} a_{kl} a_{jm}$
	t_{43}		12	1	$\sum_{K,L,M} f_K^J f_{LM}^K f^L f^M$	$\sum_{k,l,m} a_{jk} a_{kl} a_{km}$
	t_{44}		24	1	$\sum_{K,L,M} f_K^J f_L^K f_M^L f^M$	$\sum_{k,l,m} a_{jk} a_{kl} a_{lm}$
5	t_{51}		5	1	$\sum f_{KLMP}^J f^K f^L f^M f^P$	$\sum a_{jk} a_{jl} a_{jm} a_{jp}$
	t_{52}		10	6	$\sum f_{KMP}^J f_L^K f^L f^M f^P$	$\sum a_{jk} a_{kl} a_{jm} a_{jp}$
	t_{53}		15	4	$\sum f_{KP}^J f_{ML}^K f^L f^M f^P$	$\sum a_{jk} a_{kl} a_{km} a_{jp}$
	t_{54}		30	4	$\sum f_{KP}^J f_L^K f_M^L f^M f^P$	$\sum a_{jk} a_{kl} a_{lm} a_{jp}$
	t_{55}		20	3	$\sum f_{KM}^J f_L^K f^L f_P^M f^P$	$\sum a_{jk} a_{kl} a_{jm} a_{mp}$
	t_{56}		20	1	$\sum f_K^J f_{LMP}^K f^L f^M f^P$	$\sum a_{jk} a_{kl} a_{km} a_{kp}$
	t_{57}		40	3	$\sum f_K^J f_{LP}^K f_M^L f^M f^P$	$\sum a_{jk} a_{kl} a_{lm} a_{kp}$
	t_{58}		60	1	$\sum f_K^J f_L^K f_{MP}^L f^M f^P$	$\sum a_{jk} a_{kl} a_{lm} a_{lp}$
	t_{59}		120	1	$\sum f_K^J f_L^K f_M^L f_P^M f^P$	$\sum a_{jk} a_{kl} a_{lm} a_{mp}$

The Taylor Expansion of the True Solution

We can now state the general result for the qth derivative of the true solution:

Theorem 2.6. *The exact solution of (2.2) satisfies*

$$(y)^{(q)}(x_0) = \sum_{t \in LT_q} F(t)(y_0) = \sum_{t \in T_q} \alpha(t) F(t)(y_0). \qquad (2.7;q)$$

Proof. The theorem is true for $q = 1, 2, 3$ (see (2.7;1-3) above). For the computation of, say, the 4th derivative, we have to differentiate (2.7;3). This consists of two terms (corresponding to the two trees of (2.9)), each of which contains three factors $f_{\cdots}^{\cdots}$ (corresponding to the three nodes of these trees). The differentiation of these by Leibniz' rule and insertion of (2.2) for the derivatives is geometrically just the addition of a new branch with a new summation letter to *each* vertex (Fig. 2.1).

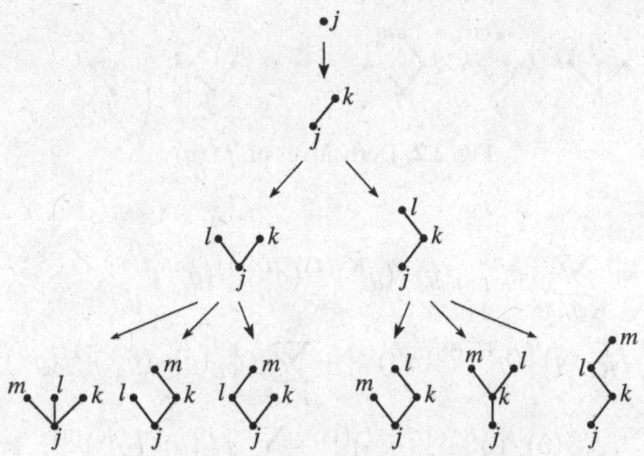

$$•j \tag{2.7;1}$$

$$\tag{2.7;2}$$

$$\tag{2.7;3}$$

$$\tag{2.7;4}$$

Fig. 2.1. Derivatives of exact solution

It is clear that by this process *all* labelled trees of order q appear for the qth derivative, each of them *exactly once*.

If we group together the terms with identical elementary differentials, we obtain the second expression of (2.7;q). □

Faà di Bruno's Formula

Our next goal will be the computation of the qth derivative of the numerical solution y_1 and of the g_j. For this, we have first to generalize the formulas (2.6;1) (the chain rule) and (2.6;2) for the qth derivative of the composition of two functions. We represent these two formulas graphically in Fig. 2.2.

Formula (2.6;2) consists of two terms; the first term contains three factors, the second contains only two. Here the node "l" is a "dummy" node, not really present in the formula, and just indicates that we have to take the second derivative. The derivation of (2.6;2) will thus lead to *five* terms which we write down for the convenience of the reader (but not for the convenience of the printer ...)

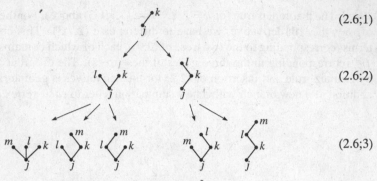

(2.6;1)

(2.6;2)

(2.6;3)

Fig. 2.2. Derivatives of $f^J(g)$

$$
\begin{aligned}
(f^J(g))^{(3)} = &\sum_{K,L,M} f^J_{KLM}(g) \cdot (g^K)^{(1)}(g^L)^{(1)}(g^M)^{(1)} \\
&+ \sum_{K,L} f^J_{KL}(g) \cdot (g^K)^{(2)}(g^L)^{(1)} + \sum_{K,L} f^J_{KL}(g) \cdot (g^K)^{(1)}(g^L)^{(2)} \\
&+ \sum_{K,M} f^J_{KM}(g) \cdot (g^K)^{(2)}(g^M)^{(1)} + \sum_{K} f^J_K(g) \cdot (g^K)^{(3)}.
\end{aligned}
\qquad (2.6;3)
$$

The corresponding trees are represented in the third line of Fig. 2.2. Each time we differentiate, we have to
i) differentiate the first factor $f^J_{K\ldots}$; i.e., we add a new branch to the root j;
ii) increase the derivative numbers of each of the g's by 1; we represent this by lengthening the corresponding branch.
Each time we add a new label. *All* trees which are obtained in this way are those "special" trees which have no ramifications except at the root.

Definition 2.7. We denote by LS_q the set of *special labelled trees of order q* which have no ramifications except at the root.

Lemma 2:8 (Faà di Bruno's formula). *For $q \geq 1$ we have*

$$
(f^J(g))^{(q-1)} = \sum_{u \in LS_q} \sum_{K_1,\ldots,K_m} f^J_{K_1,\ldots,K_m}(g) \cdot (g^{K_1})^{(\delta_1)} \ldots (g^{K_m})^{(\delta_m)} \qquad (2.6;q-1)
$$

Here, for $u \in LS_q$, m is the number of branches leaving the root and $\delta_1,\ldots,\delta_m$ are the numbers of nodes in each of these branches, such that $q = 1 + \delta_1 + \ldots + \delta_m$.
$\square$

Remark. The usual multinomial coefficients are absent here, as we use labelled trees.

The Derivatives of the Numerical Solution

> It is difficult to keep a cool head when discussing the various
> derivatives ...   (S. Gill 1956)

In order to generalize (2.5;1-3), we need the following definitions:

Definition 2.9. Let t be a labelled tree with root j; we denote by

$$\Phi_j(t) = \sum_{k,l,\ldots} a_{jk} a_{\ldots} \cdots$$

the sum over the $q-1$ remaining indices $k, l, \ldots$ (as in Definition 2.3). The summand is a product of $q-1$ a's, where all fathers stand two by two with their sons as indices. If the set A_q is written as in (2.10), we have

$$\Phi_{j_1}(t) = \sum_{j_2,\ldots,j_q} a_{t(j_2),j_2} \cdots a_{t(j_q),j_q}. \tag{2.13}$$

Definition 2.10. For $t \in LT_q$ let $\gamma(t)$ be the product of $\varrho(t)$ and all orders of the trees which appear, if the roots, one after another, are removed from t. (See Fig. 2.3 or formula (2.17)).

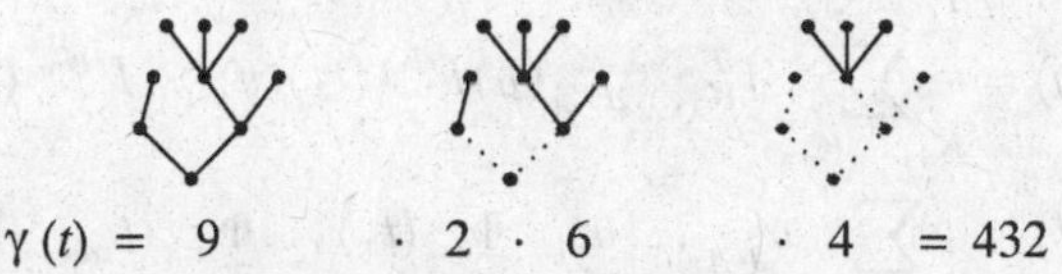

$$\gamma(t) = \quad 9 \qquad \cdot\; 2\;\cdot\;6 \qquad \cdot\;4 \quad = 432$$

Fig. 2.3. Example for the definition of $\gamma(t)$

The above expressions are of course independent of the labellings, so $\Phi_j(t)$ as well as $\gamma(t)$ also make sense in T_q. Examples are given in Table 2.2.

Theorem 2.11. *The derivatives of g_i satisfy*

$$g_i^{(q)}\big|_{h=0} = \sum_{t \in LT_q} \gamma(t) \sum_j a_{ij} \Phi_j(t) F(t)(y_0). \tag{2.5;q}$$

The numerical solution y_1 of (2.3) satisfies

$$
\begin{aligned}
y_1^{(q)}\big|_{h=0} &= \sum_{t \in LT_q} \gamma(t) \sum_j b_j \Phi_j(t) F(t)(y_0) \\
&= \sum_{t \in T_q} \alpha(t)\gamma(t) \sum_j b_j \Phi_j(t) F(t)(y_0).
\end{aligned} \tag{2.14}
$$

Proof. Because of the similarity of y_1 and g_i (see (2.3)) we only have to prove the first equation. We do this by induction on q, in exactly the same way as we obtained (2.5;1-3): we first apply Leibniz' formula (2.4) to obtain

$$(g_i^J)^{(q)}\big|_{h=0} = q \sum_j a_{ij} \big(f^J(g_j)\big)^{(q-1)}\big|_{y=y_0}. \tag{2.15}$$

Next we use Faà di Bruno's formula (Lemma 2.8). Finally we insert for the derivatives $(g_j^{K_s})^{(\delta_s)}$, which appear in (2.6;q-1) with $\delta_s < q$, the induction hypothesis (2.5;1) - (2.5;q-1) and rearrange the sums. This gives

$$(g_i^J)^{(q)}\big|_{h=0} = q \sum_{u \in LS_q} \sum_{t_1 \in LT\delta_1} \cdots \sum_{t_m \in LT\delta_m} \gamma(t_1)\ldots\gamma(t_m)\cdot$$

$$\sum_j a_{ij} \sum_{k_1} a_{jk_1} \Phi_{k_1}(t_1)\ldots\sum_{k_m} a_{jk_m} \Phi_{k_m}(t_m)\cdot \tag{2.16}$$

$$\sum_{K_1,\ldots,K_m} f_{K_1,\ldots,K_m}^J(y_0) F^{K_1}(t_1)(y_0)\ldots F^{K_m}(t_m)(y_0).$$

The main difficulty is now to understand that to each tuple

$$(u, t_1, \ldots, t_m) \qquad \text{with} \qquad u \in LS_q, \ t_s \in LT_{\delta_s}$$

there corresponds a labelled tree $t \in LT_q$ such that

$$\gamma(t) = q \cdot \gamma(t_1)\ldots\gamma(t_m) \tag{2.17}$$

$$F^J(t)(y) = \sum_{K_1,\ldots,K_m} f_{K_1,\ldots,K_m}^J(y) F^{K_1}(t_1)(y)\ldots F^{K_m}(t_m)(y) \tag{2.18}$$

$$\Phi_j(t) = \sum_{k_1,\ldots,k_m} a_{jk_1}\ldots a_{jk_m} \Phi_{k_1}(t_1)\ldots\Phi_{k_m}(t_m). \tag{2.19}$$

This labelled tree t is obtained if the branches of u are replaced by the trees $t_1, \ldots, t_m$ and the corresponding labels are taken over in a natural way, i.e., in the same order (see Fig. 2.4 for some examples).

In this way, *all* trees $t \in LT_q$ appear exactly *once*. Thus (2.16) becomes (2.5;q) after inserting (2.17), (2.18) and (2.19). $\square$

The above construction of t can also be used for a recursive definition of trees. We first observe that the equivalence class of t (in Fig. 2.4) depends only on the equivalence classes of $t_1, \ldots, t_m$.

Definition 2.12. We denote by

$$t = [t_1, \ldots, t_m] \tag{2.20}$$

the tree, which leaves over the trees $t_1, \ldots, t_m$ when its root and the adjacent branches are chopped off (Fig. 2.5).

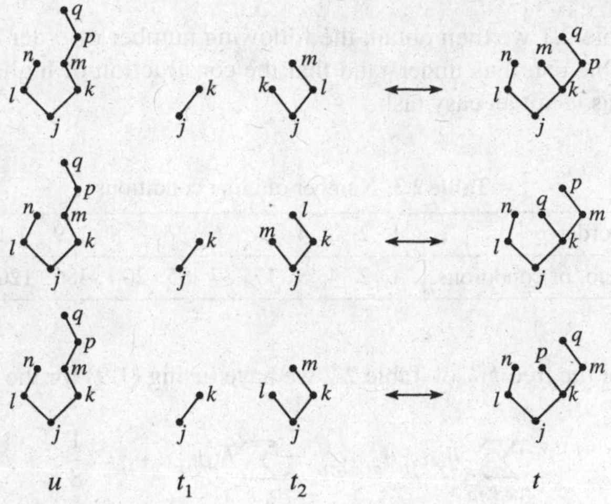

Fig. 2.4. Example for the bijection $(u, t_1, \ldots, t_m) \leftrightarrow t$

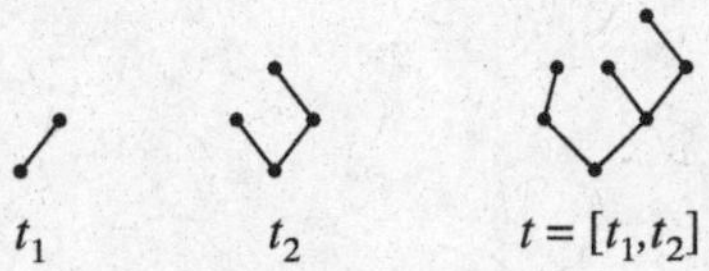

Fig. 2.5. Recursive definition of trees

With (2.20) all trees can be expressed in terms of τ; e.g., $t_{21} = [\tau]$, $t_{31} = [\tau, \tau]$, $t_{32} = [[\tau]]$, ..., etc.

The Order Conditions

Comparing Theorems 2.6 and 2.11 we now obtain:

Theorem 2.13. *A Runge-Kutta method (1.8) is of order p iff*

$$\sum_{j=1}^{s} b_j \Phi_j(t) = \frac{1}{\gamma(t)} \qquad (2.21)$$

for all trees of order $\leq p$.

Proof. While the "if" part is clear from the preceding discussion, the "only if" part needs the fact that the elementary differentials for different trees are actually *independent*. See Exercises 3 and 4 below.   □

From Table 2.1 we then obtain the following number of order conditions (see Table 2.3). One can thus understand that the construction of higher order Runge Kutta formulas is not an easy task.

Table 2.3. Number of order conditions

order p	1	2	3	4	5	6	7	8	9	10
no. of conditions	1	2	4	8	17	37	85	200	486	1205

Example. For the tree t_{42} of Table 2.2 we have (using (1.9) for the second expression)

$$\sum_{j,k,l,m} b_j a_{jk} a_{jl} a_{km} = \sum_{j,k} b_j a_{jk} c_j c_k = \frac{1}{8},$$

which is (1.11;f). All remaining conditions of (1.11) correspond to the other trees of order ≤ 4.

Exercises

1. Find all trees of order 6 and order 7.

   Hint. Search for all representations of $p-1$ as a sum of positive integers, and then insert all known trees of lower order for each term in the sum. You may also use a computer for general p.

2. (A. Cayley 1857). Denote the number of trees of order q by a_q. Prove that

   $$a_1 + a_2 x + a_3 x^2 + a_4 x^3 + \ldots = (1-x)^{-a_1}(1-x^2)^{-a_2}(1-x^3)^{-a_3}\ldots.$$

   Compare the result with Table 2.1.

3. Compute the elementary differentials of Table 2.2 for the case of the scalar non-autonomous equation (2.1), i.e., $f^1 = 1$, $f^2 = f(x,y)$. One imagines the complications met by the first authors (Kutta, Nyström, Huťa) in looking for higher order conditions. Observe also that in this case the expressions for t_{54} and t_{57} are the same, so that here Theorem 2.13 is sufficient, but not necessary for order 5.

   Hint. For, say, t_{54} we have non-zero derivatives only if $K = L = 2$. Letting M and P run from 1 to 2 we then obtain

   $$F^2(t) = (f_x + f f_y)(f_{yx} + f f_{yy})f_y$$

   (see also Butcher 1963a).

4. Show that for every $t \in T_q$ there is a system of differential equations such that $F^1(t)(y_0) = 1$ and $F^1(u)(y_0) = 0$ for all other trees u.

   Hint. For t_{54} this system would be

$$y_1' = y_2 y_5, \quad y_2' = y_3, \quad y_3' = y_4, \quad y_4' = 1, \quad y_5' = 1$$

   with all initial values $= 0$. Understand this and the general formula

$$y_{\text{father}}' = \prod y_{\text{sons}}.$$

5. Kutta (1901) claimed that the scheme given in Table 2.4 is of order 5. Was he correct in his statement? Try to correct these values.

   Result. The values for $a_{6j}(j = 1, \ldots, 5)$ should read $(6, 36, 10, 8, 0)/75$; the correct values for b_j are $(23, 0, 125, 0, -81, 125)/192$ (Nyström 1925).

Table 2.4. A method of Kutta

0						
$\frac{1}{3}$	$\frac{1}{3}$					
$\frac{2}{5}$	$\frac{4}{25}$	$\frac{6}{25}$				
1	$\frac{1}{4}$	-3	$\frac{15}{4}$			
$\frac{2}{3}$	$\frac{6}{81}$	$\frac{90}{81}$	$-\frac{50}{81}$	$\frac{8}{81}$		
$\frac{4}{5}$	$\frac{7}{30}$	$\frac{18}{30}$	$-\frac{5}{30}$	$\frac{4}{30}$	0	
	$\frac{48}{192}$	0	$\frac{125}{192}$	0	$-\frac{81}{192}$	$\frac{100}{192}$

6. Verify

$$\sum_{\varrho(t)=p} \alpha(t) = (p-1)!$$

7. Prove that a Runge-Kutta method, when applied to a linear system

$$y' = A(x)y + g(x), \tag{2.22}$$

   is of order p iff

$$\sum_j b_j c_j^{q-1} = 1/q \quad \text{for } q \leq p$$

$$\sum_{j,k} b_j c_j^{q-1} a_{jk} c_k^{r-1} = 1/((q+r)r) \quad \text{for } q+r \leq p$$

$$\sum_{j,k,l} b_j c_j^{q-1} a_{jk} c_k^{r-1} a_{kl} c_l^{s-1} = 1/((q+r+s)(r+s)s) \quad \text{for } q+r+s \leq p$$

   ... etc (write (2.22) in autonomous form and investigate which elementary differentials vanish identically; see also Crouzeix 1975).

II.3 Error Estimation and Convergence for RK Methods

Es fehlt indessen noch der Beweis dass diese Näherungs-Verfahren convergent sind oder, was practisch wichtiger ist, es fehlt ein Kriterium, um zu ermitteln, wie klein die Schritte gemacht werden müssen, um eine vorgeschriebene Genauigkeit zu erreichen.

(Runge 1905)

Since the work of Lagrange (1797) and, above all, of Cauchy, a numerically established result should be accompanied by a reliable error estimation ("... l'erreur commise sera inférieure à ..."). Lagrange gave the well-known error bounds for the Taylor polynomials and Cauchy derived bounds for the error of the Euler polygons (see Section I.7). A couple of years after the first success of the Runge-Kutta methods, Runge (1905) also required error estimates for these methods.

Rigorous Error Bounds

Runge's device for obtaining bounds for the error in one step ("local error") can be described in a few lines (free translation):

"For a method of order p consider the local error

$$e(h) = y(x_0 + h) - y_1 \tag{3.1}$$

and use its Taylor expansion

$$e(h) = e(0) + he'(0) + \ldots + \frac{h^p}{p!}e^{(p)}(\theta h) \tag{3.2}$$

with $0 < \theta < 1$ and $e(0) = e'(0) = \ldots = e^{(p)}(0) = 0$. Now compute explicitly $e^{(p)}(h)$, which will be of the form

$$e^{(p)}(h) = E_1(h) + hE_2(h), \tag{3.3}$$

where $E_1(h)$ and $E_2(h)$ contain partial derivatives of f up to order $p-1$ and p respectively. Further, because of $e^{(p)}(0) = 0$, we have $E_1(0) = 0$. Thus, if all partial derivatives of f up to order p are bounded, we have $E_1(h) = \mathcal{O}(h)$ and $E_2(h) = \mathcal{O}(1)$. So there is a constant C such that $|e^{(p)}(h)| \leq Ch$ and

$$|e(h)| \leq C\frac{h^{p+1}}{p!}. \text{"} \tag{3.4}$$

A slightly different approach is adopted by Bieberbach (1923, 1. Abschn., Kap. II, §7), explained in more detail in Bieberbach (1951): we write

$$e(h) = y(x_0 + h) - y_1 = y(x_0 + h) - y_0 - h \sum_{i=1}^{s} b_i k_i \qquad (3.5)$$

and use the Taylor expansions

$$y(x_0 + h) = y_0 + y'(x_0)h + y''(x_0)\frac{h^2}{2!} + \ldots + y^{(p+1)}(x_0 + \theta h)\frac{h^{p+1}}{(p+1)!}$$

$$k_i(h) = k_i(0) + k_i'(0)h + \ldots + k_i^{(p)}(\theta_i h)\frac{h^p}{p!}, \qquad (3.6)$$

where, for vector valued functions, the formula is valid componentwise with possibly different θ's. The first terms in the h expansion of (3.5) vanish because of the order conditions. Thus we obtain

Theorem 3.1. *If the Runge-Kutta method (1.8) is of order p and if all partial derivatives of $f(x, y)$ up to order p exist (and are continuous), then the local error of (1.8) admits the rigorous bound*

$$\|y(x_0 + h) - y_1\| \leq h^{p+1}\Big(\frac{1}{(p+1)!}\max_{t \in [0,1]}\|y^{(p+1)}(x_0 + th)\|$$
$$+ \frac{1}{p!}\sum_{i=1}^{s}|b_i|\max_{t \in [0,1]}\|k_i^{(p)}(th)\|\Big) \qquad (3.7)$$

and hence also

$$\|y(x_0 + h) - y_1\| \leq C h^{p+1}. \qquad (3.8)$$

$\square$

Let us demonstrate this result on Runge's first method (1.4), which is of order $p = 2$, applied to a scalar differential equation. Differentiating (1.1) we obtain

$$y^{(3)}(x) = \Big(f_{xx} + 2f_{xy}f + f_{yy}f^2 + f_y(f_x + f_y f)\Big)(x, y(x)) \qquad (3.9)$$

while the second derivative of $k_2(h) = f(x_0 + \frac{h}{2}, y_0 + \frac{h}{2}f_0)$ is given by

$$k_2^{(2)}(h) = \frac{1}{4}\Big(f_{xx}(x_0 + \frac{h}{2}, y_0 + \frac{h}{2}f_0) + 2f_{xy}(...)f_0 + f_{yy}(...)f_0^2\Big) \qquad (3.10)$$

(f_0 stands for $f(x_0, y_0)$). Under the assumptions of Theorem 3.1 we see that the expressions (3.9) and (3.10) are bounded by a constant independent of h, which gives (3.8).

The Principal Error Term

For higher order methods rigorous error bounds, like (3.7), become very unpractical. It is therefore much more realistic to consider the first non-zero term in the Taylor expansion of the error. For autonomous systems of equations (2.2), the error term is best obtained by subtracting the Taylor series and using (2.14) and (2.7;q).

Theorem 3.2. *If the Runge-Kutta method is of order p and if f is $(p+1)$-times continuously differentiable, we have*

$$y^J(x_0+h) - y_1^J = \frac{h^{p+1}}{(p+1)!} \sum_{t \in T_{p+1}} \alpha(t)e(t)F^J(t)(y_0) + \mathcal{O}(h^{p+2}) \qquad (3.11)$$

where

$$e(t) = 1 - \gamma(t) \sum_{j=1}^{s} b_j \Phi_j(t). \qquad (3.12)$$

$\square$

$\gamma(t)$ and $\Phi_j(t)$ are given in Definitions 2.9 and 2.10; see also formulas (2.17) and (2.19). The expressions $e(t)$ are called the *error coefficients*.

Example 3.3. For the two-parameter family of 4th order RK methods (1.17) the error coefficients for the 9 trees of Table 2.2 are ($c_2 = u$, $c_3 = v$):

$$e(t_{51}) = -\frac{1}{4} + \frac{5}{12}(u+v) - \frac{5}{6}uv, \qquad e(t_{52}) = \frac{5}{12}v - \frac{1}{4},$$

$$e(t_{53}) = \frac{5}{8}u - \frac{1}{4}, \qquad e(t_{54}) = -\frac{1}{4},$$

$$e(t_{55}) = 1 - \frac{5(b_4 + b_3(3-4v)^2)}{144 b_3 b_4 (1-v)^2}, \qquad (3.13)$$

$$e(t_{56}) = -4e(t_{51}), \qquad e(t_{57}) = -4e(t_{52}),$$

$$e(t_{58}) = -4e(t_{53}), \qquad e(t_{59}) = -4e(t_{54}).$$

Proof. The last four formulas follow from (1.12). $e(t_{59})$ is trivial, $e(t_{58})$ and $e(t_{57})$ follow from (1.11h). Further

$$e(t_{51}) = 5 \int_0^1 t(t-1)(t-u)(t-v)\, dt$$

expresses the quadrature error. For $e(t_{55})$ one best introduces $c_i' = \sum_j a_{ij}c_j$ such that $e(t_{55}) = 1 - 20 \sum_i b_i c_i' c_i$. Then from (1.11d,f) one obtains

$$c_1' = c_2' = 0, \qquad b_3 c_3' = \frac{1}{24(1-v)}, \qquad b_4 c_4' = \frac{3-4v}{24(1-v)}. \qquad \square$$

For the classical 4th order method (Table 1.2a) these error coefficients are given by Kutta (1901), p. 448 (see also Lotkin 1951) as follows

$$\left(-\frac{1}{24}, -\frac{1}{24}, \frac{1}{16}, -\frac{1}{4}, -\frac{2}{3}, \frac{1}{6}, \frac{1}{6}, -\frac{1}{4}, 1\right)$$

Kutta remarked that for the second method (Table 1.2b) ("Als besser noch erweist sich ...") the error coefficients become

$$\left(-\frac{1}{54}, \frac{1}{36}, -\frac{1}{24}, -\frac{1}{4}, -\frac{1}{9}, \frac{2}{27}, -\frac{1}{9}, \frac{1}{6}, 1\right)$$

which, with the exception of the 4th and 9th term, are all smaller than for the above method. A tedious calculation was undertaken by Ralston (1962) (and by many others) to determine optimal coefficients of (1.17). For solutions which minimize the constants (3.13), see Exercise 3 below.

Estimation of the Global Error

Das war auch eine aufregende Zeit ...   (P. Henrici 1983)

The global error is the error of the computed solution after *several* steps. Suppose that we have a one-step method which, given an initial value (x_0, y_0) and a step size h, computes a numerical solution y_1 approximating $y(x_0 + h)$. We shall denote this process by Henrici's notation

$$y_1 = y_0 + h\Phi(x_0, y_0, h) \tag{3.14}$$

and call Φ the *increment function* of the method.

The numerical solution for a point $X > x_0$ is then obtained by a step-by-step procedure

$$y_{i+1} = y_i + h_i\Phi(x_i, y_i, h_i), \qquad h_i = x_{i+1} - x_i, \qquad x_N = X \tag{3.15}$$

and our task is to estimate the *global error*

$$E = y(X) - y_N. \tag{3.16}$$

This estimate is found in a simple way, very similar to Cauchy's convergence proof for Theorem 7.3 of Chapter I: *the local errors are transported to the final point* x_N *and then added up.* This "error transport" can be done in two different ways:

a) either along the exact solution curves (see Fig. 3.1); this method can yield sharp results when sharp estimates of error propagation for the exact solutions are known, e.g., from Theorem 10.6 of Chapter I based on the logarithmic norm $\mu(\partial f/\partial y)$.

b) or along $N - i$ steps of the numerical method (see Fig. 3.2); this is the method used in the proofs of Cauchy (1824) and Runge (1905), it generalizes easily to multistep methods (see Chapter III) and will be an important tool for the existence of asymptotic expansions (see II.8).

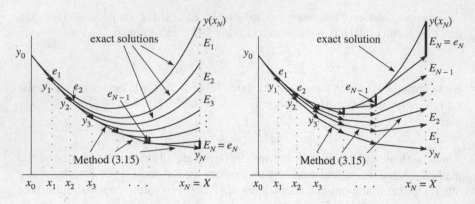

Fig. 3.1. Global error estimation, method (a)

Fig. 3.2. Global error estimation, method (b)

In both cases we first estimate the local errors e_i with the help of Theorem 3.1 to obtain

$$\|e_i\| \leq C \cdot h_{i-1}^{p+1}. \tag{3.17}$$

Warning. The e_i of Fig. 3.1 and Fig. 3.2, for $i \neq 1$, are *not* the same, but they allow similar estimates.

We then estimate the transported errors E_i: for method (a) we use the known results from Chapter I, especially Theorem I.10.6, Theorem I.10.2, or formula (I.7.17). The result is

Theorem 3.4. *Let U be a neighbourhood of $\{(x, y(x)) | x_0 \leq x \leq X\}$ where $y(x)$ is the exact solution of (1.1). Suppose that in U*

$$\left\|\frac{\partial f}{\partial y}\right\| \leq L \qquad or \qquad \mu\left(\frac{\partial f}{\partial y}\right) \leq L, \tag{3.18}$$

and that the local error estimates (3.17) are valid in U. Then the global error (3.16) can be estimated by

$$\|E\| \leq h^p \frac{C'}{L}\left(\exp\left(L(X - x_0)\right) - 1\right). \tag{3.19}$$

where $h = \max h_i$,

$$C' = \begin{cases} C & L \geq 0 \\ C\exp(-Lh) & L < 0, \end{cases}$$

and h is small enough for the numerical solution to remain in U.

Remark. For $L \to 0$ the estimate (3.19) tends to $h^p\, C\, (x_N - x_0)$.

Proof. From Theorem I.10.2 (with $\varepsilon = 0$) or Theorem I.10.6 (with $\delta = 0$) we obtain

$$\|E_i\| \leq \exp\big(L(x_N - x_i)\big)\|e_i\|. \tag{3.20}$$

We then insert this together with (3.17) into

$$\|E\| \leq \sum_{i=1}^{N} \|E_i\|.$$

Using $h_{i-1}^{p+1} \leq h^p \cdot h_{i-1}$ this leads to

$$\|E\| \leq h^p C \Big(h_0 \exp\big(L(x_N - x_1)\big) + h_1 \exp\big(L(x_N - x_2)\big) + \ldots\Big).$$

The expression in large brackets can be bounded by

$$\int_{x_0}^{x_N} \exp(L(x_N - x))dx \qquad \text{for} \qquad L \geq 0 \tag{3.21}$$

$$\int_{x_0}^{x_N} \exp(L(x_N - h - x))dx \qquad \text{for} \qquad L < 0 \tag{3.22}$$

(see Fig. 3.3). This gives (3.19).   □

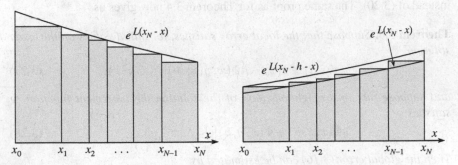

Fig. 3.3. Estimation of Riemann sums

For the second method (b) we need an estimate for $\|z_{i+1} - y_{i+1}\|$ in terms of $\|z_i - y_i\|$, where, besides (3.15),

$$z_{i+1} = z_i + h_i \Phi(x_i, z_i, h_i)$$

is a second pair of numerical solutions. For RK-methods z_{i+1} is defined by

$$\ell_1 = f(x_i, z_i),$$
$$\ell_2 = f(x_i + c_2 h_i, z_i + h_i a_{21} \ell_1), \quad \text{etc.}$$

We now subtract formulas (1.8) from this and obtain

$$\|\ell_1 - k_1\| \le L\|z_i - y_i\|,$$
$$\|\ell_2 - k_2\| \le L(1 + |a_{21}|h_i L)\|z_i - \hat{y}_i\|, \quad \text{etc.}$$

This leads to the following

Lemma 3.5. *Let L be a Lipschitz constant for f and let $h_i \le h$. Then the increment function Φ of method (1.8) satisfies*

$$\|\Phi(x_i, z_i, h_i) - \Phi(x_i, y_i, h_i)\| \le \Lambda\|z_i - y_i\| \tag{3.23}$$

where

$$\Lambda = L\Big(\sum_i |b_i| + hL \sum_{i,j} |b_i a_{ij}| + h^2 L^2 \sum_{i,j,k} |b_i a_{ij} a_{jk}| + \dots\Big). \tag{3.24}$$

□

From (3.23) we obtain

$$\|z_{i+1} - y_{i+1}\| \le (1 + h_i\Lambda)\|z_i - y_i\| \le \exp(h_i\Lambda)\|z_i - y_i\| \tag{3.25}$$

and for the errors in Fig. 3.2,

$$\|E_i\| \le \exp\big(\Lambda(x_N - x_i)\big)\|e_i\| \tag{3.26}$$

instead of (3.20). The same proof as for Theorem 3.4 now gives us

Theorem 3.6. *Suppose that the local error satisfies, for initial values on the exact solution,*

$$\|y(x + h) - y(x) - h\Phi(x, y(x), h)\| \le Ch^{p+1}, \tag{3.27}$$

and suppose that in a neighbourhood of the solution the increment function Φ satisfies

$$\|\Phi(x, z, h) - \Phi(x, y, h)\| \le \Lambda\|z - y\|. \tag{3.28}$$

Then the global error (3.16) can be estimated by

$$\|E\| \le h^p \frac{C}{\Lambda}\Big(\exp\big(\Lambda(x_N - x_0)\big) - 1\Big) \tag{3.29}$$

where $h = \max h_i$.

□

Exercises

1. (Runge 1905). Show that for explicit Runge Kutta methods with $b_i \geq 0$, $a_{ij} \geq 0$ (all i, j) of order s the Lipschitz constant Λ for Φ satisfies

$$1 + h\Lambda < \exp(hL)$$

and that (3.29) is valid with Λ replaced by L.

2. Show that $e(t_{55})$ of (3.13) becomes

$$e(t_{55}) = 1 - 5 \frac{(4v^2 - 15v + 9) - u(6v^2 - 42v + 27) - u^2(26v - 18)}{12(1 - 2u)(6uv - 4(u + v) + 3)}$$

after inserting (1.17).

3. Determine u and v in (1.17) such that in (3.13)

   a) $\max_{i=5,6,7,8} |e(t_{5i})| = \min$ b) $\sum_{i=1}^{9} |e(t_{5i})| = \min$

   c) $\max_{i=5,6,7,8} \alpha(t_{5i}) |e(t_{5i})| = \min$ d) $\sum_{i=1}^{9} \alpha(t_{5i}) |e(t_{5i})| = \min$

   Results.

   a) $u = 0.3587$, $v = 0.6346$, $\min = 0.1033$;

   b) $u = 0.3995$, $v = 0.6$, $\min = 1.55$;

   c) $u = 0.3501$, $v = 0.5839$, $\min = 0.1248$;

   d) $u = 0.3716$, $v = 0.6$, $\min = 2.53$.

   Such optimal formulas were first studied by Ralston (1962), Hull & Johnston (1964), and Hull (1967).

4. Apply an explicit Runge-Kutta method to the problem $y' = f(x, y)$, $y(0) = 0$, where

$$f(x, y) = \begin{cases} \dfrac{\lambda}{x} y + g(x) & \text{if } x > 0 \\ (1 - \lambda)^{-1} g(0) & \text{if } x = 0, \end{cases}$$

   $\lambda \leq 0$ and $g(x)$ is sufficiently differentiable (see Exercise 10 of Section I.5).

   a) Show that the error after the first step is given by

$$y(h) - y_1 = C_2 h^2 g'(0) + \mathcal{O}(h^3)$$

   where C_2 is a constant depending on λ and on the coefficients of the method. Also for high order methods we have in general $C_2 \neq 0$.

   b) Compute C_2 for the classical 4th order method (Table 1.2).

II.4 Practical Error Estimation and Step Size Selection

Ich glaube indessen, dass ein practischer Rechner sich meistens mit der geringeren Sicherheit begnügen wird, die er aus der Uebereinstimmung seiner Resultate für grössere und kleinere Schritte gewinnt.

(C. Runge 1895)

Even the simplified error estimates of Section II.3, which are content with the leading error term, are of little practical interest, because they require the computation and majorization of several partial derivatives of high orders. But the main advantage of Runge-Kutta methods, compared with Taylor series, is precisely that the computation of derivatives should be no longer necessary. However, since practical error estimates are necessary (on the one hand to ensure that the step sizes h_i are chosen sufficiently small to yield the required precision of the computed results, and on the other hand to ensure that the step sizes are sufficiently large to avoid unnecessary computational work), we shall now discuss alternative methods for error estimates.

The oldest device, used by Runge in his numerical examples, is to repeat the computations with *halved* step sizes and to compare the results: those digits which haven't changed are assumed to be correct ("... woraus ich schliessen zu dürfen glaube ...").

Richardson Extrapolation

... its usefulness for practical computations can hardly be overestimated.

(G. Birkhoff & G.C. Rota)

The idea of Richardson, announced in his classical paper Richardson (1910) which treats mainly partial differential equations, and explained in full detail in Richardson (1927), is to use more carefully the known behaviour of the error as a function of h.

Suppose that, with a given initial value (x_0, y_0) and step size h, we compute *two* steps, using a fixed Runge-Kutta method of order p, and obtain the numerical results y_1 and y_2. We then compute, starting from (x_0, y_0), *one big step* with step size $2h$ to obtain the solution w. The error of y_1 is known to be (Theorem 3.2)

$$e_1 = y(x_0 + h) - y_1 = C \cdot h^{p+1} + \mathcal{O}(h^{p+2}) \tag{4.1}$$

where C contains the error coefficients of the method and the elementary differentials $F^J(t)(y_0)$ of order $p+1$. The error of y_2 is composed of two parts: the transported error of the first step, which is

$$\left(I + h\frac{\partial f}{\partial y} + \mathcal{O}(h^2)\right)e_1,$$

and the local error of the second step, which is the same as (4.1), but with the elementary differentials evaluated at $y_1 = y_0 + \mathcal{O}(h)$. Thus we obtain

$$e_2 = y(x_0 + 2h) - y_2 = (I + \mathcal{O}(h))Ch^{p+1} + (C + \mathcal{O}(h))h^{p+1} + \mathcal{O}(h^{p+2})$$
$$= 2Ch^{p+1} + \mathcal{O}(h^{p+2}). \tag{4.2}$$

Similarly to (4.1), we have for the big step

$$y(x_0 + 2h) - w = C(2h)^{p+1} + \mathcal{O}(h^{p+2}). \tag{4.3}$$

Neglecting the terms $\mathcal{O}(h^{p+2})$, formulas (4.2) and (4.3) allow us to eliminate the unknown constant C and to "extrapolate" a better value $\widehat{y}_2$ for $y(x_0 + 2h)$, for which we obtain:

Theorem 4.1. *Suppose that y_2 is the numerical result of two steps with step size h of a Runge-Kutta method of order p, and w is the result of one big step with step size $2h$. Then the error of y_2 can be extrapolated as*

$$y(x_0 + 2h) - y_2 = \frac{y_2 - w}{2^p - 1} + \mathcal{O}(h^{p+2}) \tag{4.4}$$

and

$$\widehat{y}_2 = y_2 + \frac{y_2 - w}{2^p - 1} \tag{4.5}$$

is an approximation of order $p+1$ to $y(x_0 + 2h)$. □

Formula (4.4) is a very simple device to estimate the error of y_2 and formula (4.5) allows one to increase the precision by one additional order ("... The better theory of the following sections is complicated, and tends thereby to suggest that the practice may also be complicated; whereas it is really simple." Richardson).

Embedded Runge-Kutta Formulas

> Scraton is right in his criticism of Merson's process, although Merson did not claim as much for his process as some people expect.
> (R. England 1969)

The idea is, rather than using Richardson extrapolation, to construct Runge-Kutta formulas which themselves contain, besides the numerical approximation y_1, a second approximation $\widehat{y}_1$. The difference then yields an estimate of the local error for the less precise result and can be used for step size control (see below). Since it is at our disposal at every step, this gives more flexibility to the code and makes step rejections less expensive.

We consider two Runge-Kutta methods (one for y_1 and one for $\widehat{y}_1$) such that both use the *same* function values. We thus have to find a scheme of coefficients (see (1.8')),

$$
\begin{array}{c|ccccc}
0 & & & & & \\
c_2 & a_{21} & & & & \\
c_3 & a_{32} & a_{32} & & & \\
\vdots & \vdots & & \ddots & & \\
c_s & a_{s1} & a_{s2} & \cdots & a_{s,s-1} & \\
\hline
& b_1 & b_2 & \cdots & b_{s-1} & b_s \\
\hline
& \widehat{b}_1 & \widehat{b}_2 & \cdots & \widehat{b}_{s-1} & \widehat{b}_s
\end{array}
\tag{4.6}
$$

such that

$$
y_1 = y_0 + h(b_1 k_1 + \ldots + b_s k_s) \tag{4.7}
$$

is of order p, and

$$
\widehat{y}_1 = y_0 + h(\widehat{b}_1 k_1 + \ldots + \widehat{b}_s k_s) \tag{4.7'}
$$

is of order $\widehat{p}$ (usually $\widehat{p} = p-1$ or $\widehat{p} = p+1$). The approximation y_1 is used to continue the integration.

From Theorem 2.13, we have to satisfy the conditions

$$
\sum_{j=1}^{s} b_j \Phi_j(t) = \frac{1}{\gamma(t)} \qquad \text{for all trees of order} \le p , \tag{4.8}
$$

$$
\sum_{j=1}^{s} \widehat{b}_j \Phi_j(t) = \frac{1}{\gamma(t)} \qquad \text{for all trees of order} \le \widehat{p} . \tag{4.8'}
$$

The first methods of this type were proposed by Merson (1957), Ceschino (1962), and Zonneveld (1963). Those of Merson and Zonneveld are given in Tables 4.1 and 4.2. Here, "name $p(\widehat{p})$" means that the order of y_1 is p and the order of the error estimator $\widehat{y}_1$ is $\widehat{p}$. Merson's $\widehat{y}_1$ is of order 5 only for *linear* equations with constant coefficients; for nonlinear problems it is of order 3. This method works quite well and has been used very often, especially by NAG users. Further embedded methods were then derived by Sarafyan (1966), England (1969), and Fehlberg (1964, 1968, 1969). Let us start with the construction of some low order embedded methods.

Methods of order 3(2). It is a simple task to construct embedded formulas of order 3(2) with $s = 3$ stages. Just take a 3-stage method of order 3 (Exercise II.1.4) and put $\widehat{b}_3 = 0$, $\widehat{b}_2 = 1/2c_2$, $\widehat{b}_1 = 1 - 1/2c_2$.

Table 4.1. Merson 4("5")

0				
$\frac{1}{3}$	$\frac{1}{3}$			
$\frac{1}{3}$	$\frac{1}{6}$	$\frac{1}{6}$		
$\frac{1}{2}$	$\frac{1}{8}$	0	$\frac{3}{8}$	
1	$\frac{1}{2}$	0	$-\frac{3}{2}$	2
y_1	$\frac{1}{6}$	0	0	$\frac{2}{3}$	$\frac{1}{6}$
$\widehat{y}_1$	$\frac{1}{10}$	0	$\frac{3}{10}$	$\frac{2}{5}$	$\frac{1}{5}$

Table 4.2. Zonneveld 4(3)

0				
$\frac{1}{2}$	$\frac{1}{2}$			
$\frac{1}{2}$	0	$\frac{1}{2}$		
1	0	0	1	
$\frac{3}{4}$	$\frac{5}{32}$	$\frac{7}{32}$	$\frac{13}{32}$	$-\frac{1}{32}$
y_1	$\frac{1}{6}$	$\frac{1}{3}$	$\frac{1}{3}$	$\frac{1}{6}$
$\widehat{y}_1$	$-\frac{1}{2}$	$\frac{7}{3}$	$\frac{7}{3}$	$\frac{13}{6}$	$-\frac{16}{3}$

Methods of order 4(3). With $s = 4$ it is impossible to find a pair of order 4(3) (see Exercise 2). The idea is to add y_1 as 5th stage of the process (i.e., $a_{5i} = b_i$ for $i = 1, \ldots, 4$) and to search for a third order method which uses all five function values. Whenever the step is accepted this represents no extra work, because $f(x_0 + h, y_1)$ has to be computed anyway for the following step. This idea is called FSAL (First Same As Last). Then the order conditions (4.8') with $\widehat{p} = 3$ represent 4 linear equations for the five unknowns $\widehat{b}_1, \ldots, \widehat{b}_5$. One can arbitrarily fix $\widehat{b}_5 \neq 0$ and solve the system for the remaining parameters. With $\widehat{b}_5$ chosen such that $\widehat{b}_4 = 0$ the result is

$$\widehat{b}_1 = 2b_1 - 1/6, \qquad \widehat{b}_2 = 2(1 - c_2)b_2,$$
$$\widehat{b}_3 = 2(1 - c_3)b_3, \qquad \widehat{b}_4 = 0, \qquad \widehat{b}_5 = 1/6. \tag{4.9}$$

Automatic Step Size Control

> D'ordinaire, on se contente de multiplier ou de diviser par 2 la valeur du pas ...
> (Ceschino 1961)

We now want to write a code which automatically adjusts the step size in order to achieve a prescribed tolerance of the local error.

Whenever a starting step size h has been chosen, the program computes two approximations to the solution, y_1 and $\widehat{y}_1$. Then an estimate of the error for the less precise result is $y_1 - \widehat{y}_1$. We want this error to satisfy componentwise

$$|y_{1i} - \widehat{y}_{1i}| \le sc_i, \qquad sc_i = Atol_i + \max(|y_{0i}|, |y_{1i}|) \cdot Rtol_i \tag{4.10}$$

where $Atol_i$ and $Rtol_i$ are the desired tolerances prescribed by the user (relative errors are considered for $Atol_i = 0$, absolute errors for $Rtol_i = 0$; usually both tolerances are different from zero; they may depend on the component of the solution). As a measure of the error we take

$$err = \sqrt{\frac{1}{n} \sum_{i=1}^{n} \left(\frac{y_{1i} - \widehat{y}_{1i}}{sc_i} \right)^2}; \tag{4.11}$$

other norms, such as the max norm, are also of frequent use. Then err is compared to 1 in order to find an optimal step size. From the error behaviour $err \approx C \cdot h^{q+1}$ and from $1 \approx C \cdot h_{\text{opt}}^{q+1}$ (where $q = \min(p, \widehat{p})$) the optimal step size is obtained as ("... le procédé connu", Ceschino 1961)

$$h_{\text{opt}} = h \cdot (1/err)^{1/(q+1)}. \tag{4.12}$$

Some care is now necessary for a good code: we multiply (4.12) by a safety factor fac, usually $fac = 0.8$, 0.9, $(0.25)^{1/(q+1)}$, or $(0.38)^{1/(q+1)}$, so that the error will be acceptable the next time with high probability. Further, h is not allowed to increase nor to decrease too fast. For example, we may put

$$h_{\text{new}} = h \cdot \min\left(facmax, \max\left(facmin, fac \cdot (1/err)^{1/(q+1)}\right)\right) \tag{4.13}$$

for the new step size. Then, if $err \leq 1$, the computed step is *accepted* and the solution is advanced with y_1 and a new step is tried with h_{new} as step size. Else, the step is *rejected* and the computations are repeated with the new step size h_{new}. The maximal step size increase $facmax$, usually chosen between 1.5 and 5, prevents the code from too large step increases and contributes to its safety. It is clear that, when chosen too small, it may also unnecessarily increase the computational work. It is also advisable to put $facmax = 1$ in the steps right after a step-rejection (Shampine & Watts 1979).

Whenever y_1 is of lower order than $\widehat{y}_1$, then the difference $y_1 - \widehat{y}_1$ is (at least asymptotically) an estimate of the local error and the above algorithm keeps this estimate below the given tolerance. But isn't it more natural to continue the integration with the higher order approximation? Then the concept of "error estimation" is abandoned and the difference $y_1 - \widehat{y}_1$ is only used for the purpose of step size selection. This is justified by the fact that, due to unknown stability and instability properties of the differential system, the local errors have in general very little in common with the global errors. The procedure of continuing the integration with the higher order result is called "local extrapolation".

A modification of the above procedure (PI step size control), which is particularly interesting when applied to mildly stiff problems, is described in Section IV.2 (Volume II).

Starting Step Size

> If anything has been made foolproof, a better fool will be developed.
>
> (Heard from Dr. Pirkl, Baden)

For many years, the starting step size had to be supplied to a code. Users were assumed to have a rough idea of a good step size from mathematical knowledge or previous experience. Anyhow, a bad starting choice for h was quickly repaired by the step size control. Nevertheless, when this happens too often and when the choices are too bad, much computing time can be wasted. Therefore, several people (e.g., Watts 1983, Hindmarsh 1980) developed ideas to let the computer do this choice. We take up an idea of Gladwell, Shampine & Brankin (1987) which is based on the hypothesis that

$$\text{local error} \approx C h^{p+1} y^{(p+1)}(x_0).$$

Since $y^{(p+1)}(x_0)$ is unknown we shall replace it by approximations of the first and second derivative of the solution. The resulting algorithm is the following one:

a) Do one function evaluation $f(x_0, y_0)$ at the initial point. It is in any case needed for the first RK step. Then put $d_0 = \|y_0\|$ and $d_1 = \|f(x_0, y_0)\|$, where the norm is that of (4.11) with $sc_i = Atol_i + |y_{0i}| \cdot Rtol_i$.

b) As a first guess for the step size let

$$h_0 = 0.01 \cdot (d_0/d_1)$$

so that the increment of an explicit Euler step is small compared to the size of the initial value. If either d_0 or d_1 is smaller than 10^{-5} we put $h_0 = 10^{-6}$.

c) Perform one explicit Euler step, $y_1 = y_0 + h_0 f(x_0, y_0)$, and compute $f(x_0 + h_0, y_1)$.

d) Compute $d_2 = \|f(x_0 + h_0, y_1) - f(x_0, y_0)\|/h_0$ as an estimate of the second derivative of the solution; the norm being the same as in (a).

e) Compute a step size h_1 from the relation

$$h_1^{p+1} \cdot \max(d_1, d_2) = 0.01.$$

If $\max(d_1, d_2) \leq 10^{-15}$ we put $h_1 = \max(10^{-6}, h_0 \cdot 10^{-3})$.

f) Finally we propose as starting step size

$$h = \min(100 \cdot h_0, h_1). \tag{4.14}$$

An algorithm like the one above, or a similar one, usually gives a good guess for the initial step size (or at least avoids a very bad choice). Sometimes, more information about h is known, e.g., from previous experience or computations of similar problems.

Numerical Experiments

As a representative of 4-stage 4th order methods we consider the "3/8 Rule" of Table 1.2. We equipped it with the embedded formula (4.9) of order 3.

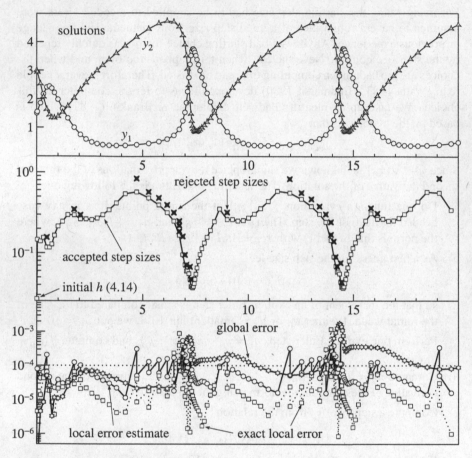

Fig. 4.1. Step size control, $Rtol = Atol = 10^{-4}$, 96 steps + 32 rejected

Step control mechanism. Fig. 4.1 presents the results of the step control mechanism (4.13) described above. As an example we choose the Brusselator (see Section I.16).

$$y_1' = 1 + y_1^2 y_2 - 4y_1$$
$$y_2' = 3y_1 - y_1^2 y_2 \tag{4.15}$$

with initial values $y_1(0) = 1.5$, $y_2(0) = 3$, integration interval $0 \le x \le 20$ and $Atol = Rtol = 10^{-4}$. The following results are plotted in this figure:

i) At the top, the solutions $y_1(x)$ and $y_2(x)$ with all accepted integration steps;

ii) then all step sizes used; the accepted ones are connected by a polygon; the rejected ones are indicated by $\times$;

iii) the third graph shows the local error estimate *err*, the exact local error and the global error; the desired tolerance is indicated by a broken horizontal line.

It can be seen that, due to the instabilities of the solutions with respect to the initial values, quite large global errors occur during the integration with small local tolerances everywhere. Further many step rejections can be observed in regions where the step size has to be decreased. This cannot easily be prevented, because right after an accepted step, the step size proposed by formula (4.13) is (apart from the safety factor) always increasing.

Numerical comparison. We are now curious to see the behaviour of the variable step size code, when compared to a fixed step size implementation. We applied both implementations to the Brusselator problem (4.15) with the initial values used there. The tolerances ($Atol = Rtol$) are chosen between 10^{-2} and 10^{-10} with ratio $\sqrt[3]{10}$. The results are then plotted in Fig. 4.2. There, the abscissa is the global error at the endpoint of integration (the "precision"), and the ordinate is the number of function evaluations (the "work"). We observe that for this problem the variable step size code is about twice as fast as the fixed step size code. There are, of course, problems (such as equation (0.1)) where variable step sizes are *much* more important than here.

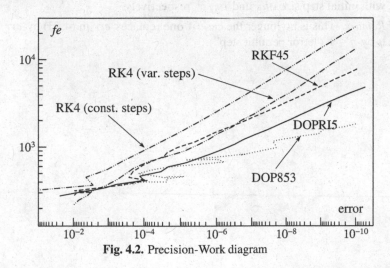

Fig. 4.2. Precision-Work diagram

In this comparison we have included some higher order methods, which will be dicussed in Section II.5. The code RKF45 (written by H.A. Watts and L.F. Shampine) is based on an embedded method of order 5(4) due to Fehlberg. The codes DOPRI5 (order 5(4)) and DOP853 (order 8(5,3)) are based on methods of

Dormand & Prince. They will be discussed in the following section. It can clearly be seen that higher order methods are, especially for higher precision, more efficient than lower order methods. We shall also understand why the 5th order method of Dormand & Prince is clearly superior to RKF45.

Exercises

1. Show that Runge's method (1.4) can be interpreted as two Euler steps (with step size $h/2$), followed by a Richardson extrapolation.

2. Prove that no 4-stage Runge-Kutta method of order 4 admits an embedded formula of order 3.

   Hint. Replace d_j by $\widehat{b}_j - b_j$ in the proof of Lemma 1.4 and deduce that $\widehat{b}_j = b_j$ for all j, which is a contradiction.

3. Show that the step size strategy (4.13) is invariant with respect to a rescaling of the independent variable. This means that it produces equivalent step size sequences when applied to the two problems

$$y' = f(x, y), \qquad y(0) = y_0, \qquad y(x_{\text{end}}) = ?$$
$$z' = \sigma \cdot f(\sigma t, z), \qquad z(0) = y_0, \qquad z(x_{\text{end}}/\sigma) = ?$$

   with initial step sizes h_0 and h_0/σ, respectively.

   Remark. This is no longer the case if one replaces *err* in (4.13) by *err*/h and q by $q - 1$ ("error per unit step").

II.5 Explicit Runge-Kutta Methods of Higher Order

Gehen wir endlich zu Näherungen von der fünften Ordnung über,
so werden die Verhältnisse etwas andere.　　　(W. Kutta 1901)

This section describes the construction of Runge-Kutta methods of higher orders,
particularly of orders $p = 5$ and $p = 8$. As can be seen from Table 2.3, the complexity and number of the order conditions to be solved increases rapidly with p.
An increasingly skilful use of simplifying assumptions will be the main tool for
this task.

The Butcher Barriers

For methods of order 5 there are 17 order conditions to be satisfied (see Table 2.2).
If we choose $s = 5$ we have 15 free parameters. Already Kutta raised the question whether there might nevertheless exist a solution ("Nun wäre es zwar möglich
..."), but he had no hope for this and turned straight away to the case $s = 6$ (see
II.2, Exercise 5). Kutta's question remained open for more than 60 years and was
answered around 1963 by three authors independently (Ceschino & Kuntzmann
1963, p. 89, Shanks 1966, Butcher 1964b, 1965b). Butcher's work is the farthest
reaching and we shall mainly follow his ideas in the following:

Theorem 5.1. *For $p \geq 5$ no explicit Runge-Kutta method exists of order p with
$s = p$ stages.*

Proof. We first treat the case $s = p = 5$: define the matrices U and V by

$$U = \begin{pmatrix} \sum_i b_i a_{i2} & \sum_i b_i a_{i3} & \sum_i b_i a_{i4} \\ \sum_i b_i a_{i2} c_2 & \sum_i b_i a_{i3} c_3 & \sum_i b_i a_{i4} c_4 \\ g_2 & g_3 & g_4 \end{pmatrix}, \quad V = \begin{pmatrix} c_2 & c_2^2 & \sum_j a_{2j} c_j - c_2^2/2 \\ c_3 & c_3^2 & \sum_j a_{3j} c_j - c_3^2/2 \\ c_4 & c_4^2 & \sum_j a_{4j} c_j - c_4^2/2 \end{pmatrix}$$

(5.1)

where

$$g_k = \sum_{i,j} b_i a_{ij} a_{jk} - \frac{1}{2} \sum_i b_i a_{ik}(1 - c_k).$$

(5.2)

Then the order conditions for order 5 imply

$$UV = \begin{pmatrix} 1/6 & 1/12 & 0 \\ 1/12 & 1/20 & 0 \\ 0 & 0 & 0 \end{pmatrix}. \tag{5.3}$$

Lemma 1.5 gives $g_4 = 0$ and consequently $c_4 = 1$ as in Lemma 1.4. Next we put in (5.1)

$$g_j = \left(\sum_i b_i a_{ij} - b_j(1 - c_j)\right)(c_j - c_5). \tag{5.4}$$

Again it can be verified by trivial computations that UV is the same as above. This time it follows that $c_4 = c_5$, hence $c_5 = 1$. Consequently, the expression

$$\sum_{i,j,k} b_i(1 - c_i)a_{ij}a_{jk}c_k \tag{5.5}$$

must be zero (because of $2 \le k < j < i$). However, by multiplying out and using two fifth-order conditions, the expression in (5.5) should be $1/120$, a contradiction.

The case $p = s = 6$ is treated by considering all "one-leg trees", i.e., the trees which consist of one leg above the root and the 5th order trees grafted on. The corresponding order conditions have the form

$$\sum_{i,j,\dots} b_i a_{ij}(a_{j\dots} \dots \text{ expressions for order 5}) = \frac{1}{\gamma(t)}.$$

If we let $b'_j = \sum_i b_i a_{ij}$ we are back in the 5th order 5-stage business and can follow the above ideas again. However, the $\gamma(t)$ values are not the same as before; as a consequence, the product UV in (5.3) now becomes

$$UV = \begin{pmatrix} \dfrac{1!}{(s-2)!} & \dfrac{2!}{(s-1)!} & 0 \\ \dfrac{2!}{(s-1)!} & \dfrac{3!}{s!} & 0 \\ 0 & 0 & 0 \end{pmatrix} \qquad (s = 6). \tag{5.3'}$$

Further, for $p = s = 7$ we use the "stork-trees" with order conditions

$$\sum_{i,j,\dots} b_i a_{ij} a_{jk}(a_{k\dots} \dots \text{ expressions for order 5}) = \frac{1}{\gamma(t)}$$

and let $b''_k = \sum_{i,j} b_i a_{ij} a_{jk}$ and so on. The general case $p = s \ge 5$ is now clear. $\qquad \Box$

6-Stage, 5th Order Processes

We now demonstrate the construction of 5th order processes with 6 stages in full detail following the ideas which allowed Butcher (1964b) to construct 7-stage, 6th order formulas.

   "In searching for such processes we are guided by the analysis of the previous section to make the following assumptions:"

$$\sum_{i=1}^{6} b_i a_{ij} = b_j(1 - c_j) \qquad j = 1, \ldots, 6, \tag{5.6}$$

$$\sum_{j=1}^{i-1} a_{ij} c_j = \frac{c_i^2}{2} \cdot \qquad i = 3, \ldots, 6, \tag{5.7}$$

$$b_2 = 0. \tag{5.8}$$

The advantage of condition (5.6) is known to us already from Section II.1 (see Lemma 1.3): we can disregard all one-leg trees other than t_{21}.

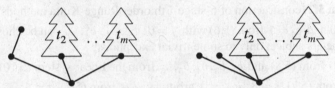

Fig. 5.1. Use of simplifying assumptions

   Condition (5.7) together with (5.8) has a similar effect: for $[[\tau], t_2, \ldots, t_m]$ and $[\tau, \tau, t_2, \ldots, t_m]$ of Fig. 5.1 (with identical but arbitrary subtrees $t_2, \ldots, t_m$) the order conditions read

$$\sum_{i,j} b_i a_{ij} c_j \Phi_i = \frac{1}{r \cdot 2} \qquad \text{and} \qquad \sum_i b_i c_i^2 \Phi_i = \frac{1}{r} \tag{5.9}$$

with known values for Φ_i and r. Since $b_2 = 0$ by (5.8) it follows from (5.7) that both conditions of (5.9) are equivalent (the condition $b_2 = 0$ is necessary for this reduction, because (5.7) cannot be satisfied for $i = 2$; otherwise we would have $c_2 = 0$ and the method would be equivalent to one of fewer stages).

   The only trees left after the above reduction are the quadrature conditions

$$\sum_{i=1}^{6} b_i c_i^{q-1} = \frac{1}{q} \qquad q = 1, 2, 3, 4, 5 \tag{5.10}$$

and the two equations

$$\sum_{i,j,k} b_i c_i a_{ij} a_{jk} c_k = \frac{1}{5 \cdot 3 \cdot 2}, \tag{5.11}$$

$$\sum_{i,j} b_i c_i a_{ij} c_j^2 = \frac{1}{5 \cdot 3} \cdot \tag{5.12}$$

We multiply (5.12) by $1/2$ and then subtract both equations to obtain

$$\sum_{i,j} b_i c_i a_{ij}\left(\sum_k a_{jk}c_k - c_j^2/2\right) = 0.$$

From (5.7) the parenthesis is zero except when $j = 2$, and therefore

$$\sum_{i=3}^{6} b_i c_i a_{i2} = 0 \tag{5.13}$$

replaces (5.11). Our last simplification is to subtract other order conditions from (5.12) to obtain

$$\sum_{i,j} b_i(1 - c_i)a_{ij}c_j(c_j - c_3) = \frac{1}{60} - \frac{c_3}{24}, \tag{5.14}$$

which has fewer terms than before, in particular because $c_6 = 1$ by (5.6) with $j = 6$. The resulting *reduced system* (5.6)-(5.8), (5.10), (5.13), (5.14) can easily be solved as follows:

Algorithm 5.2 (construction of 6-stage 5th order Runge-Kutta methods).

a) $c_1 = 0$ and $c_6 = 1$ from (5.6) with $j = 6$; c_2, c_3, c_4, c_5 can be chosen as free parameters subject only to some trivial exceptions;

b) $b_2 = 0$ from (5.8) and b_1, b_3, b_4, b_5, b_6 from the linear system (5.10);

c) a_{32} from (5.7), $i = 3$; $a_{42} = \lambda$ arbitrary; a_{43} from (5.7), $i = 4$;

d) a_{52} and a_{62} from the two linear equations (5.13) and (5.6), $j = 2$;

e) a_{54} from (5.14) and a_{53} from (5.7), $i = 5$;

f) a_{63}, a_{64}, a_{65} from (5.6), $j = 3, 4, 5$;

g)   finally a_{i1} $(i = 2, \dots, 6)$ from (1.9).

Condition (5.6) for $j = 1$ and (5.7) for $i = 6$ are automatically satisfied. This follows as in the proof of Lemma 1.4.

Embedded Formulas of Order 5

Methods of Fehlberg. The methods obtained from Algorithm 5.2 do not all possess an embedded formula of order 4. Fehlberg, interested in the construction of Runge-Kutta pairs of order 4(5), looked mainly for simplifying assumptions which depend only on c_i and a_{ij}, but not on the weights b_i. In this case the simplifying assumptions are useful for the embedded method too. Therefore Fehlberg (1969) considered (5.7), (5.8) and replaced (5.6) by

$$\sum_{j=1}^{i-1} a_{ij}c_j^2 = \frac{c_i^3}{3}, \qquad i = 3, \dots, 6. \tag{5.15}$$

As with (5.9) this allows us to disregard all trees of the form $[[\tau, \tau], t_2, \ldots, t_m]$. In order that the reduction process of Fig. 5.1 also work on a higher level, we suppose, in addition to $b_2 = 0$, that

$$\sum_i b_i a_{i2} = 0, \qquad \sum_i b_i c_i a_{i2} = 0, \qquad \sum_{i,j} b_i a_{ij} a_{j2} = 0. \qquad (5.16)$$

Then the last equations to be satisfied are

$$\sum_{i,j} b_i a_{ij} c_j^3 = \frac{1}{20} \qquad (5.17)$$

and the quadrature conditions (5.10). We remark that the equations (5.7) and (5.15) for $i = 3$ imply

$$c_3 = \frac{3}{2} c_2. \qquad (5.18)$$

We now want the method to possess an embedded formula of order 4. Analogously to (5.8) we set $\widehat{b}_2 = 0$. Then conditions (5.7) and (5.15) simplify the conditions of order 4 to 5 linear equations (the 4 quadrature conditions and $\sum_i \widehat{b}_i a_{i2} = 0$) for the 5 unknowns $\widehat{b}_1, \widehat{b}_3, \widehat{b}_4, \widehat{b}_5, \widehat{b}_6$. This system has a second solution (other than the b_i) only if it is singular, which is the case if (see Exercise 1 below)

$$c_4 = \frac{3c_2}{4 - 24c_2 + 45c_2^2}. \qquad (5.19)$$

With c_2, c_5, c_6 as free parameters, the above system can be solved and yields an embedded formula of order 4(5). The coefficients of a very popular method, constructed by Fehlberg (1969), are given in Table 5.1.

Table 5.1. Fehlberg 4(5)

0						
$\dfrac{1}{4}$	$\dfrac{1}{4}$					
$\dfrac{3}{8}$	$\dfrac{3}{32}$	$\dfrac{9}{32}$				
$\dfrac{12}{13}$	$\dfrac{1932}{2197}$	$-\dfrac{7200}{2197}$	$\dfrac{7296}{2197}$			
1	$\dfrac{439}{216}$	-8	$\dfrac{3680}{513}$	$-\dfrac{845}{4104}$		
$\dfrac{1}{2}$	$-\dfrac{8}{27}$	2	$-\dfrac{3544}{2565}$	$\dfrac{1859}{4104}$	$-\dfrac{11}{40}$	
y_1	$\dfrac{25}{216}$	0	$\dfrac{1408}{2565}$	$\dfrac{2197}{4104}$	$-\dfrac{1}{5}$	0
$\widehat{y}_1$	$\dfrac{16}{135}$	0	$\dfrac{6656}{12825}$	$\dfrac{28561}{56430}$	$-\dfrac{9}{50}$	$\dfrac{2}{55}$

All of the methods of Fehlberg are of the type $p(\widehat{p})$ with $p < \widehat{p}$. Hence, the lower order approximation is intended to be used as initial value for the next step. In order to make his methods optimal, Fehlberg tried to minimize the error coefficients for the lower order result y_1. This has the disadvantage that the local extrapolation mode (continue the integration with the higher order result) does not make sense and the estimated "error" can become substantially smaller than the true error.

> It is possible to do a lot better than the pair of Fehlberg currently regarded as "best."
>
> (L.F. Shampine 1986)

Table 5.2. Dormand-Prince 5(4) (DOPRI5)

0							
$\dfrac{1}{5}$	$\dfrac{1}{5}$						
$\dfrac{3}{10}$	$\dfrac{3}{40}$	$\dfrac{9}{40}$					
$\dfrac{4}{5}$	$\dfrac{44}{45}$	$-\dfrac{56}{15}$	$\dfrac{32}{9}$				
$\dfrac{8}{9}$	$\dfrac{19372}{6561}$	$-\dfrac{25360}{2187}$	$\dfrac{64448}{6561}$	$-\dfrac{212}{729}$			
1	$\dfrac{9017}{3168}$	$-\dfrac{355}{33}$	$\dfrac{46732}{5247}$	$\dfrac{49}{176}$	$-\dfrac{5103}{18656}$		
1	$\dfrac{35}{384}$	0	$\dfrac{500}{1113}$	$\dfrac{125}{192}$	$-\dfrac{2187}{6784}$	$\dfrac{11}{84}$	
y_1	$\dfrac{35}{384}$	0	$\dfrac{500}{1113}$	$\dfrac{125}{192}$	$-\dfrac{2187}{6784}$	$\dfrac{11}{84}$	0
$\widehat{y}_1$	$\dfrac{5179}{57600}$	0	$\dfrac{7571}{16695}$	$\dfrac{393}{640}$	$-\dfrac{92097}{339200}$	$\dfrac{187}{2100}$	$\dfrac{1}{40}$

Dormand & Prince pairs. The first efforts at minimizing the error coefficients of *the higher order result*, which is then used as numerical solution, were undertaken by Dormand & Prince (1980). Their methods of order 5 are constructed with the help of Algorithm 5.2 under the additional hypothesis (5.15). This condition is achieved by fixing the parameters c_3 and a_{42} in such a way that (5.15) holds for $i = 3$ and $i = 4$. The remaining two relations ($i = 5, 6$) are then automatically satisfied. To see this, multiply the difference $e_i = \sum_{j=1}^{i-1} a_{ij} c_j^2 - c_i^3/3$ by b_i and $b_i c_i$, respectively, sum up and deduce that all e_i must vanish.

In order to equip the method with an embedded formula, Dormand & Prince propose to use the FSAL idea (i.e., add y_1 as 7th stage). In this way the restriction (5.19) for c_4 is no longer necessary. We fix arbitrarily $\widehat{b}_7 \neq 0$, put $\widehat{b}_2 = 0$ (as in (5.8)) and compute the remaining $\widehat{b}_i$, as above for the Fehlberg case from the 4 quadrature conditions and from $\sum_i \widehat{b}_i a_{i2} = 0$.

We have thus obtained a family of 5th order Runge-Kutta methods with 4th order embedded solution with c_2, c_4, c_5 as free parameters. Dormand & Prince (1980) have undertaken an extensive search to determine these parameters in order to minimize the error coefficients for y_1 and found that $c_2 = 1/5$, $c_4 = 4/5$ and $c_5 = 8/9$ was a close rational approximation to an optimal choice. Table 5.2 presents the coefficients of this method. The corresponding code of the Appendix is called DOPRI5.

Higher Order Processes

Order 6. By Theorem 5.1 at least 7 stages are necessary for order 6. A. Huťa (1956) constructed 6th order processes with 8 stages. Finally, methods with $s = 7$, the optimal number, were derived by Butcher (1964b) along similar lines as above. He arrived at an algorithm where c_2, c_3, c_5, c_6 are free parameters.

Order 7. The existence of such a method with 8 stages is impossible by the following barrier:

Theorem 5.3 (Butcher 1965b). *For $p \geq 7$ no explicit Runge-Kutta method exists of order p with $s = p + 1$ stages.*

Since the proof of this theorem is much more complicated than that of Theorem 5.1, we do not reproduce it here.

This raises the question, whether 7th order methods with 9 stages exist. Such methods, announced by Butcher (1965b), do exist; see Verner (1971).

Order 8. As to methods of order 8, Curtis (1970) and Cooper & Verner (1972) have constructed such processes with $s = 11$. It was for a long time an open question whether there exist methods with 10 stages. John Butcher's dream of settling this difficult question before his 50th birthday did not become true. But he finally succeeded in proving the non-existence for Dahlquist's 60th birthday:

Theorem 5.4 (Butcher 1985b). *For $p \geq 8$ no explicit Runge-Kutta method exists of order p with $s = p + 2$ stages.*

For the proof, which is still more complicated, we again refer to Butcher's original paper.

Order 10. These are the highest order explicitly constructed explicit Runge-Kutta methods. Curtis (1975) constructed an 18-stage method of order 10. His construction was based solely on simplifying assumptions of the type (5.7), (5.8) and their extensions. Hairer (1978) then constructed a 17-stage method by using the complete arsenal of simplifying ideas. For more details, see the first edition, p. 189.

Embedded Formulas of High Order

It was mainly the formula manipulation genius Fehlberg who first derived high order embedded formulas. His greatest success was his 7th order formula with 8th order error estimate (Fehlberg 1968) which is of frequent use in all high precision computations, e.g., in astronomy. The coefficients are reproduced in Table 5.3.

Table 5.3. Fehlberg 7(8)

0													
$\frac{2}{27}$	$\frac{2}{27}$												
$\frac{1}{9}$	$\frac{1}{36}$	$\frac{1}{12}$											
$\frac{1}{6}$	$\frac{1}{24}$	0	$\frac{1}{8}$										
$\frac{5}{12}$	$\frac{5}{12}$	0	$-\frac{25}{16}$	$\frac{25}{16}$									
$\frac{1}{2}$	$\frac{1}{20}$	0	0	$\frac{1}{4}$	$\frac{1}{5}$								
$\frac{5}{6}$	$-\frac{25}{108}$	0	0	$\frac{125}{108}$	$-\frac{65}{27}$	$\frac{125}{54}$							
$\frac{1}{6}$	$\frac{31}{300}$	0	0	0	$\frac{61}{225}$	$-\frac{2}{9}$	$\frac{13}{900}$						
$\frac{2}{3}$	2	0	0	$-\frac{53}{6}$	$\frac{704}{45}$	$-\frac{107}{9}$	$\frac{67}{90}$	3					
$\frac{1}{3}$	$-\frac{91}{108}$	0	0	$\frac{23}{108}$	$-\frac{976}{135}$	$\frac{311}{54}$	$-\frac{19}{60}$	$\frac{17}{6}$	$-\frac{1}{12}$				
1	$\frac{2383}{4100}$	0	0	$-\frac{341}{164}$	$\frac{4496}{1025}$	$-\frac{301}{82}$	$\frac{2133}{4100}$	$\frac{45}{82}$	$\frac{45}{164}$	$\frac{18}{41}$			
0	$\frac{3}{205}$	0	0	0	0	$-\frac{6}{41}$	$-\frac{3}{205}$	$\frac{3}{41}$	$\frac{3}{41}$	$\frac{6}{41}$	0		
1	$-\frac{1777}{4100}$	0	0	$-\frac{341}{164}$	$\frac{4496}{1025}$	$-\frac{289}{82}$	$\frac{2193}{4100}$	$\frac{51}{82}$	$\frac{33}{164}$	$\frac{19}{41}$	0	1	
y_1	$\frac{41}{840}$	0	0	0	0	$\frac{34}{105}$	$\frac{9}{35}$	$\frac{9}{35}$	$\frac{9}{280}$	$\frac{9}{280}$	$\frac{41}{840}$	0	0
$\widehat{y}_1$	0	0	0	0	0	$\frac{34}{105}$	$\frac{9}{35}$	$\frac{9}{35}$	$\frac{9}{280}$	$\frac{9}{280}$	0	$\frac{41}{840}$	$\frac{41}{840}$

Fehlberg's methods suffer from the fact that they give identically zero error estimates for quadrature problems $y' = f(x)$. The first high order embedded formulas which avoid this drawback were constructed by Verner (1978). One of Verner's methods (see Table 5.4) has been implemented by T.E. Hull, W.H. Enright and K.R. Jackson as DVERK and is widely used.

Table 5.4. Verner's method of order 6(5) (DVERK)

0								
$\dfrac{1}{6}$	$\dfrac{1}{6}$							
$\dfrac{4}{15}$	$\dfrac{4}{75}$	$\dfrac{16}{75}$						
$\dfrac{2}{3}$	$\dfrac{5}{6}$	$-\dfrac{8}{3}$	$\dfrac{5}{2}$					
$\dfrac{5}{6}$	$-\dfrac{165}{64}$	$\dfrac{55}{6}$	$-\dfrac{425}{64}$	$\dfrac{85}{96}$				
1	$\dfrac{12}{5}$	-8	$\dfrac{4015}{612}$	$-\dfrac{11}{36}$	$\dfrac{88}{255}$			
$\dfrac{1}{15}$	$-\dfrac{8263}{15000}$	$\dfrac{124}{75}$	$\dfrac{643}{680}$	$-\dfrac{81}{250}$	$\dfrac{2484}{10625}$	0		
1	$\dfrac{3501}{1720}$	$-\dfrac{300}{43}$	$\dfrac{297275}{52632}$	$-\dfrac{319}{2322}$	$\dfrac{24068}{84065}$	0	$\dfrac{3850}{26703}$	
y_1	$\dfrac{3}{40}$	0	$\dfrac{875}{2244}$	$\dfrac{23}{72}$	$\dfrac{264}{1955}$	0	$\dfrac{125}{11592}$	$\dfrac{43}{616}$
$\widehat{y}_1$	$\dfrac{13}{160}$	0	$\dfrac{2375}{5984}$	$\dfrac{5}{16}$	$\dfrac{12}{85}$	$\dfrac{3}{44}$	0	0

An 8th Order Embedded Method

The first high order methods with small error constants of the *higher* order solution were constructed by Prince & Dormand (1981, Code DOPRI8 of the first edition). In the following we describe the construction of a new Dormand & Prince pair of order 8(6) which will also allow a cheap and accurate dense output (see Section II.6). This method has been announced, but not published, in Dormand & Prince (1989, p. 983). We are grateful to P. Prince for mailing us the coefficients and for his help in recovering their construction.

The essential difficulty for the construction of a high order Runge-Kutta method is to set up a "good" reduced system which implies all order conditions of Theorem 2.13. At the same time it should be simple enough to be easily solved. In extending the ideas for the construction of a 5th order process (see above), Dormand & Prince proceed as follows:

Reduced system. Suppose $s = 12$ and consider for the coefficients c_i, b_i and a_{ij} the equations:

$$\sum_{i=1}^{s} b_i c_i^{q-1} = 1/q, \qquad q = 1, \ldots, 8 \tag{5.20a}$$

$$\sum_{j=1}^{i-1} a_{ij} = c_i, \qquad i = 1, \ldots, s \tag{5.20b}$$

$$\sum_{j=1}^{i-1} a_{ij} c_j = c_i^2/2, \qquad i = 3, \ldots, s \tag{5.20c}$$

$$\sum_{j=1}^{i-1} a_{ij} c_j^2 = c_i^3/3, \qquad i = 3, \ldots, s \tag{5.20d}$$

$$\sum_{j=1}^{i-1} a_{ij} c_j^3 = c_i^4/4, \qquad i = 6, \ldots, s \tag{5.20e}$$

$$\sum_{j=1}^{i-1} a_{ij} c_j^4 = c_i^5/5, \qquad i = 6, \ldots, s \tag{5.20f}$$

$$b_2 = b_3 = b_4 = b_5 = 0 \tag{5.20g}$$

$$a_{i2} = 0 \quad \text{for} \quad i \geq 4, \qquad a_{i3} = 0 \quad \text{for} \quad i \geq 6 \tag{5.20h}$$

$$\sum_{i=j+1}^{s} b_i a_{ij} = b_j(1 - c_j), \qquad j = 4, 5, 10, 11, 12 \tag{5.20i}$$

$$\sum_{i=j+1}^{s} b_i c_i a_{ij} = 0, \qquad j = 4, 5 \tag{5.20j}$$

$$\sum_{i=j+1}^{s} b_i c_i^2 a_{ij} = 0, \qquad j = 4, 5 \tag{5.20k}$$

$$\sum_{i=k+2}^{s} b_i c_i \sum_{j=k+1}^{i-1} a_{ij} a_{jk} = 0, \qquad k = 4, 5 \tag{5.20l}$$

$$\sum_{i=1}^{s} b_i c_i \sum_{j=1}^{i-1} a_{ij} c_j^5 = 1/48. \tag{5.20m}$$

Verification of the order conditions. The equations (5.20a) are the order conditions for the bushy trees $[\tau, \ldots, \tau]$ and (5.20m) is that for the tree $[\tau, [\tau, \tau, \tau, \tau, \tau]]$. For the verification of further order conditions we shall show that the reduced system implies

$$\sum_{i=j+1}^{s} b_i a_{ij} = b_j(1 - c_j) \quad \text{for all } j. \tag{5.21}$$

If we denote the difference by $d_j = \sum_{i=j+1}^{s} b_i a_{ij} - b_j(1 - c_j)$ then $d_2 = d_3 = 0$ by (5.20g,h) and $d_4 = d_5 = d_{10} = d_{11} = d_{12} = 0$ by (5.20i). The conditions (5.20a-g) imply

$$\sum_{j=1}^{s} d_j c_j^{q-1} = 0 \quad \text{for} \quad q = 1, \ldots, 5. \tag{5.22}$$

Hence, the remaining 5 values must also vanish if c_1, c_6, c_7, c_8, c_9 are distinct. The significance of condition (5.21) is already known from Lemma 1.3 and from formula (5.6). It implies that all one-leg trees $t = [t_1]$ can be disregarded.

Fig. 5.2. Use of simplifying assumptions

Conditions (5.20c-f) are an extension of (5.6) and (5.15). Their importance will be, once more, demonstrated on an example. Consider the two trees of Fig. 5.2 and suppose that their encircled parts are identical. Then the corresponding order conditions are

$$\sum_{i,j=1}^{s} \Theta_i a_{ij} c_j^3 = \frac{1}{r \cdot 5 \cdot 4} \qquad \text{and} \qquad \sum_{i=1}^{s} \Theta_i c_i^4 = \frac{1}{r \cdot 5} \qquad (5.23)$$

with known values for Θ_i and r. If (5.20e) is satisfied and if

$$\Theta_2 = \Theta_3 = \Theta_4 = \Theta_5 = 0 \qquad (5.24)$$

then both conditions are equivalent so that the left-hand tree can be neglected. The conditions (5.20g,i-l) correspond to (5.24) for certain trees. Finally the assumption (5.20h) together with (5.20g,i-k) implies that for arbitrary Φ_i, Ψ_j and for $q \in \{1, 2, 3\}$,

$$\sum_i b_i \Phi_i a_{i2} = 0$$
$$\sum_{i,j} b_i \Phi_i a_{ij} \Psi_j a_{j2} = 0 \qquad \text{and} \qquad \sum_i b_i \Phi_i a_{i3} = 0$$
$$\sum_{i,j,k} b_i c_i^{q-1} a_{ij} \Phi_j a_{jk} \Psi_k a_{k2} = 0 \qquad \qquad \sum_{i,j} b_i c_i^{q-1} a_{ij} \Phi_j a_{j3} = 0$$

which are again conditions of type (5.24). Using these relations the verification of the order conditions (order 8) is straightforward; all trees are reduced to those corresponding to (5.20a) and (5.20m).

Solving the reduced system. Compared to the original 200 order conditions of Theorem 2.13 for the 78 coefficients b_i, a_{ij} (the c_i are defined by (5.20b)), the 74 conditions of the reduced system present a considerable simplification. We can hope for a solution with 4 degrees of freedom.

We start by expressing the coefficients b_i, a_{ij} in terms of the c_i. Because of (5.20g), condition (5.20a) represents a linear system for $b_1, b_6, \ldots, b_{12}$, which has a unique solution if $c_1, c_6, \ldots, c_{12}$ are distinct. For a fixed i $(1 \le i \le 8)$ conditions (5.20b-f) represent a linear system for $a_{i1}, \ldots, a_{i,i-1}$. Since there are sometimes less unknowns than equations (mainly due to (5.20h)) restrictions have to be imposed on the c_i. One verifies (similarly to (5.18)) that the relations

$$c_1 = 0, \qquad c_2 = \frac{2}{3} c_3, \qquad c_3 = \frac{2}{3} c_4,$$
$$c_4 = \frac{6 - \sqrt{6}}{10} c_6, \qquad c_5 = \frac{6 + \sqrt{6}}{10} c_6, \qquad c_6 = \frac{4}{3} c_7 \qquad (5.25a)$$

allow the computation of the a_{ij} with $i \le 8$ (Step 1 in Fig. 5.3).

If $b_{12} \ne 0$ (which will be assumed in our construction), condition (5.20i) for $j = 12$ implies

$$c_{12} = 1, \qquad (5.25b)$$

and for $j = 11$ it yields the value for $a_{12,11}$. We next compute the expressions

$$e_j = \sum_{i=j+1}^{s} b_i c_i a_{ij} - \frac{b_j}{2} (1 - c_j^2), \qquad j = 1, \ldots, s. \qquad (5.26)$$

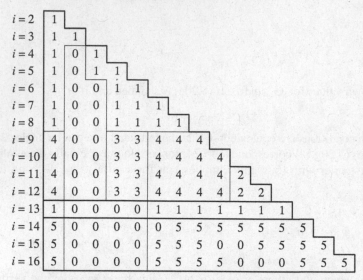

	$j{=}1$	2	3	4	5	6	7	8	9	10	11	12	13	14	15
$i=2$	1														
$i=3$	1	1													
$i=4$	1	0	1												
$i=5$	1	0	1	1											
$i=6$	1	0	0	1	1										
$i=7$	1	0	0	1	1	1									
$i=8$	1	0	0	1	1	1	1								
$i=9$	4	0	0	3	3	4	4	4							
$i=10$	4	0	0	3	3	4	4	4	4						
$i=11$	4	0	0	3	3	4	4	4	4	2					
$i=12$	4	0	0	3	3	4	4	4	4	2	2				
$i=13$	1	0	0	0	0	1	1	1	1	1	1	1			
$i=14$	5	0	0	0	0	0	5	5	5	5	5	5	5		
$i=15$	5	0	0	0	0	5	5	5	0	0	5	5	5	5	
$i=16$	5	0	0	0	0	5	5	5	5	0	0	0	5	5	5

Fig. 5.3. Steps in the construction of an 8th order RK method;
the entries 0 indicate vanishing coefficients;
the stages $i = 14, 15, 16$ will be used for dense output, see II.6.

We have $e_{12} = 0$ by (5.25b), $e_{11} = b_{12}a_{12,11} - b_{11}(1 - c_{11}^2)/2$ is known and $e_2 = e_3 = e_4 = e_5 = 0$ by (5.20g,h,j). The remaining 6 values are determined by the system

$$\sum_{j=1}^{s} e_j c_j^{q-1} = 0, \qquad q = 1, \dots, 6 \tag{5.27}$$

which follows from (5.20a-f,m). The conditions (5.20i) and (5.26) for $j = 10$ then yield $a_{12,10}$ and $a_{11,10}$ (Step 2 in Fig. 5.3).

We next compute a_{ij} ($i = 9, 10, 11, 12; j = 4, 5$) from the remaining 8 equations of (5.20i-l). This is indicated as Step 3 in Fig. 5.3. Finally, we use the conditions (5.20b-f) with $i \geq 9$ for the computation of the remaining coefficients (Step 4). A difficulty still arises from the case $i = 9$, where only 4 parameters for five equations are at our disposal. A tedious computation shows that this system has a solution if (see Exercise 6 below)

$$2c_9 = \frac{3\sigma_1 - 28\sigma_2 + 189\sigma_3 + 14\sigma_1\sigma_2 - 168\sigma_1\sigma_3 + 98\sigma_2\sigma_3}{6 - 21\sigma_1 + 35\sigma_2 - 42\sigma_3 + 21\sigma_1^2 + 98\sigma_2^2 + 735\sigma_3^2 - 84\sigma_1\sigma_2 + 168\sigma_1\sigma_3 - 490\sigma_2\sigma_3} \tag{5.25c}$$

where

$$\sigma_1 = c_6 + c_7 + c_8, \qquad \sigma_2 = c_6c_7 + c_6c_8 + c_7c_8, \qquad \sigma_3 = c_6c_7c_8. \tag{5.28}$$

The reduced system (5.20) leaves c_7, c_8, c_{10}, c_{11} as free parameters. Dormand

& Prince propose the following numerical values:

$$c_7 = 1/4, \qquad c_8 = 4/13, \qquad c_{10} = 3/5, \qquad c_{11} = 6/7 .$$

All remaining coefficients are then determined by the above procedure. Since c_4 and c_5 (see (5.25a)) are not rational, there is no easy way to present the coefficients in a tableau.

Embedded method. We look for a second method with the same c_i, a_{ij} but with different weights, say $\widehat{b}_i$. If we require that

$$\sum_{i=1}^{s} \widehat{b}_i c_i^{q-1} = 1/q, \qquad q = 1, \ldots, 6 \qquad (5.29a)$$
$$\widehat{b}_2 = \widehat{b}_3 = \widehat{b}_4 = \widehat{b}_5 = 0 \qquad (5.29b)$$
$$\sum_{i=j+1}^{s} \widehat{b}_i a_{ij} = 0, \qquad j = 4, 5 \qquad (5.29c)$$

then one can verify (similarly as above for the 8th order method) that the corresponding Runge-Kutta method is of order 6. The system (5.29) consists of 12 linear equations for 12 unknowns. A comparison with (5.20) shows that $b_1, \ldots, b_{12}$ is a solution of (5.29). Furthermore, the corresponding homogeneous system has the nontrivial solution $e_1, \ldots, e_{12}$ (see (5.27) and (5.20l)). Therefore

$$\widehat{b}_i = b_i + \alpha e_i \qquad (5.30)$$

is a solution of (5.29) for all values of α. Dormand & Prince suggest taking α in such a way that $\widehat{b}_6 = 2$.

A program based on this method (with a different error estimator, see Section II.10) has been written and is called DOP853. It is documented in the Appendix. The performance of this code, compared to methods of lower order, is impressive. See for example the results for the Brusselator in Fig. 4.2.

Exercises

1. Consider a Runge-Kutta method with s stages that satisfies (5.7)-(5.8), (5.15), (5.17) and the first two relations of (5.16).

   a) If the relation (5.19) holds, then the method possesses an embedded formula of order 4.

   b) The condition (5.19) implies that the last relation of (5.16) is automatically satisfied.

   Hint. The order conditions for the embedded method constitute a linear system for the $\widehat{b}_i$ which has to be singular. This implies that

   $$a_{i2} = \alpha c_i + \beta c_i^2 + \gamma c_i^3 \qquad \text{for} \qquad i \neq 2. \qquad (5.31)$$

Multiplying (5.31) with b_i and $b_i c_i$ and summing up, yields two relations for $\alpha, \beta, J\gamma$. These together with (5.31) for $i = 3, 4$ yield (5.19).

2. Construct a 6-stage 5th order formula with $c_3 = 1/3$, $c_4 = 1/2$, $c_5 = 2/3$ possessing an embedded formula of order 4.

3. (Butcher). Show that for any Runge-Kutta method of order 5,

$$\sum_i b_i \left(\sum_j a_{ij} c_j - \frac{c_i^2}{2} \right)^2 = 0.$$

Consequently, there exists no explicit Runge-Kutta method of order 5 with all $b_i > 0$.

Hint. Multiply out and use order conditions.

4. Write a code with a high order Runge-Kutta method (or take one) and solve numerically the Arenstorf orbit of the restricted three body problem (0.1) (see the introduction) with initial values

$$y_1(0) = 0.994, \qquad y_1'(0) = 0, \qquad y_2(0) = 0,$$
$$y_2'(0) = -2.0317326295573368357302057924,$$

Compute the solutions for

$$x_{\text{end}} = 11.124340337266085134999734047.$$

The initial values are chosen such that the solution is periodic to this precision. The plotted solution curve has one loop less than that of the introduction.

5. (Shampine 1979). Show that the storage requirement of a Runge-Kutta method can be substantially decreased if s is large.

Hint. Suppose, for example, that $s = 15$.
After computing (see (1.8)) $k_1, k_2, \ldots, k_9$, compute the sums

$$\sum_{j=1}^{9} a_{ij} k_j \quad \text{for } i = 10, 11, 12, 13, 14, 15, \quad \sum_{j=1}^{9} b_j k_j, \quad \sum_{j=1}^{9} \widehat{b}_j k_j;$$

then the memories occupied by $k_2, k_3, \ldots, k_9$ are not needed any longer. Another possibility for reducing the memory requirement is offered by the zero-pattern of the coefficients.

6. Show that the reduced system (5.20) implies (5.25c).

Hint. The equations (5.20b-f) imply that for $i \in \{1, 6, 7, 8, 9\}$

$$\alpha a_{i4} + \beta a_{i5} = \sigma_3 \frac{c_i^2}{2} - \sigma_2 \frac{c_i^3}{3} + \sigma_1 \frac{c_i^4}{4} - \frac{c_i^5}{5} \tag{5.32}$$

with σ_j given by (5.28). The constants α and β are not important. Further, for the same values of i one has

$$0 = c_i(c_i - c_6)(c_i - c_7)(c_i - c_8)(c_i - c_9) \qquad (5.33)$$
$$= \sigma_3 c_9 c_i - (\sigma_3 + c_9 \sigma_2)c_i^2 + (\sigma_2 + c_9 \sigma_1)c_i^3 - (\sigma_1 + c_9)c_i^4 + c_i^5.$$

Multiplying (5.32) and (5.33) by $e_i, b_i, b_i c_i, b_i c_i^2$, summing up from $i = 1$ to s and using (5.20) gives the relation

$$\begin{pmatrix} \times & \times & \times \\ \times & \times & \times \\ 0 & 0 & b_{12}^{-1} \end{pmatrix} \begin{pmatrix} e_{10} & b_{10} & b_{10}c_{10} & b_{10}c_{10}^2 \\ e_{11} & b_{11} & b_{11}c_{11} & b_{11}c_{11}^2 \\ 0 & b_{12} & b_{12} & b_{12} \end{pmatrix} = \begin{pmatrix} 0 & \gamma_1 & \gamma_2 & \gamma_3 \\ 0 & \delta_1 & \delta_2 & \delta_3 \\ 0 & 1 & 1 & 1 \end{pmatrix}$$
$$(5.34)$$

where

$$\gamma_j = \frac{\sigma_3}{2 \cdot (j+2)} - \frac{\sigma_2}{3 \cdot (j+3)} + \frac{\sigma_1}{4 \cdot (j+4)} - \frac{1}{5 \cdot (j+5)}$$

$$\delta_j = \frac{\sigma_3 c_9}{j+1} - \frac{\sigma_3 + c_9 \sigma_2}{j+2} + \frac{\sigma_2 + c_9 \sigma_1}{j+3} - \frac{\sigma_1 + c_9}{j+4} + \frac{1}{j+5}$$

and the "$\times$" indicate certain values. Deduce from (5.34) and $e_{11} \neq 0$ that the most left matrix of (5.34) is singular. This implies that the right-hand matrix of (5.34) is of rank 2 and yields equation (5.25c).

7. Prove that the 8th order method given by (5.20; $s = 12$) does not possess a 6th order embedding with $\hat{b}_{12} \neq b_{12}$, not even if one adds the numerical result y_1 as 13th stage (FSAL).

II.6 Dense Output, Discontinuities, Derivatives

> ... providing "interpolation" for Runge-Kutta methods. ... this
> capability and the features it makes possible will be the hallmark
> of the next generation of Runge-Kutta codes.
>
> (L.F. Shampine 1986)

The present section is mainly devoted to the construction of dense output formulas for Runge-Kutta methods. This is important for many practical questions such as graphical output, event location or the treatment of discontinuities in differential equations. Further, the numerical computation of derivatives with respect to initial values and parameters is discussed, which is particularly useful for the integration of boundary value problems.

Dense Output

Classical Runge-Kutta methods are inefficient, if the number of output points becomes very large (Shampine, Watts & Davenport 1976). This motivated the construction of dense output formulas (Horn 1983). These are Runge-Kutta methods which provide, in addition to the numerical result y_1, cheap numerical approximations to $y(x_0 + \theta h)$ for the *whole* integration interval $0 \leq \theta \leq 1$. "Cheap" means without or, at most, with only a few additional function evaluations.

We start from an s-stage Runge-Kutta method with given coefficients c_i, a_{ij} and b_j, eventually add $s^* - s$ new stages, and consider formulas of the form

$$u(\theta) = y_0 + h \sum_{i=1}^{s^*} b_i(\theta) k_i, \tag{6.1}$$

where

$$k_i = f\left(x_0 + c_i h, y_0 + h \sum_{j=1}^{i-1} a_{ij} k_j\right), \qquad i = 1, \ldots, s^* \tag{6.2}$$

and $b_i(\theta)$ are polynomials to be determined such that

$$u(\theta) - y(x_0 + \theta h) = \mathcal{O}(h^{p^*+1}). \tag{6.3}$$

Usually $s^* \geq s + 1$ since we include (at least) the first function evaluation of the subsequent step $k_{s+1} = hf(x_0 + h, y_1)$ in the formula with $a_{s+1,j} = b_j$ for all j. A Runge-Kutta method, provided with a formula (6.1), will be called a *continuous* Runge-Kutta method.

Theorem 6.1. *The error of the approximation (6.1) is of order p^* (i.e., the local error satisfies (6.3)), if and only if*

$$\sum_{j=1}^{s^*} b_j(\theta)\Phi_j(t) = \frac{\theta^{\varrho(t)}}{\gamma(t)} \qquad for \qquad \varrho(t) \le p^* \tag{6.4}$$

with $\Phi_j(t)$, $\varrho(t)$, $\gamma(t)$ given in Section II.2.

Proof. The qth derivative (with respect to h) of the numerical approximation is given by (2.14) with b_j replaced by $b_j(\theta)$; that of the exact solution $y(x_0 + \theta h)$ is $\theta^q y^{(q)}(x_0)$. The statement thus follows as in Theorem 2.13.   □

Corollary 6.2. *Condition (6.4) implies that the derivatives of (6.1) approximate the derivatives of the exact solution as*

$$h^{-k}u^{(k)}(\theta) - y^{(k)}(x_0 + \theta h) = \mathcal{O}(h^{p^* - k + 1}). \tag{6.5}$$

Proof. Comparing the qth derivative (with respect to h) of $u'(\theta)$ with that of $hy'(x_0 + \theta h)$ we find that (6.5) (for $k = 1$) is equivalent to

$$\sum_{j=1}^{s^*} b_j'(\theta)\Phi_j(t) = \frac{\varrho(t)\theta^{\varrho(t)-1}}{\gamma(t)} \qquad for \qquad \varrho(t) \le p^*.$$

This, however, follows from (6.4) by differentiation. The case $k > 1$ is obtained similarly.   □

We write the polynomials $b_j(\theta)$ as

$$b_j(\theta) = \sum_{q=1}^{p^*} b_{jq}\theta^q , \tag{6.6}$$

so that the equations (6.4) become a system of simultaneous linear equations of the form

$$\underbrace{\begin{pmatrix} 1 & 1 & \cdots & 1 \\ \Phi_1(t_{21}) & \Phi_2(t_{21}) & \cdots & \Phi_{s^*}(t_{21}) \\ \Phi_1(t_{31}) & \Phi_2(t_{31}) & \cdots & \Phi_{s^*}(t_{31}) \\ \vdots & \vdots & & \vdots \end{pmatrix}}_{\Phi} \underbrace{\begin{pmatrix} b_{11} & b_{12} & b_{13} & \cdots \\ b_{21} & b_{22} & b_{23} & \cdots \\ \vdots & \vdots & \vdots & \\ b_{s^*1} & b_{s^*2} & b_{s^*3} & \cdots \end{pmatrix}}_{B} = \underbrace{\begin{pmatrix} 1 & 0 & 0 & \cdots \\ 0 & \frac{1}{2} & 0 & \cdots \\ 0 & 0 & \frac{1}{3} & \cdots \\ 0 & 0 & \frac{1}{6} & \cdots \\ \vdots & \vdots & \vdots & \vdots \end{pmatrix}}_{G} \tag{6.4'}$$

where the $\Phi_j(t)$ are known numbers depending on a_{ij} and c_i. Using standard linear algebra the solution of this system can easily be discussed. It may happen, however, that the order p^* of the dense output is smaller than the order p of the underlying method.

Example. For "the" Runge-Kutta method of Table 1.2 (with $s^* = s = 4$) equations (6.4') with $p^* = 3$ produce a unique solution

$$b_1(\theta) = \theta - \frac{3\theta^2}{2} + \frac{2\theta^3}{3}, \qquad b_2(\theta) = b_3(\theta) = \theta^2 - \frac{2\theta^3}{3}, \qquad b_4(\theta) = -\frac{\theta^2}{2} + \frac{2\theta^3}{3}$$

which constitutes a dense output solution which is globally continuous but not C^1.

Hermite interpolation. A much easier way (than solving (6.4')) and more efficient for low order dense output formulas is the use of Hermite interpolation (Shampine 1985). Whatever the method is, we have two function values y_0, y_1 and two derivatives $f_0 = f(x_0, y_0)$, $f_1 = f(x_0 + h, y_1)$ at our disposal and can thus do cubic polynomial interpolation. The resulting formula is

$$u(\theta) = (1 - \theta)y_0 + \theta y_1 + \theta(\theta - 1)\Big((1 - 2\theta)(y_1 - y_0) + (\theta - 1)hf_0 + \theta hf_1\Big). \quad (6.7)$$

Inserting the definition of y_1 into (6.7) shows that Hermite interpolation is a special case of (6.1). Whenever the underlying method is of order $p \geq 3$ we thus obtain a continuous Runge-Kutta method of order 3.

Since the function and derivative values on the right side of the first interval coincide with those on the left side of the second interval, Hermite interpolation leads to a globally C^1 approximation of the solution.

The 4-stage 4th order methods of Section II.1 do not possess a dense output of order 4 without any additional function evaluations (see Exercise 1). Therefore the question arises whether it is really important to have a dense output of the same order. Let us consider an interval far away from the initial value, say $[x_n, x_{n+1}]$, and denote by $z(x)$ the local solution, i.e., the solution of the differential equation which passes through (x_n, y_n). Then the error of the dense output is composed of two terms:

$$u(\theta) - y(x_n + \theta h) = \big(u(\theta) - z(x_n + \theta h)\big) + \big(z(x_n + \theta h) - y(x_n + \theta h)\big).$$

The term to the far right reflects the global error of the method and is of size $\mathcal{O}(h^p)$. In order that both terms be of the same order of magnitude it is thus sufficient to require $p^* = p - 1$.

The situation changes, if we also need accurate values of the derivative $y'(x_n + \theta h)$ (see Section 5 of Enright, Jackson, Nørsett & Thomsen (1986) for a discussion of problems where this is important). We have

$$h^{-1}u'(\theta) - y'(x_n + \theta h) = \big(h^{-1}u'(\theta) - z'(x_n + \theta h)\big) + \big(z'(x_n + \theta h) - y'(x_n + \theta h)\big)$$

and the term to the far right is of size $\mathcal{O}(h^p)$ if $f(x, y)$ satisfies a Lipschitz condition. A comparison with (6.5) shows that we need $p^* = p$ in order that both error terms be of comparable size.

Boot-strapping process (Enright, Jackson, Nørsett & Thomsen 1986). This is a general procedure for increasing iteratively the order of dense output formulas.

Suppose that we already have a 3rd order dense output at our disposal (e.g., from Hermite interpolation). We then fix arbitrarily an $\alpha \in (0,1)$ and denote the 3rd order approximation at $x_0 + \alpha h$ by y_α. The idea is now that $hf(x_0 + \alpha h, y_\alpha)$ is a 4th order approximation to $hy'(x_0 + \alpha h)$. Consequently, the 4th degree polynomial $u(\theta)$ defined by

$$u(0) = y_0, \qquad u'(0) = hf(x_0, y_0)$$
$$u(1) = y_1, \qquad u'(1) = hf(x_0 + h, y_1) \qquad (6.8)$$
$$u'(\alpha) = hf(x_0 + \alpha h, y_\alpha)$$

(which exists uniquely for $\alpha \neq 1/2$) yields the desired formula. The interpolation error is $\mathcal{O}(h^5)$ and each quantity of (6.8) approximates the corresponding exact solution value with an error of $\mathcal{O}(h^5)$.

The extension to arbitrary order is straightforward. Suppose that a dense output formula $u_0(\theta)$ of order $p^* < p$ is known. We then evaluate this polynomial at $p^* - 2$ distinct points $\alpha_i \in (0,1)$ and compute the values $f(x_0 + \alpha_i h, u_0(\alpha_i))$. The interpolation polynomial $u_1(\theta)$ of degree $p^* + 1$, defined by

$$u_1(0) = y_0, \qquad u_1'(0) = hf(x_0, y_0)$$
$$u_1(1) = y_1, \qquad u_1'(1) = hf(x_0 + h, y_1) \qquad (6.9)$$
$$u_1'(\alpha_i) = hf(x_0 + \alpha_i h, u_0(\alpha_i)), \qquad i = 1, \ldots p^* - 2,$$

yields an interpolation formula of order $p^* + 1$. Obviously, the α_i in (6.9) have to be chosen such that the corresponding interpolation problem admits a solution.

Continuous Dormand & Prince Pairs

The method of Dormand & Prince (Table 5.2) is of order 5(4) so that we are mainly interested in dense output formulas with $p^* = 4$ and $p^* = 5$.

Order 4. A continuous formula of order 4 can be obtained without any additional function evaluation. Since the coefficients satisfy (5.7), it follows from the difference of the order conditions for the trees t_{31} and t_{32} (notation of Table 2.2) that

$$b_2(\theta) = 0 \qquad (6.10)$$

is necessary. This condition together with (5.7) and (5.15) then implies that the order conditions are equivalent for the following pairs of trees: t_{31} and t_{32}, t_{41} and t_{42}, t_{41} and t_{43}. Hence, for order 4, only 5 conditions have to be considered (the four quadrature conditions and $\sum_i b_i(\theta)a_{i2} = 0$). We can arbitrarily choose $b_7(\theta)$ and the coefficients $b_1(\theta), b_3(\theta), \ldots, b_6(\theta)$ are then uniquely determined.

As for the choice of $b_7(\theta)$, Shampine (1986) proposed minimizing, for each θ, the error coefficients (Theorem 3.2)

$$e(t) = \theta^5 - \gamma(t) \sum_{j=1}^{7} b_j(\theta) \Phi_j(t) \qquad \text{for} \qquad t \in T_5, \qquad (6.11)$$

weighted by $\alpha(t)$ of Definition 2.5, in the square norm. These expressions can be seen to depend linearly on $b_7(\theta)$,

$$\alpha(t)e(t) = \zeta(t, \theta) - b_7(\theta)\eta(t),$$

thus the minimal value is found for

$$b_7(\theta) = \sum_{t \in T_5} \zeta(t, \theta)\eta(t) \Big/ \sum_{t \in T_5} \eta^2(t).$$

The resulting formula, given by Dormand & Prince (1986), is

$$b_7(\theta) = \theta^2(\theta - 1) + \theta^2(\theta - 1)^2 10 \cdot (7414447 - 8293050\theta)/29380423. \qquad (6.12)$$

The other coefficients, written in a fashion which makes the Hermite-part clearly visible, are then given by

$$b_1(\theta) = \theta^2(3 - 2\theta) \cdot b_1 + \theta(\theta - 1)^2$$
$$\qquad - \theta^2(\theta - 1)^2 5 \cdot (2558722523 - 3140301 60\theta)/11282082432$$
$$b_3(\theta) = \theta^2(3 - 2\theta) \cdot b_3 + \theta^2(\theta - 1)^2 100 \cdot (882725551 - 157015080\theta)/32700410799$$
$$b_4(\theta) = \theta^2(3 - 2\theta) \cdot b_4 - \theta^2(\theta - 1)^2 25 \cdot (443332067 - 3140301 60\theta)/1880347072$$
$$b_5(\theta) = \theta^2(3 - 2\theta) \cdot b_5 + \theta^2(\theta - 1)^2 32805 \cdot (23143187 - 34892240\theta)/199316789632$$
$$b_6(\theta) = \theta^2(3 - 2\theta) \cdot b_6 - \theta^2(\theta - 1)^2 55 \cdot (29972135 - 70767360\theta)/822651844. \qquad (6.13)$$

It can be directly verified that the interpolation polynomial $u(\theta)$ defined by (6.10), (6.12) and (6.13) satisfies

$$
\begin{aligned}
u(0) &= y_0, & u'(0) &= hf(x_0, y_0), \\
u(1) &= y_1, & u'(1) &= hf(x_0 + h, y_1),
\end{aligned}
\qquad (6.14)
$$

so that it produces globally a $\mathcal{C}^1$ approximation of the solution.

Instead of using the above 5th degree polynomial $u(\theta)$, Shampine (1986) suggests evaluating it only at the midpoint, $y_{1/2} = u(1/2)$, and then doing quartic polynomial interpolation with the five values y_0, $hf(x_0, y_0)$, y_1, $hf(x_0 + h, y_1)$, $y_{1/2}$. This dense output is also $\mathcal{C}^1$, is easier to implement and the difference to the above formula "... is not significant" (Dormand & Prince 1986).

We have implemented Shampine's dense output in the code DOPRI5 (see Appendix). The advantages of such a dense output for graphical representations of the solution can already be seen from Fig. 0.1 of the introduction to Chapter II. For a more thorough study we have applied DOPRI5 to the Brusselator (4.15) with initial values $y_1(0) = 1.5$, $y_2(0) = 3$, integration interval $0 \leq x \leq 10$ and error tolerance $Atol = Rtol = 10^{-4}$. The global error of the above 4th order continuous solution is displayed in Fig. 6.1 for both components. The error shows the same quality throughout; the grid points, which are represented by the symbols $\square$ and $\bigcirc$, are by no means outstanding.

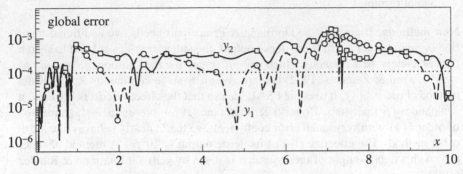

Fig. 6.1. Error of dense output of DOPRI5

Order 5. For a dense output of order $p^* = 5$ for the Dormand & Prince method the linear system (6.4') has no solution since

$$\text{rank}\,(\Phi|G) = 9 \qquad \text{and} \qquad \text{rank}\,(\Phi) = 7 \qquad (6.15)$$

as can be verified by Gaussian elimination. Such a linear system has a solution if and only if the two ranks in (6.15) are *equal* . So we must append additional stages to the method. Each new stage adds a new column to the matrix Φ, thus may increase the rank of Φ by one without changing rank $(\Phi|G)$. Therefore we obtain

Lemma 6.3 (Owren & Zennaro 1991). *Consider a Runge-Kutta method of order* p. *For the construction of a continuous extension of order* $p^* = p$ *one has to add at least*

$$\delta := \text{rank}\,(\Phi|G) - \text{rank}\,(\Phi) \qquad (6.16)$$

stages. $\qquad\qquad\qquad\qquad\qquad\qquad\qquad\qquad\qquad\qquad\qquad\qquad\qquad\qquad\qquad$ $\square$

For the Dormand & Prince method we thus need at least two additional stages. There are several possibilities for constructing such dense output formulas:

a) Shampine (1986) shows that one new function evaluation allows one to compute a 5th order approximation at the midpoint $x_0 + h/2$. If one evaluates anew the function at this point to get an approximation of $y'(x_0 + h/2)$, one can do quintic Hermite interpolation to get a dense output of order 5.

b) Use the 4th order formula constructed above at two different output points and do boot-strapping. This has been done by Calvé & Vaillancourt (1990).

c) Add two arbitrary new stages and solve the order conditions. This leads to methods with 10 free parameters (Calvo, Montijano & Rández 1992) which can then be used to minimize the error terms. This seems to give the best output formulas.

New methods. If anyhow the Dormand & Prince pair needs two additional function evaluations for a 5th order dense output, the suggestion lies at hand to search for completely new methods which use *all* stages for the solution y_1 and $\widehat{y}_1$ as well. Owren & Zennaro (1992) constructed an 8-stage continuous Runge-Kutta method of order $5(4)$. It uses the FSAL idea so that the effective cost is 7 function evaluations (fe) per step. Bogacki & Shampine (1989) present a 7-stage method of order $5(4)$ with very small error coefficients, so that it nearly behaves like a 6th order method. The effective cost of its dense output is 10 fe. A method of order $6(5)$ with a dense output of order $p^* = 5$ is given by Calvo, Montijano & Rández (1990).

Dense Output for DOP853

We are interested in a continuous extension of the 8th order method of Section II.5 (formula (5.20)). A dense output of order 6 can be obtained for free (add y_1 as 13th stage and solve the linear system (6.19a-c) below with $s^* = s + 1 = 13$). Following Dormand & Prince we shall construct a dense output of order $p^* = 7$. We add three further stages (by Lemma 6.3 this is the minimal number of additional stages). The values for c_{14}, c_{15}, c_{16} are chosen arbitrarily as

$$c_{14} = 0.1, \qquad c_{15} = 0.2, \qquad c_{16} = 7/9 \qquad (6.17)$$

and the coefficients a_{ij} are assumed to satisfy, for $i \in \{14, 15, 16\}$,

$$\sum_{j=1}^{i-1} a_{ij} c_j^{q-1} = c_i^q/q, \qquad q = 1, \ldots, 6 \qquad (6.18a)$$

$$a_{i2} = a_{i3} = a_{i4} = a_{i5} = 0 \qquad (6.18b)$$

$$\sum_{j=k+1}^{i-1} a_{ij} a_{jk} = 0, \qquad k = 4, 5. \qquad (6.18c)$$

This system can easily be solved (step 5 of Fig. 5.3). We are still free to set some coefficients equal to 0 (see Fig. 5.3).

We next search for polynomials $b_i(\theta)$ such that the conditions (6.4) are satisfied for all trees of order ≤ 7. We find the following necessary conditions ($s^* = 16$)

$$\sum_{i=1}^{s^*} b_i(\theta) c_i^{q-1} = \theta^q/q, \qquad q = 1, \ldots, 7 \qquad (6.19a)$$

$$b_2(\theta) = b_3(\theta) = b_4(\theta) = b_5(\theta) = 0 \qquad (6.19b)$$

$$\sum_{i=j+1}^{s^*} b_i(\theta) a_{ij} = 0, \qquad j = 4, 5 \qquad (6.19c)$$

$$\sum_{i=j+1}^{s^*} b_i(\theta) c_i a_{ij} = 0, \qquad\qquad j = 4, 5 \tag{6.19d}$$

$$\sum_{i,j=1}^{s^*} b_i(\theta) a_{ij} c_j^5 = \theta^7/42. \tag{6.19e}$$

Here (6.19a,e) are order conditions for $[\tau, \ldots, \tau]$ and $[[\tau, \tau, \tau, \tau, \tau]]$. The property $b_2(\theta) = 0$ follows from $0 = \sum_i b_i(\theta)(\sum_j a_{ij} c_j - c_i^2/2) = -b_2(\theta) c_2^2/2$ and the other three conditions of (6.19b) are a consequence of the relations $0 = \sum_i b_i(\theta) c_i^{q-1}(\sum_j a_{ij} c_j^3 - c_i^4/4) = 0$ for $q = 1, 2, 3$. The necessity of the conditions (6.19c,d) is seen similarly.

On the other hand, the conditions (6.19) are also sufficient for the dense output to be of order 7. We first remark that (6.19), (6.18) and (5.20) imply

$$\sum_{i,j=k+1}^{s^*} b_i(\theta) a_{ij} a_{jk} = 0, \qquad k = 4, 5 \tag{6.20}$$

(see Exercise 3). The verification of the order conditions (6.4) is then possible without difficulty.

System (6.19) consists of 16 linear equations for 16 unknowns which possess a unique solution. An interesting property of the continuous solution (6.1) obtained in this manner is that it yields a global C^1-approximation to the solution, i.e.,

$$u(0) = y_0, \qquad u(1) = y_1, \qquad u'(0) = hf(y_0), \qquad u'(1) = hf(y_1). \tag{6.21}$$

For the verification of this property we define a polynomial $q(\theta)$ of degree 7 by the relations (6.21) and by $q(\theta_i) = u(\theta_i)$ for 4 distinct values θ_i which are different from 0 and 1. Obviously, $q(\theta)$ is of the form (6.1) and defines a dense output of order 7. Due to the uniqueness of the $b_i(\theta)$ we must have $q(\theta) \equiv u(\theta)$ so that (6.21) is verified.

Event Location

Often the output value x_{end} for which the solutions are wanted is not known in advance, but depends implicitly on the computed solutions. An example of such a situation is the search for periodic solutions and limit cycles discussed in Section I.16, where we wanted to know when the solution reaches the Poincaré-section for the first time.

Such problems are very easily treated when a dense output $u(x)$ is available. Suppose we want to determine x such that

$$g(x, y(x)) = 0. \tag{6.22}$$

Algorithm 6.4. Compute the solution step-by-step until a sign change appears between $g(x_i, y_i)$ and $g(x_{i+1}, y_{i+1})$ (this is, however, not completely safe because g may change sign twice in an integration interval; use the dense output at intermediate values if more safety is needed). Then replace $y(x)$ in (6.22) by the approximation $u(x)$ and solve the resulting equation numerically, e.g. by bisection or Newton iterations.

This algorithm can be conveniently done in the subroutine SOLOUT, which is called after every accepted step (see Appendix). If the value of x, satisfying (6.22), has been found, the integration is stopped by setting IRTRN $= -1$.

Whenever the function g of (6.22) also depends on $y'(x)$, it is advisable to use a dense output of order $p^* = p$.

Discontinuous Equations

> If you write some software which is half-way useful, sooner or later someone will use it on discontinuities. You have to scope about ...
>
> (A.R. Curtis 1986)

In many applications the function defining a differential equation is not analytic or continuous everywhere. A common example is a problem which (at least locally) can be written in the form

$$y' = \begin{cases} f_I(y) & \text{if} \quad g(y) > 0 \\ f_{II}(y) & \text{if} \quad g(y) < 0 \end{cases} \tag{6.23}$$

with sufficiently differentiable functions g, f_I and f_{II}. The derivative of the solution is thus in general discontinuous on the surface

$$S = \{y; \ g(y) = 0\}.$$

The function $g(y)$ is called a *switching function*.

In order to understand the situations which can occur when the solution of (6.23) meets the surface S in a point y_0 (i.e., $g(y_0) = 0$), we consider the scalar products

$$\begin{aligned} a_I &= \langle \operatorname{grad} g(y_0), f_I(y_0) \rangle \\ a_{II} &= -\langle \operatorname{grad} g(y_0), f_{II}(y_0) \rangle \end{aligned} \tag{6.24}$$

which can be approximated numerically by $a_I \approx g(y_0 + \delta f_I(y_0))/\delta$ with small enough δ. Since the vector $\operatorname{grad} g(y_0)$ points towards the domain of f_I, the inequality $a_I < 0$ tells us that the flow for f_I is "pushing" against S, while for $a_I > 0$ the flow is "pulling". The same argument holds for a_{II} and the flow for f_{II}. Therefore, apart from degenerate cases where either a_I or a_{II} vanishes, we can distinguish the following four cases (see Fig. 6.2):

1) $a_I > 0, a_{II} < 0$: the flow traverses S from $g < 0$ to $g > 0$.

2) $a_I < 0, a_{II} > 0$: the flow traverses S from $g > 0$ to $g < 0$.

3) $a_I > 0, a_{II} > 0$: the flow "pulls" on both sides; the solution is not unique; except in the case of an unhappily chosen initial value, this situation would normally not occur.

4) $a_I < 0, a_{II} < 0$: here *both* flows push against S; the solution is trapped in S and the problem no longer has a classical solution.

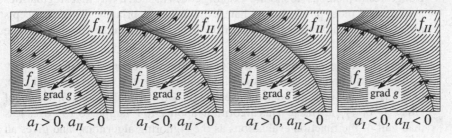

| $a_I > 0, a_{II} < 0$ | $a_I < 0, a_{II} > 0$ | $a_I > 0, a_{II} > 0$ | $a_I < 0, a_{II} < 0$ |

Fig. 6.2. Solutions near the surface of discontinuity

Crossing a discontinuity. The *numerical* computation of a solution crossing a discontinuity (cases 1 and 2) can be performed as follows:

a) *Ignoring the discontinuity:* apply a variable step size code with local error control (such as DOPRI5) and hope that the step size mechanism would handle the discontinuity appropriately. Consider the example (which represents the flow of the second picture of Fig. 6.2)

$$y' = \begin{cases} x^2 + 2y^2 & \text{if } (x+0.05)^2 + (y+0.15)^2 \le 1 \\ 2x^2 + 3y^2 - 2 & \text{if } (x+0.05)^2 + (y+0.15)^2 > 1 \end{cases} \qquad (6.25)$$

with initial value $y(0) = 0.3$. The discontinuity for this problem occurs at $x \approx 0.6234$ and the code, applied with $Atol = Rtol = 10^{-5}$, detects the discontinuity fairly well by means of numerous rejected steps (see Fig. 6.3; this figure, however, is much less dramatic than an analogous drawing (see Gear & Østerby 1984) for multistep methods). The numerical solution for $x = 1$ then has an error of $5.9 \cdot 10^{-4}$.

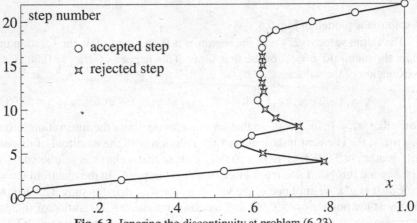

Fig. 6.3. Ignoring the discontinuity at problem (6.23)

b) *Singularity detecting codes.* Concepts have been developed (Gear & Østerby (1984) for multistep methods, Enright, Jackson, Nørsett & Thomsen (1988) for Runge-Kutta methods) to modify existing codes in such a way that singularities are detected more precisely and handled more appropriately. These concepts are mainly based on the behaviour of the local error estimate compared to the step size.

c) *Use the switching function:* stop the computation at the surface of discontinuity using Algorithm 6.4 and restart the integration with the new right-hand side. One has to take care that during one integration step only function values of either f_I or f_{II} are used. This algorithm, applied to Example (6.25), uses less than half of the function evaluations as the "ignoring algorithm" and gives an error of $6.6 \cdot 10^{-6}$ at the point $x = 1$. It is thus not only faster, but also much more reliable.

Example 6.5. Coulomb's law of friction (Coulomb 1785), which states that the force of friction is *independent* of the speed, gives rise to many situations with discontinuous differential equations. Consider the example (see Den Hartog 1930, Reissig 1954, Taubert 1976)

$$y'' + 2Dy' + \mu \operatorname{sign} y' + y = A \cos(\omega x). \tag{6.26}$$

where the Coulomb-force $\mu \operatorname{sign} y'$ is accompanied by a viscosity term Dy'. We fix the parameters as $D = 0.1$, $\mu = 4$, $A = 2$ and $\omega = \pi$, and choose the initial values

$$y(0) = 3, \qquad y'(0) = 4. \tag{6.27}$$

Equation (6.26), written in the form (6.23), is

$$y' = v$$
$$v' = -0.2v - y + 2\cos(\pi x) - \begin{cases} 4 & \text{if } v > 0 \\ -4 & \text{if } v < 0. \end{cases} \tag{6.28}$$

Its solution is plotted in Fig. 6.4.

The initial value (6.27) is in the region $v > 0$ and we follow the solution until it hits the manifold $v = 0$ for the first time. This happens for $x_1 \approx 0.5628$. An investigation of the values

$$a_I = -y(x_1) + 2\cos(\pi x_1) - 4, \qquad a_{II} = y(x_1) - 2\cos(\pi x_1) - 4 \tag{6.29}$$

shows that $a_I < 0$, $a_{II} > 0$, so that we have to continue the integration into the region $v < 0$. The next intersection of the solution with the manifold of discontinuity is at $x_2 \approx 2.0352$. Here $a_I < 0$, $a_{II} < 0$, so that a classical solution does not exist beyond this point and the solution remains "trapped" in the manifold ($v = 0$, $y = Const = y(x_2)$) until one of the values a_I or a_{II} changes sign. This happens for a_{II} at the point $x_3 \approx 2.6281$ and we can continue the integration of (6.28) in the region $v < 0$ (see Fig. 6.4). The same situation then repeats periodically.

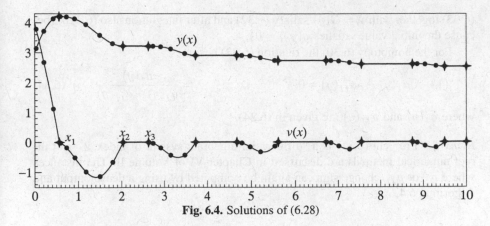

Fig. 6.4. Solutions of (6.28)

Solutions in the manifold. In the case $a_I < 0$, $a_{II} < 0$ the solution of (6.23) can neither be continued along the flow of $y' = f_I(y)$ nor along that of $y' = f_{II}(y)$. However, the physical process, described by the differential equation (6.23), possesses a solution (see Example 6.5). Early papers on this subject studied the convergence of Euler polygons, pushed across the border again and again by the conflicting vector fields (see, e.g., Taubert 1976). Later it became clear that it is much more advantageous to pursue the solution *in* the manifold S, i.e., solve a so-called differential algebraic problem. This approach is advocated by Eich (1992), who attributes the ideas to the thesis of G. Bock, by Eich, Kastner-Maresch & Reich (unpublished manuscript, 1991), and by Stewart (1990). We must decide, however, *which* vector field in S should determine the solution. Several motivations (see Exercises 8 and 9 below) suggest to search this field in the convex hull

$$f(y, \lambda) = (1 - \lambda)f_I(y) + \lambda f_{II}(y), \tag{6.30}$$

of f_I and f_{II}. This coincides, for the special problem (6.23), with Filippov's "generalized solution" (Filippov 1960); but other homotopies may be of interest as well. The value of λ must be chosen in such a way that the solution remains in S. This means that we have to solve the problem

$$y' = f(y, \lambda) \tag{6.31a}$$
$$0 = g(y). \tag{6.31b}$$

Differentiating (6.31b) with respect to time yields

$$0 = \operatorname{grad} g(y)y' = \operatorname{grad} g(y)f(y, \lambda). \tag{6.32}$$

If this relation allows λ to be expressed as a function of y, say as $\lambda = G(y)$, then (6.31a) becomes the ordinary differential equation

$$y' = f(y, G(y)) \tag{6.33}$$

which can be solved by standard integration methods. Obviously, the solution of

(6.33) together with $\lambda = G(y)$ satisfy (6.32) and after integration also (6.31b) (because the initial value satisfies $g(y_0) = 0$).

For the homotopy (6.30) the relation (6.32) becomes

$$(1 - \lambda)a_I(y) - \lambda a_{II}(y) = 0, \qquad \text{i.e.,} \qquad \lambda = \frac{a_I(y)}{a_I(y) + a_{II}(y)}, \qquad (6.34)$$

where $a_I(y)$ and $a_{II}(y)$ are given in (6.24).

Remark. Problem (6.31) is a "differential-algebraic system of index 2" and direct numerical methods are discussed in Chapter VI of Volume II. The instances where a_I or a_{II} change sign can again be computed by using a dense output and Algorithm 6.4.

Numerical Computation of Derivatives with Respect to Initial Values and Parameters

For the efficient computation of boundary value problems by a shooting technique as explained in Section I.15, we need to compute the derivatives of the solutions with respect to (the missing) initial values. Also, if we want to adjust unknown parameters from given data, say by a nonlinear least squares procedure, we have to compute the derivatives of the solutions with respect to parameters in the differential equation.

We shall restrict our discussion to the problem

$$y' = f(x, y, B), \qquad y(x_0) = y_0(B) \qquad (6.35)$$

where the right-hand side function and the initial values depend on a real parameter B. The generalization to more than one parameter is straightforward. There are several possibilities for computing the derivative $\partial y/\partial B$.

External differentiation. Denote the numerical solution, obtained by a variable step size code with a fixed tolerance, by $y_{Tol}(x_{\text{end}}, x_0, B)$. Then the most simple device is to approximate the derivative by a finite difference

$$\frac{1}{\Delta B} \Big(y_{Tol}(x_{\text{end}}, x_0, B + \Delta B) - y_{Tol}(x_{\text{end}}, x_0, B) \Big). \qquad (6.36)$$

However, due to the error control mechanism with its IF's and THEN's and step rejections, the function $y_{Tol}(x_{\text{end}}, x_0, B)$ is by no means a smooth function of the parameter B. Therefore, the errors of the two numerical results in (6.36) are not correlated, so that the error of (6.36) as an approximation to $\partial y/\partial B(x_{\text{end}}, x_0, B)$ is of size $\mathcal{O}(Tol/\Delta B) + \mathcal{O}(\Delta B)$, the second term coming from the discretization (6.36). This suggests taking for ΔB something like $\sqrt{Tol}$, and the error of (6.36) becomes of size $\mathcal{O}(\sqrt{Tol})$.

Internal differentiation. We know from Section I.14 that $\Psi = \partial y/\partial B$ is the solution of the variational equation

$$\Psi' = \frac{\partial f}{\partial y}(x, y, B)\Psi + \frac{\partial f}{\partial B}(x, y, B), \qquad \Psi(x_0) = \frac{\partial y_0}{\partial B}(B). \qquad (6.37)$$

Here y is the solution of (6.35). Hence, (6.35) and (6.37) together constitute a differential system for y and Ψ, which can be solved simultaneously by any code. If the partial derivatives $\partial f/\partial y$ and $\partial f/\partial B$ are available analytically, then the error of $\partial y/\partial B$, obtained by this procedure, is obviously of size Tol. This algorithm is equivalent to "internal differentiation" as introduced by Bock (1981).

If $\partial f/\partial y$ and $\partial f/\partial B$ are not available one can approximate them by finite differences so that (6.37) becomes

$$\Psi' = \frac{1}{\Delta B}\Big(f(x, y + \Delta B \cdot \Psi, B + \Delta B) - f(x, y, B)\Big). \qquad (6.38)$$

The solution of (6.38), when inserted into (6.37), gives raise to a defect of size $\mathcal{O}(\Delta B) + \mathcal{O}(eps/\Delta B)$, where eps is the precision of the computer (independent of Tol). By Theorem I.10.2, the difference of the solutions of (6.38) and (6.37) is of the same size. Choosing $\Delta B \approx \sqrt{eps}$ the error of the approximation to $\partial y/\partial B$, obtained by solving (6.35), (6.38), will be of order $Tol + \sqrt{eps}$, so that for $Tol \geq \sqrt{eps}$ the result is as precise as that obtained by integration of (6.37). Observe that external differentiation and the numerical solution of (6.35), (6.38) need about the same number of function evaluations.

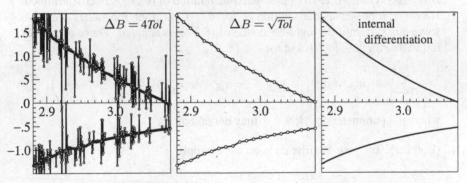

Fig. 6.5. Derivatives of the solution of (6.39) with respect to B

As an example we consider the Brusselator

$$\begin{aligned}
y_1' &= 1 + y_1^2 y_2 - (B+1)y_1 & y_1(0) &= 1.3 \\
y_2' &= By_1 - y_1^2 y_2 & y_2(0) &= B
\end{aligned} \qquad (6.39)$$

and compute $\partial y/\partial B$ at $x = 20$ for various B ranging from $B = 2.88$ to $B = 3.08$. We applied the code DOPRI5 with $Atol = Rtol = Tol = 10^{-4}$. The numerical result is displayed in Fig. 6.5. External differentiation has been applied, once with $\Delta B = \sqrt{Tol}$ and a second time with $\Delta B = 4 Tol$. This numerical example clearly demonstrates that internal differentiation is to be preferred.

Exercises

1. (Owren & Zennaro 1991, Carnicer 1991). The 4-stage 4th order methods of Section II.1 do not possess a dense output of order 4 (also if the numerical solution y_1 is included as 5th stage). Prove this statement.

2. Consider a Runge-Kutta method of order p and use Richardson extrapolation for step size control. Besides the numerical solution y_0, y_1, y_2 we consider the extrapolated values (see Section II.4)

$$\widehat{y}_1 = y_1 + \frac{y_2 - w}{(2^p - 1)2}, \qquad \widehat{y}_2 = y_2 + \frac{y_2 - w}{2^p - 1}$$

and do quintic polynomial interpolation based on $y_0, f(x_0, y_0), \widehat{y}_1, f(x_0 + h, y_1), \widehat{y}_2, f(x_0 + 2h, \widehat{y}_2)$. Prove that the resulting dense output formula is of order $p^* = \min(5, p+1)$.

Remark. It is not necessary to evaluate f at $\widehat{y}_1$.

3. Prove that the conditions (6.19), (6.18) and (5.20) imply (6.20).

Hint. The system (6.19) together with one relation of (6.20) is overdetermined. However, it possesses the solution b_i for $\theta = 1$. Further, the values $b_i c_i$ also solve this system if the right-hand side of (6.19a) is adapted. These properties imply that for $k \in \{4, 5\}$ and for $i \in \{1, 6, \ldots, 16\}$

$$\sum_{j=k+1}^{i-1} a_{ij} a_{jk} = \alpha a_{i4} + \beta a_{i5} + \gamma c_i a_{i4} + \delta c_i a_{i5} + \varepsilon \left(\sum_{j=1}^{i-1} a_{ij} c_j^5 - \frac{c_i^6}{6} \right),$$

where the parameters $\alpha, \beta, \gamma, \delta, \varepsilon$ may depend on k.

4. (Butcher). Try your favorite code on the example

$$\begin{aligned} y_1' &= f_1(y_1, y_2), & y_1(0) &= 1 \\ y_2' &= f_2(y_1, y_2), & y_2(0) &= 0 \end{aligned}$$

where f is defined as follows.

    If $(|y_1| > |y_2|)$ then
        $f_1 = 0, \quad f_2 = \text{sign}\,(y_1)$
    Else
        $f_2 = 0, \quad f_1 = -\text{sign}\,(y_2)$
    End If .

Compute $y_1(8), y_2(8)$. Show that the exact solution is periodic.

5. Do numerical computations for the problem $y' = f(y)$, $y(0) = 1$, $y(3) = ?$
   where

$$f(y) = \begin{cases} y^2 & \text{if } 0 \le y \le 2 \\ \left. \begin{array}{l} \text{a) } 1 \\ \text{b) } 4 \\ \text{c) } -4 + 4y \end{array} \right\} & \text{if } 2 < y \end{cases}$$

Remark. The correct answer would be (a) 4.5, (b) 12, (c) $\exp(10) + 1$.

6. Consider an s-stage Runge-Kutta method and denote by $\tilde{s}$ the number of distinct c_i. Prove that the order of any continuous extension is $\le \tilde{s}$.

   Hint. Let $q(x)$ be a polynomial of degree $\tilde{s}$ satisfying $q(c_i) = 0$ (for $i = 1, \ldots, s$) and investigate the expression $\sum_i b_i(\theta) q(c_i)$.

7. (Step size freeze). Consider the following algorithm for the computation of $\partial y / \partial B$: first compute numerically the solution of (6.35) and denote it by $y_h(x_{\text{end}}, B)$. At the same time memorize all the selected step sizes. This step size sequence is then used to solve (6.35) with B replaced by $B + \Delta B$. The result is denoted by $y_h(x_{\text{end}}, B + \Delta B)$. Then approximate the derivative $\partial y / \partial B$ by

$$\frac{1}{\Delta B}\Big(y_h(x_{\text{end}}, B + \Delta B) - y_h(x_{\text{end}}, B)\Big).$$

   Prove that this algorithm is equivalent to the solution of the system (6.35), (6.38), if only the components of y are considered for error control and step size selection.

   Remark. For large systems this algorithm needs less storage requirements than internal differentiation, in particular if the derivative with respect to several parameters is computed.

8. (Taubert 1976). Show that for the discontinuous problem (6.23) the Euler polygons converge to Filippov's solution (6.30), (6.31).

   Hint. The difference quotient of a piece of the Euler polygon lies in the convex hull of points $f_I(y)$ and $f_{II}(y)$.

   Remark. This result can either be interpreted as pleading for myriads of Euler steps, or as a motivation for the homotopy (6.30).

9. Another motivation for formula (6.30): suppose that a small particle of radius ε is transported in a possibly discontinuous flow. Then its movement might be described by the mean of f

$$f_\varepsilon(y) = \int_{B_\varepsilon(y)} f(z)\, dz \Big/ \int_{B_\varepsilon(y)} dz$$

   which is continuous in y. Show that the solution of $y'_\varepsilon = f_\varepsilon(y)$ becomes, for $\varepsilon \to 0$, that of (6.33) and (6.34).

II.7 Implicit Runge-Kutta Methods

It has been traditional to consider only explicit processes
(J.C. Butcher 1964a)

The high speed computing machines make it possible to enjoy
the advantage of intricate methods
(P.C. Hammer & J.W. Hollingsworth 1955)

The first *implicit* RK methods were used by Cauchy (1824) for the sake of — you
have guessed correctly — error estimation (Méthodes diverses qui peuvent être
employées au Calcul numérique ...; see Exercise 5). Cauchy inserted the mean
value theorem into the integral studied in Sections I.8 and II.1,

$$y(x_1) = y(x_0) + \int_{x_0}^{x_1} f(x, y(x))\, dx, \tag{7.1}$$

to obtain

$$y_1 = y_0 + hf(x_0 + \theta h, y_0 + \Theta(y_1 - y_0)) \tag{7.2}$$

with $0 \leq \theta, \Theta \leq 1$ (the "θ-method"). The extreme cases are $\theta = \Theta = 0$ (the explicit
Euler method) and $\theta = \Theta = 1$

$$y_1 = y_0 + hf(x_1, y_1), \tag{7.3}$$

which we call the *implicit* or *backward Euler method*.

For the sake of more efficient numerical processes, we apply, as we did in
Section II.1, the midpoint rule $(\theta = \Theta = 1/2)$ and obtain from (7.2) by setting
$k_1 = (y_1 - y_0)/h$:

$$\begin{aligned} k_1 &= f\left(x_0 + \frac{h}{2},\, y_0 + \frac{h}{2} k_1\right), \\ y_1 &= y_0 + hk_1. \end{aligned} \tag{7.4}$$

This method is called the *implicit midpoint rule*.

Still another possibility is to approximate (7.1) by the *trapezoidal rule* and to
obtain

$$y_1 = y_0 + \frac{h}{2}\left(f(x_0, y_0) + f(x_1, y_1)\right). \tag{7.5}$$

Let us also look at the Radau scheme

$$\begin{aligned} y(x_1) - y(x_0) &= \int_{x_0}^{x_0+h} f(x, y(x))\, dx \\ &\approx \frac{h}{4}\left(f(x_0, y_0) + 3f\left(x_0 + \frac{2}{3} h, y(x_0 + \frac{2}{3} h)\right)\right). \end{aligned}$$

Here we need to approximate $y(x_0 + 2h/3)$. One idea would be the use of quadratic interpolation based on y_0, y_0' and $y(x_1)$,

$$y\left(x_0 + \frac{2}{3}h\right) \approx \frac{5}{9}y_0 + \frac{4}{9}y(x_1) + \frac{2}{9}hf(x_0, y_0).$$

The resulting method, given by Hammer & Hollingsworth (1955), is

$$
\begin{aligned}
k_1 &= f(x_0, y_0) \\
k_2 &= f\left(x_0 + \frac{2}{3}h, y_0 + \frac{h}{3}(k_1 + k_2)\right) \\
y_1 &= y_0 + \frac{h}{4}(k_1 + 3k_2).
\end{aligned}
\tag{7.6}
$$

All these schemes are of the form (1.8) if the summations are extended up to "s".

Definition 7.1. Let b_i, a_{ij} $(i, j = 1, \ldots, s)$ be real numbers and let c_i be defined by (1.9). The method

$$
\begin{aligned}
k_i &= f\left(x_0 + c_i h, y_0 + h \sum_{j=1}^{s} a_{ij} k_j\right) \qquad i = 1, \ldots, s \\
y_1 &= y_0 + h \sum_{i=1}^{s} b_i k_i
\end{aligned}
\tag{7.7}
$$

is called an *s-stage Runge-Kutta method*. When $a_{ij} = 0$ for $i \leq j$ we have an explicit (ERK) method. If $a_{ij} = 0$ for $i < j$ and at least one $a_{ii} \neq 0$, we have a *diagonal implicit Runge-Kutta method* (DIRK). If in addition all diagonal elements are identical ($a_{ii} = \gamma$ for $i = 1, \ldots, s$), we speak of a *singly diagonal* implicit (SDIRK) method. In all other cases we speak of an *implicit* Runge-Kutta method (IRK).

The tableau of coefficients used above for ERK-methods is obviously extended to include all the other non-zero a_{ij}'s above the diagonal. For methods (7.3), (7.4) and (7.6) it is given in Table 7.1.

Renewed interest in implicit Runge-Kutta methods arose in connection with *stiff* differential equations (see Volume II).

Table 7.1. Implicit Runge-Kutta methods

1	1		$1/2$	$1/2$		0	0	0
	1			1		$2/3$	$1/3$	$1/3$
							$1/4$	$3/4$

Implicit Euler   Implicit midpoint rule   Hammer & Hollingsworth

Existence of a Numerical Solution

For implicit methods, the k_i's can no longer be evaluated successively, since (7.7) constitutes a system of implicit equations for the determination of k_i. For DIRK-methods we have a sequence of implicit equations of dimension n for k_1, then for k_2, etc. For fully implicit methods $s \cdot n$ unknowns (k_i, $i = 1, \ldots, s$; each of dimension n) have to be determined simultaneously, which still increases the difficulty. A natural question is therefore (the reason for which the original version of Butcher (1964a) was returned by the editors): do equations (7.7) possess a solution at all?

Theorem 7.2. *Let* $f : \mathbb{R} \times \mathbb{R}^n \to \mathbb{R}^n$ *be continuous and satisfy a Lipschitz condition with constant L (with respect to y). If*

$$h < \frac{1}{L \, \max_i \sum_j |a_{ij}|} \tag{7.8}$$

there exists a unique solution of (7.7), which can be obtained by iteration. If $f(x, y)$ *is p times continuously differentiable, the functions* k_i *(as functions of h) are also in* C^p .

Proof. We prove the existence by iteration ("... on la résoudra facilement par des approximations successives ...", Cauchy 1824)

$$k_i^{(m+1)} = f\left(x_0 + c_i h, y_0 + h \sum_{j=1}^{s} a_{ij} k_j^{(m)}\right).$$

We define $K \in \mathbb{R}^{sn}$ as $K = (k_1, \ldots, k_s)^T$ and use the norm $\|K\| = \max_i(\|k_i\|)$. Then (7.7) can be written as $K = F(K)$ where

$$F_i(K) = f\left(x_0 + c_i h, y_0 + h \sum_{j=1}^{s} a_{ij} k_j\right), \quad i = 1, \ldots, s.$$

The Lipschitz condition and a repeated use of the triangle inequality then show that

$$\|F(K_1) - F(K_2)\| \leq hL \max_{i=1,\ldots,s} \sum_{j=1}^{s} |a_{ij}| \cdot \|K_1 - K_2\|$$

which from (7.8) is a contraction. The contraction mapping principle then ensures the existence and uniqueness of the solution and the convergence of the fixed-point iteration.

The differentiability result is ensured by the Implicit Function Theorem of classical analysis: (7.7) is written as $\Phi(h, K) = K - F(K) = 0$. The matrix of partial derivatives $\partial \Phi / \partial K$ for $h = 0$ is the identity matrix and therefore the solution of $\Phi(h, K) = 0$, which for $h = 0$ is $k_i = f(x_0, y_0)$, is continuously differentiable in a neighbourhood of $h = 0$. $\qquad\square$

If the assumptions on f in Theorem 7.2 are only satisfied in a neighbourhood of the initial value, then further restrictions on h are needed in order that the argument of f remains in this neighbourhood. Uniqueness is then only of local nature.

The step size restriction (7.8) becomes useless for stiff problems (L large). We return to this question in Vol. II, Sections IV.8 and IV.14.

The definition of *order* is the same as for explicit methods and the order conditions are derived in precisely the same way as in Section II.2.

Example 7.3. Let us study implicit two-stage methods of order 3: the order conditions become (see Theorem 2.1)

$$b_1 + b_2 = 1, \qquad b_1 c_1 + b_2 c_2 = \frac{1}{2}, \qquad b_1 c_1^2 + b_2 c_2^2 = \frac{1}{3}$$

$$b_1(a_{11}c_1 + a_{12}c_2) + b_2(a_{21}c_1 + a_{22}c_2) = \frac{1}{6}. \tag{7.9}$$

The first three equations imply the following orthogonality relation (from the theory of Gaussian integration):

$$\int_0^1 (x - c_1)(x - c_2)\, dx = 0, \quad \text{i.e., } c_2 = \frac{2 - 3c_1}{3 - 6c_1} \quad (c_1 \neq 1/2) \tag{7.10}$$

and

$$b_1 = \frac{c_2 - 1/2}{c_2 - c_1}, \qquad b_2 = \frac{c_1 - 1/2}{c_1 - c_2}.$$

In the fourth equation we insert $a_{21} = c_2 - a_{22}$, $a_{11} = c_1 - a_{12}$ and consider a_{12} and c_1 as free parameters. This gives

$$a_{22} = \frac{1/6 - b_1 a_{12}(c_2 - c_1) - c_1/2}{b_2(c_2 - c_1)}. \tag{7.11}$$

For $a_{12} = 0$ we obtain a one-parameter family of DIRK-methods of order 3. An SDIRK-method is obtained if we still require $a_{11} = a_{22}$ (Nørsett 1974b, Crouzeix 1975, see Table 7.2). For order 4 we have 4 additional conditions, with only two free parameters left. Nevertheless there exists a unique solution (see Table 7.3).

Table 7.2. SDIRK method, order 3

γ	γ	0
$1 - \gamma$	$1 - 2\gamma$	γ
	1/2	1/2

$$\gamma = \frac{3 \pm \sqrt{3}}{6}$$

Table 7.3. Hammer & Hollingsworth, order 4

$\dfrac{1}{2} - \dfrac{\sqrt{3}}{6}$	$\dfrac{1}{4}$	$\dfrac{1}{4} - \dfrac{\sqrt{3}}{6}$
$\dfrac{1}{2} + \dfrac{\sqrt{3}}{6}$	$\dfrac{1}{4} + \dfrac{\sqrt{3}}{6}$	$\dfrac{1}{4}$
	1/2	1/2

The Methods of Kuntzmann and Butcher of Order 2s

It is clear that formula (7.4) and the method of Table 7.3 extend the one-point and two-point Gaussian quadrature formulas, respectively. Kuntzmann (1961) (see Ceschino & Kuntzmann 1963, p. 106) and Butcher (1964a) then discovered that for all s there exist IRK-methods of order $2s$. The main tools of proof are the following *simplifying assumptions*

$$B(p): \quad \sum_{i=1}^{s} b_i c_i^{q-1} = \frac{1}{q} \qquad q = 1, \ldots, p,$$

$$C(\eta): \quad \sum_{j=1}^{s} a_{ij} c_j^{q-1} = \frac{c_i^q}{q} \qquad i = 1, \ldots, s, \ q = 1, \ldots, \eta,$$

$$D(\zeta): \quad \sum_{i=1}^{s} b_i c_i^{q-1} a_{ij} = \frac{b_j}{q}(1 - c_j^q) \qquad j = 1, \ldots, s, \ q = 1, \ldots, \zeta.$$

Condition $B(p)$ simply means that the quadrature formula (b_i, c_i) is of order p or, equivalently, that the order conditions (2.21) are satisfied for the bushy trees $[\tau, \ldots, \tau]$ up to order p.

The assumption $C(\eta)$ implies that the pairs of trees in Fig. 7.1 give identical order conditions for $q \leq \eta$. In contrast to explicit Runge-Kutta methods (see (5.7) and (5.15)) there is no need to require conditions such as $b_2 = 0$ (see (5.8)), because $\sum_j a_{ij} c_j^{q-1} = c_i^q/q$ is valid for all i.

The assumption $D(\zeta)$ is an extension of (1.12). It means that the order condition of the left-hand tree of Fig. 7.2 is implied by those of the two right-hand trees if $q \leq \zeta$.

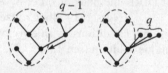   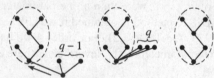

Fig. 7.1. Reduction with $C(q)$ **Fig. 7.2.** Reduction with $D(q)$

Theorem 7.4 (Butcher 1964a). *If $B(p)$, $C(\eta)$ and $D(\zeta)$ are satisfied with $p \leq 2\eta + 2$ and $p \leq \zeta + \eta + 1$, then the method is of order p.*

Proof. The above reduction by $C(\eta)$ implies that it is sufficient to consider trees $t = [t_1, \ldots, t_m]$ of order $\leq p$, where the subtrees $t_1, \ldots, t_m$ are either equal to τ or of order $\geq \eta + 1$. Since $p \leq 2\eta + 2$ either all subtrees are equal to τ or there is exactly one subtree different from τ. In the second case the number of τ's is $\leq \zeta - 1$ by $p \leq \eta + \zeta + 1$ and the reduction by $D(\zeta)$ can be applied. Therefore, after all these reductions, only the bushy trees are left and they are satisfied by $B(p)$.   □

To obtain the formulas of order $2s$, Butcher assumed $B(2s)$ (i.e., the c_i and b_i are the coefficients of the Gaussian quadrature formula) and $C(s)$. This implies $D(s)$ (see Exercise 7) so that Theorem 7.4 can be applied with $p = 2s$, $\eta = s$ and $\zeta = s$. Hence the method, obtained in this way, is of order $2s$. For $s = 3$ and 4 the coefficients are given in Tables 7.4 and 7.5. They can still be expressed by radicals for $s = 5$ and are given in Butcher (1964a), p. 57.

Impressive numerical results from celestial mechanics for these methods were first reported in the thesis of D. Sommer (see Sommer 1965).

Table 7.4. Kuntzmann & Butcher method, order 6

$\dfrac{1}{2} - \dfrac{\sqrt{15}}{10}$	$\dfrac{5}{36}$	$\dfrac{2}{9} - \dfrac{\sqrt{15}}{15}$	$\dfrac{5}{36} - \dfrac{\sqrt{15}}{30}$
$\dfrac{1}{2}$	$\dfrac{5}{36} + \dfrac{\sqrt{15}}{24}$	$\dfrac{2}{9}$	$\dfrac{5}{36} - \dfrac{\sqrt{15}}{24}$
$\dfrac{1}{2} + \dfrac{\sqrt{15}}{10}$	$\dfrac{5}{36} + \dfrac{\sqrt{15}}{30}$	$\dfrac{2}{9} + \dfrac{\sqrt{15}}{15}$	$\dfrac{5}{36}$
	$\dfrac{5}{18}$	$\dfrac{4}{9}$	$\dfrac{5}{18}$

Table 7.5. Kuntzmann & Butcher method, order 8

$\frac{1}{2} - \omega_2$	ω_1	$\omega_1' - \omega_3 + \omega_4'$	$\omega_1' - \omega_3 - \omega_4'$	$\omega_1 - \omega_5$
$\frac{1}{2} - \omega_2'$	$\omega_1 - \omega_3' + \omega_4$	ω_1'	$\omega_1' - \omega_5'$	$\omega_1 - \omega_3' - \omega_4$
$\frac{1}{2} + \omega_2'$	$\omega_1 + \omega_3' + \omega_4$	$\omega_1' + \omega_5'$	ω_1'	$\omega_1 + \omega_3' - \omega_4$
$\frac{1}{2} + \omega_2$	$\omega_1 + \omega_5$	$\omega_1' + \omega_3 + \omega_4'$	$\omega_1' + \omega_3 - \omega_4'$	ω_1
	$2\omega_1$	$2\omega_1'$	$2\omega_1'$	$2\omega_1$

$$\omega_1 = \frac{1}{8} - \frac{\sqrt{30}}{144}, \qquad \omega_1' = \frac{1}{8} + \frac{\sqrt{30}}{144},$$

$$\omega_2 = \frac{1}{2}\sqrt{\frac{15 + 2\sqrt{30}}{35}}, \qquad \omega_2' = \frac{1}{2}\sqrt{\frac{15 - 2\sqrt{30}}{35}},$$

$$\omega_3 = \omega_2\left(\frac{1}{6} + \frac{\sqrt{30}}{24}\right), \qquad \omega_3' = \omega_2'\left(\frac{1}{6} - \frac{\sqrt{30}}{24}\right),$$

$$\omega_4 = \omega_2\left(\frac{1}{21} + \frac{5\sqrt{30}}{168}\right), \qquad \omega_4' = \omega_2'\left(\frac{1}{21} - \frac{5\sqrt{30}}{168}\right),$$

$$\omega_5 = \omega_2 - 2\omega_3, \qquad \omega_5' = \omega_2' - 2\omega_3'.$$

An important interpretation of the assumption $C(\eta)$ is the following:

Lemma 7.5. *The assumption* $C(\eta)$ *implies that the internal stages*

$$g_i = y_0 + h \sum_{j=1}^{s} a_{ij} k_j, \qquad k_j = f(x_0 + c_j h, g_j) \qquad (7.12)$$

satisfy for $i = 1, \ldots, s$

$$g_i - y(x_0 + c_i h) = \mathcal{O}(h^{\eta+1}). \qquad (7.13)$$

Proof. Because of $C(\eta)$ the exact solution satisfies (Taylor expansion)

$$y(x_0 + c_i h) = y_0 + h \sum_{j=1}^{s} a_{ij} y'(x_0 + c_j h) + \mathcal{O}(h^{\eta+1}). \qquad (7.14)$$

Subtracting (7.14) from (7.12) yields

$$g_i - y(x_0 + c_i h) = h \sum_{j=1}^{s} a_{ij} \Big(f\big(x_0 + c_j h, g_j\big) - f\big(x_0 + c_j h, y(x_0 + c_j h)\big) \Big)$$
$$+ \mathcal{O}(h^{\eta+1})$$

and Lipschitz continuity of f proves (7.13). $\qquad\qquad\square$

IRK Methods Based on Lobatto Quadrature

Lobatto quadrature rules (Lobatto 1852, Radau 1880, p. 307) modify the idea of Gaussian quadrature by requiring that the first and the last node coincide with the interval ends, i.e., $c_1 = 0$, $c_s = 1$. These points are easier to handle and, in a step-by-step procedure, can be used twice. The remaining c's are then adjusted optimally, i.e., as the zeros of the Jacobi orthogonal polynomial $P_{s-2}^{(1,1)}(x)$ or of $P'_{s-1}(x)$ (see e.g., Abramowitz & Stegun 1964, 25.4.32 for the interval [-1,1]) and lead to formulas of order $2s - 2$.

J.C. Butcher (1964a, p. 51, 1964c) then found that Lobatto quadrature rules can be extended to IRK-methods whose coefficient matrix is zero in the first line and the last column. The first and the last stage then become *explicit* and the number of implicit stages reduces to $s - 2$. The methods are characterized by $B(2s - 2)$ and $C(s - 1)$. As in Exercise 7 this implies $D(s - 1)$ so that by Theorem 7.4 the method is of order $2s - 2$. For $s = 3$ and 4, the coefficients are given in Table 7.6.

We shall see in Volume II (Section IV.3, Table 3.1) that these methods, although preferable as concerns the relation between order and implicit stages, are not sufficiently stable for stiff differential equations.

Table 7.6. Butcher's Lobatto formulas of orders 4 and 6

0	0	0	0
0	0	0	0
$\dfrac{5-\sqrt5}{10}$	$\dfrac{5+\sqrt5}{60}$	$\dfrac{1}{6}$	$\dfrac{15-7\sqrt5}{60}$	0
$\dfrac{5+\sqrt5}{10}$	$\dfrac{5-\sqrt5}{60}$	$\dfrac{15+7\sqrt5}{60}$	$\dfrac{1}{6}$	0
1	$\dfrac{1}{6}$	$\dfrac{5-\sqrt5}{12}$	$\dfrac{5+\sqrt5}{12}$	0

Left tableau:

0	0	0	0
$\dfrac{1}{2}$	$\dfrac{1}{4}$	$\dfrac{1}{4}$	0
1	0	1	0
	$\dfrac{1}{6}$	$\dfrac{2}{3}$	$\dfrac{1}{6}$

Right tableau (bottom row): $\dfrac{1}{12}\quad \dfrac{5}{12}\quad \dfrac{5}{12}\quad \dfrac{1}{12}$

Collocation Methods

> Es ist erstaunlich dass die Methode trotz ihrer Primitivität und der geringen Rechenarbeit in vielen Fällen ... sogar gute Ergebnisse liefert.
> (L. Collatz 1951)

> Nous allons montrer l'équivalence de notre définition avec la définition traditionnelle de certaines formules de Runge Kutta implicites.
> (Guillou & Soulé 1969)

The concept of collocation is old and universal in numerical analysis (see e.g., pp. 28,29,32,181,411,453,483,495 of Collatz 1960, Frazer, Jones & Skan 1937). For ordinary differential equations it consists in searching for a polynomial of degree s whose derivative coincides ("co-locates") at s given points with the vector field of the differential equation (Guillou & Soulé 1969, Wright 1970). Still another approach is to combine Galerkin's method with numerical quadrature (see Hulme 1972).

Definition 7.6. For s a positive integer and $c_1,\dots,c_s$ distinct real numbers (typically between 0 and 1), the corresponding *collocation polynomial* $u(x)$ of degree s is defined by

$$u(x_0) = y_0 \qquad \text{(initial value)} \tag{7.15a}$$
$$u'(x_0+c_ih) = f\big(x_0+c_ih, u(x_0+c_ih)\big), \qquad i=1,\dots,s. \tag{7.15b}$$

The numerical solution is then given by

$$y_1 = u(x_0+h). \tag{7.15c}$$

If some of the c_i coincide, the collocation condition (7.15b) will contain higher derivatives and lead to multi-derivative methods (see Section II.13). Accordingly, for the moment, we suppose them all distinct.

Theorem 7.7 (Guillou & Soulé 1969, Wright 1970). *The collocation method (7.15) is equivalent to the s-stage IRK-method (7.7) with coefficients*

$$a_{ij} = \int_0^{c_i} \ell_j(t)\, dt, \qquad b_j = \int_0^1 \ell_j(t)\, dt \qquad i,j = 1,\ldots,s, \qquad (7.16)$$

where the $\ell_j(t)$ are the Lagrange polynomials

$$\ell_j(t) = \prod_{k \neq j} \frac{(t - c_k)}{(c_j - c_k)}. \qquad (7.17)$$

Proof. Put $u'(x_0 + c_i h) = k_i$, so that

$$u'(x_0 + th) = \sum_{j=1}^s k_j \cdot \ell_j(t) \qquad \text{(Lagrange)}.$$

Then integrate

$$u(x_0 + c_i h) = y_0 + h \int_0^{c_i} u'(x_0 + th)\, dt \qquad (7.18)$$

and insert into (7.15b) together with (7.16). The IRK-method (7.7) then comes out. $\qquad \square$

As a consequence of this result, the existence and uniqueness of the collocation polynomial (for sufficiently small h) follows from Theorem 7.2.

Theorem 7.8. *An implicit Runge-Kutta method with all c_i different and of order at least s is a collocation method iff $C(s)$ is true.*

Proof. $C(s)$ determines the a_{ij} uniquely. We write it as

$$\sum_{j=1}^s a_{ij} p(c_j) = \int_0^{c_i} p(t)\, dt \qquad (7.19)$$

for all polynomials p of degree $\leq s - 1$. The a_{ij} given by (7.16) satisfy this relation, because (7.16) inserted into (7.19) is just the Lagrange interpolation formula. $\qquad \square$

Theorem 7.9. *Let $M(t) = \prod_{i=1}^s (t - c_i)$ and suppose that M is orthogonal to polynomials of degree $r - 1$,*

$$\int_0^1 M(t) t^{q-1}\, dt = 0, \qquad q = 1,\ldots,r, \qquad (7.20)$$

then method (7.15) has order $p = s + r$.

Proof. The following proof uses the Gröbner & Alekseev Formula, which gives nice insight in the background of the result. An alternative proof is indicated in Exercise 7 below. One can also linearize the equation, apply the *linear* variation-of-constants formula and estimate the error (Guillou & Soulé 1969).

The orthogonality condition (7.20) means that the quadrature formula

$$\int_{x_0}^{x_0+h} g(t)\,dt = h\sum_{j=1}^{s} b_j g(x_0 + c_j h) + err(g) \qquad (7.21)$$

is of order $s + r = p$, and its error is bounded by

$$|err(g)| \leq Ch^{p+1}\cdot\max|g^{(p)}(x)|. \qquad (7.22)$$

The principal idea of the proof is now the following: we consider

$$u'(x) = f\big(x, u(x)\big) + \big(u'(x) - f(x, u(x))\big)$$

as a perturbation of

$$y'(x) = f\big(x, y(x)\big)$$

and integrate the Gröbner & Alekseev Formula (I.14.18) with the quadrature formula (7.21). Due to (7.15b), the result is identically zero, since at the collocation points the defect is zero. Thus from (7.21) and (7.22)

$$\|y(x_0 + h) - u(x_0 + h)\| = \|err(g)\| \leq C\cdot h^{p+1}\cdot\max_{x_0 \leq t \leq x_0+h}\|g^{(p)}(t)\|, \quad (7.23)$$

where

$$g(t) = \frac{\partial y}{\partial y_0}\big(x, t, u(t)\big)\cdot\big(u'(t) - f(t, u(t))\big),$$

and we see that the local error behaves like $\mathcal{O}(h^{p+1})$.

There remains, however, a small technical detail: to show that the derivatives of $g(t)$ remain bounded for $h \to 0$. These derivatives contain partial derivatives of $f(t, y)$ and derivatives of $u(t)$. We shall see in the next theorem that these derivatives remain bounded for $h \to 0$. $\qquad\square$

Theorem 7.10. *The collocation polynomial $u(x)$ gives rise to a continuous IRK method of order s, i.e., for all $x_0 \leq x \leq x_0 + h$ we have*

$$\|y(x) - u(x)\| \leq C\cdot h^{s+1}. \qquad (7.24)$$

Moreover, for the derivatives of $u(x)$ we have

$$\|y^{(k)}(x) - u^{(k)}(x)\| \leq C\cdot h^{s+1-k} \qquad k = 0,\ldots,s. \qquad (7.25)$$

Proof. The exact solution $y(x)$ satisfies the collocation condition everywhere, hence *also* at the points $x_0 + c_i h$. So, in exactly the same way as in the proof of Theorem 7.7, we apply the Lagrange interpolation formula to $y'(x)$:

$$y'(x_0 + th) = \sum_{j=1}^{s} f(x_0 + c_j h, y(x_0 + c_j h)) \ell_j(t) + h^s R(t, h)$$

where $R(t, h)$ is a smooth function of both variables. Integration and subtraction from (7.18) gives

$$y(x_0 + th) - u(x_0 + th) = h \sum_{j=1}^{s} \Delta f_j \cdot \int_0^t \ell_j(\tau) \, d\tau + h^{s+1} \int_0^t R(\tau, h) \, d\tau, \quad (7.26)$$

where

$$\Delta f_j = f(x_0 + c_j h, y(x_0 + c_j h)) - f(x_0 + c_j h, u(x_0 + c_j h)).$$

The kth derivative of (7.26) with respect to t is

$$h^k \left(y^{(k)}(x_0 + th) - u^{(k)}(x_0 + th) \right) = h \sum_{j=1}^{s} \Delta f_j \cdot \ell_j^{(k-1)}(t) + h^{s+1} \frac{\partial^{k-1} R}{\partial t^{k-1}}(t, h),$$

so that the result follows from the boundedness of the derivatives of $R(t, h)$ and from $\Delta f_j = \mathcal{O}(h^{s+1})$ which is a consequence of Lemma 7.5.   □

Remark. Only *some* IRK methods are collocation methods. An extension of the collocation idea ("Perturbed Collocation", see Nørsett & Wanner 1981) applies to *all* IRK methods.

Exercises

1. Compute the one-point collocation method $(s = 1)$ with $c_i = \theta$ and compare with (7.2). Determine its order in dependence of θ.

2. Compute all collocation methods with $s = 2$ of order 2 in dependence of c_1 and c_2.

3. Specify in the method of Exercise 2 $c_1 = 1/3$, $c_2 = 1$ as well as $c_1 = 0$, $c_2 = 2/3$. Determine the orders of the obtained methods and explain.

4. Interpret the implicit midpoint rule (7.4) and the explicit Euler method as collocation methods. Is method (7.5) a collocation method? Method (7.6)?

5. (Cauchy 1824). Find from equation (7.2) conditions for the function $f(x, y)$ such that for scalar differential equations

$$y_1(\text{explicit Euler}) \geq y(x_1) \geq y_1(\text{implicit Euler}).$$

Compute five steps with $h = 0.2$ with both methods to obtain upper and lower bounds for $y(1)$, the solution of

$$y' = \cos \frac{x+y}{5}, \qquad y(0) = 0.$$

Cauchy's result: $0.9659 \le y(1) \le 0.9810$. For one single step with $h = 1$ he obtained $0.926 \le y(1) \le 1$.

Compute the exact solution by elementary integration.

6. Determine the orders of the methods of Table 7.7. Generalize to arbitrary s (Ehle 1968).

   Hint. Use Theorems 7.8 and 7.9.

Table 7.7. Methods of Ehle

Radau IIA, order 5

$\dfrac{4-\sqrt{6}}{10}$	$\dfrac{88-7\sqrt{6}}{360}$	$\dfrac{296-169\sqrt{6}}{1800}$	$\dfrac{-2+3\sqrt{6}}{225}$
$\dfrac{4+\sqrt{6}}{10}$	$\dfrac{296+169\sqrt{6}}{1800}$	$\dfrac{88+7\sqrt{6}}{360}$	$\dfrac{-2-3\sqrt{6}}{225}$
1	$\dfrac{16-\sqrt{6}}{36}$	$\dfrac{16+\sqrt{6}}{36}$	$\dfrac{1}{9}$
	$\dfrac{16-\sqrt{6}}{36}$	$\dfrac{16+\sqrt{6}}{36}$	$\dfrac{1}{9}$

Lobatto IIIA, order 4

0	0	0	0
$\dfrac{1}{2}$	$\dfrac{5}{24}$	$\dfrac{1}{3}$	$-\dfrac{1}{24}$
1	$\dfrac{1}{6}$	$\dfrac{2}{3}$	$\dfrac{1}{6}$
	$\dfrac{1}{6}$	$\dfrac{2}{3}$	$\dfrac{1}{6}$

7. (Butcher 1964a). Give an algebraic proof of Theorem 7.9.

   Hint. From Theorem 7.8 we have $C(s)$.

   Next the condition $B(p)$ with $p = s + r$ (theory of Gaussian quadrature formulas) implies $D(r)$. To see this, multiply the two vectors $u_j = \sum_i b_i c_i^{q-1} a_{ij}$ and $v_j = b_j(1 - c_j^q)/q$ $(j = 1, \dots, s)$ by the Vandermonde matrix

$$V = \begin{pmatrix} 1 & 1 & \cdots & 1 \\ c_1 & c_2 & \cdots & c_s \\ \vdots & \vdots & & \vdots \\ c_1^{s-1} & c_2^{s-1} & \cdots & c_s^{s-1} \end{pmatrix}.$$

   Finally apply Theorem 7.4.

II.8 Asymptotic Expansion of the Global Error

Mein Verzicht auf das Restglied war leichtsinnig ...

(W. Romberg 1979)

Our next goal will be to perfect Richardson's extrapolation method (see Section II.4) by doing *repeated* extrapolation and eliminating more and more terms Ch^{p+k} of the error. A sound theoretical basis for this procedure is given by the study of the asymptotic behaviour of the global error. For problems of the type $y' = f(x)$, which lead to integration, the answer is given by the Euler-Maclaurin formula and has been exploited by Romberg (1955) and his successors. The first rigorous treatments for differential equations are due to Henrici (1962) and Gragg (1964) (see also Stetter 1973). We shall follow here the successive elimination of the error terms given by Hairer & Lubich (1984), which also generalizes to multistep methods.

Suppose we have a one-step method which we write, in Henrici's notation, as

$$y_{n+1} = y_n + h\Phi(x_n, y_n, h). \tag{8.1}$$

If the method is of order p, it possesses at each point of the solution $y(x)$ a *local error* of the form

$$
\begin{aligned}
y(x+h) - y(x) - h\Phi(x, y(x), h) = \\
d_{p+1}(x)h^{p+1} + \ldots + d_{N+1}(x)h^{N+1} + \mathcal{O}(h^{N+2})
\end{aligned}
\tag{8.2}
$$

whenever the differential equation is sufficiently differentiable. For Runge-Kutta methods these error terms were computed in Section II.2 (see also Theorem 3.2).

The Global Error

Let us now set $y_n =: y_h(x)$ for the numerical solution at $x = x_0 + nh$. We then know from Theorem 3.6 that the global error behaves like h^p. We shall search for a function $e_p(x)$ such that

$$y(x) - y_h(x) = e_p(x)h^p + o(h^p). \tag{8.3}$$

The idea is to consider

$$y_h(x) + e_p(x)h^p =: \widehat{y}_h(x) \tag{8.4a}$$

as the numerical solution of a new method

$$\widehat{y}_{n+1} = \widehat{y}_n + h\widehat{\Phi}(x_n, \widehat{y}_n, h). \tag{8.4b}$$

By comparison with (8.1), we see that the increment function for the new method is

$$\widehat{\Phi}(x, \widehat{y}, h) = \Phi\big(x, \widehat{y} - e_p(x)h^p, h\big) + \big(e_p(x+h) - e_p(x)\big)h^{p-1}. \tag{8.5}$$

Our task is to find a function $e_p(x)$, with $e_p(x_0) = 0$, such that the method with increment function $\widehat{\Phi}$ is of order $p+1$.

Expanding the local error of the one-step method $\widehat{\Phi}$ into powers of h we obtain

$$y(x+h) - y(x) - h\widehat{\Phi}(x, y(x), h)$$
$$= \Big(d_{p+1}(x) + \frac{\partial f}{\partial y}\big(x, y(x)\big)e_p(x) - e'_p(x)\Big)h^{p+1} + \mathcal{O}(h^{p+2}) \tag{8.6}$$

where we have used

$$\frac{\partial \Phi}{\partial y}(x, y, 0) = \frac{\partial f}{\partial y}(x, y). \tag{8.7}$$

The term in h^{p+1} vanishes if $e_p(x)$ is defined as the solution of

$$e'_p(x) = \frac{\partial f}{\partial y}\big(x, y(x)\big)e_p(x) + d_{p+1}(x), \qquad e_p(x_0) = 0. \tag{8.8}$$

By Theorem 3.6, applied to the method $\widehat{\Phi}$, we now have

$$y(x) - y_h(x) = e_p(x)h^p + \mathcal{O}(h^{p+1}) \tag{8.9}$$

and the first term of the desired asymptotic expansion has been determined.

We now repeat the procedure with the method with increment function $\widehat{\Phi}$. It is of order $p+1$ and again satisfies condition (8.7). The final result of this procedure is the following

Theorem 8.1 (Gragg 1964). *Suppose that a given method with sufficiently smooth increment function Φ satisfies the consistency condition $\Phi(x, y, 0) = f(x, y)$ and possesses an expansion (8.2) for the local error. Then the global error has an asymptotic expansion of the form*

$$y(x) - y_h(x) = e_p(x)h^p + \ldots + e_N(x)h^N + E_h(x)h^{N+1} \tag{8.10}$$

where the $e_j(x)$ are solutions of inhomogeneous differential equations of the form (8.8) with $e_j(x_0) = 0$ and $E_h(x)$ is bounded for $x_0 \leq x \leq x_{\mathrm{end}}$ and $0 \leq h \leq h_0$.
□

The differentiability properties of the $e_j(x)$ depend on those of f and Φ (see (8.8) and (8.2)). The expansion (8.10) will be the theoretical basis for all discussions of extrapolation methods.

Examples. 1. For the equation $y' = y$ and Euler's method we have with $h = 1/n$ and $x = 1$, using the binomial theorem,

$$y_h(1) = \left(1 + \frac{1}{n}\right)^n = 1 + 1 + \left(1 - \frac{1}{n}\right)\frac{1}{2!} + \left(1 - \frac{1}{n}\right)\left(1 - \frac{2}{n}\right)\frac{1}{3!} + \ldots$$

By multiplying out, this gives

$$y(1) - y_h(1) = -\sum_{i=1}^{\infty} h^i \sum_{j=1}^{\infty} \frac{S_{i+j}^{(j)}}{(i+j)!} = 1.359h - 1.246h^2 \pm \ldots$$

where the $S_i^{(j)}$ are the Stirling numbers of the first kind (1730, see Abramowitz & Stegun 1964, Section 24.1.3). This is, of course, the Taylor series for the function

$$e - (1+h)^{1/h} = e - \exp\left(1 - \frac{h}{2} + \frac{h^2}{3} \pm \ldots\right) = e\left(\frac{1}{2}h - \frac{11}{24}h^2 + \frac{7}{16}h^3 \pm \ldots\right)$$

with *convergence radius* $r = 1$.

2. For the differential equation $y' = f(x)$ and the trapezoidal rule (7.5), the expansion (8.10) becomes

$$\int_0^1 f(x)\,dx - y_h(1) = -\sum_{k=1}^{N} \frac{h^{2k}}{(2k)!} B_{2k}\left(f^{(2k-1)}(1) - f^{(2k-1)}(0)\right) + \mathcal{O}(h^{2N+1}),$$

the well known Euler-Maclaurin formula (1736). For $N \to \infty$, the series will usually diverge, due to the fast growth of the Bernoulli numbers for large k. It may, however, be useful for small values of N and we call it an *asymptotic expansion* (Poincaré 1893).

Variable h

Theorem 8.1 is not only valid for equal step sizes. A reasonable assumption for the case of variable step sizes is the existence of a function $\tau(x) > 0$ such that the step sizes depend as

$$x_{n+1} - x_n = \tau(x_n)h \tag{8.11}$$

on a parameter h. Then the local error expansion (8.2) becomes

$$y(x + \tau(x)h) - y(x) - h\tau(x)\Phi(x, y(x), \tau(x)h) = d_{p+1}(x)\tau^{p+1}(x)h^{p+1} + \ldots$$

and instead of (8.5) we have

$$\widehat{\Phi}(x, \widehat{y}, \tau(x)h) = \Phi(x, \widehat{y} - e_p(x)h^p, \tau(x)h) + \frac{h^p}{h\tau(x)}\left(e_p(x + \tau(x)h) - e_p(x)\right).$$

With this the local error expansion for the new method becomes, instead of (8.6),

$$y(x + \tau(x)h) - y(x) - h\tau(x)\widehat{\Phi}(x, y(x), \tau(x)h)$$

$$= \tau(x)\Big(d_{p+1}(x)\tau^p(x) + \frac{\partial f}{\partial y}(x, y(x))e_p(x) - e'_p(x)\Big)h^{p+1} + \mathcal{O}(h^{p+2})$$

and the proof of Theorem 8.1 generalizes with slight modifications.

Negative h

The most important extrapolation algorithms will use asymptotic expansions with *even* powers of h. In order to provide a theoretical basis for these methods, we need to explain the meaning of $y_h(x)$ for h *negative*.

Motivation. We write (8.1) as

$$y_h(x+h) = y_h(x) + h\Phi(x, y_h(x), h) \tag{8.1'}$$

and replace h by $-h$ to obtain

$$y_{-h}(x-h) = y_{-h}(x) - h\Phi(x, y_{-h}(x), -h).$$

Next we replace x by $x+h$ which gives

$$y_{-h}(x) = y_{-h}(x+h) - h\Phi(x+h, y_{-h}(x+h), -h). \tag{8.12}$$

This is an implicit equation for $y_{-h}(x+h)$, which possesses a unique solution for sufficiently small h (by the implicit function theorem). We write this solution in the form

$$y_{-h}(x+h) = y_{-h}(x) + h\Phi^*(x, y_{-h}(x), h). \tag{8.13}$$

The comparison of (8.12) and (8.13) (with $A = y_{-h}(x+h)$, $B = y_{-h}(x)$) leads us to the following definition.

Definition 8.2. Let $\Phi(x, y, h)$ be the increment function of a method. Then we define the increment function $\Phi^*(x, y, h)$ of the *adjoint method* by the pair of formulas

$$\begin{aligned} B &= A - h\Phi(x+h, A, -h) \\ A &= B + h\Phi^*(x, B, h). \end{aligned} \tag{8.14}$$

Example. The adjoint method of explicit Euler is implicit Euler.

Theorem 8.3. *Let Φ be the Runge-Kutta method (7.7) with coefficients a_{ij}, b_j, c_i $(i, j = 1, \ldots, s)$. Then the adjoint method Φ^* is equivalent to a Runge-Kutta method with s stages and with coefficients*

$$\begin{aligned} c_i^* &= 1 - c_{s+1-i} \\ a_{ij}^* &= b_{s+1-j} - a_{s+1-i,s+1-j} \\ b_j^* &= b_{s+1-j}. \end{aligned}$$

Proof. The formulas (8.14) indicate that for the definition of the adjoint method we have, starting from (7.7), to exchange $y_0 \leftrightarrow y_1$, $h \leftrightarrow -h$ and replace $x_0 \to x_0 + h$. This then leads to

$$k_i = f\left(x_0 + (1 - c_i)h, y_0 + h \sum_{j=1}^{s}(b_j - a_{ij})k_j\right)$$

$$y_1 = y_0 + h \sum_{j=1}^{s} b_j k_j.$$

In order to preserve the usual natural ordering of $c_1, \ldots, c_s$, we also permute the k_i-values and replace all indices i by $s + 1 - i$. □

Properties of the Adjoint Method

Theorem 8.4. $\Phi^{**} = \Phi$.

Proof. This property, which is the reason for the name "adjoint", is seen by replacing $h \to -h$ and then $x \to x + h$, $B \to A$, $A \to B$ in (8.14). □

Theorem 8.5. *The adjoint method has the same order as the original method. Its principal error term is the error term of the first method multiplied by $(-1)^p$.*

Proof. We replace h by $-h$ in (8.2), then $x \to x + h$ and rearrange the terms. This gives (using $d_{p+1}(x + h) = d_{p+1}(x) + \mathcal{O}(h)$)

$$y(x) + d_{p+1}(x)h^{p+1}(-1)^p + \mathcal{O}(h^{p+2})$$
$$= y(x + h) - h\Phi(x + h, y(x + h), -h).$$

Here we let B be the left-hand side of this identity, $A = y(x + h)$, and use (8.14). This leads to

$$y(x + h) = y(x) + d_{p+1}(x)h^{p+1}(-1)^p + h\Phi^*(x, y(x), h) + \mathcal{O}(h^{p+2}),$$

which expresses the statement of the theorem. □

Theorem 8.6. *The adjoint method has exactly the same asymptotic expansion (8.10) as the original method, with h replaced by $-h$.*

Proof. We repeat the procedure which led to the proof of Theorem 8.1, with h negative. The first separated term corresponding to (8.9) will be

$$y(x) - y_{-h}(x) = e_p(x)(-h)^p + \mathcal{O}(h^{p+1}). \tag{8.9'}$$

This is true because the solution of (8.8) with initial value $e_p(x_0) = 0$ has the same sign change as the inhomogenity $d_{p+1}(x)$. This settles the first term. To continue, we prove that the transformation (8.4b) commutes with the adjunction operation, i.e., that

$$(\widehat{\Phi})^* = (\Phi^*)\widehat{}. \tag{8.15}$$

In order to prove (8.15), we obtain from (8.4a) and the definition of $\widehat{\Phi}$

$$y_h(x+h) + e_p(x+h)h^p = y_h(x) + e_p(x)h^p + h\widehat{\Phi}\big(x, y_h(x) + e_p(x)h^p, h\big).$$

Here again, we substitute $h \to -h$ followed by $x \to x+h$. Finally, we apply (8.14) with $B = y_{-h}(x) + e_p(x)(-h)^p$ and $A = y_{-h}(x+h) + e_p(x+h)(-h)^p$ to obtain

$$\begin{aligned}
&y_{-h}(x+h) + e_p(x+h)(-h)^p \\
&= y_{-h}(x) + e_p(x)(-h)^p + h(\widehat{\Phi})^*\big(x, y_{-h}(x) + e_p(x)(-h)^p, h\big).
\end{aligned} \tag{8.16}$$

On the other hand, if we perform the transformation (see Theorem 8.5)

$$\widehat{y}_{-h}(x) = y_{-h}(x) + e_p(x)(-h)^p \tag{8.4'}$$

and insert this into (8.13), we obtain (8.16) again, but this time with $(\Phi^*)\widehat{}$ instead of $(\widehat{\Phi})^*$. This proves (8.15). $\qquad\square$

Symmetric Methods

Definition 8.7. A method is *symmetric* if $\Phi = \Phi^*$.

Example. The trapezoidal rule (7.5) and the implicit mid-point rule (7.4) are symmetric: the exchanges $y_1 \leftrightarrow y_0$, $h \leftrightarrow -h$ and $x_0 \leftrightarrow x_0 + h$ leave these methods invariant. The following two theorems (Wanner 1973) characterize symmetric IRK methods.

Theorem 8.8. *If*

$$a_{s+1-i,s+1-j} + a_{ij} = b_{s+1-j} = b_j, \qquad i,j = 1,\ldots,s, \tag{8.17}$$

then the corresponding Runge-Kutta method is symmetric. Moreover, if the b_i are nonzero and the c_i distinct and ordered as $c_1 < c_2 < \ldots < c_s$, then condition (8.17) is also necessary for symmetry.

Proof. The sufficiency of (8.17) follows from Theorem 8.3. The condition $c_i = 1 - c_{s+1-i}$ can be verified by adding up (8.17) for $j = 1,\ldots,s$.

Symmetry implies that the original method (with coefficients c_i, a_{ij}, b_j) and the adjoint method (c_i^*, a_{ij}^*, b_j^*) give identical numerical results. If we apply both methods to $y' = f(x)$ we obtain

$$\sum_{i=1}^{s} b_i f(c_i) = \sum_{i=1}^{s} b_i^* f(c_i^*)$$

for all $f(x)$. Our assumption on b_i and c_i thus yields

$$b_i^* = b_i, \qquad c_i^* = c_i \qquad \text{for all} \quad i.$$

We next apply both methods to $y_1' = f(x)$, $y_2' = x^q y_1$ and obtain

$$\sum_{i,j=1}^{s} b_i c_i^q a_{ij} f(c_j) = \sum_{i,j=1}^{s} b_i^* c_i^{*q} a_{ij}^* f(c_j^*).$$

This implies $\sum_i b_i c_i^q a_{ij} = \sum_i b_i c_i^q a_{ij}^*$ for $q = 0, 1, \ldots$ and hence also $a_{ij}^* = a_{ij}$ for all i, j.   □

Theorem 8.9. *A collocation method based on symmetrically distributed collocation points is symmetric.*

Proof. If $c_i = 1 - c_{s+1-i}$, the Lagrange polynomials satisfy $\ell_i(t) = \ell_{s+1-i}(1-t)$. Condition (8.17) is then an easy consequence of (7.19).   □

The following important property of symmetric methods, known intuitively for many years, now follows from the above results.

Theorem 8.10. *If in addition to the assumptions of Theorem 8.1 the underlying method is symmetric, then the asymptotic expansion (8.10) contains only even powers of h:*

$$y(x) - y_h(x) = e_{2q}(x)h^{2q} + e_{2q+2}(x)h^{2q+2} + \ldots \tag{8.18}$$

with $e_{2j}(x_0) = 0$.

Proof. If $\Phi^* = \Phi$, we have $y_{-h}(x) = y_h(x)$ from (8.13) and the result follows from Theorem 8.6.   □

Exercises

1. Assume the one-step method (8.1) to be of order $p \geq 2$ and in addition to $\Phi(x, y, 0) = f(x, y)$ assume

$$\frac{\partial \Phi}{\partial h}(x, y, 0) = \frac{1}{2}\left(\frac{\partial f}{\partial x}(x, y) + \frac{\partial f}{\partial y}(x, y) \cdot f(x, y)\right). \qquad (8.19)$$

   Show that the principal local error term of the method $\widehat{\Phi}$ defined in (8.5) is then given by

$$\widehat{d}_{p+2}(x) = d_{p+2}(x) - \frac{1}{2}\frac{\partial f}{\partial y}(x, y(x))d_{p+1}(x) - \frac{1}{2}d'_{p+1}(x).$$

   Verify that (8.19) is satisfied for all RK-methods of order ≥ 2.

2. Consider the second order method

$$\begin{array}{c|cc}
0 & & \\
1 & 1 & \\
\hline
 & 1/2 & 1/2
\end{array}$$

   applied to the problem $y' = y$, $y(0) = 1$. Show that

$$d_3(x) = \frac{1}{6}e^x, \quad d_4(x) = \frac{1}{24}e^x, \quad e_2(x) = \frac{1}{6}xe^x, \quad \widehat{d}_4(x) = -\frac{1}{8}e^x.$$

3. Consider the second order method

$$\begin{array}{c|ccc}
0 & & & \\
1/2 & 1/2 & & \\
1 & 0 & 1 & \\
\hline
 & 1/4 & 1/2 & 1/4
\end{array}$$

   Show that for this method

$$d_3(x) = \frac{1}{24}\left(F(t_{32})(y(x)) - \frac{1}{2}F(t_{31})(y(x))\right)$$

$$d_4(x) = \frac{1}{24}\left(F(t_{44})(y(x)) + \frac{1}{4}F(t_{43})(y(x)) - \frac{1}{4}F(t_{41})(y(x))\right)$$

   in the notation of Table 2.2. Show that this implies

$$\widehat{d}_4(x) = 0 \qquad \text{and} \qquad e_3(x) = 0,$$

   so that one step of Richardson extrapolation increases the order of the method by two. Find a connection between this method and the GBS-algorithm of Section II.9.

4. Discuss the symmetry of the IRK methods of Section II.7.

II.9 Extrapolation Methods

> The following method of approximation may or may not be new, but as I believe it to be of practical importance ...
>
> (S.A. Corey 1906)

> The h^2-extrapolation was discovered by a hint from theory followed by arithmetical experiments, which gave pleasing results.
>
> (L.F. Richardson 1927)

> Extrapolation constitutes a powerful means ...
>
> (R. Bulirsch & J. Stoer 1966)

> Extrapolation does not appear to be a particularly effective way ..., our tests raise the question as to whether there is any point to pursuing it as a separate method.
>
> (L.F. Shampine & L.S. Baca 1986)

Definition of the Method

Let $y' = f(x, y)$, $y(x_0) = y_0$ be a given differential system and $H > 0$ a basic step size. We choose a sequence of positive integers

$$n_1 < n_2 < n_3 < \ldots \tag{9.1}$$

and define the corresponding step sizes $h_1 > h_2 > h_3 > \ldots$ by $h_i = H/n_i$. We then choose a numerical method of order p and compute the numerical results of our initial value problem by performing n_i steps with step size h_i to obtain

$$y_{h_i}(x_0 + H) =: T_{i,1} \tag{9.2}$$

(the letter "T" stands historically for "trapezoidal rule"). We then eliminate as many terms as possible from the asymptotic expansion (8.10) by computing the interpolation polynomial

$$p(h) = \widehat{y} - e_p h^p - e_{p+1} h^{p+1} - \ldots - e_{p+k-2} h^{p+k-2} \tag{9.3}$$

such that

$$p(h_i) = T_{i,1} \qquad i = j,\ j-1, \ldots, j-k+1. \tag{9.4}$$

Finally we *"extrapolate to the limit"* $h \to 0$ and use

$$p(0) = \widehat{y} =: T_{j,k}$$

as numerical result. Conditions (9.4) consist of k linear equations for the k unknowns $\widehat{y}, e_p, \ldots, e_{p+k-2}$.

Example. For $k = 2$, $n_1 = 1$, $n_2 = 2$ the above definition is identical to Richardson's extrapolation discussed in Section II.4.

Theorem 9.1. *The value $T_{j,k}$ represents a numerical method of order $p + k - 1$.*

Proof. We compare (9.4) and (9.3) with the asymptotic expansion (8.10) which we write in the form (with $N = p + k - 1$)

$$T_{i,1} = y(x_0+H) - e_p(x_0+H)h_i^p - \ldots - e_{p+k-2}(x_0+H)h_i^{p+k-2} - \Delta_i, \quad (9.4')$$

where

$$\Delta_i = e_{p+k-1}(x_0+H)h_i^{p+k-1} + E_{h_i}(x_0+H)h_i^{p+k} = \mathcal{O}(H^{p+k})$$

because $e_{p+k-1}(x_0) = 0$ and $h_i \leq H$. This is a linear system for the unknowns $y(x_0+H)$, $H^p e_p(x_0+H), \ldots, H^{p+k-2}e_{p+k-2}(x_0+H)$ with the Vandermonde-like matrix

$$A = \begin{pmatrix} 1 & \dfrac{1}{n_j^p} & \cdots & \dfrac{1}{n_j^{p+k-2}} \\ \vdots & \vdots & & \vdots \\ 1 & \dfrac{1}{n_{j-k+1}^p} & \cdots & \dfrac{1}{n_{j-k+1}^{p+k-2}} \end{pmatrix}.$$

It is the same as (9.4), just with the right-hand side perturbed by the $\mathcal{O}(H^{p+k})$-terms Δ_i. The matrix A is invertible (see Exercise 6). Therefore by subtraction we obtain

$$|y(x_0+H) - \widehat{y}| \leq \|A^{-1}\|_\infty \cdot \max|\Delta_i| = \mathcal{O}(H^{p+k}). \qquad \square$$

Remark. The case $p = 1$ (as well as $p = 2$ with expansions in h^2) can also be treated by interpreting the difference $y(x_0 + H) - \widehat{y}$ as an interpolation error (see (9.21)).

A great advantage of the method is that it provides a complete table of numerical results

$$\begin{array}{lllll} T_{11} \\ T_{21} & T_{22} \\ T_{31} & T_{32} & T_{33} & & (9.5) \\ T_{41} & T_{42} & T_{43} & T_{44} \\ \cdots & \cdots & \cdots & \cdots & \cdots \end{array}$$

which form a sequence of embedded methods and allow easy estimates of the local error and strategies for variable order. Several step-number sequences are in use for (9.1):

The *"Romberg sequence"* (Romberg 1955):

$$1, 2, 4, 8, 16, 32, 64, 128, 256, 512, \ldots \qquad (9.6)$$

The *"Bulirsch sequence"* (see also Romberg 1955):

$$1, 2, 3, 4, 6, 8, 12, 16, 24, 32, \ldots \tag{9.7}$$

alternating powers of 2 with 1.5 times 2^k. This sequence needs fewer function evaluations for higher orders than the previous one and became prominent through the success of the "Gragg-Bulirsch-Stoer algorithm" (Bulirsch & Stoer 1966).

The above sequences have the property that for integration problems $y' = f(x)$ many function values can be saved and re-used for smaller h_i. Further, $\lim\inf(n_{i+1}/n_i)$ remains bounded away from 1 ("Toeplitz condition") which allows convergence proofs for $j = k \to \infty$ (Bauer, Rutishauser & Stiefel 1963). However, if we work with differential equations and with fixed or bounded order, the most economic sequence is the *"harmonic sequence"* (Deuflhard 1983)

$$1, 2, 3, 4, 5, 6, 7, 8, 9, 10, \ldots . \tag{9.8}$$

The Aitken - Neville Algorithm

For the case $p = 1$, (9.3) and (9.4) become a classical interpolation problem and we can compute the values of $T_{j,k}$ economically by the use of classical methods. Since we need only the values of the interpolation polynomials at the point $h = 0$, the most economical algorithm is that of "Aitken - Neville" (Aitken 1932, Neville 1934, based on ideas of Jordan 1928) which leads to

$$T_{j,k+1} = T_{j,k} + \frac{T_{j,k} - T_{j-1,k}}{(n_j/n_{j-k}) - 1}. \tag{9.9}$$

If the basic method used is *symmetric,* we know that the underlying asymptotic expansion is in powers of h^2 (Theorem 8.9), and each extrapolation eliminates *two* powers of h. We may thus simply replace in (9.3) h by h^2 and for $p = 2$ (i.e., $q = 1$ in (8.18)) also use the Aitken - Neville algorithm with this modification. This leads to

$$T_{j,k+1} = T_{j,k} + \frac{T_{j,k} - T_{j-1,k}}{(n_j/n_{j-k})^2 - 1} \tag{9.10}$$

instead of (9.9).

Numerical example. We solve the problem

$$y' = (-y\sin x + 2\tan x)y, \qquad y(\pi/6) = 2/\sqrt{3} \tag{9.11}$$

with true solution $y(x) = 1/\cos x$ and basic step size $H = 0.2$ by Euler's method. Fig. 9.1 represents, for each of the entries $T_{j,k}$ of the extrapolation tableau, the *numerical work* $(1 + n_j - 1 + n_{j-1} - 1 + \ldots + n_{j-k+1} - 1)$ compared to the *precision* $(|T_{j,k} - y(x_0 + H)|)$ in double logarithmic scale. The first picture is for the Romberg sequence (9.6), the second for the Bulirsch sequence (9.7), and the last

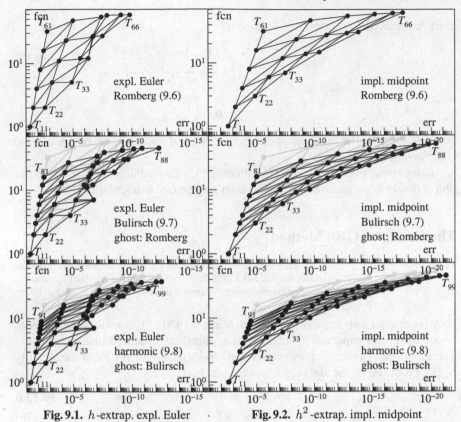

Fig. 9.1. h-extrap. expl. Euler

Fig. 9.2. h^2-extrap. impl. midpoint

for the harmonic sequence (9.8). In pictures 2 and 3 the results of the foregoing graphics are repeated as a shaded "ghost" (... of Canterville) in order to demonstrate how the results are better than those for the predecessor. Nobody is perfect, however. The "best" method in these comparisons, the harmonic sequence, suffers for high orders from a strong influence of rounding errors (see Exercise 5 below; the computations of Fig. 9.1, 9.2 and 9.4 have been made in quadruple precision).

The analogous results for the symmetric implicit mid-point rule (7.4) are presented in Fig. 9.2. Although implicit, this method is easy to implement for this particular example. We again use the same basic step size $H = 0.2$ as above and the same step-number sequences (9.6), (9.7), (9.8). Here, the "numerical work" $(n_j + n_{j-1} + \ldots + n_{j-k+1})$ represents *implicit* stages and therefore can not be compared to the values of the explicit method. The precisions, however, show a drastic improvement.

Rational Extrapolation. Many authors in the sixties claimed that it is better to use rational functions instead of polynomials in (9.3). In this case the formula (9.9) must be replaced by (Bulirsch & Stoer 1964)

$$T_{j,k+1} = T_{j,k} + \frac{T_{j,k} - T_{j-1,k}}{\left(\frac{n_j}{n_{j-k}}\right)\left(1 - \frac{T_{j,k}-T_{j-1,k}}{T_{j,k}-T_{j-1,k-1}}\right) - 1} \tag{9.12}$$

where

$$T_{j,0} = 0.$$

For systems of differential equations the division of vectors is to be understood componentwise.

Later numerical experiments (Deuflhard 1983) showed that rational extrapolation is nearly never more advantageous than polynomial extrapolation.

The Gragg or GBS Method

> Since it is fully explicit GRAGG's algorithm is so ideally suited as a basis for RICHARDSON extrapolation that no other symmetric two-step algorithm can compete with it.   (H.J. Stetter 1970)

Here we can not do better than quote from Stetter (1970): "Expansions in powers of h^2 are extremely important for an efficient application of Richardson extrapolation. Therefore it was a great achievement when Gragg proved in 1963 that the quantity $S_h(x)$ produced by the algorithm ($x = x_0 + 2nh$, $x_i = x_0 + ih$)

$$y_1 = y_0 + hf(x_0, y_0) \tag{9.13a}$$

$$y_{i+1} = y_{i-1} + 2hf(x_i, y_i) \qquad i = 1, 2, \ldots, 2n \tag{9.13b}$$

$$S_h(x) = \frac{1}{4}\left(y_{2n-1} + 2y_{2n} + y_{2n+1}\right) \tag{9.13c}$$

possesses an asymptotic expansion in even powers of h and has satisfactory stability properties. This led to the construction of the very powerful G(ragg)-B(ulirsch)-S(toer)-extrapolation algorithm ...".

Gragg's *proof* of this property was very long and complicated and it was again "a great achievement" that Stetter had the elegant idea of interpreting (9.13b) as a *one-step* algorithm by rewriting (9.13) in terms of odd and even indices: for this purpose we define

$$h^* = 2h, \qquad x_k^* = x_0 + kh^*, \qquad u_0 = v_0 = y_0,$$

$$u_k = y_{2k}, \qquad v_k = y_{2k+1} - hf(x_{2k}, y_{2k}) = \frac{1}{2}\left(y_{2k+1} + y_{2k-1}\right). \tag{9.14}$$

Then the method (9.13) can be rewritten as (see Fig. 9.3)

$$\begin{pmatrix} u_{k+1} \\ v_{k+1} \end{pmatrix} = \begin{pmatrix} u_k \\ v_k \end{pmatrix} + h^* \begin{pmatrix} f\left(x_k^* + \frac{h^*}{2}, v_k + \frac{h^*}{2}f(x_k^*, u_k)\right) \\ \frac{1}{2}\left(f(x_k^* + h^*, u_{k+1}) + f(x_k^*, u_k)\right) \end{pmatrix}. \tag{9.15}$$

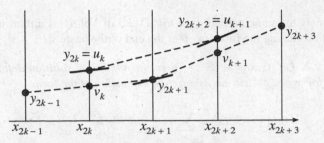

Fig. 9.3. Symmetry of the Gragg method

This method, which maps the pair (u_k, v_k) to (u_{k+1}, v_{k+1}), can be seen from Fig. 9.3 to be *symmetric*. The symmetry can also be checked analytically (see Definition 8.7) by exchanging $u_{k+1} \leftrightarrow u_k$, $v_{k+1} \leftrightarrow v_k$, $h^* \leftrightarrow -h^*$, $x_k^* \leftrightarrow x_k^* + h^*$. A trivial calculation then shows that this leaves formula (9.15) invariant. Method (9.15) is consistent with the differential equation (let $h^* \to 0$ in the increment function)

$$u' = f(x, v) \qquad u(x_0) = y_0$$
$$v' = f(x, u) \qquad v(x_0) = y_0, \qquad\qquad (9.16)$$

whose exact solution is simply $u(x) = v(x) = y(x)$. Therefore, we have from Theorem 8.10 that

$$y(x) - u_{h^*}(x) = \sum_{j=1}^{\ell} a_{2j}(x)(h^*)^{2j} + (h^*)^{2\ell+2} A(x, h^*) \qquad (9.17a)$$

$$y(x) - v_{h^*}(x) = \sum_{j=1}^{\ell} b_{2j}(x)(h^*)^{2j} + (h^*)^{2\ell+2} B(x, h^*) \qquad (9.17b)$$

and $a_{2j}(x_0) = b_{2j}(x_0) = 0$. We see from (9.14) and (9.17a) that $y_h(x)$ possesses an expansion in even powers of h, provided that the number of steps is even; i.e., for $x = x_0 + 2nh$,

$$y(x) - y_h(x) = \sum_{j=1}^{\ell} \widehat{a}_{2j}(x)h^{2j} + h^{2\ell+2}\widehat{A}(x, h) \qquad (9.18)$$

where $\widehat{a}_{2j}(x) = 2^{2j} a_{2j}(x)$ and $\widehat{A}(x, h) = 2^{2\ell+2} A(x, 2h)$.

The so-called *smoothing step*, i.e., formula

$$S_h(x_0 + 2nh) = \frac{1}{4}\left(y_{2n-1} + 2y_{2n} + y_{2n+1}\right) = \frac{1}{2}\left(u_n + v_n\right)$$

(see (9.13c) and (9.14)) had its historical origin in the "weak stability" of the explicit midpoint rule (9.13b) (see also Fig. III.9.2). However, since the method is anyway followed by extrapolation, this step is not of great importance (Shampine & Baca 1983). It is a little more costly and increases the "stability domain" by approximately the same amount (see Fig. IV.2.3 of Vol. II). Further, it has the advantage of evaluating the function f at the end of the basic step.

Theorem 9.2. *Let* $f(x, y) \in C^{2\ell+2}$, *then the numerical solution defined in (9.13) possesses for* $x = x_0 + 2nh$ *an asymptotic expansion of the form*

$$y(x) - S_h(x) = \sum_{j=1}^{\ell} e_{2j}(x)h^{2j} + h^{2\ell+2}C(x, h) \qquad (9.19)$$

with $e_{2j}(x_0) = 0$ *and* $C(x, h)$ *bounded for* $x_0 \leq x \leq \overline{x}$ *and* $0 \leq h \leq h_0$.

Proof. By adding (9.17a) and (9.17b) and using $h^* = 2h$ we obtain (9.19) with $e_{2j}(x) = (a_{2j}(x) + b_{2j}(x))2^{2j-1}$. $\qquad \qquad \qquad \square$

This method can thus be used for Richardson extrapolation in the same way as symmetric methods above: we choose a step-number sequence, with the condition that the n_j are even, i.e.,

$$2, 4, 8, 16, 32, 64, 128, 256, \ldots \qquad (9.6')$$
$$2, 4, 6, 8, 12, 16, 24, 32, 48, \ldots \qquad (9.7')$$
$$2, 4, 6, 8, 10, 12, 14, 16, 18, \ldots \qquad (9.8')$$

set

$$T_{i,1} := S_{h_i}(x_0 + H)$$

and compute the extrapolated expressions $T_{i,j}$, based on the h^2-expansion, by the Aitken-Neville formula (9.10).

Numerical example. Fig. 9.4 represents the numerical results of this algorithm applied to Example (9.11) with step size $H = 0.2$. The step size sequences are Romberg (9.6') (above), Bulirsch (9.7') (middle), and harmonic (9.8') (below). The algorithm *with* smoothing step (numerical work $= 1 + n_j + n_{j-1} + \ldots + n_{j-k+1}$) is represented *left*, the results *without* smoothing step (numerical work $= 1 + n_j - 1 + n_{j-1} - 1 + \ldots + n_{j-k+1} - 1$) are on the *right*.

The results are nearly identical to those for the implicit midpoint rule (Fig. 9.2), but much more valuable, since here the method is explicit. In the pictures on the left the values for extrapolated Euler (from Fig. 9.1) are repeated as a "ghost" and demonstrate clearly the importance of the h^2-expansion, especially in the diagonal T_{kk} for large values of k. The ghost in the pictures on the right are the values *with* smoothing step from the left; the differences are seen to be tiny.

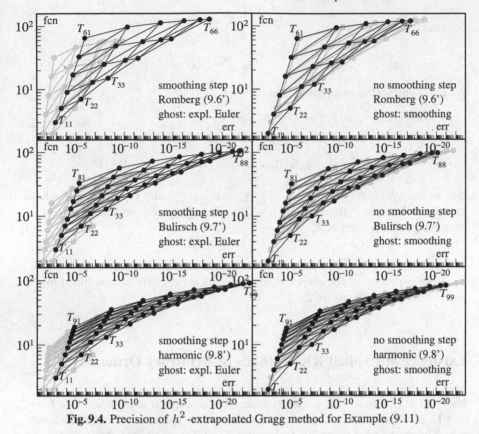

Fig. 9.4. Precision of h^2-extrapolated Gragg method for Example (9.11)

Asymptotic Expansion for Odd Indices

For completeness, we still want to derive the existence of an h^2 expansion for y_{2k+1} from (9.17b), although this is of no practical importance for the numerical algorithm described above.

Theorem 9.3 (Gragg 1964). *For $x = x_0 + (2k+1)h$ we have*

$$y(x) - y_h(x) = \sum_{j=1}^{\ell} \widehat{b}_{2j}(x)h^{2j} + h^{2\ell+2}\widehat{B}(x,h) \qquad (9.20)$$

where the coefficients $\widehat{b}_{2j}(x)$ are in general different from those for even indices and $\widehat{b}_{2j}(x_0) \neq 0$.

Proof. y_{2k+1} can be computed (see Fig. 9.3) either from v_k by a forward step or from v_{k+1} by a backward step. For the sake of symmetry, we take the mean of both expressions and write

$$y_{2k+1} = \frac{1}{2}\left(v_k + v_{k+1}\right) + \frac{h}{2}\left(f(x_k^*, u_k) - f(x_{k+1}^*, u_{k+1})\right).$$

We now subtract the exact solution and obtain

$$
\begin{aligned}
2\big(y_h(x) - y(x)\big) &= v_{2h}(x - h) - y(x - h) \\
&\quad + v_{2h}(x + h) - y(x + h) + y(x - h) - 2y(x) + y(x + h) \\
&\quad + h\Big(f\big(x - h, u_{2h}(x - h)\big) - f\big(x + h, u_{2h}(x + h)\big)\Big).
\end{aligned}
$$

Due to the symmetry of $u_{2h}(x)$ $(u_{2h}(\xi) = u_{-2h}(\xi))$ and of $v_{2h}(x)$ the whole expression becomes symmetric in h. Thus the asymptotic expansion for y_{2k+1} contains no odd powers of h. □

Both expressions, for even and for odd indices, can still be combined into a single formula (see Exercise 2).

Existence of Explicit RK Methods of Arbitrary Order

Each of the expressions $T_{j,k}$ clearly represents an explicit RK-method (see Exercise 1). If we apply the well-known error formula for polynomial interpolation (see e.g., Abramowitz & Stegun 1964, formula 25.2.27) to (9.19), we obtain

$$y(x_0 + H) - T_{j,k} = \frac{(-1)^k}{n_j^2 \cdot \ldots \cdot n_{j-k+1}^2} e_{2k}(x_0 + H)H^{2k} + \mathcal{O}(H^{2k+2}). \quad (9.21)$$

Since $e_k(x_0) = 0$, we have

$$y(x_0 + H) - T_{j,k} = \frac{(-1)^k}{n_j^2 \cdot \ldots \cdot n_{j-k+1}^2} e_{2k}'(x_0)H^{2k+1} + \mathcal{O}(H^{2k+2}). \quad (9.22)$$

This shows that $T_{j,k}$ represents an explicit Runge-Kutta method of order $2k$. As an application of this result we have:

Theorem 9.4 (Gragg 1964). *For p even, there exists an explicit RK-method of order p with $s = p^2/4 + 1$ stages.*

Proof. This result is obtained by counting the number of necessary function evaluations of the GBS-algorithm using the harmonic sequence and without the final smoothing step. □

Remark. The extrapolated Euler method leads to explicit Runge-Kutta methods with $s = p(p-1)/2 + 1$ stages. This shows once again the importance of the h^2 expansion.

Order and Step Size Control

Extrapolation methods have the advantage that in addition to the step size also the order (i.e., number of columns) can be changed at each step. Because of this double freedom, the *practical implementation* in an optimal way is more complicated than for fixed-order RK-methods. The first codes were developed by Bulirsch & Stoer (1966) and their students. Very successful extrapolation codes due to P. Deuflhard and his collaborators are described in Deuflhard (code DIFEX1, 1983).

The choice of the *step size* can be performed in exactly the same way as for fixed-order embedded methods (see Section II.4). If the first k lines of the extrapolation tableau are computed, we have T_{kk} as the highest-order approximation (of order $2k$ by (9.22)) and in addition $T_{k,k-1}$ of order $2k - 2$. It is therefore natural to use the expression

$$err_k = \|T_{k,k-1} - T_{k,k}\| \tag{9.23}$$

for step size control. The norm is the same as in (4.11). As in (4.12) we get for the optimal step size the formula

$$H_k = H \cdot 0.94 \cdot (0.65/err_k)^{1/(2k-1)} \tag{9.24}$$

where this time we have chosen a safety factor depending partly on the order.

For the choice of an *optimal order* we need a measure of work, which allows us to compare different methods. The work for computing T_{kk} can be measured by the number A_k of function evaluations. For the GBS-algorithm it is given recursively by

$$\begin{align} A_1 &= n_1 + 1 \\ A_k &= A_{k-1} + n_k. \end{align} \tag{9.25}$$

However, a large number of function evaluations can be compensated by a large step size H_k, given by (9.24). We therefore consider

$$W_k = \frac{A_k}{H_k}, \tag{9.26}$$

the *work per unit step,* as a measure of work. The idea is now to choose the order (i.e., the index k) in such a way that W_k is minimized.

Let us describe the *combined order and step size control* in some more detail. We assume that at some point of integration the step size H and the index k $(k > 2)$ are proposed. The step is then realized in the following way: we first compute $k - 1$ lines of the extrapolation tableau and also the values H_{k-2}, W_{k-2}, err_{k-1}, H_{k-1}, W_{k-1}.

a) *Convergence in line* $k - 1$. If $err_{k-1} \leq 1$, we accept $T_{k-1,k-1}$ as numerical solution and continue the integration with the new proposed quantities

$$k_{\text{new}} = \begin{cases} k & \text{if } W_{k-1} < 0.9 \cdot W_{k-2} \\ k-1 & \text{else} \end{cases}$$

$$H_{\text{new}} = \begin{cases} H_{k_{\text{new}}} & \text{if } k_{\text{new}} \leq k-1 \\ H_{k-1}(A_k/A_{k-1}) & \text{if } k_{\text{new}} = k. \end{cases} \tag{9.27}$$

In (9.27), the only non-trivial formula is the choice of the step size H_{new} in the case of an order-increase $k_{\text{new}} = k$. In this case we want to avoid the computation of err_k, so that H_k and W_k are unknown. However, since our k is assumed to be close to the optimal value, we have $W_k \approx W_{k-1}$ which leads to the proposed step size increase.

b) *Convergence monitor.* If $err_{k-1} > 1$, we first decide whether we may expect convergence at least in line $k + 1$. It follows from (9.22) that, asymptotically,

$$\|T_{k,k-2} - T_{k,k-1}\| \approx \left(\frac{n_2}{n_k}\right)^2 err_{k-1} \tag{9.28}$$

with err_{k-1} given by (9.23). Unfortunately, err_k cannot be compared with (9.28), since different factors (depending on the differential equation to be solved) are involved in the asymptotic formula (cf. (9.22)). If we nevertheless assume that err_k is $(n_2/n_1)^2$ times smaller than (9.28) we obtain $err_k \approx (n_1/n_k)^2 err_{k-1}$. We therefore already reject the step at this point, if

$$err_{k-1} > \left(\frac{n_{k+1}n_k}{n_1 n_1}\right)^2 \tag{9.29}$$

and restart with $k_{\text{new}} \leq k - 1$ and H_{new} according to (9.27). If the contrary of (9.29) holds, we compute the next line of the extrapolation tableau, i.e., $T_{k,k}$, err_k, H_k and W_k.

c) *Convergence in line* k. If $err_k \leq 1$, we accept T_{kk} as numerical solution and continue the integration with the new proposed values

$$k_{\text{new}} = \begin{cases} k-1 & \text{if } W_{k-1} < 0.9 \cdot W_k \\ k+1 & \text{if } W_k < 0.9 \cdot W_{k-1} \\ k & \text{in all other cases} \end{cases}$$

$$H_{\text{new}} = \begin{cases} H_{k_{\text{new}}} & \text{if } k_{\text{new}} \leq k \\ H_k(A_{k+1}/A_k) & \text{if } k_{\text{new}} = k+1. \end{cases} \tag{9.30}$$

d) *Second convergence monitor.* If $err_k > 1$, we check, as in (b), the relation

$$err_k > \left(\frac{n_{k+1}}{n_1}\right)^2. \tag{9.31}$$

If (9.31) is satisfied, the step is rejected and we restart with $k_{\text{new}} \leq k$ and H_{new} of (9.30). Otherwise we continue.

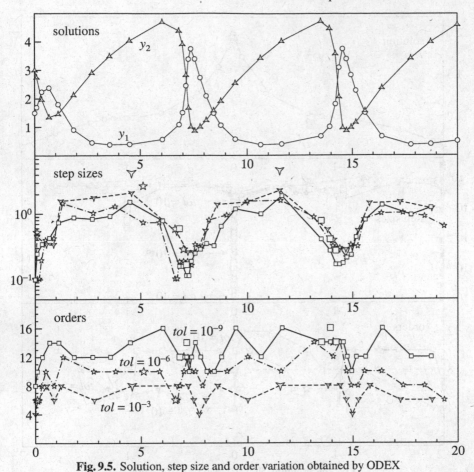

Fig. 9.5. Solution, step size and order variation obtained by ODEX

e) *Hope for convergence in line* $k+1$. We compute err_{k+1}, H_{k+1} and W_{k+1}. If $err_{k+1} \leq 1$, we accept $T_{k+1,k+1}$ as numerical solution and continue the integration with the new proposed order

$$
\begin{aligned}
&k_{\text{new}} := k \\
&\text{if } (W_{k-1} < 0.9 \cdot W_k) \qquad k_{\text{new}} := k-1 \\
&\text{if } (W_{k+1} < 0.9 \cdot W_{k_{\text{new}}}) \qquad k_{\text{new}} := k+1.
\end{aligned}
\tag{9.32}
$$

If $err_{k+1} > 1$ the step is rejected and we restart with $k_{\text{new}} \leq k$ and H_{new} of (9.24).

The following slight modifications of the above algorithm are recommended:

i) Storage considerations lead to a limitation of the number of columns of the extrapolation tableau, say by k_{max} (e.g., $k_{\text{max}} = 9$). For the proposed index k_{new} we require $2 \leq k_{\text{new}} \leq k_{\text{max}} - 1$. This allows us to activate (e) at each step.

ii) After a step-rejection the step size and the order may not be increased.

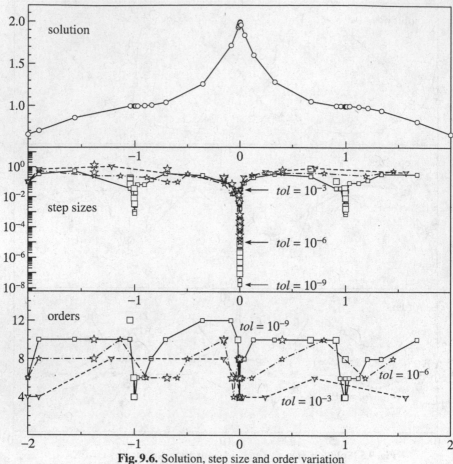

Fig. 9.6. Solution, step size and order variation
obtained by ODEX at the discontinuous example (9.33)

Numerical study of the combined step size and order control. We show in the following examples how the step size and the order vary for the above algorithm. For this purpose we have written the FORTRAN-subroutine ODEX (see Appendix).

As a first example we again take the *Brusselator* (cf. Section II.4). As in Fig. 4.1, the first picture of Fig. 9.5 shows the two components of the solution (obtained with $Atol = Rtol = 10^{-9}$). In the remaining two pictures we have plotted the step sizes and orders for the three tolerances 10^{-3} (broken line), 10^{-6} (dashes and dots) and 10^{-9} (solid line). One can easily observe that the extrapolation code automatically chooses a suitable order (depending essentially on *Tol*). Step-rejections are indicated by larger symbols.

We next study the behaviour of the order control near discontinuities. In the example

$$y' = -\text{sign}(x) \, |1 - |x|| \, y^2, \qquad y(-2) = 2/3, \qquad -2 \le x \le 2 \qquad (9.33)$$

we have a discontinuity in the first derivative of $y(x)$ at $x = 0$ and two discontinuities in the second derivative (at $x = \pm 1$). The numerical results are shown in Fig. 9.6 for three tolerances. In all cases the error at the endpoint is about $10 \cdot Tol$. The discontinuities at $x = \pm 1$ are not recognized in the computations with $Tol = 10^{-3}$ and $Tol = 10^{-6}$. Whenever a discontinuity is detected, the order drops to 4 (lowest possible) in its neighbourhood, so that these points are passed rather efficiently.

Dense Output for the GBS Method

Extrapolation methods are methods best suited for high precision which typically take very large (basic) step sizes during integration. The reasons for the need of a dense output formula (discussed in Section II.6) are therefore particularly important here. First attempts to provide extrapolation methods with a dense output are due to Lindberg (1972) for the implicit trapezoidal rule, and to Shampine, Baca & Bauer (1983) who constructed a 3rd order dense output for the GBS method. We present here the approach of Hairer & Ostermann (1990) (see also Simonsen 1990).

It turned out that the existence of high order dense output is only possible if the step number sequence satisfies some restrictions such as

$$n_{j+1} - n_j = 0 \, (\text{mod } 4) \qquad \text{for} \quad j = 1, 2, 3, \ldots \qquad (9.34)$$

which, for example, is fulfilled by the sequence

$$\{2, 6, 10, 14, 18, 22, 26, 30, 34, \ldots\}. \qquad (9.35)$$

The idea is, once again, to do Hermite interpolation. To begin with, high order approximations are as usual at our disposal for the values y_0, y_0', y_1, y_1' by using y_0, $f(x_0, y_0)$, T_{kk}, $f(x_0 + H, T_{kk})$, where T_{kk} is supposed to be the highest order approximation computed and used for continuation of the solution.

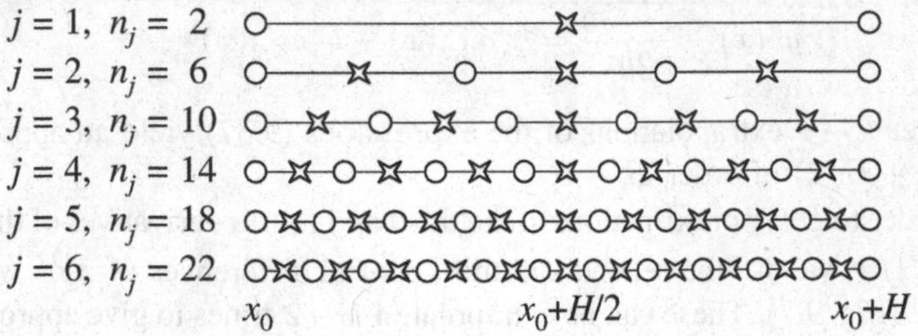

Fig. 9.7. Evaluation points for a GBS step

For more inspiration, we represent in Fig. 9.7 the steps taken by Gragg's midpoint rule for the step number sequence (9.35). The symbols $\circ$ and $\times$ indicate that the even steps and the odd steps possess a *different* asymptotic expansion (see Theorem 9.3) and must not be blended. We see that, owing to condition (9.34), the midpoint values $y_{n_j/2}^{(j)}$, obtained during the computation of T_{j1}, all have the same parity and can therefore also be extrapolated to yield an approximation for $y(x_0 + H/2)$ of order $2k - 1$ (remember that in Theorem 9.3, $\widehat{b}_{2j}(x_0) \neq 0$).

We next insert (9.20) for $x = x_0 + H/2$ into $f(x, y)$

$$f_{n_j/2}^{(j)} := f(x, y_{n_j/2}^{(j)}) = f\left(x, y(x) - h_j^2 \widehat{b}_2(x) - h_j^4 \widehat{b}_4(x) \ldots\right)$$

and develop in powers of h_j to obtain

$$y'(x) - f_{n_j/2}^{(j)} = h_j^2 a_{2,1}(x) + h_j^4 a_{4,1}(x) + \ldots . \qquad (9.36)$$

This shows that the f-values at the midpoint $x_0 + H/2$ (for $j = 1, 2, \ldots k$) possess an asymptotic expansion and can be extrapolated $k - 1$ times to yield an approximation to $y'(x_0 + H/2)$ of order $2k - 1$.

But this is not enough. We now consider, similar to an idea which goes back to the papers of Deuflhard & Nowak (1987) and Lubich (1989), the central differences $\delta f_i = f_{i+1} - f_{i-1}$ at the midpoint which, by Fig. 9.7, are available for $j = 1, 2, \ldots, k$ and are based on even parity. By using (9.18) and by developing into powers of h_j we obtain

$$\frac{\delta f_{n_j/2}^{(j)}}{2h_j} = \frac{f(x+h_j, y_{n_j/2+1}^{(j)}) - f(x-h_j, y_{n_j/2-1}^{(j)})}{2h_j}$$

$$= \left(f\left(x+h_j, y(x+h_j) - h_j^2 \widehat{a}_2(x+h_j) - h_j^4 \widehat{a}_4(x+h_j) - \ldots\right) - \right.$$

$$\left. f\left(x-h_j, y(x-h_j) - h_j^2 \widehat{a}_2(x-h_j) - h_j^4 \widehat{a}_4(x-h_j) - \ldots\right)\right) \Big/ 2h_j$$

$$= \frac{y'(x+h_j) - y'(x-h_j)}{2h_j} - h_j^2 c_2(x) - h_j^4 c_4(x) - \ldots .$$

Finally we insert the Taylor series for $y'(x+h)$ and $y'(x-h)$ to obtain an expansion

$$y''(x) - \frac{\delta f_{n_j/2}^{(j)}}{2h_j} = h_j^2 a_{2,2}(x) + h_j^4 a_{4,2}(x) + \ldots . \qquad (9.38)$$

Therefore, $k - 1$ extrapolations of the expressions (9.37) yield an approximation to $y''(x_0 + H/2)$ of order $2k - 1$.

In order to get approximations to the third and fourth derivatives of the solution at $x_0 + H/2$, we use the second and third central differences of $f_i^{(j)}$ which exist for $j \geq 2$ (Fig. 9.7). These can be extrapolated $k - 2$ times to give approximations of order $2k - 3$.

The continuation of this process yields the following algorithm:

Step 1. For each $j \in \{1, \ldots, k\}$, compute approximations to the derivatives of $y(x)$ at $x_0 + H/2$ by:

$$d_j^{(0)} = y_{n_j/2}^{(j)}, \qquad d_j^{(\kappa)} = \frac{\delta^{\kappa-1} f_{n_j/2}^{(j)}}{(2h_j)^{\kappa-1}} \qquad \text{for} \quad \kappa = 1, \ldots, 2j. \tag{9.39}$$

Step 2. Extrapolate $d_j^{(0)}$ $(k-1)$ times and $d_j^{(2\ell-1)}$, $d_j^{(2\ell)}$ $(k-\ell)$ times to obtain improved approximations $d^{(\kappa)}$ to $y^{(\kappa)}(x_0 + H/2)$.

Step 3. For given μ $(-1 \le \mu \le 2k)$ define the polynomial $P_\mu(\theta)$ of degree $\mu+4$ by

$$P_\mu(0) = y_0, \qquad\qquad P_\mu'(0) = Hf(x_0, y_0),$$
$$P_\mu(1) = T_{kk}, \qquad\qquad P_\mu'(1) = Hf(x_0+H, T_{kk}) \tag{9.40}$$
$$P_\mu^{(\kappa)}(1/2) = H^\kappa d^{(\kappa)} \qquad \text{for} \quad \kappa = 0, \ldots, \mu.$$

This computation of $P_\mu(\theta)$ does not need any further function evaluation since $f(x_0 + H, T_{kk})$ has to be computed anyway for the next step. Further, $P_\mu(\theta)$ gives a global $\mathcal{C}^1$ approximation to the solution.

Theorem 9.5 (Hairer & Ostermann 1990). *If the step number sequence satisfies (9.34), then the error of the dense output polynomial $P_\mu(\theta)$ satisfies*

$$y(x_0+\theta H) - P_\mu(\theta) = \begin{cases} \mathcal{O}(H^{2k+1}) & \text{if } n_1 = 4 \text{ and } \mu \ge 2k-4 \\ \mathcal{O}(H^{2k}) & \text{if } n_1 = 2 \text{ and } \mu \ge 2k-5. \end{cases} \tag{9.40}$$

Proof. Since $P_\mu(\theta)$ is a polynomial of degree $\mu+4$ the error due to interpolation is of size $\mathcal{O}(H^{\mu+5})$. This explains the restriction on μ in (9.40). As explained above, the function value and derivative data used for Hermite interpolation have the required precision

$$H^\kappa y^{(\kappa)}(x_0+H/2) - H^\kappa d^{(\kappa)} = \begin{cases} \mathcal{O}(H^{2k}) & \text{if } \kappa = 0, \\ \mathcal{O}(H^{2k+1}) & \text{if } \kappa \text{ is odd}, \\ \mathcal{O}(H^{2k+2}) & \text{if } \kappa \ge 2 \text{ is even}. \end{cases}$$

In the case $n_1 = 4$ the parity of the central point $x_0+H/2$ is *even* (in contrary to Fig. 9.7), we therefore apply (9.18) and gain one order because then the functions $a_{i,0}(x)$ vanish at x_0. $\square$

Control of the Interpolation Error

> At one time ... every young mathematician was familiar with $\operatorname{sn} u$, $\operatorname{cn} u$, and $\operatorname{dn} u$, and algebraic identities between these functions figured in every examination.
>
> (E.H. Neville, Jacobian Elliptic Functions, 1944)

Numerical example. We apply the above dense output formula with $\mu = 2k - 3$ (as is standard in ODEX) to the differential equations of the Jacobian elliptic functions sn, cn, dn (see Abramowitz & Stegun 1964, 16.16):

$$
\begin{aligned}
y_1' &= y_2 y_3   &\qquad y_1(0) &= 0 \\
y_2' &= -y_1 y_3   &\qquad y_2(0) &= 1 \\
y_3' &= -0.51 \cdot y_1 y_2 &\qquad y_3(0) &= 1
\end{aligned}
\tag{9.41}
$$

with integration interval $0 \le x \le 10$ and error tolerance $Atol = Rtol = 10^{-9}$. The error for the three components of the obtained continuous solution is displayed in Fig. 9.8 (upper picture; the ghosts are the solution curves) and gives a quite disappointing impression when compared with the precision at the grid points. We shall now see that these horrible bumps are nothing else than interpolation errors.

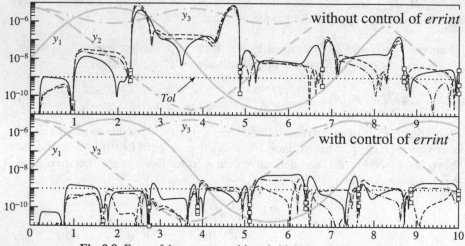

Fig. 9.8. Error of dense output without/with interpolation control

Assume that in the definition of $P_\mu(\theta)$ the basic function and derivative values are replaced by the exact values $y(x_0 + H)$, $y'(x_0 + H)$, and $y^{(\kappa)}(x_0 + H/2)$. Then the error of $P_\mu(\theta)$ is given by

$$
\theta^2 (1 - \theta)^2 \left(\theta - \frac{1}{2}\right)^{\mu+1} \frac{y^{(\mu+5)}(\xi)}{(\mu+5)!} H^{\mu+5}
\tag{9.42}
$$

where $\xi \in (x_0, x_0 + H)$ (possibly different for each component). The function $\theta^2(1-\theta)^2(\theta - 1/2)^{\mu+1}$ has its maximum at

$$\theta_{\mu+1} = \frac{1}{2} \pm \frac{1}{2}\sqrt{\frac{\mu+1}{\mu+5}} \tag{9.43}$$

which, for large μ, are close to the ends of the integration intervals and indicate precisely the locations of the large bumps in Fig. 9.8. This demonstrates the need for a code which not only controls the error at the grid points, but also takes care of the interpolation error. To this end we denote by a_μ the coefficient of $\theta^{\mu+4}$ in the polynomial $P_\mu(\theta)$ and consider (Hairer & Ostermann 1992)

$$P_\mu(\theta) - P_{\mu-1}(\theta) = \theta^2(1-\theta)^2\left(\theta - \frac{1}{2}\right)^\mu a_\mu \tag{9.44}$$

as an approximation for the interpolation error for $P_{\mu-1}(\theta)$ and use

$$errint = \|P_\mu(\theta_\mu) - P_{\mu-1}(\theta_\mu)\| \tag{9.45}$$

as error estimator (the norm is again that of (4.11)). Then, if $errint > 10$ the step is rejected and recomputed with

$$H_{\text{int}} = H(1/errint)^{1/(\mu+4)}$$

because $errint = \mathcal{O}(H^{\mu+4})$. Otherwise the subsequent step is computed subject to the restriction $H \leq H_{\text{int}}$.

This modified step size strategy makes the code, together with its dense output, more robust. The corresponding numerical results for the problem (9.41) are presented in the lower graph of Fig. 9.8.

Exercises

1. Show that the extrapolated Euler methods $T_{3,1}, T_{3,2}, T_{3,3}$ (with step-number sequence (9.8)) are equivalent to the Runge-Kutta methods of Table 9.1. Compute also the Runge-Kutta schemes corresponding to the first elements of the GBS algorithm.

Table 9.1. Extrapolation methods as Runge-Kutta methods

					0					0			
0					1/2	1/2				1/2	1/2		
1/3	1/3				1/3	1/3	0			1/3	1/3	0	
2/3	1/3	1/3			2/3	1/3	0	1/3		2/3	1/3	0	1/3
	1/3	1/3	1/3			0	−1	1	1		0	−2	3/2 3/2
	$T_{3,1}$ order 1					$T_{3,2}$ order 2					$T_{3,3}$ order 3		

2. Combine (9.18) and (9.19) into the formula ($x = x_0 + kh$)

$$y(x) - y_k = \sum_{j=1}^{\ell} \Big(\alpha_{2j}(x) + (-1)^k \beta_{2j}(x)\Big) h^{2j} + h^{2\ell+2} E(x, h)$$

for the asymptotic expansion of the Gragg method defined by (9.13a,b).

3. (Stetter 1970). Prove that for every real b (generally between 0 and 1) the method

$$y_1 = y_0 + h\Big(bf(x_0, y_0) + (1-b)f(x_1, y_1)\Big)$$

$$y_{i+1} = y_{i-1} + h\Big((1-b)f(x_{i-1}, y_{i-1}) + 2bf(x_i, y_i) + (1-b)f(x_{i+1}, y_{i+1})\Big)$$

possesses an expansion in powers of h^2. Prove the same property for the smoothing step

$$S_h(x) = \frac{1}{2}\Big(y_{2n} + y_{2n-1} + h(1-b)f(x_{2n-1}, y_{2n-1}) + hbf(x_{2n}, y_{2n})\Big).$$

4. (Stetter 1970). Is the Euler step (9.13a) essential for an h^2-expansion? Prove that a first order starting procedure

$$y_1 = y_0 + h\Phi(x_0, y_0, h)$$

for (9.13a) produces an h^2-expansion if the quantities
$y_{-1} = y_0 - h\Phi(x_0, y_0, -h)$, y_0, and y_1 satisfy (9.13b) for $i = 0$.

5. Study the numerical instability of the extrapolation scheme for the harmonic sequence, i.e., suppose that the entries T_{11}, T_{21}, T_{31} ... are disturbed with rounding errors ε, $-\varepsilon$, ε, ... and compute the propagation of these errors into the extrapolation tableau (9.5).

Result. Due to the linearity of the extrapolation scheme, we suppose the T_{ik} equal zero and $\varepsilon = 1$. Then the results for sequence (9.8') are

```
   1.
  -1.  -1.67
   1.    2.60     3.13
  -1.  -3.57    -5.63    -6.21
   1.    4.56     9.13    11.94    12.69
  -1.  -5.55   -13.63   -21.21   -25.35   -26.44
   1.    6.54    19.13    35.01    47.65    54.14    55.82
  -1.  -7.53   -25.63   -54.31   -84.09  -105.64  -116.30  -119.03
   1.    8.53    33.13    80.13   140.14   195.34   232.96   251.10  255.73
```

hence, for order 18, we lose approximately two digits due to roundoff errors.

6. (Laguerre 1883[*]). If $a_1, a_2, \ldots, a_n$ are distinct positive real numbers and $r_1, r_2, \ldots, r_n$ are distinct reals, then

$$A = \begin{pmatrix} a_1^{r_1} & a_1^{r_2} & \cdots & a_1^{r_n} \\ a_2^{r_1} & a_2^{r_2} & \cdots & a_2^{r_n} \\ \vdots & \vdots & & \vdots \\ a_n^{r_1} & a_n^{r_2} & \cdots & a_n^{r_n} \end{pmatrix}$$

is invertible.

Hint (Pólya & Szegö 1925, Vol. II, Abschn. V, Problems 76-77[*]). Show by induction on n that, if the function $g(t) = \sum_{i=1}^{n} \alpha_i t^{r_i}$ has n distinct positive zeros, then $g(t) \equiv 0$. By Rolle's theorem the function

$$\frac{d}{dt}\left(t^{-r_1} g(t)\right) = \sum_{i=2}^{n} \alpha_i (r_i - r_1) t^{r_i - r_1 - 1}$$

has $n - 1$ positive distinct zeros and the induction hypothesis can be applied.

---

[*] We are grateful to our colleague J. Steinig for these references.

II.10 Numerical Comparisons

> The Pleiades seem to be among the first stars mentioned in astronomical liter-
> ature, appearing in Chinese annals of 2357 B.C. ...
> > (R.H. Allen, Star names, their love and meaning, 1899, Dover 1963)

> If you enjoy fooling around making pictures, instead of typesetting ordinary
> text, TEX will be a source of endless frustration/amusement for you, ...
> > (D. Knuth, The TEXbook, p. 389)

Problems

EULR — Euler's equation of rotation of a rigid body ("Diese merkwürdig sym-
metrischen und eleganten Formeln ...", A. Sommerfeld 1942, vol. I, § 26.1, Euler
1758)

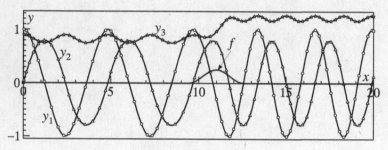

Fig. 10.1. Solutions of Euler's equations (10.1)

$$I_1 \, y_1' = (I_2 - I_3) \, y_2 y_3$$
$$I_2 \, y_2' = (I_3 - I_1) \, y_3 y_1$$
$$I_3 \, y_3' = (I_1 - I_2) \, y_1 y_2 + f(x) \tag{10.1}$$

where y_1, y_2, y_3 are the coordinates of $\vec{\omega}$, the rotation vector, and I_1, I_2, I_3 are the
principal moments of inertia. The third coordinate has an additional exterior force

$$f(x) = \begin{cases} 0.25 \cdot \sin^2 x & \text{if } 3\pi \le x \le 4\pi \\ 0 & \text{otherwise} \end{cases} \tag{10.1'}$$

which is discontinuous in its second derivative. We choose the constants and initial
values as

$$I_1 = 0.5, \;\; I_2 = 2, \;\; I_3 = 3, \;\; y_1(0) = 1, \;\; y_2(0) = 0, \;\; y_3(0) = 0.9$$

(see Fig. 10.1) and check the numerical precision at the output points

$$x_{\text{end}} = 10 \quad \text{and} \quad x_{\text{end}} = 20 .$$

AREN — the Arenstorf orbit (0.1) for the restricted three body problem with initial values (0.2) integrated óver one period $0 \leq x \leq x_{\text{end}}$ (see Fig. 0.1). The precision is checked at the endpoint, here the solution is most sensitive to errors of the initial phase.

LRNZ — the solution of the Saltzman-Lorenz equations (I.16.17) displayed in Fig. I.16.8, i.e., with constants and initial values

$$\sigma = 10, \quad r = 28, \quad b = \frac{8}{3}, \quad y_1(0) = -8, \quad y_2(0) = 8, \quad y_3(0) = 27 . \tag{10.2}$$

The solution is, for large values of x, *extremely* sensitive to the errors of the first integration steps (see Fig. I.16.10 and its discussion). For example, at $x = 50$ the numerical solution becomes totally wrong, even if the computations are performed in quadruple precision with $Tol = 10^{-20}$. Hence the numerical results of *all* methods would be equally useless and no comparison makes any sense. Therefore we choose

$$x_{\text{end}} = 16$$

and check the numerical solution at this point. Even here, all computations with $Tol \geq 10^{-7}$, say, fall into a chaotic cloud of meaningless results (see Fig. 10.5).

PLEI — a celestial mechanics problem (which we call "the Pleiades"): seven stars in the plane with coordinates x_i, y_i and masses $m_i = i$ $(i = 1, \ldots, 7)$:

$$x_i'' = \sum_{j \neq i} m_j(x_j - x_i)/r_{ij}$$
$$y_i'' = \sum_{j \neq i} m_j(y_j - y_i)/r_{ij} \tag{10.3}$$

where

$$r_{ij} = \left((x_i - x_j)^2 + (y_i - y_j)^2\right)^{3/2}, \qquad i, j = 1, \ldots, 7.$$

The initial values are

$$
\begin{aligned}
&x_1(0) = 3, &&x_2(0) = 3, &&x_3(0) = -1, &&x_4(0) = -3, \\
&x_5(0) = 2, &&x_6(0) = -2, &&x_7(0) = 2, \\
&y_1(0) = 3, &&y_2(0) = -3, &&y_3(0) = 2, &&y_4(0) = 0, \\
&y_5(0) = 0, &&y_6(0) = -4, &&y_7(0) = 4, \\
&x_i'(0) = y_i'(0) = 0, &&\text{for all } i \text{ with the exception of} \\
&x_6'(0) = 1.75, &&x_7'(0) = -1.5, &&y_4'(0) = -1.25, &&y_5'(0) = 1,
\end{aligned}
\tag{10.4}
$$

and we integrate for $0 \leq t \leq t_{\text{end}} = 3$. Fig. 10.2a represents the movement of these 7 bodies in phase coordinates. The initial value is marked by an "i", the final value at $t = t_{\text{end}}$ is marked by an "f". Between these points, 19 time-equidistant output points are plotted and connected by a dense output formula. There occur several quasi-collisions which are displayed in Table 10.1.

Table 10.1. Quasi-collisions in the PLEI problem

Body$_1$	1	1	3	1	2	5
Body$_2$	7	3	5	7	6	7
r_{ij}^2	0.0129	0.0193	0.0031	0.0011	0.1005	0.0700
time	1.23	1.46	1.63	1.68	1.94	2.14

The resulting violent shapes of the derivatives $x_i'(t), y_i'(t)$ are displayed in Fig. 10.2b and show that automatic step size control is essential for this example.

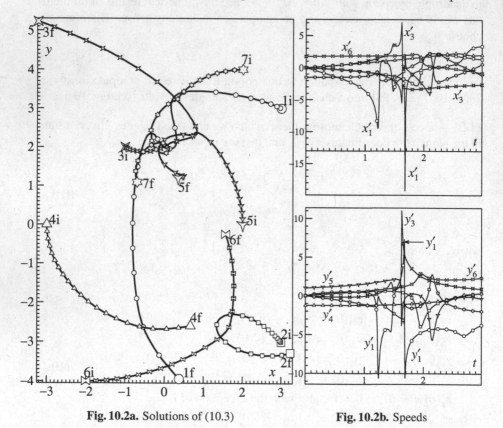

Fig. 10.2a. Solutions of (10.3)   **Fig. 10.2b.** Speeds

ROPE — the movement of a hanging rope (see Fig. 10.3a) of length 1 under gravitation and under the influence of a horizontal force

$$F_y(t) = \left(\frac{1}{\cosh(4t - 2.5)}\right)^4 \tag{10.5a}$$

acting at the point $s = 0.75$ as well as a vertical force

$$F_x(t) = 0.4 \tag{10.5b}$$

acting at the endpoint $s = 1$.

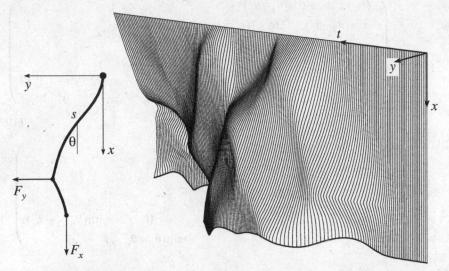

Fig. 10.3a. Hanging rope   **Fig. 10.3b.** Solution for $0 \leq t \leq 3.723$.

If this problem is discretized, then Lagrange theory (see (I.6.18); see also Exercises IV.1.2 and IV.1.4 of Volume II) leads to the following equations for the unknown angles θ_k:

$$\sum_{k=1}^{n} a_{lk} \ddot{\theta}_k = -\sum_{k=1}^{n} b_{lk} \dot{\theta}_k^2 - n\left(n + \frac{1}{2} - l\right) \sin \theta_l \tag{10.6}$$

$$- n^2 \sin \theta_l \cdot F_x(t) + \begin{cases} n^2 \cos \theta_l \cdot F_y(t) & \text{if } l \leq 3n/4 \\ 0 & \text{if } l > 3n/4, \end{cases} \quad l = 1, \ldots, n$$

where

$$a_{lk} = g_{lk} \cos(\theta_l - \theta_k), \qquad b_{lk} = g_{lk} \sin(\theta_l - \theta_k), \qquad g_{lk} = n + \frac{1}{2} - \max(l, k). \tag{10.7}$$

We choose

$$n = 40, \qquad \theta_l(0) = \dot{\theta}_l(0) = 0, \qquad 0 \leq t \leq 3.723. \tag{10.8}$$

The resulting system is of dimension 80. The special structure of G^{-1} (see (IV.1.16–18) of Volume II) allows one to evaluate $\ddot{\theta}_l$ with the following algorithm:

a) Let $v_l = -n\left(n+\frac{1}{2}-l\right)\sin\theta_l - n^2\sin\theta_l \cdot F_x + \begin{cases} n^2\cos\theta_l \cdot F_y \\ 0 \end{cases}$

b) Compute $w = Dv + \dot{\theta}^2$,

c) Solve the tridiagonal system $Cu = w$,

d) Compute $\ddot{\theta} = Cv + Du$,

where

$$
C = \begin{pmatrix}
1 & -\cos(\theta_1-\theta_2) & & & & \\
-\cos(\theta_2-\theta_1) & 2 & -\cos(\theta_2-\theta_3) & & & \\
& -\cos(\theta_3-\theta_2) & \ddots & & \ddots & \\
& & \ddots & & 2 & -\cos(\theta_{n-1}-\theta_n) \\
& & & & -\cos(\theta_n-\theta_{n-1}) & 3
\end{pmatrix}.
$$

$$(10.9)$$

$$
D = \begin{pmatrix}
0 & -\sin(\theta_1-\theta_2) & & & & \\
-\sin(\theta_2-\theta_1) & 0 & -\sin(\theta_2-\theta_3) & & & \\
& -\sin(\theta_3-\theta_2) & \ddots & & \ddots & \\
& & \ddots & & 0 & -\sin(\theta_{n-1}-\theta_n) \\
& & & & -\sin(\theta_n-\theta_{n-1}) & 0
\end{pmatrix}.
$$

BRUS — the reaction-diffusion equation (Brusselator with diffusion)

$$
\begin{aligned}
\frac{\partial u}{\partial t} &= 1 + u^2v - 4.4u + \alpha\left(\frac{\partial^2 u}{\partial x^2} + \frac{\partial^2 u}{\partial y^2}\right) \\
\frac{\partial v}{\partial t} &= 3.4u - u^2v + \alpha\left(\frac{\partial^2 v}{\partial x^2} + \frac{\partial^2 v}{\partial y^2}\right)
\end{aligned}
$$

$$(10.10)$$

for $0 \le x \le 1$, $0 \le y \le 1$, $t \ge 0$, $\alpha = 2 \cdot 10^{-3}$ together with the Neumann boundary conditions

$$
\frac{\partial u}{\partial \mathbf{n}} = 0, \qquad \frac{\partial v}{\partial \mathbf{n}} = 0, \tag{10.11}
$$

and the initial conditions

$$
u(x,y,0) = 0.5 + y, \qquad v(x,y,0) = 1 + 5x. \tag{10.12}
$$

By the method of lines (cf. Section I.6) this problem becomes a system of ordinary differential equations. We put

$$
x_i = \frac{i-1}{N-1}, \qquad y_j = \frac{j-1}{N-1}, \qquad i,j = 1,\ldots,N
$$

and define
$$U_{ij}(t) = u(x_i, y_j, t), \qquad V_{ij}(t) = v(x_i, y_j, t). \tag{10.13}$$

Discretizing the derivatives in (10.10) with respect to the space variables we obtain for $i, j = 1, \ldots, N$

$$U'_{ij} = 1 + U_{ij}^2 V_{ij} - 4.4 U_{ij} + \alpha(N-1)^2 \Big(U_{i+1,j} + U_{i-1,j} + U_{i,j+1} + U_{i,j-1} - 4U_{ij} \Big)$$

$$V'_{ij} = 3.4 U_{ij} - U_{ij}^2 V_{ij} + \alpha(N-1)^2 \Big(V_{i+1,j} + V_{i-1,j} + V_{i,j+1} + V_{i,j-1} - 4V_{ij} \Big),$$
$$\tag{10.14}$$

an ODE of dimension $2N^2$. Because of the boundary condition (10.11) we have

$$U_{0,j} = U_{2,j}, \quad U_{N+1,j} = U_{N-1,j}, \quad U_{i,0} = U_{i,2}, \quad U_{i,N+1} = U_{i,N-1}$$

and similarly for the V_{ij}-quantities. We choose $N = 21$ so that the system is of dimension 882 and check the numerical solutions at the output point $t_{\text{end}} = 7.5$. The solution of (10.14) (in the (x, y)-space) is represented in Fig. 10.4a and Fig. 10.4b for u and v respectively.

Performance of the Codes

Several codes were applied to each of the test problems with $Tol = 10^{-3}$, $Tol = 10^{-3-1/8}$, $Tol = 10^{-3-2/8}$, $Tol = 10^{-3-3/8}, \ldots$ (for the large problems with $Tol = 10^{-3}$, $Tol = 10^{-3-1/4}$, $Tol = 10^{-3-2/4}, \ldots$) up to, in general, $Tol = 10^{-14}$, then the numerical result at the output points were compared with an "exact solution" (computed very precisely in quadruple precision). Each of these results then corresponds to one point of Fig. 10.5, where this precision is compared (in double logarithmic scale) to the number of function evaluations. The "integer" tolerances 10^{-3}, 10^{-4}, $10^{-5}, \ldots$ are distinguishable as enlarged symbols. All codes were applied with complete "standard" parameter settings and were not at all "tuned" to these particular problems.

A comparison of the *computing time* (instead of the number of function evaluations) gave no significant difference. Therefore, only one representative of the small problems (LRNZ) and one large problem (BRUS) are displayed in Fig. 10.6. All computations have been performed in REAL*8 ($Uround = 1.11 \cdot 10^{-16}$) on a Sun Workstation (SunBlade 100).

The codes used are the following:

RKF45 — symbol ⋈ — a product of Shampine and Watts' programming art based on Fehlberg's pair of orders 4 and 5 (Table 5.1). The method is used in the "local extrapolation mode", i.e., the numerical solution is advanced with the 5th order result. The code is usually, except for low precision, the slowest of all, which is explained by its low order. The results of the "time"-picture Fig. 10.6 for this

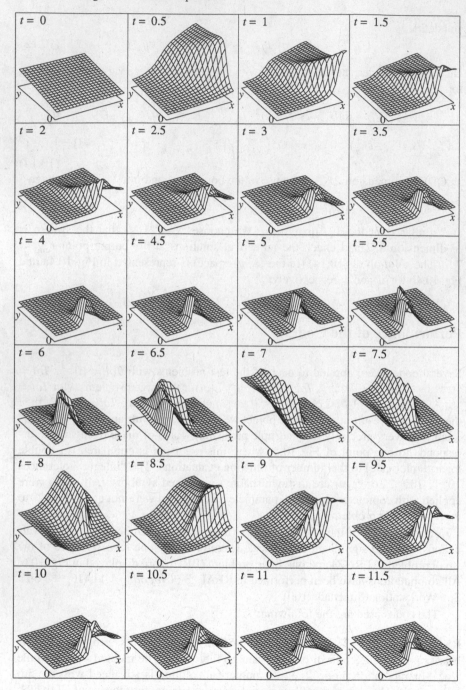

Fig. 10.4a. Solution $u(x, y, t)$ for the BRUS problem

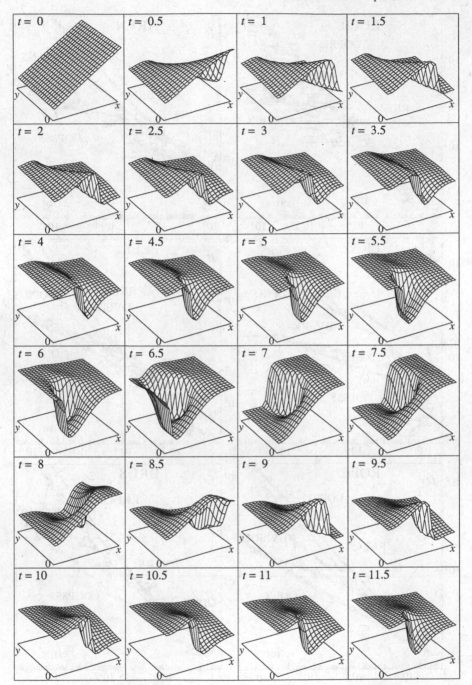

Fig. 10.4b. Solution $v(x, y, t)$ for the BRUS problem

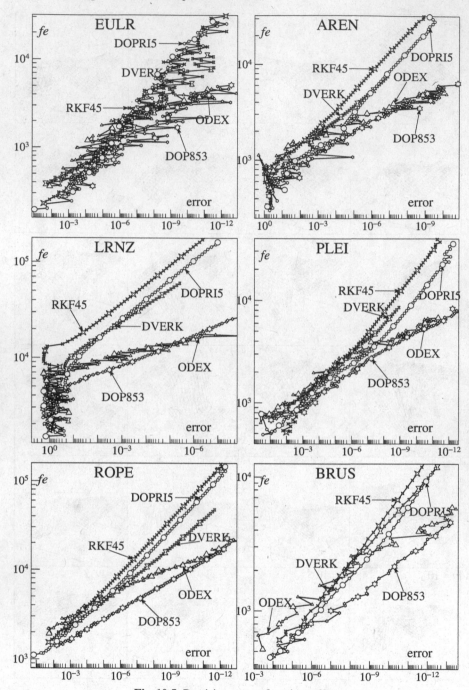

Fig. 10.5. Precision versus function calls

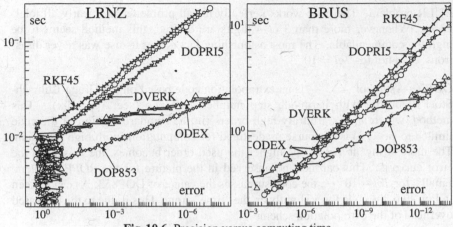

Fig. 10.6. Precision versus computing time

code are relatively better than those on the "function calls" front (Fig. 10.5). This indicates that the code has particularly small overhead.

DOPRI5 — symbol $\bigcirc$ — the method of Dormand & Prince of order 5 with embedded error estimator of order 4 (see Table 5.2). The code is explained in the Appendix. The method has precisely the same order as that used in RKF45, but the error constants are much more optimized. Therefore the "error curves" in Fig. 10.5 are nicely parallel to those of RKF45, but appear translated to the side of higher precision. One usually gains between a half and one digit of numerical precision for comparable numerical work. The code performs specially well between $Tol = 10^{-3}$ and $Tol = 10^{-8}$ in the AREN problem. This is simply due to an accidental sign change of the error for the most sensitive solution component.

DVERK — symbol $\boxtimes$ — this widely known code implements Verner's 6th order method of Table 5.4 and was written by Hull, Enright & Jackson. It has been included in the IMSL library for many years and the source code is available through na-net. The corresponding error curves in Fig. 10.5 appear to be less steep than those of DOPRI5, which illustrates the higher order of the method. However, the error constants seem to be less optimal so that this code surpasses the performance of DOPRI5 only for very stringent tolerances. It is significantly better than DOPRI5 solely in problems EULR and ROPE. The code, as it was, failed at the BRUS problem for $Tol = 10^{-3}$ and $Tol = 10^{-4}$. Therefore these computations were started with $Tol = 10^{-5}$.

DOP853 — symbol $\stackrel{\wedge}{\leftrightarrow}$ — is the method of Dormand & Prince of order 8 explained in Section II.5 (formulas (5.20) – (5.30), see Appendix). The 6th order error estimator (5.29), (5.30) has been replaced by a 5th order estimator with 3rd order correction (see below). This was necessary to make the code robust for the

EULR problem. The code works perfectly for all problems and nearly all tolerances. Whenever more than 3 or 4 digits are desired, this method seems to be
highly recommendable. The most astonishing fact is that its use was never disas-
trous, even not for $Tol = 10^{-3}$.

ODEX — symbol $\triangle$ — is an extrapolation code based on the Gragg-Bulirsch-
Stoer algorithm with harmonic step number sequence (see Appendix). This
method, which allows arbitrary high orders (in the standard version of the code
limited to $p \leq 18$) is of course predestined for computations with high precision.
The more stringent Tol is, the higher the used order becomes, the less steep the
error curve is. This can best be observed in the picture for the ROPE problem.
Finally, for $Tol \approx 10^{-12}$, the code surpasses the values of DOP853. As can be seen
in Fig. 10.6, the code loses slightly on the "time"-front. This is due to the increased
overhead of the extrapolation scheme.

The numerical results of ODEX behave very similarly to those of DIFEX1
(Deuflhard 1983).

A "Stretched" Error Estimator for DOP853

In preliminary stages of our numerical tests we had written a code "DOPR86"
based on the method of order 8 of Dormand & Prince with the 6th order error
estimator described in Section II.5. For most problems the results were excellent.
However, there are some situations in which the error control of DOPR86 did not
work safely:

When applied to the BRUS problem with $Tol = 10^{-3}$ or $Tol = 10^{-4}$ the code
stopped with an overflow message. The reason was the following: when the step
size is too large, the internal stages are too far away from the solution and their
modulus increases at each stage (e.g., by a factor 10^5 between stage 11 and stage
12). Due to the fact that $\widehat{b}_{12} = b_{12}$ (see (5.30) (5.26) and (5.25b)) the difference
$\widehat{y}_1 - y_1$ is not influenced by the last stage and is smaller (by a factor of 10^5) than
the modulus of y_1. Hence, the error estimator scaled by (4.10) is $\leq 10^{-5}$ and a
completely wrong step will be accepted.

The code DOPR86 also had severe difficulties when applied to problems with
discontinuities such as EULR. The worst results were obtained for the problem

$$
\begin{aligned}
y_1' &= y_2 y_3 & y_1(0) &= 0 \\
y_2' &= -y_3 y_1 & y_2(0) &= 1 \\
y_3' &= -0.51 \cdot y_1 y_2 + f(x) & y_3(0) &= 1
\end{aligned}
\tag{10.15}
$$

where $f(x)$, given in (10.1'), has a discontinuous second derivative. The results for this problem and the code DOPR86 for very many different Tol values
($Tol = 10^{-3}, 10^{-3-1/24}, 10^{-3-2/24}, \ldots, 10^{-14}$) are displayed in Fig. 10.7. There,

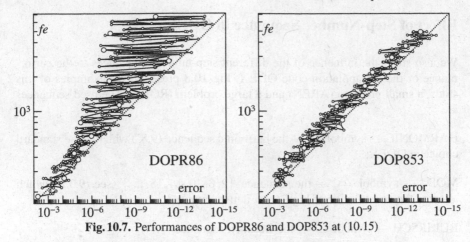

Fig. 10.7. Performances of DOPR86 and DOP853 at (10.15)

the (dotted) diagonal is of exact slope $1/8$ and represents the theoretical convergence speed of the method of order 8. It can be observed that this convergence is well attained by *some* results, but others lose precision of up to 8 digits from the desired tolerance. We explain this disappointing behaviour by the fact that $\widehat{b}_{12} = b_{12}$ and that the 12th stage is the only one where the function is evaluated at the endpoint of the step. Whenever the discontinuity of f'' is by accident slightly to the left of a grid point, the error estimator ignores it and the code reports a wrong value.

Unfortunately, the basic 8th order method does not possess a 6th order embedding with $\widehat{b}_{12} \neq b_{12}$ (unless additional function evaluations are used). Therefore, we decided to construct a 5th order approximation $\widehat{y}_1$. It can be obtained by taking $\widehat{b}_6, \widehat{b}_7, \widehat{b}_{12}$ as free parameters, e.g.,

$$\widehat{b}_6 = b_6/2 + 1, \qquad \widehat{b}_7 = b_7/2 + 0.45, \qquad \widehat{b}_{12} = b_{12}/2,$$

by putting $\widehat{b}_2 = \widehat{b}_3 = \widehat{b}_4 = \widehat{b}_5 = 0$ and by determining the remaining coefficients such that this quadrature formula has order 5. Due to the simplifying assumptions (5.20) all conditions for order 5 are then satisfied. In order to prevent a serious *over*-estimation of the error, we consider a second embedded method $\widetilde{y}_1$ of order 3 based on the nodes $c_1 = 0$, c_9 and $c_{12} = 1$ so that two error estimators

$$err_5 = \|\widehat{y}_1 - y_1\| = \mathcal{O}(h^6), \qquad err_3 = \|\widetilde{y}_1 - y_1\| = \mathcal{O}(h^4) \qquad (10.16)$$

are available. Similarly to a procedure which is common for quadrature formulas (R. Piessens, E. de Doncker-Kapenga, C.W. Überhuber & D.K. Kahaner 1983, Berntsen & Espelid 1991) we consider

$$err = err_5 \cdot \frac{err_5}{\sqrt{err_5^2 + 0.01 \cdot err_3^2}} = \mathcal{O}(h^8) \qquad (10.17)$$

as error estimator. It behaves asymptotically like the global error of the method. The corresponding code DOP853 gives satisfactory results for all the above problems (see right picture in Fig. 10.7).

Effect of Step-Number Sequence in ODEX

We also study the influence of the different step-number sequences to the performance of the extrapolation code ODEX. Fig. 10.8 presents two examples of this study, a small problem (AREN) and a large problem (ROPE). The used sequences are

HARMONIC — symbol ○ — the harmonic sequence (9.8') which is the standard choice in ODEX;

MOD4 — symbol △ — the sequence $\{2, 6, 10, 14, 18, \ldots\}$ (see (9.35)) which allowed the construction of high-order dense output;

BULIRSCH — symbol □ — the Bulirsch sequence (9.7');

ROMBERG — symbol ⬠ — the Romberg sequence (9.6');

DNSECTRL — symbol ✳ — the error control for the MOD4 sequence taking into account the interpolation error of the dense output solution (9.42). This is included only in the small problem, since (complete) dense output on large problems would need too much memory.

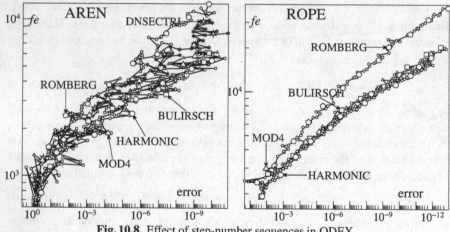

Fig. 10.8. Effect of step-number sequences in ODEX

Discussion. With the exception of the clear inferiority of the Romberg sequence, especially for high precision, and a certain price to be paid for the dense output error control, there is not much difference between the first three sequences. Although the harmonic sequence appears to be slightly superior, the difference is statistically not very significant.

II.11 Parallel Methods

> We suppose that we have a computer with a number of arithmetic
> processors capable of simultaneous operation and seek to devise
> parallel integration algorithms for execution on such a computer.
>
> (W.L. Miranker & W. Liniger 1967)
>
> "PARALYSING ODES"   (K. Burrage, talk in Helsinki 1990)

Parallel machines are computers with more than one processor and this facility
might help us to speed up the computations in ordinary differential equations. This
is particularly interesting for very large problems, for very costly function evaluation, or for fast real-time simulations. A second motivation is the desire to make a
code, with the help of parallel computations, not necessarily faster but more robust
and reliable.

Early attempts for finding parallel methods are Nievergelt (1964) and Miranker
& Liniger (1967). See also the survey papers Miranker (1971) and Jackson (1991).

We distinguish today essentially between two types of parallel architectures:

SIMD (single instruction multiple data): all processors execute the same in-
structions with possibly different input data.

MIMD (multiple instruction multiple data): the different processors can act
independently.

The exploitation of parallelism for an ordinary differential equation

$$y' = f(x, y), \qquad y(x_0) = y_0 \tag{11.1}$$

can be classified into two main categories (Gear 1987, 1988):

Parallelism across the system. Often the problem itself offers more or less trivial
applications for parallelism, e.g.,

> if several solutions are required for various initial or parameter values;

> if the right-hand side of (11.1) is very costly, but structured in such a way that
  the computation of *one* function evaluation can be split efficiently across the
  various processors;

> space discretizations of partial differential equations (such as the Brusselator
  problem (10.14)) whose function evaluation can be done simultaneously for all
  components on an SIMD machine with thousands of processors;

> the solution of boundary value problems with the multiple shooting method
  (see Section I.15) where all computations on the various sub-intervals can be
  done in parallel;

> doing all the high-dimensional linear algebra in the Runge-Kutta method (11.2) in parallel;

> parallelism in the linear algebra for Newton's method for *implicit* Runge-Kutta methods (see Section IV.8).

These types of parallelism, of course, depend strongly on the problem and on the type of the computer.

Parallelism across the method. This is problem-independent and means that, due to a special structure of the method, several function values can be evaluated in parallel within one integration step. This will be discussed in this section in more detail.

Parallel Runge-Kutta Methods

> ... it seems that *explicit* Runge-Kutta methods are not facilitated much by parallelism at the method level.
>
> (Iserles & Nørsett 1990)

Consider an explicit Runge-Kutta method

$$k_i = f\left(x_0 + c_i h, y_0 + h \sum_{j=1}^{i-1} a_{ij} k_j\right), \quad i = 1, \ldots, s$$

$$y_1 = y_0 + h \sum_{i=1}^{s} b_i k_i. \tag{11.2}$$

Suppose, for example, that the coefficients have the zero-pattern indicated in Fig. 11.1.

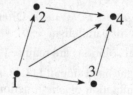

Fig. 11.1. Parallel method   **Fig. 11.2.** Production graph

Each arrow in the corresponding "production graph" G (Fig. 11.2), pointing from vertex "i" to vertex "j", stands for a non-zero a_{ji}. Here the vertices 2 and 3 are independent and can be evaluated in parallel. We call the number of vertices in the longest chain of successive arrows (here 3) the *number of sequential function evaluations* σ.

In general, if the Runge-Kutta matrix A can be partitioned (possibly after a permutation of the stages) as

$$A = \begin{pmatrix} 0 & & & & \\ A_{21} & 0 & & & \\ A_{31} & A_{32} & 0 & & \\ \vdots & \vdots & & \ddots & \\ A_{\sigma 1} & A_{\sigma 2} & \cdots & A_{\sigma, \sigma-1} & 0 \end{pmatrix}, \tag{11.3}$$

where A_{ij} is a matrix of size $\mu_i \times \mu_j$, then the derivatives $k_1, \ldots, k_{\mu_1}$ as well as $k_{\mu_1+1}, \ldots, k_{\mu_1+\mu_2}$, and so on, can be computed in parallel and one step of the method is executed in σ sequential function evaluations (if $\mu = \max_i \mu_i$ processors are at disposal). The following theorem is a severe restriction on parallel methods. It appeared in hand-written notes by K. Jackson & S. Nørsett around 1986. For a publication see Jackson & Nørsett (1992) and Iserles & Nørsett (1990).

Theorem 11.1. *For an explicit Runge-Kutta method with σ sequential stages the order p satisfies*

$$p \le \sigma, \tag{11.4}$$

for any number μ of available processors.

Proof. Each non-zero term of the expressions $\Phi_i(t)$ for the "tall" trees t_{21}, t_{32}, t_{44}, $t_{59}, \ldots$ (see Table 2.2 and Definition 2.9) $\sum a_{ij} a_{jk} a_{k\ell} a_{\ell m} \cdots$ corresponds to a connected chain of arrows in the production graph. Since their length is limited by σ, these terms are all zero for $\varrho(t) > \sigma$. $\square$

Methods with $p = \sigma$ will be called *P-optimal methods*. The Runge-Kutta methods of Section II.1 for $p \le 4$ are all P-optimal. Only for $p > 4$ does the subsequent construction of P-optimal methods allow one to increase the order with the help of parallelism.

Remark. The fact that the "stability function" (see Section IV.2) of an explicit parallel Runge-Kutta method is a polynomial of degree $\le \sigma$ allows a second proof of Theorem 11.1. Further, P-optimal methods all have the same stability function $1 + z + z^2/2! + \ldots + z^\sigma/\sigma!$.

Parallel Iterated Runge-Kutta Methods

One possibility of constructing P-optimal methods is by fixed point iteration. Consider an arbitrary (explicit or implicit) Runge-Kutta method with coefficients

$$c = (c_1, \ldots, c_s)^T, \qquad A = (a_{ij})_{i,j=1}^s, \qquad b^T = (b_1, \ldots, b_s)$$

and define $\widehat{y}_1$ by

$$k_i^{(0)} = 0$$

$$k_i^{(\ell)} = f\left(x_0 + c_i h, y_0 + h \sum_{j=1}^s a_{ij} k_j^{(\ell-1)}\right), \qquad \ell = 1, \ldots, \sigma \tag{11.5}$$

$$\widehat{y}_1 = y_0 + h \sum_{i=1}^s b_i k_i^{(\sigma)}.$$

This algorithm can be interpreted as an explicit Runge-Kutta method with scheme

$$
\begin{array}{c|cccccc}
0 & 0 \\
c & A & 0 \\
c & 0 & A & 0 \\
\vdots & \vdots & & \ddots & \ddots \\
c & 0 & \cdots & 0 & A & 0 \\
\hline
& 0 & \cdots & 0 & 0 & b^T
\end{array}
\tag{11.6}
$$

It has σ sequential stages if s processors are available. To compute its order we use a Lipschitz condition for $f(x, y)$ and obtain

$$\max_i \|k_i^{(\ell)} - k_i\| \le Ch \cdot \max_i \|k_i^{(\ell-1)} - k_i\|$$

where k_i are the stage-vectors of the basic method. Since $k_i^{(0)} - k_i = \mathcal{O}(1)$ this implies $k_i^{(\sigma)} - k_i = \mathcal{O}(h^\sigma)$ and consequently the difference to the solution of the basic method satisfies $\widehat{y}_1 - y_1 = \mathcal{O}(h^{\sigma+1})$.

Theorem 11.2. *The parallel iterated Runge-Kutta method (11.5) is of order*

$$p = \min(p_0, \sigma), \tag{11.7}$$

if p_0 denotes the order of the basic method.

Proof. The statement follows from

$$\widehat{y}_1 - y(x_0 + h) = \widehat{y}_1 - y_1 + y_1 - y(x_0 + h) = \mathcal{O}(h^{\sigma+1}) + \mathcal{O}(h^{p_0+1}). \qquad \square$$

This theorem shows that the choice $\sigma = p_0$ in (11.5) yields P-optimal explicit Runge-Kutta methods (i.e., $\sigma = p$). If we take as basic method the s-stage collocation method based on the Gaussian quadrature ($p_0 = 2s$) then we obtain a method of order $p = 2s$ which is P-optimal on s processors. P.J. van der Houwen

& B.P. Sommeijer (1990) have done extensive numerical experiments with this method.

Extrapolation Methods

It turns out that the GBS-algorithm (Section II.9) without smoothing step is also P-optimal. Indeed, all the values T_{j1} can be computed independently of each other. If we choose the step number sequence $\{2, 4, 6, 8, 10, 12, \ldots\}$ then the computation of T_{k1} requires $2k$ sequential function evaluations. Hence, if k processors are available (one for each T_{j1}), the numerical approximation T_{kk}, which is of order $p = 2k$, can be computed with $\sigma = 2k$ sequential stages. When the processors are of type MIMD we can compute T_{11} and $T_{k-1,1}$ on one processor $(2 + 2(k-1) = 2k$ function evaluations). Similarly, T_{21} and $T_{k-2,1}$ occupy another processor, etc. In this way, the number of necessary processors is reduced by a factor close to 2 without increasing the number of sequential stages.

The order and step size strategy, discussed in Section II.9, should, of course, be adapted for an implementation on parallel computers. The "hope for convergence in line $k+1$" no longer makes sense because this part of the algorithm is now as costly as the whole step. Similarly, there is no reason to accept already $T_{k-1,k-1}$ as numerical approximation, because T_{kk} is computed on the same time level as $T_{k-1,k-1}$. Moreover, the numbers A_k of (9.25) should be replaced by $A_k = n_k$ which will in general increase the order used by the code.

Increasing Reliability

> ... using parallelism to improve *reliability* and *functionality* rather than efficiency. (W.H. Enright & D.J. Higham 1991)

For a given Runge-Kutta method parallel computation can be used to give a reliable error estimate or an accurate dense output. This has been advocated by Enright & Higham (1991) and will be the subject of this subsection.

Consider a Runge-Kutta method of order p, choose distinct numbers $0 = \sigma_0 < \sigma_1 < \ldots < \sigma_k = 1$ and apply the Runge-Kutta method in parallel with step sizes $\sigma_1 h, \ldots, \sigma_{k-1} h, \sigma_k h = h$. This gives approximations

$$y_{\sigma_i} \approx y(x_0 + \sigma_i h). \tag{11.8}$$

Then compute $f(x_0 + \sigma_i h, y_{\sigma_i})$ and do Hermite interpolation with the values

$$y_{\sigma_i}, \ hf(x_0 + \sigma_i h, y_{\sigma_i}), \qquad i = 0, 1, \ldots, k, \tag{11.9}$$

i.e., compute

$$u(\theta) = \sum_{i=0}^{k} v_i(\theta) y_{\sigma_i} + h \sum_{i=0}^{k} w_i(\theta) f(x_0 + \sigma_i h, y_{\sigma_i}) \tag{11.10}$$

where $v_i(\theta)$ and $w_i(\theta)$ are the scalar polynomials

$$\left. \begin{aligned} v_i(\theta) &= \ell_i^2(\theta) \cdot \left(1 - 2\ell_i'(\sigma_i)(\theta - \sigma_i)\right) \\ w_i(\theta) &= \ell_i^2(\theta) \cdot (\theta - \sigma_i) \end{aligned} \right\} \quad \text{with} \quad \ell_i(\theta) = \prod_{\substack{j=0 \\ j \neq i}}^{k} \frac{(\theta - \sigma_j)}{(\sigma_i - \sigma_j)} \, . \quad (11.11)$$

The interpolation error, which is $\mathcal{O}(h^{2k+2})$, may be neglected if $2k + 2 > p + 1$.

As to the choice of σ_i we denote the local error of the method by $le = y_1 - y(x_0 + h)$. It follows from Taylor expansion (see Theorem 3.2) that

$$y_{\sigma_i} - y(x_0 + \sigma_i h) = \sigma_i^{p+1} \cdot le + \mathcal{O}(h^{p+2})$$

and consequently the error of (11.10) satisfies (for $2k + 2 > p + 1$)

$$u(\theta) - y(x_0 + \theta h) = \left(\sum_{i=1}^{k} \sigma_i^{p+1} v_i(\theta) \right) \cdot le + \mathcal{O}(h^{p+2}). \quad (11.12)$$

The coefficient of le is equal to 1 for $\theta = 1$ and it is natural to search for suitable σ_i such that

$$\left| \sum_{i=1}^{k} \sigma_i^{p+1} v_i(\theta) \right| \leq 1 \quad \text{for all} \quad \theta \in [0, 1] \, . \quad (11.13)$$

Indeed, under the assumption $2k - 1 \leq p < 2k + 1$, it can be shown that numbers $0 = \sigma_0 < \sigma_1 < \ldots < \sigma_{k-1} < \sigma_k = 1$ exist satisfying (11.13) (see Exercise 1). Selected values of σ_i proposed by Enright & Higham (1991), which satisfy this condition are given in Table 11.1. For such a choice of σ_i the error (11.12) of the dense output is bounded (at least asymptotically) by the local error le at the endpoint of integration. This implementation of a dense output provides a simple way to estimate le. Since $u(\theta)$ is an $\mathcal{O}(h^{p+1})$-approximation of $y(x_0 + \theta h)$, the defect of $u(\theta)$ satisfies

$$u'(\theta) - hf\left(x_0 + \theta h, u(\theta)\right) = \left(\sum_{i=1}^{k} \sigma_i^{p+1} v_i'(\theta) \right) \cdot le + \mathcal{O}(h^{p+2}) \, . \quad (11.14)$$

If we take a σ^* different from σ_i such that $\sum_{i=1}^{k} \sigma_i^{p+1} v_i'(\sigma^*) \neq 0$ (see Table 11.1) then only one function evaluation, namely $f(x_0 + \sigma^* h, u(\sigma^*))$, allows the computation of an asymptotically correct approximation of le from (11.14). This error estimate can be used for step size selection and for improving the numerical result (local extrapolation). In the local extrapolation mode one then loses the C^1 continuity of the dense output.

With the use of an additional processor the quantities y_{σ^*} and $f(x_0 + \sigma^* h, y_{\sigma^*})$ can be computed simultaneously with y_{σ_i} and $f(x_0 + \sigma_i h, y_{\sigma_i})$. If the polynomial $u(\theta)$ is required to satisfy $u(\sigma^*) = y_{\sigma^*}$, but not $u'(\sigma^*) = hf(x_0 + \sigma^* h, y_{\sigma^*})$, then the estimate (11.14) of the local error le does not need any further evaluation of f.

Table 11.1. Good values for σ_i

p	k	$\sigma_1, \ldots, \sigma_{k-1}$	σ^*
5	3	0.2, 0.4	0.88
6	3	0.2, 0.4	0.88
7	4	0.2, 0.4, 0.7	0.94
8	4	0.2, 0.4, 0.6	0.93

Exercises

1. Let the positive integers k and p satisfy $2k - 1 \le p < 2k + 1$. Then show that there exist numbers $0 = \sigma_0 < \sigma_1 < \ldots < \sigma_{k-1} < \sigma_k = 1$ such that (11.13) is true for all $\theta \in [0, 1]$.

   Hint. Put $\sigma_j = j\varepsilon$ for $j = 1, \ldots, k - 1$ and show that (11.13) is verified for sufficiently small $\varepsilon > 0$. Of course, in a computer program, one should use σ_j which satisfy (11.13) and are well separated in order to avoid roundoff errors.

II.12 Composition of B-Series

> At the Dundee Conference in 1969, a paper by J. Butcher was read
> which contained a surprising result. (H.J. Stetter 1971)

We shall now derive a theorem on the composition of what we call B-series (in honour of J. Butcher). This will have many applications and will lead to a better understanding of order conditions for all general classes of methods (composition of methods, multiderivative methods of Section II.13, general linear methods of Section III.8, Rosenbrock methods in Exercise 2 of Section IV.7).

Composition of Runge-Kutta Methods

There is no five-stage explicit Runge-Kutta method of order 5 (Section II.5). This led Butcher (1969) to the idea of searching for different five-stage methods such that a certain *composition* of these methods produces a fifth-order result ("effective order"). Although not of much practical interest (mainly due to the problem of changing step size), this was the starting point of a fascinating algebraic theory of numerical methods.

Suppose we have two methods, say of three stages,

$$
\begin{array}{c|ccc}
0 & & & \\
\widehat{c}_2 & \widehat{a}_{21} & & \\
\widehat{c}_3 & \widehat{a}_{31} & \widehat{a}_{32} & \\
\hline
& \widehat{b}_1 & \widehat{b}_2 & \widehat{b}_3
\end{array}
\qquad
\begin{array}{c|ccc}
0 & & & \\
\widetilde{c}_2 & \widetilde{a}_{21} & & \\
\widetilde{c}_3 & \widetilde{a}_{31} & \widetilde{a}_{32} & \\
\hline
& \widetilde{b}_1 & \widetilde{b}_2 & \widetilde{b}_3
\end{array}
\qquad (12.1)
$$

which are applied one after the other to a starting value y_0 with the same step size:

$$
g_i = y_0 + h \sum_j \widehat{a}_{ij} f(g_j), \qquad y_1 = y_0 + h \sum_j \widehat{b}_j f(g_j) \qquad (12.2)
$$

$$
\ell_i = y_1 + h \sum_j \widetilde{a}_{ij} f(\ell_j), \qquad y_2 = y_1 + h \sum_j \widetilde{b}_j f(\ell_j). \qquad (12.3)
$$

If we insert y_1 from (12.2) into (12.3) and group all g_i, ℓ_i together, we see that the composition can be interpreted as a large Runge-Kutta method with coefficients

$$
\begin{array}{c|cccccc}
0 \\
\widehat{c}_2 & \widehat{a}_{21} \\
\widehat{c}_3 & \widehat{a}_{31} & \widehat{a}_{32} \\
\sum \widehat{b}_i & \widehat{b}_1 & \widehat{b}_2 & \widehat{b}_3 \\
\sum \widehat{b}_i + \widetilde{c}_2 & \widehat{b}_1 & \widehat{b}_2 & \widehat{b}_3 & \widetilde{a}_{21} \\
\sum \widehat{b}_i + \widetilde{c}_3 & \widehat{b}_1 & \widehat{b}_2 & \widehat{b}_3 & \widetilde{a}_{31} & \widetilde{a}_{32} \\
\hline
& \widehat{b}_1 & \widehat{b}_2 & \widehat{b}_3 & \widetilde{b}_1 & \widetilde{b}_2 & \widetilde{b}_3
\end{array}
\quad \equiv \quad
\begin{array}{c|cccccc}
0 \\
c_2 & a_{21} \\
c_3 & a_{31} & a_{32} \\
c_4 & a_{41} & a_{42} & a_{43} \\
c_5 & a_{51} & a_{52} & a_{53} & a_{54} \\
c_6 & a_{61} & a_{62} & a_{63} & a_{64} & a_{65} \\
\hline
& b_1 & b_2 & b_3 & b_4 & b_5 & b_6
\end{array}
\tag{12.4}
$$

It is now of interest to study the *order conditions* of the new method. For this, we have to compute the expressions (see Table 2.2)

$$
\sum b_i, \quad 2 \sum b_i c_i, \quad 3 \sum b_i c_i^2, \quad 6 \sum b_i a_{ij} c_j, \quad \text{etc.}
$$

If we insert the values from the left tableau of (12.4), a computation, which for low orders is still not too difficult, shows that these expressions can be written in terms of the corresponding expressions for the two methods (12.1). We shall denote these expressions for the *first* method by $\mathbf{a}(t)$, for the *second* method by $\mathbf{b}(t)$, and for the *composite* method by $\mathbf{ab}(t)$:

$$
\mathbf{a}(\,.\,) = \sum \widehat{b}_i, \quad \mathbf{a}(\nearrow) = 2 \cdot \sum \widehat{b}_i \widehat{c}_i, \quad \mathbf{a}(\vee) = 3 \cdot \sum \widehat{b}_i \widehat{c}_i^2, \quad \ldots \tag{12.5a}
$$

$$
\mathbf{b}(\,.\,) = \sum \widetilde{b}_i, \quad \mathbf{b}(\nearrow) = 2 \cdot \sum \widetilde{b}_i \widetilde{c}_i, \quad \mathbf{b}(\vee) = 3 \cdot \sum \widetilde{b}_i \widetilde{c}_i^2, \quad \ldots \tag{12.5b}
$$

$$
\mathbf{ab}(\,.\,) = \sum b_i, \quad \mathbf{ab}(\nearrow) = 2 \cdot \sum b_i c_i, \quad \mathbf{ab}(\vee) = 3 \cdot \sum b_i c_i^2, \quad \ldots \tag{12.5c}
$$

The above mentioned formulas are then

$$
\begin{aligned}
\mathbf{ab}(\,.\,) &= \mathbf{a}(\,.\,) + \mathbf{b}(\,.\,) \\
\mathbf{ab}(\nearrow) &= \mathbf{a}(\nearrow) + 2\mathbf{b}(\,.\,)\mathbf{a}(\,.\,) + \mathbf{b}(\nearrow) \\
\mathbf{ab}(\vee) &= \mathbf{a}(\vee) + 3\mathbf{b}(\,.\,)\mathbf{a}(\,.\,)^2 + 3\mathbf{b}(\nearrow)\mathbf{a}(\,.\,) + \mathbf{b}(\vee) \\
\mathbf{ab}(\curlyvee) &= \mathbf{a}(\curlyvee) + 3\mathbf{b}(\,.\,)\mathbf{a}(\nearrow) + 3\mathbf{b}(\nearrow)\mathbf{a}(\,.\,) + \mathbf{b}(\curlyvee)
\end{aligned}
\tag{12.6}
$$

etc.

It is now, of course, of interest to have a general understanding of these formulas for arbitrary trees. This, however, is not easy in the above framework ("... a tedious calculation shows that ..."). Further, there are problems of identifying different methods with identical numerical results (see Exercise 1 below). Also, we want the theory to include more general processes than Runge-Kutta methods, for example the exact solution or multi-derivative methods.

B-Series

All these difficulties can be avoided if we consider directly the composition of the series appearing in Section II.2. We define by

$$T = \{\emptyset\} \cup T_1 \cup T_2 \cup \dots, \qquad LT = \{\emptyset\} \cup LT_1 \cup LT_2 \cup \dots$$

the sets of all trees and labelled trees, respectively.

Definition 12.1 (Hairer & Wanner 1974). Let $\mathbf{a}(\emptyset)$, $\mathbf{a}(\;.\;)$, $\mathbf{a}(\nearrow)$, $\mathbf{a}(\vee)$, ... be a sequence of real coefficients defined for all trees $\mathbf{a} : T \to \mathbb{R}$. Then we call the series (see Theorem 2.11, Definitions 2.2, 2.3)

$$
\begin{aligned}
B(\mathbf{a}, y) &= \mathbf{a}(\emptyset)y + h\mathbf{a}(\;.\;)f(y) + \frac{h^2}{2!}\,\mathbf{a}(\nearrow)F(\nearrow)(y) + \dots \\
&= \sum_{t \in LT} \frac{h^{\varrho(t)}}{\varrho(t)!}\,\mathbf{a}(t)F(t)(y) = \sum_{t \in T} \frac{h^{\varrho(t)}}{\varrho(t)!}\,\alpha(t)\,\mathbf{a}(t)F(t)(y)
\end{aligned}
\tag{12.7}
$$

a *B-series*.

We have seen in Theorems 2.11 and 2.6 that the numerical solution of a Runge-Kutta method as well as the exact solution are B-series. The coefficients of the latter are all equal to 1.

Usually we are only interested in a finite number of terms of these series (only as high as the orders of the methods under consideration, or as far as f is differentiable) and all subsequent results are valid modulo error terms $\mathcal{O}(h^{k+1})$.

Definition 12.2. Let $t \in LT$ be a labelled tree of order $q = \varrho(t)$ and $0 \le i \le q$ be a fixed integer. Then we denote by $s_i(t) = s$ the *subtree* formed by the first i indices and by $d_i(t)$ (the *difference set*) the set of subtrees formed by the remaining indices. In the graphical representation we distinguish the subtree s by fat nodes and doubled lines.

Example 12.3. For the labelled tree $t = \overset{\displaystyle{}}{\underset{j}{P \vee k}}^{\,l\; m}$ we have:

$i = 0$:	$s_0(t) = \emptyset$,	$d_0(t) = \{\vee\}$
$i = 1$:	$s_1(t) = \,.\,$,	$d_1(t) = \{\,.\,,\,.\,,\nearrow\}$
$i = 2$:	$s_2(t) = \nearrow$,	$d_2(t) = \{\,.\,,\,.\,,\,.\,\}$
$i = 3$:	$s_3(t) = \vee$,	$d_3(t) = \{\,.\,,\,.\,\}$
$i = 4$:	$s_4(t) = \vee$,	$d_4(t) = \{\,.\,\}$
$i = 5$:	$s_5(t) = t = \vee$,	$d_5(t) = \emptyset$

Definition 12.4. Let $\mathbf{a} : T \to \mathbb{R}$ and $\mathbf{b} : T \to \mathbb{R}$ be two sequences of coefficients such that $\mathbf{a}(\emptyset) = 1$. Then for a tree t of order $q = \varrho(t)$ we define the *composition*

$$\mathbf{ab}(t) = \frac{1}{\alpha(t)} \sum \Big(\sum_{i=0}^{q} \binom{q}{i} \mathbf{b}\big(s_i(t)\big) \prod_{z \in d_i(t)} \mathbf{a}(z) \Big) \tag{12.8}$$

where the first summation is over all $\alpha(t)$ different labellings of t (see Definition 2.5).

Example 12.5. It is easily seen that the formulas of (12.6) are special cases of (12.8). The tree t of Example 12.3 possesses 6 different labellings

These lead to

$$\mathbf{ab}(\mathbf{\Psi}) = \mathbf{b}(\emptyset)\mathbf{a}(\mathbf{\Psi}) + 5\mathbf{b}(\,.\,)\mathbf{a}(\,.\,)^2\mathbf{a}(\mathbf{\nearrow})$$

$$+ 10\Big(\frac{1}{2}\,\mathbf{b}(\mathbf{\nearrow})\mathbf{a}(\,.\,)\mathbf{a}(\mathbf{\nearrow}) + \frac{1}{2}\,\mathbf{b}(\mathbf{\nearrow})\mathbf{a}(\,.\,)^3\Big)$$

$$+ 10\Big(\frac{1}{6}\,\mathbf{b}(\mathbf{v})\mathbf{a}(\mathbf{\nearrow}) + \frac{4}{6}\,\mathbf{b}(\mathbf{v})\mathbf{a}(\,.\,)^2 + \frac{1}{6}\,\mathbf{b}(\mathbf{>})\mathbf{a}(\,.\,)^2\Big) \tag{12.9}$$

$$+ 5\Big(\frac{1}{2}\,\mathbf{b}(\mathbf{\Psi})\mathbf{a}(\,.\,) + \frac{1}{2}\,\mathbf{b}(\mathbf{\diamondsuit})\mathbf{a}(\,.\,)\Big) + \mathbf{b}(\mathbf{\Psi}).$$

Here is the main theorem of this section:

Theorem 12.6 (Hairer & Wanner 1974). *As above, let* $\mathbf{a} : T \to \mathbb{R}$ *and* $\mathbf{b} : T \to \mathbb{R}$ *be two sequences of coefficients such that* $\mathbf{a}(\emptyset) = 1$. *Then the composition of the two corresponding B-series is again a B-series*

$$B(\mathbf{b}, B(\mathbf{a}, y)) = B(\mathbf{ab}, y) \tag{12.10}$$

where the "product" $\mathbf{ab} : T \to \mathbb{R}$ *is that of Definition 12.4.*

Proof. We denote the inner series by

$$B(\mathbf{a}, y) = g(h). \tag{12.11}$$

Then the proof is similar to the development of Section II.2 (see Fig. 2.2), with the difference that, instead of $f(g)$, we now start from

$$B(\mathbf{b}, g) = \sum_{s \in LT} \frac{h^{\varrho(s)}}{\varrho(s)!}\,\mathbf{b}(s)F(s)(g) \tag{12.12}$$

and have to compute the derivatives of this function: let us select the term $s = \mathbf{>}$

of this series,

$$\frac{h^3}{3!}\, \mathbf{b}(\text{\Large\texthorizontalsplit}) \sum_{L,M} f_L^K(g) f_M^L(g) f^M(g).$$ (12.13)

The qth derivative of this expression, for $h=0$, is by Leibniz' formula

$$\binom{q}{3} \mathbf{b}(\text{\Large\texthorizontalsplit}) \sum_{L,M} \big(f_L^K(g) f_M^L(g) f^M(g) \big)^{(q-3)} \Big|_{h=0}.$$ (12.14)

We now compute, as we did in Lemma 2.8, the derivatives of

$$f_L^K(g) f_M^L(g) f^M(g)$$ (12.15)

using the classical rules of differential calculus; this gives for the first derivative

$$\sum_N f_{LN}^K \cdot (g^N)' f_M^L f^M + \sum_N f_L^K f_{MN}^L \cdot (g^N)' f^M + \sum_N f_L^K f_M^L f_N^M \cdot (g^N)'$$

and so on. We again represent this in graphical form in Fig. 12.1.

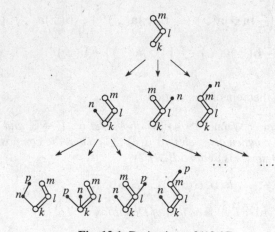

Fig. 12.1. Derivatives of (12.15)

We see that we arrive at trees u of order q such that $s_3(u)=s$ (where $3=\varrho(s)$) and the elements of $d_3(u)$ have no ramifications. The corresponding expressions are similar to (2.6;q-1) in Lemma 2.8. We finally have to insert the derivatives of g (see (12.11)) and rearrange the terms. Then, as in Fig. 2.4, the tall branches of $d_3(u)$ are replaced by trees z of order δ, multiplied by $\mathbf{a}(z)$. Thus the coefficient which we obtain for a given tree t is just given by (12.8).

The factor $1/\alpha(t)$ is due to the fact that in $B(\mathbf{ab},y)$ the term with $\mathbf{ab}(t)F(t)$ appears $\alpha(t)$ times. $\square$

Since $hf(y) = B(\mathbf{b}, y)$ is a special B-series with $\mathbf{b}(\,\centerdot\,) = 1$ and all other $\mathbf{b}(t) = 0$, we have the following

Corollary 12.7. *If* $\mathbf{a} : T \to \mathbb{R}$ *with* $\mathbf{a}(\emptyset) = 1$, *then*

$$hf(B(\mathbf{a}, y)) = B(\mathbf{a}', y)$$

with

$$\mathbf{a}'(\emptyset) = 0, \quad \mathbf{a}'(\,\centerdot\,) = 1$$

$$\mathbf{a}'([t_1, \ldots, t_m]) = \varrho(t)\, \mathbf{a}(t_1) \cdot \ldots \cdot \mathbf{a}(t_m) \tag{12.16}$$

where $t = [t_1, \ldots, t_m]$ *means that* $d_1(t) = \{t_1, t_2, \ldots, t_m\}$ *(Definition 2.12).*

Proof. We obtain (12.16) from (12.8) with $i = 1$, $q = \varrho(t)$ and the fact that the expression in brackets is independent of the labelling of t. □

Order Conditions for Runge-Kutta Methods

As an application of Corollary 12.7, we demonstrate the derivation of order conditions for Runge-Kutta methods: we write method (2.3) as

$$g_i = y_0 + \sum_{j=1}^{s} a_{ij} k_j, \qquad k_i = hf(g_i), \qquad y_1 = y_0 + \sum_{j=1}^{s} b_j k_j. \tag{12.17}$$

If we assume g_i, k_i and y_1 to be B-series, whose coefficients we denote by $\mathbf{g}_i, \mathbf{k}_i, \mathbf{y}_1$

$$g_i = B(\mathbf{g}_i, y_0), \qquad k_i = B(\mathbf{k}_i, y_0), \qquad y_1 = B(\mathbf{y}_1, y_0),$$

then Corollary 12.7 immediately allows us to transcribe formulas (12.17) as

$$\mathbf{g}_i(\emptyset) = 1, \qquad\qquad \mathbf{k}_i(\,\centerdot\,) = 1, \qquad\qquad\qquad \mathbf{y}_1(\emptyset) = 1,$$

$$\mathbf{g}_i(t) = \sum_{j=1}^{s} a_{ij} \mathbf{k}_j(t), \quad \mathbf{k}_i(t) = \varrho(t)\, \mathbf{g}_i(t_1) \cdot \ldots \cdot \mathbf{g}_i(t_m), \quad \mathbf{y}_1(t) = \sum_{j=1}^{s} b_j \mathbf{k}_j(t)$$

which leads easily to formulas (2.17), (2.19) and Theorem 2.11.

Also, if we put $y(h) = B(\mathbf{y}, y_0)$ for the *true solution*, and compare the derivative $hy'(h)$ of the series (12.7) with $hf\big(y(h)\big)$ from Corollary 12.7, we immediately obtain $\mathbf{y}(t) = 1$ for all t, so that Theorem 2.6 drops out. The order conditions are then obtained as in Theorem 2.13 by comparing the coefficients of the B-series $B(\mathbf{y}, y_0)$ and $B(\mathbf{y}_1, y_0)$.

Butcher's "Effective Order"

We search for a 5-stage Runge-Kutta method $\mathbf{a}$ and for a method $\mathbf{d}$, such that $\mathbf{dad}^{-1}$ represents a fifth order method $\mathbf{u}$. This means that we have to satisfy

$$\mathbf{da}(t) = \mathbf{yd}(t) \qquad \text{for} \qquad \varrho(t) \leq 5, \tag{12.18}$$

where $\mathbf{y}(t) = 1$ represents the B-series of the exact solution. Then

$$(\mathbf{dad}^{-1})^k = \mathbf{da}^k \mathbf{d}^{-1} = (\mathbf{da})\mathbf{a}^{k-2}(\mathbf{ad}^{-1}). \tag{12.19}$$

If now two Runge-Kutta methods $\mathbf{b}$ and $\mathbf{c}$ are constructed such that $\mathbf{b} = \mathbf{da}$ and $\mathbf{c} = \mathbf{ad}^{-1}$ up to order 5, then applying one step of $\mathbf{b}$ followed by $k-2$ steps of $\mathbf{a}$ and a final step of $\mathbf{c}$ is equivalent (up to order 5) to k steps of the 5th order method $\mathbf{dad}^{-1}$ (see Fig. 12.2). A possible set of coefficients, computed by Butcher (1969), is given in Table 12.1 (method $\mathbf{a}$ has classical order 4).

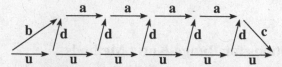

Fig. 12.2. Effective increase of order

Stetter's approach. Soon after the appearance of Butcher's purely algebraic proof, Stetter (1971) gave an elegant analytic explanation. Consider the principal global error term $e_p(x)$ which satisfies the variational equation (8.8). The question is, under which conditions on the local error $d_{p+1}(x)$ (see (8.8)) this equation can be solved, for special initial values, without effort. We write equation (8.8) as

$$e'(x) - \frac{\partial f}{\partial y}(y(x)) \cdot e(x) = d(x) \tag{12.20}$$

and want $e(x)$ to possess an expansion of the form

$$e(x) = \sum_{t \in T_p} \alpha(t)\, \mathbf{e}(t)\, F(t)(y(x)) \tag{12.21}$$

with constant coefficients $\mathbf{e}(t)$. Simply inserting (12.21) into (12.20) yields

$$d(x) = \sum_{t \in T_p} \alpha(t)\, \mathbf{e}(t) \Big(\frac{d}{dx}\big(F(t)(y(x))\big) - f'(y(x)) \cdot F(t)(y(x)) \Big). \tag{12.22}$$

Thus, *(12.21) is the exact solution of the variational equation, if the local error $d(x)$ has the symmetric form (12.22)*. Then, if we replace the initial value y_0 by the "starting procedure"

$$\widehat{y}_0 := y_0 - h^p e(x_0) = y_0 - h^p \sum_{t \in T_p} \alpha(t)\, \mathbf{e}(t)\, F(t)(y_0) \tag{12.23}$$

Table 12.1. Butcher's method of effective order 5

Method **a**

0				
$\frac{1}{5}$	$\frac{1}{5}$			
$\frac{2}{5}$	0	$\frac{2}{5}$		
$\frac{1}{2}$	$\frac{3}{16}$	0	$\frac{5}{16}$	
1	$\frac{1}{4}$	0	$-\frac{5}{4}$	2
	$\frac{1}{6}$	0	0	$\frac{2}{3}$	$\frac{1}{6}$

Method **b**

0				
$\frac{1}{5}$	$\frac{1}{5}$			
$\frac{2}{5}$	0	$\frac{2}{5}$		
$\frac{3}{4}$	$\frac{75}{64}$	$-\frac{9}{4}$	$\frac{117}{64}$	
1	$-\frac{37}{36}$	$\frac{7}{3}$	$-\frac{3}{4}$	$\frac{4}{9}$
	$\frac{19}{144}$	0	$\frac{25}{48}$	$\frac{2}{9}$	$\frac{1}{8}$

Method **c**

0				
$\frac{1}{5}$	$\frac{1}{5}$			
$\frac{2}{5}$	0	$\frac{2}{5}$		
$\frac{3}{4}$	$\frac{161}{192}$	$-\frac{19}{12}$	$\frac{287}{192}$	
1	$-\frac{27}{28}$	$\frac{19}{7}$	$-\frac{291}{196}$	$\frac{36}{49}$
	$\frac{7}{48}$	0	$\frac{475}{1008}$	$\frac{2}{7}$	$\frac{7}{72}$

(or by a Runge-Kutta method equivalent to this up to order $p+1$; this would represent "method **d**" in Fig. 12.2), its error satisfies $y(x_0) - \widehat{y}_0 = h^p e(x_0) + \mathcal{O}(h^{p+1})$. By Theorem 8.1 the numerical solution $\widehat{y}_n$ of the Runge-Kutta method applied to $\widehat{y}_0$ satisfies $y(x_n) - \widehat{y}_n = h^p e(x_n) + \mathcal{O}(h^{p+1})$. Therefore the "finishing procedure"

$$y_n := \widehat{y}_n + h^p e(x_n) = \widehat{y}_n + h^p \sum_{t \in T_p} \alpha(t)\, \mathbf{e}(t)\, F(t)(\widehat{y}_n) + \mathcal{O}(h^{p+1}) \qquad (12.24)$$

(or some equivalent Runge-Kutta method) gives a $(p+1)$th order approximation to the solution.

Example. Butcher's method **a** of Table 12.1 has the local error

$$d_6(x) = \frac{1}{6!}\left(-\frac{1}{24}F(\Psi) - \frac{1}{4}F(\psi) - \frac{1}{8}F(\diamondsuit) + \frac{1}{6}F(\Upsilon) + \frac{1}{2}F(\Upsilon)\right). \qquad (12.25)$$

The right-hand side of (12.22) would be (the derivation $\frac{d}{dx}F$ attaches a new twig to each of the nodes, the product $f'(y) \cdot F$ lifts the tree on a stilt)

$$
e(\psi)\Big(F(\psi) + 3F(\psi) - F(Y)\Big)
$$
$$
+3e(\vee)\Big(F(\psi) + F(\psi) + F(\psi) + F(\psi) - F(Y)\Big)
$$
$$
+e(Y)\Big(F(\psi) + F(Y) + 2F(Y) - F(Y)\Big) \tag{12.26}
$$
$$
+e(\psi)\Big(F(\psi) + F(Y) + F(Y) + F(\psi) - F(\psi)\Big).
$$

Comparison of (12.25) and (12.26) shows that this method does indeed have the desired symmetry if

$$
e(\psi) = e(\vee) = -\frac{1}{6!} \cdot \frac{1}{24}, \qquad e(Y) = e(\psi) = \frac{1}{6!} \cdot \frac{1}{8}.
$$

This allows one to construct a Runge-Kutta method as starting procedure corresponding to (12.23) up to the desired order.

Exercises

1. Show that the pairs of methods given in Tables 12.2 - 12.4 produce, at least for h sufficiently small, identical numerical results.

   Result. a) is seen by permutation of the stages, b) by neglecting superfluous stages (Dahlquist & Jeltsch 1979), c) by identifying equal stages (Stetter 1973, Hundsdorfer & Spijker 1981). See also the survey on "The Runge-Kutta space" by Butcher (1984).

2. Extend formulas (12.6) by computing the composition $\mathbf{ab}(t)$ for all trees of order 4 and 5.

3. Verify that the methods given in Table 12.1 satisfy the stated order properties.

4. Prove, using Theorem 12.6, that the set

$$
G = \{\mathbf{a} : T \to \mathbb{R} \mid \mathbf{a}(\emptyset) = 1\}
$$

   together with the composition law of Definition 12.4 is a (non-commutative) group.

5. (Equivalence of Butcher's and Stetter's approach). Let $\mathbf{a} : T \to \mathbb{R}$ represent a Runge-Kutta method of classical order p and effective order $p+1$, i.e., $\mathbf{a}(t) = 1$ for $\varrho(t) \leq p$ and

$$
\mathbf{da}(t) = \mathbf{yd}(t) \qquad \text{for} \qquad \varrho(t) \leq p+1 \tag{12.27}
$$

   for some $\mathbf{d} : T \to \mathbb{R}$ and with $\mathbf{y}(t)$ as in (12.18). Prove that then the local error $h^{p+1}d(x) + \mathcal{O}(h^{p+2})$ of the method $\mathbf{a}$ has the symmetric form (12.22). This

Table 12.2. Equivalent methods a)

$$
\begin{array}{c|cc}
0 & & \\
1 & 1 & 0 \\
\hline
 & 1/4 & 3/4
\end{array}
\qquad
\begin{array}{c|cc}
1 & 0 & 1 \\
0 & 0 & 0 \\
\hline
 & 3/4 & 1/4
\end{array}
$$

Table 12.3. Equivalent methods b)

$$
\begin{array}{c|cccc}
1 & 2 & 0 & 0 & -1 \\
3 & 0 & 1 & 2 & 0 \\
7 & 0 & 3 & 4 & 0 \\
2 & 1 & 0 & 0 & 1 \\
\hline
 & 1/2 & 0 & 0 & 1/2
\end{array}
\qquad
\begin{array}{c|cc}
1 & 2 & -1 \\
2 & 1 & 1 \\
\hline
 & 1/2 & 1/2
\end{array}
$$

Table 12.4. Equivalent methods c)

$$
\begin{array}{c|cccc}
1 & 1 & 1 & 1 & -2 \\
1 & 2 & 2 & -1 & -2 \\
1 & -1 & -1 & 5 & -2 \\
-1 & -1 & 2 & 1 & -3 \\
\hline
 & 1/4 & 1/4 & 1/4 & 1/4
\end{array}
\qquad
\begin{array}{c|cc}
1 & 3 & -2 \\
-1 & 2 & -3 \\
\hline
 & 3/4 & 1/4
\end{array}
$$

means that, in this situation, Butcher's effective order is equivalent to Stetter's approach.

Hint. Start by expanding condition (12.27) (using (12.8)) for the first trees. Possible simplifications are then best seen if the second sum $\sum_{i=0}^{q}$ (for $\mathbf{yd}$) is arranged *downwards* $(i = q, q-1, \ldots, 0)$. One then arrives recursively at the result

$$
\mathbf{d}(t) = \mathbf{d}(\,.\,)^{\varrho(t)} \qquad \text{for } \varrho(t) \le p-1.
$$

Then express the error coefficients $\mathbf{a}(t) - 1$ for $\varrho(t) = p+1$ in terms of $\mathbf{d}(s) - \mathbf{d}(\,.\,)^{\varrho(s)}$ where $\varrho(s) = p$. Formula (12.22) then becomes visible.

6. Prove that for $t = [t_1, \ldots, t_m]$ the coefficient $\alpha(t)$ of Definition 2.5 satisfies the recurrence relation

$$
\alpha(t) = \binom{\varrho(t) - 1}{\varrho(t_1), \ldots, \varrho(t_m)} \alpha(t_1) \cdot \ldots \cdot \alpha(t_m) \cdot \frac{1}{\mu_1! \mu_2! \ldots}. \tag{12.28}
$$

The integers $\mu_1, \mu_2, \ldots$ count the equal trees among $t_1, \ldots, t_m$.

Hint. The multinomial coefficient in (12.28) counts the possible partitionnings of the labels $2, \ldots, \varrho(t)$ to the m subtrees $t_1, \ldots, t_m$. Equal subtrees lead to equal labellings. Hence the division by $\mu_1! \mu_2! \ldots$.

II.13 Higher Derivative Methods

In Section I.8 we studied the computation of higher derivatives of solutions of

$$(y^J)' = f^J(x, y^1, \ldots, y^n), \qquad J = 1, \ldots, n. \tag{13.1}$$

The chain rule

$$(y^J)'' = \frac{\partial f^J}{\partial x}(x, y) + \frac{\partial f^J}{\partial y^1}(x, y) \cdot f^1(x, y) + \ldots + \frac{\partial f^J}{\partial y^n}(x, y) \cdot f^n(x, y) \tag{13.2}$$

leads to the differential operator D which, when applied to a function $\Psi(x, y)$, is given by

$$(D\Psi)(x, y) = \frac{\partial \dot{\Psi}}{\partial x}(x, y) + \frac{\partial \Psi}{\partial y^1}(x, y) \cdot f^1(x, y) + \ldots + \frac{\partial \Psi}{\partial y^n}(x, y) \cdot f^n(x, y). \tag{13.2'}$$

Since $Dy^J = f^J$, we see by extending (13.2) that

$$(y^J)^{(\ell)} = (D^\ell y^J)(x, y), \qquad \ell = 0, 1, 2, \ldots. \tag{13.3}$$

This notation allows us to define a new class of methods which combine features of Runge-Kutta methods as well as Taylor series methods:

Definition 13.1. Let $a_{ij}^{(r)}$, $b_j^{(r)}$, $(i, j = 1, \ldots, s, \ r = 1, \ldots, q)$ be real coefficients. Then the method

$$k_i^{(\ell)} = \frac{h^\ell}{\ell!}(D^\ell y)\left(x_0 + c_i h, y_0 + \sum_{r=1}^q \sum_{j=1}^s a_{ij}^{(r)} k_j^{(r)}\right)$$

$$y_1 = y_0 + \sum_{r=1}^q \sum_{j=1}^s b_j^{(r)} k_j^{(r)} \tag{13.4}$$

is called an *s-stage q-derivative Runge-Kutta method*. If $a_{ij}^{(r)} = 0$ for $i \leq j$, the method is *explicit*, otherwise *implicit*.

A natural extension of (1.9) is here, because of $Dx = 1$, $D^\ell x = 0$ $(\ell \geq 2)$,

$$c_i = \sum_{j=1}^s a_{ij}^{(1)}. \tag{13.5}$$

Definition 13.1 is from Kastlunger & Wanner (1972), but special methods of this type have been considered earlier in the literature. In particular, the very successful methods of Fehlberg (1958, 1964) have this structure.

Collocation Methods

A natural way of obtaining s-stage q-derivative methods is to use the collocation idea with *multiple nodes*, i.e., to replace (7.15b) by

$$u^{(\ell)}(x_0 + c_i h) = (D^\ell y)\big(x_0 + c_i h, u(x_0 + c_i h)\big) \quad i = 1, \ldots, s, \quad \ell = 1, \ldots, q_i \tag{13.6}$$

where $u(x)$ is a polynomial of degree $q_1 + q_2 + \ldots + q_s$ and $q_1, \ldots, q_s$, the "multiplicities" of the nodes $c_1, \ldots, c_s$, are given integers. For example $q_1 = m$, $q_2 = \ldots = q_s = 1$ leads to Fehlberg-type methods.

In order to generalize the results and ideas of Section II.7, we have to replace the Lagrange interpolation of Theorem 7.7 by *Hermite* interpolation (Hermite 1878: "Je me suis proposé de trouver un polynôme ..."). The reason is that (13.6) can be interpreted as an ordinary collocation condition with clusters of q_i nodes "infinitely" close together (Rolle's theorem). We write Hermite's formula as

$$p(t) = \sum_{j=1}^{s} \sum_{r=1}^{q_j} \frac{1}{r!} \ell_{jr}(t) p^{(r-1)}(c_j) \tag{13.7}$$

for polynomials $p(t)$ of degree $\sum q_j - 1$. Here the "basis" polynomials $\ell_{jr}(t)$ of degree $\sum q_j - 1$ must satisfy

$$l_{jr}^{(k)}(c_i) = \begin{cases} r! & \text{if} \quad i = j \quad \text{and} \quad k = r - 1 \\ 0 & \text{else} \end{cases} \tag{13.8}$$

and are best obtained from Newton's interpolation formula (with multiple nodes). We now use this formula, as we did in Section II.7, for $p(t) = hu'(x_0 + th)$:

$$hu'(x_0 + th) = \sum_{j=1}^{s} \sum_{r=1}^{q_j} \ell_{jr}(t) k_j^{(r)}, \tag{13.9}$$

with

$$k_j^{(r)} = \frac{h^r}{r!} u^{(r)}(x_0 + c_j h). \tag{13.10}$$

If we insert

$$u(x_0 + c_i h) = y_0 + \int_0^{c_i} hu'(x_0 + th) \, dt$$

together with (13.9) into (13.6), we get:

Theorem 13.2. *The collocation method (13.6) is equivalent to an s-stage q-derivative implicit Runge-Kutta method (13.4) with*

$$a_{ij}^{(r)} = \int_0^{c_i} \ell_{jr}(t)\, dt, \qquad b_j^{(r)} = \int_0^1 \ell_{jr}(t)\, dt. \qquad (13.11)$$

□

Theorems 7.8, 7.9, and 7.10 now generalize immediately to the case of "confluent" quadrature formulas; i.e., the q-derivative Runge-Kutta method possesses the *same order* as the underlying quadrature formula

$$\int_0^1 p(t)\, dt \approx \sum_{j=1}^s \sum_{r=1}^{q_j} b_j^{(r)} p^{(r-1)}(c_j).$$

The "algebraic" proof of this result (extending Exercise 7 of Section II.7) is more complicated and is given, for the case $q_j = q$, in Kastlunger & Wanner (1972b).

The formulas corresponding to condition $C(\eta)$ are given by

$$\sum_{j=1}^s \sum_{r=1}^{q_j} a_{ij}^{(r)} \binom{\varrho}{r} c_j^{\varrho-r} = c_i^\varrho, \qquad \varrho = 1, 2, \ldots, \sum_{j=1}^s q_j. \qquad (13.12)$$

These equations uniquely determine the $a_{ij}^{(r)}$, once the c_i have been chosen, by a linear system with a "confluent" Vandermonde matrix (see e.g., Gautschi 1962). Formula (13.12) is obtained by setting $p(t) = t^{\varrho-1}$ in (13.7) and then integrating from 0 to c_i.

Examples of methods. "Gaussian" quadrature formulas with multiple nodes exist for *odd q* (Stroud & Stancu 1965) and extend to q-derivative implicit Runge-Kutta methods (Kastlunger & Wanner 1972b): for $s = 1$ we have, of course, $c_1 = 1/2$ which yields

$$b_1^{(2k)} = 0, \qquad b_1^{(2k+1)} = 2^{-2k}, \qquad a_{11}^{(k)} = (-1)^{k+1} 2^{-k}.$$

We give also the coefficients for the case $s = 2$ and $q_1 = q_2 = 3$. The nodes c_i and the weights $b_i^{(k)}$ are those of Stroud & Stancu. The method has order 8:

$$c_1 = 0.185394435825045 \qquad c_2 = 1 - c_1$$
$$b_1^{(1)} = 0.5 \qquad b_2^{(1)} = b_1^{(1)}$$
$$b_1^{(2)}/2! = 0.0240729420844974 \qquad b_2^{(2)} = -b_1^{(2)}$$
$$b_1^{(3)}/3! = 0.00366264960671727 \qquad b_2^{(3)} = b_1^{(3)}$$

$$a_{ij}^{(1)} = \begin{pmatrix} 0.201854115831005 & -0.0164596800059598 \\ 0.516459680005959 & 0.298145884168994 \end{pmatrix}$$

$$a_{ij}^{(2)} = \begin{pmatrix} -0.0223466569080541 & 0.00868878773082417 \\ 0.0568346718998190 & -0.0704925410770490 \end{pmatrix}$$

$$a_{ij}^{(3)} = \begin{pmatrix} 0.0116739668400997 & -0.00215351251065784 \\ 0.0241294101509615 & 0.0103019308002039 \end{pmatrix}$$

Hermite-Obreschkoff Methods

We now consider the special case of collocation methods with $s=2$, $c_1=0$, $c_2=1$. These methods can be obtained in closed form by repeated partial integration as follows (Darboux 1876, Hermite 1878):

Lemma 13.3. *Let m be a given positive integer and $P(t)$ a polynomial of exact degree m. Then*

$$\sum_{j=0}^{m} h^j (D^j y)(x_1,y_1) P^{(m-j)}(0) = \sum_{j=0}^{m} h^j (D^j y)(x_0,y_0) P^{(m-j)}(1) \qquad (13.13)$$

defines a multiderivative method (13.4) of order m.

Proof. We let $y(x)$ be the exact solution and start from

$$h^{m+1} \int_0^1 y^{(m+1)}(x_0 + ht)P(1-t)\, dt = \mathcal{O}(h^{m+1}).$$

This integral is now transformed by repeated partial integration until all derivatives of the polynomial $P(1-t)$ are used up. This leads to

$$\sum_{j=0}^{m} h^j y^{(j)}(x_1)P^{(m-j)}(0) = \sum_{j=0}^{m} h^j y^{(j)}(x_0)P^{(m-j)}(1) + \mathcal{O}(h^{m+1}).$$

If this is subtracted from (13.13) we find the difference of the left-hand sides to be $\mathcal{O}(h^{m+1})$, which shows by the implicit function theorem that (13.13) determines y_1 to this order if $P^{(m)}$, which is a constant, is $\neq 0$.   □

The argument $1-t$ in P (instead of the more natural t) avoids the sign changes in the partial integrations.

A good choice for $P(t)$ is, of course, a polynomial for which most derivatives disappear at $t = 0$ and $t = 1$. Then the method (13.13) is, by keeping the same order m, most economical. We write

$$P(t) = \frac{t^k(t-1)^\ell}{(k+\ell)!}$$

and obtain

$$y_1 - \frac{\ell}{(k+\ell)} \frac{h}{1!} (Dy)(x_1, y_1) + \frac{\ell(\ell-1)}{(k+\ell)(k+\ell-1)} \frac{h^2}{2!} (D^2 y)(x_1, y_1) \pm \ldots$$

$$= y_0 + \frac{k}{(k+\ell)} \frac{h}{1!} (Dy)(x_0, y_0) + \frac{k(k-1)}{(k+\ell)(k+\ell-1)} \frac{h^2}{2!} (D^2 y)(x_0, y_0) + \ldots$$

(13.14)

which is a method of order $m = k + \ell$. After the ℓth term in the first line and the kth term in the second line, the coefficients automatically become zero. Special cases of this method are:

$$\begin{array}{lll}
k = 1, & \ell = 0: & \text{explicit Euler} \\
k \geq 1, & \ell = 0: & \text{Taylor series} \\
k = 0, & \ell = 1: & \text{implicit Euler} \\
k = 1, & \ell = 1: & \text{trapezoidal rule.}
\end{array}$$

Darboux and Hermite advocated the use of this formula for the approximations of functions, Obreschkoff (1940) for the computation of integrals, Loscalzo & Schoenberg (1967), Loscalzo (1969) as well as Nørsett (1974a) for the solution of differential equations.

Fehlberg Methods

Another class of multiderivative methods is due to Fehlberg (1958, 1964): the idea is to subtract from the solution of $y' = f(x, y)$, $y(x_0) = y_0$ m terms of the Taylor series (see Section I.8)

$$\widehat{y}(x) := y(x) - \sum_{i=0}^{m} Y_i (x - x_0)^i,$$

(13.15)

and to solve the resulting differential equation $\widehat{y}'(x) = \widehat{f}(x, \widehat{y}(x))$, where

$$\widehat{f}(x, \widehat{y}(x)) = f\left(x, \widehat{y} + \sum_{i=0}^{m} Y_i (x - x_0)^i\right) - \sum_{i=1}^{m} Y_i \, i \, (x - x_0)^{i-1},$$

(13.16)

by a Runge-Kutta method. Thus, knowing that the solution of (13.16) and its first m derivatives are zero at the initial value, we can achieve much higher orders.

In order to understand this, we develop the Taylor series of the solution for the non-autonomous case, as we did at the beginning of Section II.1. We thereby omit the hats and suppose the transformation (13.15) already carried out. We then have from (1.6) (see also Exercise 3 of Section II.2)

$$f = 0,$$

$$f_x + f_y f = 0,$$

$$f_{xx} + 2 f_{xy} f + f_{yy} f^2 + f_y (f_x + f_y f) = 0, \text{ etc.}$$

These formulas recursively imply that $f = 0$, $f_x = 0$, ..., $\partial^{m-1}f/\partial x^{m-1} = 0$. All elementary differentials of order $\leq m$ and most of those of higher orders then become *zero* and the corresponding order conditions can be omitted. The first non-zero terms are

$$\frac{\partial^m f}{\partial x^m} \qquad \text{for order} \quad m+1,$$

$$\frac{\partial^{m+1} f}{\partial x^{m+1}} \quad \text{and} \quad \frac{\partial f}{\partial y} \cdot \frac{\partial^m f}{\partial x^m} \quad \text{for order} \quad m+2,$$

and so on. The corresponding order conditions are then

$$\sum_{i=1}^{s} b_i c_i^m = \frac{1}{m+1}$$

for order $m+1$,

$$\sum_{i=1}^{s} b_i c_i^{m+1} = \frac{1}{m+2} \quad \text{and} \quad \sum_{i,j} b_i a_{ij} c_j^m = \frac{1}{(m+1)(m+2)}$$

for order $m+2$, and so on.

The condition $\sum a_{ij} = c_i$, which usually allows several terms of (1.6) to be grouped together, is not necessary, because all these other terms are zero.

A complete insight is obtained by considering the method as being *partitioned* applied to the *partitioned system* $y' = f(x,y)$, $x' = 1$. This will be explained in Section II.15 (see Fig. 15.3).

Example 13.4. A solution with $s = 3$ stages of the (seven) conditions for order $m+3$ is given by Fehlberg (1964). The choice $c_1 = c_3 = 1$ minimizes the numerical work for the evaluation of (13.16) and the other coefficients are then uniquely determined (see Table 13.1).

Fehlberg (1964) also derived an embedded method with two additional stages of orders $m+3$ $(m+4)$. These methods were widely used in the sixties for scientific computations.

Table 13.1. Fehlberg, order $m+3$

1			
θ	$\dfrac{\theta^m}{m+3}$		
1	$-\dfrac{1}{m+1}$	$\dfrac{2}{(m+1)\theta^m}$	
	0	$\dfrac{m+3}{2(m+1)(m+2)\theta^m}$	$\dfrac{1}{2(m+2)}$

$$\theta = \frac{m+1}{m+3}$$

General Theory of Order Conditions

For the same reason as in Section II.2 we assume that (13.1) is autonomous. The general form of the order conditions for method (13.4) was derived in the thesis of Kastlunger (see Kastlunger & Wanner 1972). It later became a simple application of the composition theorem for B-series (Hairer & Wanner 1974). The point is that from Theorem 2.6,

$$\frac{h^i}{i!}(D^iy)(y_0) = \sum_{t\in LT,\varrho(t)=i}\frac{h^i}{i!}F(t)(y_0) = B(\mathbf{y}^{(i)}, y_0) \tag{13.17}$$

is a B-series with coefficients

$$\mathbf{y}^{(i)}(t) = \begin{cases} 1 & \text{if } \varrho(t) = i \\ 0 & \text{otherwise.} \end{cases} \tag{13.18}$$

Thus, in extension of Corollary 12.7, we have

$$\frac{h^i}{i!}(D^iy)(B(\mathbf{a}, y_0)) = B(\mathbf{a}^{(i)}, y_0) \tag{13.19}$$

where, from formula (12.8) with $q = \varrho(t)$,

$$\mathbf{a}^{(i)}(t) = (\mathbf{a}\mathbf{y}^{(i)})(t) = \frac{1}{\alpha(t)}\binom{q}{i}\sum\prod_{z\in d_i(t)}\mathbf{a}(z), \tag{13.20}$$

and the sum is over all $\alpha(t)$ different labellings of t. This allows us to compute recursively the coefficients of the B-series which appear in (13.4).

Example 13.5. The tree $t = \vee\!\!\!\vee$ sketched in Fig. 13.1 possesses three different labellings, two of which produce the same difference set $d_2(t)$, so that formula (13.20) becomes

$$\mathbf{a}''(\vee\!\!\!\vee) = 2\big(2(\mathbf{a}(\,\boldsymbol{.}\,))^2 + \mathbf{a}(\boldsymbol{\nearrow})\big). \tag{13.21}$$

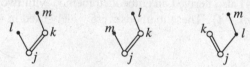

Fig. 13.1. Different labellings of $\vee\!\!\!\vee$

For all other trees of order ≤ 4 we have $\alpha(t) = 1$ and (13.20) leads to the following table of second derivatives

$$\begin{aligned}
&\mathbf{a}''(\,\boldsymbol{.}\,) = 0 && \mathbf{a}''(\boldsymbol{\nearrow}) = 1 \\
&\mathbf{a}''(\vee) = 3\mathbf{a}(\,\boldsymbol{.}\,) && \mathbf{a}''(\boldsymbol{>}) = 3\mathbf{a}(\,\boldsymbol{.}\,) \\
&\mathbf{a}''(\vee\!\!\!\vee) = 6(\mathbf{a}(\,\boldsymbol{.}\,))^2 && \mathbf{a}''(\diamondsuit) = 4(\mathbf{a}(\,\boldsymbol{.}\,))^2 + 2\mathbf{a}(\boldsymbol{\nearrow}) \\
&\mathbf{a}''(Y) = 6(\mathbf{a}(\,\boldsymbol{.}\,))^2 && \mathbf{a}''(\diamondsuit\!\!\!\diamond) = 6\mathbf{a}(\boldsymbol{\nearrow}).
\end{aligned} \tag{13.22}$$

Once these expressions have been established, we write formulas (13.4) in the form

$$k_i^{(\ell)} = \frac{h^\ell}{\ell!} \, (D^\ell y)(g_i)$$

$$g_i = y_0 + \sum_{r=1}^{q} \sum_{j=1}^{s} a_{ij}^{(r)} k_j^{(r)}, \qquad y_1 = y_0 + \sum_{r=1}^{q} \sum_{j=1}^{s} b_j^{(r)} k_j^{(r)} \tag{13.23}$$

and suppose the expressions $k_i^{(\ell)}$, g_i, y_1 to be B-series

$$k_i^{(\ell)} = B(\mathbf{k}_i^{(\ell)}, y_0), \qquad g_i = B(\mathbf{g}_i, y_0), \qquad y_1 = B(\mathbf{y}_1, y_0).$$

Then equations (13.23) can be translated into

$$\mathbf{k}_i^{(1)}(t) = \varrho(t)\mathbf{g}_i(t_1)\cdot \ldots \cdot \mathbf{g}_i(t_m), \qquad \mathbf{k}_i^{(1)}(\tau) = 1 \qquad \text{(see (12.16))}$$

$$\mathbf{k}_i^{(2)}(t) = \mathbf{g}_i''(t) \qquad \text{from (13.22)}$$

$$\mathbf{k}_i^{(3)}(t) = \mathbf{g}_i'''(t) \qquad \text{from Exercise 1 or Exercise 2, etc.}$$

$$\mathbf{g}_i(t) = \sum_{r=1}^{q} \sum_{j=1}^{s} a_{ij}^{(r)} \mathbf{k}_j^{(r)}(t), \qquad \mathbf{y}_1(t) = \sum_{r=1}^{q} \sum_{j=1}^{s} b_j^{(r)} \mathbf{k}_j^{(r)}(t).$$

These formulas recursively determine all the coefficients. Method (13.4) (together with (13.5)) is then of order p if, as usual,

$$\mathbf{y}_1(t) = 1 \quad \text{for all } t \text{ with } \varrho(t) \le p. \tag{13.24}$$

More details and special methods are given in Kastlunger & Wanner (1972); see also Exercise 3.

Exercises

1. Extend Example 13.5 and obtain formulas for $\mathbf{a}^{(3)}(t)$ for all trees of order ≤ 4.

2. (Kastlunger). Prove the following variant form of formula (13.20) which extends (12.16) more directly and can also be used to obtain the formulas of Example 13.5. If $t = [t_1, \ldots, t_m]$ then

$$\mathbf{a}^{(i)}(t) = \frac{\varrho(t)}{i} \sum_{\substack{\lambda_1 + \ldots + \lambda_m = i-1 \\ \lambda_1, \ldots, \lambda_m \ge 0}} \mathbf{a}^{(\lambda_1)}(t_1) \ldots \mathbf{a}^{(\lambda_m)}(t_m)$$

Hint. See Kastlunger & Wanner (1972); Hairer & Wanner (1973), Section 5.

3. Show that the conditions for order 3 of method (13.4) are given by

$$\sum_i b_i^{(1)} = 1$$

$$2 \sum_i b_i^{(1)} c_i + \sum_i b_i^{(2)} = 1$$

$$3 \sum_i b_i^{(1)} c_i^2 + 3 \sum_i b_i^{(2)} c_i + \sum_i b_i^{(3)} = 1$$

$$6 \sum_{i,j} b_i^{(1)} a_{ij}^{(1)} c_j + 3 \sum_i b_i^{(1)} e_i + 3 \sum_i b_i^{(2)} c_i + \sum_i b_i^{(3)} = 1,$$

where $c_i = \sum_j a_{ij}^{(1)}$, $e_i = \sum_j a_{ij}^{(2)}$.

4. (Zurmühl 1952, Albrecht 1955). Differentiate a given first order system of differential equations $y' = f(x, y)$ to obtain

$$y'' = (D^2 y)(x, y), \qquad y(x_0) = y_0, \qquad y'(x_0) = f_0.$$

Apply to this equation a special method for higher order systems (see the following Section II.14) to obtain higher-derivative methods. Show that the following method is of order six

$$k_1 = h^2 g(x_0, y_0)$$

$$k_2 = h^2 g\left(x_0 + \frac{h}{4}, y_0 + \frac{h}{4} f_0 + \frac{1}{32} k_1\right)$$

$$k_3 = h^2 g\left(x_0 + \frac{h}{2}, y_0 + \frac{h}{2} f_0 + \frac{1}{24} (-k_1 + 4k_2)\right)$$

$$k_4 = h^2 g\left(x_0 + \frac{3h}{4}, y_0 + \frac{3h}{4} f_0 + \frac{1}{32} (3k_1 + 4k_2 + 2k_3)\right)$$

$$y_1 = y_0 + h f_0 + \frac{1}{90} (7k_1 + 24k_2 + 6k_3 + 8k_4)$$

where $g(x, y) = (D^2 y)(x, y) = D f(x, y) = f_x(x, y) + f_y(x, y) \cdot f(x, y)$.

II.14 Numerical Methods
for Second Order Differential Equations

Mutationem motus proportionalem esse vi motrici impressae
(Newton's Lex II, 1687)

Many differential equations which appear in practice are systems of the *second order*

$$y'' = f(x, y, y').$$ (14.1)

This is mainly due to the fact that the forces are proportional to acceleration, i.e., to second derivatives. As mentioned in Section I.1, such a system can be transformed into a first order differential equation of doubled dimension by considering the vector (y, y') as the new variable:

$$\begin{pmatrix} y \\ y' \end{pmatrix}' = \begin{pmatrix} y' \\ f(x, y, y') \end{pmatrix} \qquad \begin{array}{l} y(x_0) = y_0 \\ y'(x_0) = y_0'. \end{array}$$ (14.2)

In order to solve (14.1) numerically, one can for instance apply a Runge-Kutta method (explicit or implicit) to (14.2). This yields

$$k_i = y_0' + h \sum_{j=1}^{s} a_{ij} k_j'$$

$$k_i' = f\left(x_0 + c_i h,\ y_0 + h \sum_{j=1}^{s} a_{ij} k_j,\ y_0' + h \sum_{j=1}^{s} a_{ij} k_j'\right)$$ (14.3)

$$y_1 = y_0 + h \sum_{i=1}^{s} b_i k_i, \qquad y_1' = y_0' + h \sum_{i=1}^{s} b_i k_i'.$$

If we insert the first formula of (14.3) into the others we obtain (assuming (1.9) and an order ≥ 1)

$$k_i' = f\left(x_0 + c_i h,\ y_0 + c_i h y_0' + h^2 \sum_{j=1}^{s} \bar{a}_{ij} k_j',\ y_0' + h \sum_{j=1}^{s} a_{ij} k_j'\right)$$

$$y_1 = y_0 + h y_0' + h^2 \sum_{i=1}^{s} \bar{b}_i k_i', \qquad y_1' = y_0' + h \sum_{i=1}^{s} b_i k_i'$$ (14.4)

where

$$\bar{a}_{ij} = \sum_{k=1}^{s} a_{ik} a_{kj}, \qquad \bar{b}_i = \sum_{j=1}^{s} b_j a_{ji}.$$ (14.5)

For an implementation the representation (14.4) is preferable to (14.3), since about half of the storage can be saved. This may be important, in particular if the dimension of equation (14.1) is large.

Nyström Methods

> R.H. Merson: "... I have not seen the paper by Nyström. Was it in English?"
> J.M. Bennett: "In German actually, not Finnish."
> (From the discussion following a talk of Merson 1957)

E.J. Nyström (1925) was the first to consider methods of the form (14.4) in which the coefficients do not necessarily satisfy (14.5) ("Da bis jetzt die *direkte* Anwendung der Rungeschen Methode auf den wichtigen Fall von Differentialgleichungen zweiter Ordnung nicht behandelt war ..." Nyström, 1925). Such direct methods are called *Nyström methods*.

Definition 14.1. A Nyström method (14.4) has *order p* if for sufficiently smooth problems (14.1)

$$y(x_0 + h) - y_1 = \mathcal{O}(h^{p+1}), \qquad y'(x_0 + h) - y_1' = \mathcal{O}(h^{p+1}). \qquad (14.6)$$

An example of an explicit Nyström method where condition (14.5) is violated is given in Table 14.1. Nyström claimed that this method would be simpler to apply than "Runge-Kutta's" and reduce the work by about 25%. This is, of course, not true if the Runge-Kutta method is applied as in (14.4) (see also Exercise 2).

Table 14.1. Nyström, order 4

c_i		$\overline{a}_{ij}$				a_{ij}			
0									
$\frac{1}{2}$	$\frac{1}{8}$				$\frac{1}{2}$				
$\frac{1}{2}$	$\frac{1}{8}$	0			0	$\frac{1}{2}$			
1	0	0	$\frac{1}{2}$		0	0	1		
$\overline{b}_i \rightarrow$	$\frac{1}{6}$	$\frac{1}{6}$	$\frac{1}{6}$	0	$\frac{1}{6}$	$\frac{2}{6}$	$\frac{2}{6}$	$\frac{1}{6}$	$\leftarrow b_i$

A *real* improvement can be achieved in the case where the right-hand side of (14.1) does not depend on y', i.e.,

$$y'' = f(x, y). \qquad (14.7)$$

Here the Nyström method becomes

$$k_i' = f(x_0 + c_i h,\; y_0 + c_i h y_0' + h^2 \sum_{j=1}^{s} \bar{a}_{ij} k_j')$$

$$y_1 = y_0 + h y_0' + h^2 \sum_{i=1}^{s} \bar{b}_i k_i', \qquad y_1' = y_0' + h \sum_{i=1}^{s} b_i k_i', \tag{14.8}$$

and the coefficients a_{ij} are no longer needed. Some examples are given in Table 14.2. The fifth-order method of Table 14.2 needs only four evaluations of f. This is a considerable improvement compared to Runge-Kutta methods where at least six evaluations are necessary (cf. Theorem 5.1).

Table 14.2. Methods for $y'' = f(x, y)$

Nyström, order 4

c_i		$\bar{a}_{ij}$	
0			
$\frac{1}{2}$	$\frac{1}{8}$		
1	0	$\frac{1}{2}$	
$\bar{b}_i$	$\frac{1}{6}$	$\frac{1}{3}$	0
b_i	$\frac{1}{6}$	$\frac{4}{6}$	$\frac{1}{6}$

Nyström, order 5

		$\bar{a}_{ij}$		
0				
$\frac{1}{5}$	$\frac{1}{50}$			
$\frac{2}{3}$	$\frac{-1}{27}$	$\frac{7}{27}$		
1	$\frac{3}{10}$	$\frac{-2}{35}$	$\frac{9}{35}$	
$\bar{b}_i$	$\frac{14}{336}$	$\frac{100}{336}$	$\frac{54}{336}$	0
b_i	$\frac{14}{336}$	$\frac{125}{336}$	$\frac{162}{336}$	$\frac{35}{336}$

Global convergence. Introducing the variable $z_n = (y_n, y_n')^T$, a Nyström method (14.4) can be written in the form

$$z_1 = z_0 + h\Phi(x_0, z_0, h) \tag{14.9}$$

where

$$\Phi(x_0, z_0, h) = \begin{pmatrix} y_0' + h \sum_i \bar{b}_i k_i' \\ \sum_i b_i k_i' \end{pmatrix}.$$

(14.9) is just a special one-step method for the differential equation (14.2). For a pth order Nyström method the local error $(y(x_0 + h) - y_1,\; y'(x_0 + h) - y_1')^T$ can be bounded by Ch^{p+1} (Definition 14.1), which is in agreement with formula (3.27). The convergence theorems of Section II.3 and the results on asymptotic expansions of the global error (Section II.8) are also valid here.

Our next aim is to derive the order conditions for Nyström methods. For this purpose we extend the theory of Section II.2 to second order differential equations (Hairer & Wanner 1976).

The Derivatives of the Exact Solution

As for first order equations we may restrict ourselves to systems of autonomous differential equations

$$(y^J)'' = f^J(y^1, \ldots, y^n, y'^1, \ldots, y'^n) \tag{14.10}$$

(if necessary, add $x'' = 0$). The superscript index J denotes the Jth component of the corresponding vector. We now calculate the derivatives of the exact solution of (14.10). The second derivative is given by (14.10):

$$(y^J)^{(2)} = f^J(y, y'). \tag{14.11;2}$$

A repeated differentiation of this equation, using (14.10), leads to

$$(y^J)^{(3)} = \sum_K \frac{\partial f^J}{\partial y^K}(y, y') \cdot y'^K + \sum_K \frac{\partial f^J}{\partial y'^K}(y, y') f^K(y, y') \tag{14.11;3}$$

$$(y^J)^{(4)} = \sum_{K,L} \frac{\partial^2 f^J}{\partial y^K \partial y^L}(y, y') \cdot y'^K \cdot y'^L \tag{14.11;4}$$

$$+ \sum_{K,L} \frac{\partial^2 f^J}{\partial y^K \partial y'^L}(y, y') \cdot y'^K \cdot f^L(y, y') + \sum_K \frac{\partial f^J}{\partial y^K}(y, y') f^K(y, y')$$

$$+ \sum_{K,L} \frac{\partial^2 f^J}{\partial y'^K \partial y^L}(y, y') f^K(y, y') \cdot y'^L$$

$$+ \sum_{K,L} \frac{\partial^2 f^J}{\partial y'^K \partial y'^L}(y, y') f^K(y, y') f^L(y, y')$$

$$+ \sum_{K,L} \frac{\partial f^J}{\partial y'^K}(y, y') \frac{\partial f^K}{\partial y^L}(y, y') y'^L$$

$$+ \sum_{K,L} \frac{\partial f^J}{\partial y'^K}(y, y') \frac{\partial f^K}{\partial y'^L}(y, y') f^L(y, y')$$

The continuation of this process becomes even more complex than for first order differential equations. A graphical representation of the above formulas will therefore be very helpful. In order to distinguish the derivatives with respect to y and y' we need two kinds of vertices: "meagre" and "fat". Fig. 14.1 shows the graphs that correspond to the above formulas.

Definition 14.2. A *labelled N-tree of order q* is a labelled tree (see Definition 2.2)

$$t : A_q \setminus \{j\} \to A_q$$

together with a mapping

$$t' : A_q \to \{\text{"meagre", "fat"}\}$$

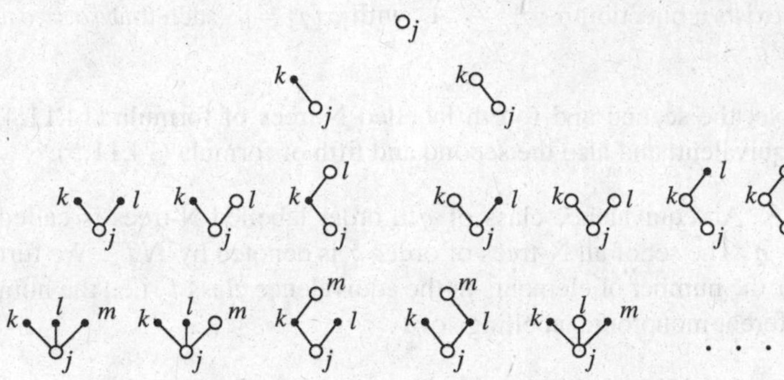

$$(14.11;2)$$

$$(14.11;3)$$

$$(14.11;4)$$

$$(14.11;5)$$

Fig. 14.1. The derivatives of the exact solution

which satisfies:

a) the root of t is always fat; i.e., $t'(j) =$ "fat";

b) a meagre vertex has at most one son and this son has to be fat.

We denote by LNT_q the set of all labelled N-trees of order q.

The reason for condition (b is that all derivatives of $g(y, y') = y'$ vanish identically with the exception of the first derivative with respect to y'.

In the sequel we use the notation *end-vertex* for a vertex which has no son. If no confusion is possible, we write t instead of (t, t') for a labelled N-tree.

Definition 14.3. For a labelled N-tree t we denote by

$$F^J(t)(y, y')$$

the expression which is a *sum* over the indices of all fat vertices of t (without "j", the index of the root) and over the indices of all meagre end-vertices. The *general term* of this sum is a product of expressions

$$\frac{\partial^r f^K}{\partial y^L \dots \partial y'^M \dots}(y, y') \qquad \text{and} \qquad y'^K. \qquad (14.12)$$

A factor of the first type appears if the fat vertex k is connected via a meagre son with $l, \dots$ and directly with a fat son $m, \dots$; a factor y'^K appears if "k" is the index of a meagre end-vertex. The vector $F(t)(y, y')$ is again called an *elementary differential*.

For some examples see Table 14.3 below. Observe that the indices of the meagre vertices, which are not end-vertices, play no role in the above definition. In analogy to Definition 2.4 we have

Definition 14.4. Two labelled N-trees (t, t') and (u, u') are *equivalent,* if they differ only by a permutation of their indices; i.e., if they have the same order, say

q, and if there exists a bijection $\sigma : A_q \to A_q$ with $\sigma(j) = j$, such that $t\sigma = \sigma u$ on $A_q \setminus \{j\}$ and $t'\sigma = u'$.

For example, the second and fourth labelled N-trees of formula (14.11;4) in Fig. 14.1 are equivalent; and also the second and fifth of formula (14.11;5).

Definition 14.5. An equivalence class of qth order labelled N-trees is called an *N-tree of order* q. The set of all N-trees of order q is denoted by NT_q. We further denote by $\alpha(t)$ the number of elements in the equivalence class t, i.e., the number of possible different monotonic labellings of t.

Representatives of N-trees up to order 5 are shown in Table 14.3. We are now able to give a closed formula for the derivatives of the exact solution of (14.10).

Theorem 14.6. *The exact solution of (14.10) satisfies*

$$y^{(q)} = \sum_{t \in LNT_{q-1}} F(t)(y, y') = \sum_{t \in NT_{q-1}} \alpha(t) F(t)(y, y'). \tag{14.11;q}$$

Proof. The general formula is obtained by continuing the computation for (14.11;2-4) as in Section II.2.   □

The Derivatives of the Numerical Solution

We first rewrite (14.4) as

$$g_i = y_0 + c_i h y_0' + \sum_{j=1}^{s} \bar{a}_{ij} h^2 f(g_j, g_j'), \qquad g_i' = y_0' + \sum_{j=1}^{s} a_{ij} h f(g_j, g_j')$$

$$y_1 = y_0 + h y_0' + \sum_{i=1}^{s} \bar{b}_i h^2 f(g_i, g_i'), \qquad y_1' = y_0' + \sum_{i=1}^{s} b_i h f(g_i, g_i')$$

$$\tag{14.13}$$

so that the intermediate values g_i, g_i' are treated in the same way as y_1, y_1'. In (14.13) there appear expressions of the form $h^2 \varphi(h)$ and $h\varphi(h)$. Therefore we have to use in addition to (2.4) the formula

$$\left(h^2 \varphi(h)\right)^{(q)}\Big|_{h=0} = q \cdot (q-1) \cdot \left(\varphi(h)\right)^{(q-2)}\Big|_{h=0}. \tag{14.14}$$

We now compute successively the derivatives of g_i^J and $g_i'^J$ at $h = 0$:

$$\left(g_i^J\right)^{(1)}\Big|_{h=0} = c_i y_0'^J \tag{14.15;1}$$

$$\left(g_i'^J\right)^{(1)}\Big|_{h=0} = \sum_j a_{ij} f^J\Big|_{y_0, y_0'} \tag{14.16;1}$$

$$\left. (g_i^J)^{(2)} \right|_{h=0} = 2 \sum_j \overline{a}_{ij} f^J \Big|_{y_0, y_0'}. \tag{14.15;2}$$

For a further differentiation we need

$$(f^J(g_j, g_j'))^{(1)} = \sum_K \frac{\partial f^J}{\partial y^K}(g_j, g_j')(g_j^K)^{(1)} + \sum_K \frac{\partial f^J}{\partial y'^K}(g_j, g_j')(g_j'^K)^{(1)}. \tag{14.17}$$

With this formula we then obtain

$$\left. (g_i'^J)^{(2)} \right|_{h=0} = 2 \sum_j a_{ij} c_j \sum_K \frac{\partial f^J}{\partial y^K} \cdot y'^K \Big|_{y_0, y_0'}$$
$$+ 2 \sum_{j,k} a_{ij} a_{jk} \sum_K \frac{\partial f^J}{\partial y'^K} \cdot f^K \Big|_{y_0, y_0'} \tag{14.16;2}$$

$$\left. (g_i^J)^{(3)} \right|_{h=0} = 3 \cdot 2 \sum_j \overline{a}_{ij} c_j \sum_K \frac{\partial f^J}{\partial y^K} \cdot y'^K \Big|_{y_0, y_0'}$$
$$+ 3 \cdot 2 \sum_{j,k} \overline{a}_{ij} a_{jk} \sum_K \frac{\partial f^J}{\partial y'^K} \cdot f^K \Big|_{y_0, y_0'}. \tag{14.15;3}$$

To write down a general formula we need

Definition 14.7. For a labelled N-tree we denote by $\Phi_j(t)$ the expression which is a sum over the indices of all fat vertices of t (without "j", the index of the root). The general term of the sum is a product of

a_{kl} if the fat vertex "k" has a fat son "l";

$\overline{a}_{kl}$ if the fat vertex "k" is connected via a meagre son with "l"; and

c_k^m if the fat vertex "k" is connected with m meagre end-vertices.

Theorem 14.8. *The g_i, g_i' of (14.13) satisfy*

$$\left. (g_i)^{(q+1)} \right|_{h=0} = (q+1) \sum_{t \in LNT_q} \gamma(t) \sum_{j=1}^s \overline{a}_{ij} \Phi_j(t) F(t)(y_0, y_0') \tag{14.15;q+1}$$

$$\left. (g_i')^{(q)} \right|_{h=0} = \sum_{t \in LNT_q} \gamma(t) \sum_{j=1}^s a_{ij} \Phi_j(t) F(t)(y_0, y_0') \tag{14.16;q}$$

where $\gamma(t)$ is given in Definition 2.10.

Proof. For small values of q these formulas were obtained above; for general values of q they are proved like Theorem 2.11. System (14.2) is a special case of what will later be treated as a *partitioned system* (see Section II.15). Theorem 14.8 will then appear again in a new light. □

Because of the similarity of the formulas for g_i and y_1, g_i' and y_1' we have

Theorem 14.9. *The numerical solution y_1, y_1' of (14.13) satisfies*

$$(y_1)^{(q)}\big|_{h=0} = q \sum_{t \in LNT_{q-1}} \gamma(t) \sum_{i=1}^{s} \overline{b}_i\, \Phi_i(t)\, F(t)(y_0, y_0') \qquad (14.18;q)$$

$$(y_1')^{(q-1)}\big|_{h=0} = \sum_{t \in LNT_{q-1}} \gamma(t) \sum_{i=1}^{s} b_i\, \Phi_i(t)\, F(t)(y_0, y_0') . \qquad (14.19;q\text{-}1)$$

$\square$

The Order Conditions

For the study of the order of a Nyström method (Definition 14.1) one has to compare the Taylor series of y_1, y_1' with that of the true solution $y(x_0 + h), y'(x_0 + h)$.

Theorem 14.10. *A Nyström method (14.4) is of order p iff*

$$\sum_{i=1}^{s} \overline{b}_i \Phi_i(t) = \frac{1}{(\varrho(t)+1) \cdot \gamma(t)} \quad \text{for N-trees } t \text{ with } \varrho(t) \leq p-1, \qquad (14.20)$$

$$\sum_{i=1}^{s} b_i \Phi_i(t) = \frac{1}{\gamma(t)} \quad \text{for N-trees } t \text{ with } \varrho(t) \leq p . \qquad (14.21)$$

Here $\varrho(t)$ denotes the order of the N-tree t, $\Phi_i(t)$ and $\gamma(t)$ are given by Definition 14.7 and formula (2.17).

Proof. The "if" part is an immediate consequence of Theorems 14.6 and 14.9. The "only if" part can be shown in the same way as for first order equations (cf. Exercise 4 of Section II.2). $\square$

Let us briefly discuss whether the extra freedom in the choice of the parameters of (14.4) (by discarding the assumption (14.5)) can lead to a considerable improvement. Since the order conditions for Runge-Kutta methods (Theorem 2.13) are a subset of (14.21) (see Exercise 3 below), it is impossible to gain order with this extra freedom. Only some (never all) error coefficients can be made smaller. Therefore we shall turn to Nyström methods (14.8) for special second order differential equations (14.7).

For the study of the order conditions for (14.8) we write (14.7) in autonomous form

$$y'' = f(y). \qquad (14.22)$$

This special form implies that those elementary differentials which contain derivatives with respect to y' vanish identically. Consequently, only the following subset of N-trees has to be considered:

Definition 14.11. An N-tree t is called a *special N-tree* or *SN-tree,* if the fat vertices have only meagre sons.

Theorem 14.12. *A Nyström method (14.8) for the special differential equation (14.7) is of order p, iff*

$$\sum_{i=1}^{s} \overline{b}_i \Phi_i(t) = \frac{1}{(\varrho(t)+1) \cdot \gamma(t)} \quad \text{for SN-trees } t \text{ with } \varrho(t) \le p-1, \quad (14.23)$$

$$\sum_{i=1}^{s} b_i \Phi_i(t) = \frac{1}{\gamma(t)} \quad \text{for SN-trees } t \text{ with } \varrho(t) \le p . \quad (14.24)$$
$$\square$$

All SN-trees up to order 5, together with the elementary differentials and the expressions Φ_j, ϱ, α, and γ, which are needed for the order conditions, are given in Table 14.3.

Higher order systems. The extension of the ideas of this section to *higher order* systems

$$y^{(n)} = f(x, y, y', \ldots, y^{(n-1)}) \quad (14.25)$$

is now more or less straightforward. Again, a real improvement is only possible in the case when the right-hand side of (14.25) depends only on x and y. A famous paper on this subject is the work of Zurmühl (1948). Tables of order conditions and methods are given in Hebsacker (1982).

On the Construction of Nyström Methods

The following simplifying assumptions are useful for the construction of Nyström methods.

Lemma 14.13. *Under the assumption*

$$\overline{b}_i = b_i(1 - c_i) \qquad i = 1, \ldots, s \quad (14.26)$$

the condition (14.24) implies (14.23).

Proof. Let t be an SN-tree of order $\le p-1$ and denote by u the SN-tree of order $\varrho(t)+1$ obtained from t by attaching a new branch with a meagre vertex to the root of t. By Definition 14.7 we have $\Phi_i(u) = c_i \Phi_i(t)$ and from formula (2.17) it

Table 14.3. SN-trees, elementary differentials and order conditions

t	graph	$\varrho(t)$	$\alpha(t)$	$\gamma(t)$	$F^J(t)(y,y')$	$\Phi_j(t)$
t_1		1	1	1	f^J	1
t_2		2	1	2	$\sum_K f_K^J y'^K$	c_j
t_3		3	1	3	$\sum_{K,L} f_{KL}^J y'^K y'^L$	c_j^2
t_4		3	1	6	$\sum_L f_L^J f^L$	$\sum_l \bar{a}_{jl}$
t_5		4	1	4	$\sum_{K,L,M} f_{KLM}^J y'^K y'^L y'^M$	c_j^3
t_6		4	3	8	$\sum_{L,M} f_{LM}^J y'^L f^M$	$\sum_m c_j \bar{a}_{jm}$
t_7		4	1	24	$\sum_{L,M} f_L^J f_M^L y'^M$	$\sum_l \bar{a}_{jl} c_l$
t_8		5	1	5	$\sum_{K,L,M,P} f_{KLMP}^J y'^K y'^L y'^M y'^P$	c_j^4
t_9		5	6	10	$\sum_{L,M,P} f_{LMP}^J y'^L y'^M f^P$	$\sum_p c_j^2 \bar{a}_{jp}$
t_{10}		5	3	20	$\sum_{M,P} f_{MP}^J f^M f^P$	$\sum_{m,p} \bar{a}_{jm} \bar{a}_{jp}$
t_{11}		5	4	30	$\sum_{L,M,P} f_{LP}^J f_M^L y'^M y'^P$	$\sum_l c_j \bar{a}_{jl} c_l$
t_{12}		5	1	60	$\sum_{L,M,P} f_L^J f_{MP}^L y'^M y'^P$	$\sum_l \bar{a}_{jl} c_l^2$
t_{13}		5	1	120	$\sum_{L,P} f_L^J f_P^L f^P$	$\sum_{l,p} \bar{a}_{jl} \bar{a}_{lp}$

follows that $\gamma(u) = (\varrho(t)+1)\gamma(t)/\varrho(t)$. The conclusion now follows since

$$\sum_{i=1}^s \bar{b}_i \Phi_i(t) = \sum_{i=1}^s b_i \Phi_i(t) - \sum_{i=1}^s b_i \Phi_i(u) = \frac{1}{\gamma(t)} - \frac{1}{\gamma(u)} = \frac{1}{(\varrho(t)+1)\gamma(t)}.$$

$\square$

Lemma 14.14. *Let t and u be two SN-trees as sketched in Fig. 14.2, where the encircled parts are assumed to be identical. Then under the assumption*

$$\sum_{j=1}^s \bar{a}_{ij} = \frac{c_i^2}{2} \qquad i=1,\ldots,s \tag{14.27}$$

the order conditions for t and u are the same.

Proof. It follows from Definition 14.7 and (14.27) that $\Phi_i(t) = \Phi_i(u)/2$ and from formula (2.17) that $\gamma(t) = 2\gamma(u)$. Both order conditions are thus identical. $\square$

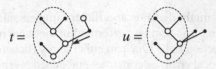

$$t = \qquad\qquad u =$$

Fig. 14.2. Trees of Lemma 14.14

Condition (14.26) allows us to neglect the equations (14.23), while condition (14.27) plays a similar role to that of (1.9) for Runge-Kutta methods. It expresses the fact that the g_i of (14.13) approximate $y(x_0 + c_i h)$ up to $\mathcal{O}(h^3)$. As a consequence of Lemma 14.14, SN-trees which have at least one fat end-vertex can be left out (i.e., $t_4, t_6, t_9, t_{10}, t_{13}$ of Table 14.3).

With the help of (14.26) and (14.27) *explicit Nyström methods* (14.8) *of order 5* with $s = 4$ can now easily be constructed: the order conditions for the trees t_1, t_2, t_3, t_5 and t_8 just indicate that the quadrature formula with nodes $c_1 = 0$, c_2, c_3, c_4 and weights b_1, b_2, b_3, b_4 is of order 5. Thus the nodes c_i have to satisfy the orthogonality relation

$$\int_0^1 x(x - c_2)(x - c_3)(x - c_4)\, dx = 0$$

and we see that two degrees of freedom are still left in the choice of the quadrature formula. The $\bar{a}_{ij}$ are now uniquely determined and can be computed as follows: $\bar{a}_{21}$ is given by (14.27) for $i = 2$. The order conditions for t_7 and t_{11} constitute two linear equations for the unknowns

$$\sum_{j=1}^2 \bar{a}_{3j} c_j \quad \text{and} \quad \sum_{j=1}^3 \bar{a}_{4j} c_j .$$

Together with (14.27, $i = 3$) one now obtains $\bar{a}_{31}$ and $\bar{a}_{32}$. Finally, the order condition for t_{12} leads to $\sum_j \bar{a}_{4j} c_j^2$ and the remaining coefficients $\bar{a}_{41}, \bar{a}_{42}, \bar{a}_{43}$ can be computed from a Vandermonde-type linear system. The method of Table 14.2 is obtained in this way.

For still higher order methods it is helpful to use further simplifying assumptions; for example

$$\sum_{j=1}^s \bar{a}_{ij} c_j^q = \frac{c_i^{q+2}}{(q+2)(q+1)} \tag{14.28}$$

which, for $q = 0$, reduces to (14.27), and

$$\sum_{i=1}^s b_i c_i^q \bar{a}_{ij} = b_j \left(\frac{c_j^{q+2}}{(q+2)(q+1)} - \frac{c_j}{q+1} + \frac{1}{q+2} \right) \tag{14.29}$$

which can be considered a generalization of condition $D(\zeta)$ of Section II.7. For more details we refer to Hairer & Wanner (1976) and also to Albrecht (1955), Battin (1976), Beentjes & Gerritsen (1976), Hairer (1977, 1982), where Nyström methods of higher order are presented.

Embedded Nyström methods. For an efficient implementation we need a step size control mechanism. This can be performed in the same manner as for Runge-Kutta methods (see Section II.4). One can either apply Richardson extrapolation in order to estimate the local error, or construct embedded Nyström methods.

A series of embedded Nyström methods has been constructed by Fehlberg (1972). These methods use a $(p+1)$-st order approximation to $y(x_0 + h)$ for step size control. A $(p+1)$-st order approximation to $y'(x_0 + h)$ is not needed, since the lower order approximations are used for step continuation.

As for first order differential equations, local extrapolation — to use the higher order approximations for step continuation — turns out to be superior. Bettis (1973) was apparently the first to use this technique. His proposed method is of order 5(4). A method of order 7(6) has been constructed by Dormand & Prince (1978), methods of order 8(7), 9(8), 10(9) and 11(10) are given by Filippi & Gräf (1986) and further methods of order 8(6) and 12(10) are presented by Dormand, El-Mikkawy & Prince (1987).

In certain situations (see Section II.6) it is important that a Nyström method be equipped with a dense output formula. Such procedures are given by Dormand & Prince (1987) and, for general initial value problems $y'' = f(x, y, y')$, by Fine (1987).

An Extrapolation Method for $y'' = f(x, y)$

Les calculs originaux, comprenant environt 3.000 pages in-folio avec 358 grandes planches, et encore 3.800 pages de développements mathématiques correspondants, appartiennent maintenant à la collection de manuscrits de la Bibliothèque de l'Université, Christiania.   (Störmer 1921)

If we rewrite the differential equation (14.7) as a first order system

$$\begin{pmatrix} y \\ y' \end{pmatrix}' = \begin{pmatrix} y' \\ f(x, y) \end{pmatrix}, \qquad \begin{pmatrix} y \\ y' \end{pmatrix}(x_0) = \begin{pmatrix} y_0 \\ y_0' \end{pmatrix} \qquad (14.30)$$

we can apply the GBS-algorithm (9.13) directly to (14.30); this yields

$$y_1 = y_0 + h y_0' \qquad (14.31a)$$
$$y_1' = y_0' + h f(x_0, y_0)$$
$$y_{i+1} = y_{i-1} + 2 h y_i' \qquad (14.31b)$$
$$y_{i+1}' = y_{i-1}' + 2 h f(x_i, y_i) \qquad i = 1, 2, \ldots, 2n$$
$$S_h(x) = (y_{2n-1} + 2 y_{2n} + y_{2n+1})/4 \qquad (14.31c)$$
$$S_h'(x) = (y_{2n-1}' + 2 y_{2n}' + y_{2n+1}')/4.$$

Here, $S_h(x)$ and $S_h'(x)$ are the numerical approximations to $y(x)$ and $y'(x)$ at $x = x_0 + H$, where $H = 2nh$ and $x_i = x_0 + ih$. We now make the following important observation: for the computation of $y_0, y_2, y_4, \ldots, y_{2n}$ (even indices) and of $y_1', y_3', \ldots, y_{2n+1}'$ (odd indices) only the function values $f(x_0, y_0)$, $f(x_2, y_2)$ $, \ldots, f(x_{2n}, y_{2n})$ have to be calculated. Furthermore, we know from (9.17) that y_{2n} and $(y_{2n-1}' + y_{2n+1}')/2$ each possess an asymptotic expansion in even powers of h. It is therefore obvious that (14.31c) should be replaced by (Gragg 1965)

$$S_h(x) = y_{2n}$$
$$S_h'(x) = (y_{2n-1}' + y_{2n+1}')/2. \tag{14.31c'}$$

Using this final step, the number of function evaluations is reduced by a factor of two. These numerical approximations can now be used for extrapolation. We take the harmonic sequence (9.8'), put

$$T_{i1} = S_h(x_0 + H), \qquad T_{i1}' = S_h'(x_0 + H)$$

and compute the extrapolated expressions $T_{i,j}$ and $T_{i,j}'$ by the Aitken & Neville formula (9.10).

Remark. Eliminating the y_j'-values in (14.31b) we obtain the equivalent formula

$$y_{i+2} - 2y_i + y_{i-2} = (2h)^2 f(x_i, y_i), \tag{14.32}$$

which is often called *Störmer's rule*. For the implementation the formulation (14.31b) is to be preferred, since it is more stable with respect to round-off errors (see Section III.10).

Dense output. As for the derivation of Section II.9 for the GBS algorithm we shall do Hermite interpolation based on derivatives of the solution at x_0, $x_0 + H$ and $x_0 + H/2$. At the endpoints of the considered interval we have $y_0, y_0', y_0'' = f(x_0, y_0)$ and y_1, y_1', y_1'' at our disposal. The derivatives at the midpoint can be obtained by extrapolation of suitable differences of function values. However, one has to take care of the fact that y_i and $f(x_i, y_i)$ are available only for even indices, whereas y_i' is available for odd indices only. For the same reason as for the GBS method, the step number sequence has to satisfy (9.34). For notational convenience, the following description is restricted to the sequence (9.35).

We suppose that T_{kk} and T_{kk}' are accepted approximations to the solution. Then the construction of a dense output formula can be summarized as follows:

Step 1. For each $j \in \{1, \ldots, k\}$ compute the approximations to the derivatives of $y(x)$ at $x_0 + H/2$ by (δ is the central difference operator):

$$d_j^{(0)} = \frac{1}{2}\Big(y_{n_j/2-1} + y_{n_j/2+1}\Big), \qquad d_j^{(1)} = y_{n_j/2}',$$

$$d_j^{(\kappa)} = \frac{1}{2} \cdot \frac{1}{(2h_j)^{\kappa-2}}\Big(\delta^{\kappa-2} f_{n_j/2-1}^{(j)} + \delta^{\kappa-2} f_{n_j/2+1}^{(j)}\Big), \qquad \kappa = 2, 4, \ldots, 2j,$$

$$d_j^{(\kappa)} = \frac{\delta^{\kappa-2} f_{n_j/2}^{(j)}}{(2h_j)^{\kappa-2}}, \qquad \kappa = 3, 5, \ldots, 2j+1. \tag{14.33}$$

Step 2. Extrapolate $d_j^{(0)}$, $d_j^{(1)}$ $(k-1)$ times and $d_j^{(2\ell)}$, $d_j^{(2\ell+1)}$ $(k-\ell)$ times to obtain improved approximations $d^{(\kappa)}$ to $y^{(\kappa)}(x_0 + H/2)$.

Step 3. For given μ $(-1 \leq \mu \leq 2k+1)$ define the polynomial $P_\mu(\theta)$ of degree $\mu + 6$ by

$$P_\mu(0) = y_0, \qquad P'_\mu(0) = y'_0, \qquad P''_\mu(0) = f(x_0, y_0)$$
$$P_\mu(1) = T_{kk}, \qquad P'_\mu(1) = T'_{kk}, \qquad P''_\mu(1) = f(x_0 + H, T_{kk}) \qquad (14.34)$$
$$P_\mu^{(\kappa)}(1/2) = H^\kappa d^{(\kappa)} \qquad \text{for} \quad \kappa = 0, 1, \ldots, \mu.$$

Since T_{kk}, T'_{kk} are the initial values for the next step, the dense output obtained by the above algorithm is a global $\mathcal{C}^2$ approximation to the solution. It satisfies

$$y(x_0 + \theta H) - P_\mu(\theta) = \mathcal{O}(H^{2k}) \qquad \text{if} \quad \mu \geq 2k - 7 \qquad (14.35)$$

(compare Theorem 9.5). In the code ODEX2 of the Appendix the value $\mu = 2k - 5$ is suggested as standard choice.

Problems for Numerical Comparisons

PLEI — the celestial mechanics problem (10.3) which is the only problem of Section II.10 already in the special form (14.7).

ARES — the ARENstorf orbit in Second order form (14.7). This is the restricted three body problem (0.1) with initial values (0.2) integrated over one period $0 \leq x \leq x_{\text{end}}$ (see Fig. 0.1) in a *fixed* coordinate system. Then the equations of motion become

$$y''_1 = \mu' \frac{a_1(x) - y_1}{D_1} + \mu \frac{b_1(x) - y_1}{D_2}$$
$$y''_2 = \mu' \frac{a_2(x) - y_2}{D_1} + \mu \frac{b_2(x) - y_2}{D_2} \qquad (14.36)$$

where

$$D_1 = \left((y_1 - a_1(x))^2 + (y_2 - a_2(x))^2\right)^{3/2}, \quad D_2 = \left((y_1 - b_1(x))^2 + (y_2 - b_2(x))^2\right)^{3/2}$$

and the movement of sun and moon are described by

$$a_1(x) = -\mu \cos x \quad a_2(x) = -\mu \sin x \quad b_1(x) = \mu' \cos x \quad b_2(x) = \mu' \sin x.$$

The initial values

$$y_1(0) = 0.994, \qquad y_1'(0) = 0, \qquad y_2(0) = 0,$$

$$y_2'(0) = -2.00158510637908252240537862224 + 0.994,$$

$$x_{\text{end}} = 17.06521656015796255889172062 49,$$

are those of (0.2) enlarged by the speed of the rotation. The exact solution values are the initial values transformed by the rotation of the coordinate system.

CPEN — the nonlinear Coupled PENdulum (see Fig. 14.3).

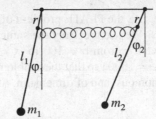

Fig. 14.3. Coupled pendulum

The kinetic as well as potential energies

$$T = \frac{m_1 l_1^2 \dot{\varphi}_1^2}{2} + \frac{m_2 l_2^2 \dot{\varphi}_2^2}{2}$$

$$V = -m_1 l_1 \cos \varphi_1 - m_2 l_2 \cos \varphi_2 + \frac{c_0 r^2 (\sin \varphi_1 - \sin \varphi_2)^2}{2}$$

lead by Lagrange theory (equations (I.6.21)) to

$$\ddot{\varphi}_1 = -\frac{\sin \varphi_1}{l_1} - \frac{c_0 r^2}{m_1 l_1^2} (\sin \varphi_1 - \sin \varphi_2) \cos \varphi_1 + f(t)$$

$$\ddot{\varphi}_2 = -\frac{\sin \varphi_2}{l_2} - \frac{c_0 r^2}{m_2 l_1^2} (\sin \varphi_2 - \sin \varphi_1) \cos \varphi_2.$$

(14.37)

We choose the parameters

$$l_1 = l_2 = 1, \quad m_1 = 1, \quad m_2 = 0.99, \quad r = 0.1, \quad c_0 = 0.01, \quad t_{\text{end}} = 496$$

and all initial values and speeds for $t = 0$ equal to zero. The first pendulum is then pushed into movement by a (somewhat idealized) hammer as

$$f(t) = \begin{cases} \sqrt{1 - (1 - t)^2} & \text{if } |t - 1| \le 1; \\ 0 & \text{otherwise.} \end{cases}$$

The resulting solutions are displayed in Fig. 14.4. The nonlinearities in this problem produce quite different sausages (cf. "Mon Oncle" de Jacques Tati 1958) from those people are accustomed to from linear problems (cf. Sommerfeld 1942, §20).

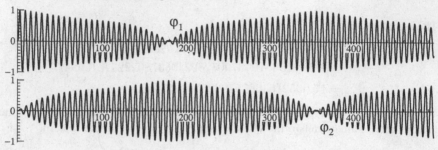

Fig. 14.4. Movement of the coupled pendulum (14.37)

WPLT — the Weak PLaTe, i.e., the PLATE problem of Section IV.10 (see Volume II) with weakened stiffness. We use precisely the same equations as (IV.10.6) and reduce the stiffness parameter σ from $\sigma = 100$ to $\sigma = 1/16$. We also remove the friction ($\omega = 0$ instead of $\omega = 1000$) so that the problem becomes purely of second order. It is linear, nonautonomous, and of dimension 40.

Performance of the Codes

Several codes were applied to each of the above four problems with 89 different tolerances between $Tol = 10^{-3}$ and $Tol = 10^{-14}$ (exactly as in Section II.10). The number of function evaluations (Fig. 14.5) and the computer time (Fig. 14.6) on a Sun Workstation (SunBlade 100) are plotted as a function of the global error at the endpoint of the integration interval. The codes used are the following:

RKN6 — symbol ⋈ — is the low order option of the Runge-Kutta-Nyström code presented in Brankin, Gladwell, Dormand, Prince & Seward (1989). It is based on a fixed-order embedded Nyström method of order 6(4), whose coefficients are given in Dormand & Prince (1987). This code is provided with a dense output.

RKN12 — symbol ⋈ — is the high order option of the Runge-Kutta-Nyström code presented in Brankin & al. (1989). It is based on the method of order 12(10), whose coefficients are given in Dormand, El-Mikkawy & Prince (1987). This code is not equipped with a dense output.

ODEX2 — symbol ◯ — is the extrapolation code based on formula (14.31a,b,c') and uses the harmonic step number sequence (see Appendix). It is implemented in the same way as ODEX (the extrapolation code for first order differential equations). In particular, the order and step size strategy is that of Section II.9. A dense output is available. Similar results are obtained by the code DIFEX2 of Deuflhard & Bauer (see Deuflhard 1985).

In order to demonstrate the superiority of the special methods for $y'' = f(x, y)$, we have included the results obtained by DOP853 (symbol ✩) and ODEX (symbol

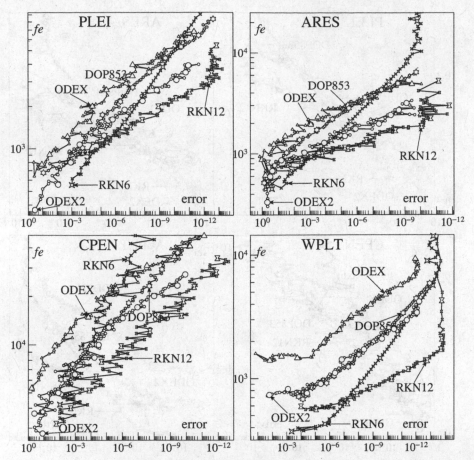

Fig. 14.5. Precision versus function evaluations

$\triangle$) which were already described in Section II.10. For their application we had to rewrite the four problems as a first order system by introducing the first derivatives as new variables. The code ODEX2 is nearly twice as efficient as ODEX which is in agreement with the theoretical considerations. Similarly the Runge-Kutta-Nyström codes RKN6 and RKN12 are a real improvement over DOP853.

A comparison of Fig. 14.5 and 14.6 shows a significant difference. The extrapolation codes ODEX and ODEX2 are relatively better on the "time"-pictures than for the function evaluation counts. With the exception of problem WPLT the performance of the code ODEX2 then becomes comparable to that of RKN12. As can be observed especially at the WPLT problem, the code RKN12 overshoots, for stringent tolerances, significantly the desired precision. It becomes less efficient if *Tol* is chosen too close to *Uround*.

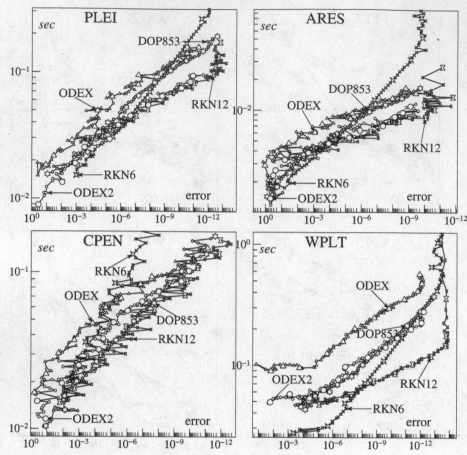

Fig. 14.6. Precision versus computing time

Exercises

1. Verify that the methods of Table 14.2 are of order 4 and 5, respectively.

2. The error coefficients of a pth order Nyström method are defined by

$$e(t) = 1 - (\varrho(t) + 1)\gamma(t) \sum_i \bar{b}_i \Phi_i(t) \qquad \text{for } \varrho(t) = p,$$
$$e'(t) = 1 - \gamma(t) \sum_i b_i \Phi_i(t) \qquad \text{for } \varrho(t) = p + 1. \qquad (14.38)$$

a) The assumption (14.26) implies that

$$e(t) = -\varrho(t)e'(u) \qquad \text{for } \varrho(t) = p,$$

where u is the N-tree obtained from t by adding a branch with a meagre vertex to the root of t.

b) Compute the error coefficients of Nyström's method (Table 14.1) and compare them to those of the classical Runge-Kutta method.

3. Show that the order conditions for Runge-Kutta methods (Theorem 2.13) are a subset of the conditions (14.21). They correspond to the N-trees, all of whose vertices are fat.

4. Sometimes the definition of order of Nyström methods (14.8) is relaxed to

$$y(x_0+h) - y_1 = \mathcal{O}(h^{p+1})$$
$$y'(x_0+h) - y_1' = \mathcal{O}(h^p)$$

(14.39)

(see Nyström 1925). Show that the conditions (14.39) are not sufficient to obtain global convergence of order p.

Hint. Investigate the asymptotic expansion of the global error with the help of Theorem 8.1 and formula (8.8).

5. The numerical solutions T_{kk} and T_{kk}' of the extrapolation method of this section are equivalent to a Nyström method of order $p = 2k$ with $s = p^2/8 + p/4 + 1$ stages.

6. A *collocation method* for $y'' = f(x, y, y')$ (or $y'' = f(x, y)$) can be defined as follows: let $u(x)$ be a polynomial of degree $s+1$ defined by

$$u(x_0) = y_0, \qquad u'(x_0) = y_0'$$ (14.40)
$$u''(x_0 + c_i h) = f\big(x_0 + c_i h, u(x_0 + c_i h), u'(x_0 + c_i h)\big), \quad i = 1, \ldots, s,$$

then the numerical solution is given by $y_1 = u(x_0 + h)$, $y_1' = u'(x_0 + h)$.

a) Prove that this collocation method is equivalent to the Nyström method (14.4) where

$$a_{ij} = \int_0^{c_i} \ell_j(t)\, dt, \qquad \bar{a}_{ij} = \int_0^{c_i} (c_i - t)\ell_j(t)\, dt,$$
$$b_i = \int_0^1 \ell_i(t)\, dt, \qquad \bar{b}_i = \int_0^1 (1 - t)\ell_i(t)\, dt,$$

(14.41)

and $\ell_j(t)$ are the Lagrange polynomials of (7.17).

b) The a_{ij} satisfy $C(s)$ (see Theorem 7.8) and the $\bar{a}_{ij}$ satisfy (14.28) for $q = 0, 1, \ldots, s-1$. These equations uniquely define a_{ij} and $\bar{a}_{ij}$.

c) In general, a_{ij} and $\bar{a}_{ij}$ do not satisfy (14.5).

d) If $M(t) = \prod_{i=1}^s (t - c_i)$ is orthogonal to all polynomials of degree $r - 1$,

$$\int_0^1 M(t) t^{q-1}\, dt = 0, \qquad q = 1, \ldots, r,$$

then the collocation method (14.40) has order $p = s + r$.

e) The polynomial $u(x)$ yields an approximation to the solution $y(x)$ on the whole interval $[x_0, x_0 + h]$. The following estimates hold:

$$y(x) - u(x) = \mathcal{O}(h^{s+2}), \qquad y'(x) - u'(x) = \mathcal{O}(h^{s+1}).$$

II.15 P-Series for Partitioned Differential Equations

> Divide ut regnes   (N. Machiavelli 1469-1527)

In the previous section we considered direct methods for second order differential equations $y'' = f(y, y')$. The idea was to write the equation as a partitioned differential system

$$\begin{pmatrix} y \\ y' \end{pmatrix}' = \begin{pmatrix} y' \\ f(y, y') \end{pmatrix} \tag{15.1}$$

and to discretize the two components, y and y', by different formulas. There are many other situations where the problem possesses a natural partitioning. Typical examples are the Hamiltonian equations (I.6.26, I.14.26) and singular perturbation problems (see Chapter VI of Volume II). It may also be of interest to separate linear and nonlinear parts or the "non-stiff" and "stiff" components of a differential equation.

We suppose that the differential system is partitioned as

$$\begin{pmatrix} y_a \\ y_b \end{pmatrix}' = \begin{pmatrix} f_a(y_a, y_b) \\ f_b(y_a, y_b) \end{pmatrix} \tag{15.2}$$

where the solution vector is separated into two components y_a, y_b, each of which may itself be a vector. An extension to more components is straight-forward.

For the numerical solution of (15.2) we consider the *partitioned method*

$$k_i = f_a\Big(y_{a0} + h \sum_{j=1}^{s} a_{ij}k_j, \; y_{b0} + h \sum_{j=1}^{s} \widehat{a}_{ij}\ell_j\Big)$$

$$\ell_i = f_b\Big(y_{a0} + h \sum_{j=1}^{s} a_{ij}k_j, \; y_{b0} + h \sum_{j=1}^{s} \widehat{a}_{ij}\ell_j\Big) \tag{15.3}$$

$$y_{a1} = y_{a0} + h \sum_{i=1}^{s} b_i k_i, \qquad y_{b1} = y_{b0} + h \sum_{i=1}^{s} \widehat{b}_i \ell_i$$

where the coefficients a_{ij}, b_i and $\widehat{a}_{ij}, \widehat{b}_i$ represent two different Runge-Kutta schemes. The first methods of this type are due to Hofer (1976) and Griepentrog (1978) who apply an explicit method to the nonstiff part and an implicit method to the stiff part of a differential equation. Later Rentrop (1985) modified this idea by combining explicit Runge-Kutta methods with Rosenbrock-type methods (Section IV.7). Recent interest for partitioned methods came up when solving Hamiltonian systems (see Section II.16 below).

The subject of this section is the derivation of the order conditions for method (15.3). For order p it is necessary that each of the two Runge-Kutta schemes under consideration be of order p. This can be seen by applying the method to $y'_a = f_a(y_a)$, $y'_b = f_b(y_b)$. But this is not sufficient, the coefficients have to satisfy certain *coupling conditions*. In order to understand this, we first look at the derivatives of the exact solution of (15.2). Then we generalize the theory of B-series (see Section II.12) to the new situation (Hairer 1981) and derive the order conditions in the same way as in II.12 for Runge-Kutta methods.

Derivatives of the Exact Solution, P-Trees

In order to avoid sums and unnecessary indices we assume that y_a and y_b in (15.2) are scalar quantities. All subsequent formulas remain valid for vectors if the derivatives are interpreted as multi-linear mappings. Differentiating (15.2) and inserting (15.2) again for the derivatives we obtain for the first component y_a

$$y_a^{(1)} = f_a \tag{15.4;1}$$

$$y_a^{(2)} = \frac{\partial f_a}{\partial y_a} f_a + \frac{\partial f_a}{\partial y_b} f_b \tag{15.4;2}$$

$$y_a^{(3)} = \frac{\partial^2 f_a}{\partial y_a^2}(f_a, f_a) + \frac{\partial^2 f_a}{\partial y_b \partial y_a}(f_b, f_a) + \frac{\partial f_a}{\partial y_a}\frac{\partial f_a}{\partial y_a} f_a + \frac{\partial f_a}{\partial y_a}\frac{\partial f_a}{\partial y_b} f_b \tag{15.4;3}$$

$$+ \frac{\partial^2 f_a}{\partial y_a \partial y_b}(f_a, f_b) + \frac{\partial^2 f_a}{\partial y_b^2}(f_b, f_b) + \frac{\partial f_a}{\partial y_b}\frac{\partial f_b}{\partial y_a} f_a + \frac{\partial f_a}{\partial y_b}\frac{\partial f_b}{\partial y_b} f_b.$$

Similar formulas hold for the derivatives of y_b.

For a graphical representation of these formulas we need two different kinds of vertices. As in Section II.14 we use "meagre" and "fat" vertices, which will correspond to f_a and f_b, respectively. Formulas (15.4) can then be represented as shown in Fig. 15.1.

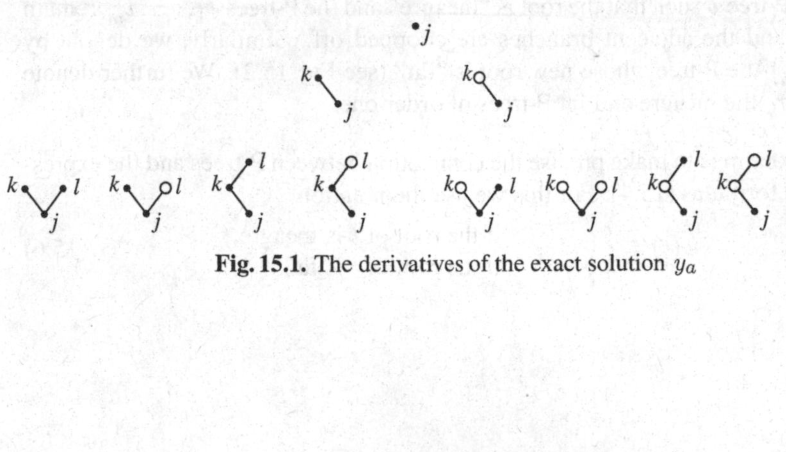

$$(15.4;1)$$
$$(15.4;2)$$
$$(15.4;3)$$

Fig. 15.1. The derivatives of the exact solution y_a

Definition 15.1. A *labelled P-tree* of *order* q is a labelled tree (see Definition 2.2)

$$t : A_q \setminus \{j\} \to A_q$$

together with a mapping

$$t' : A_q \to \{\text{"meagre"}, \text{"fat"}\}.$$

We denote by LTP_q^a the set of those labelled P-trees of order q, whose root is meagre (i.e., $t'(j) = $"meagre"). Similarly, LTP_q^b is the set of qth order labelled P-trees with a "fat" root.

Due to the symmetry of the second derivative the 2nd and 5th expressions in (15.4;3) are equal. We therefore define:

Definition 15.2. Two labelled P-trees (t, t') and (u, u') are *equivalent*, if they have the same order, say q, and if there exists a bijection $\sigma : A_q \to A_q$ such that $\sigma(j) = j$ and the following diagram commutes:

$$
\begin{array}{ccc}
A_q \setminus \{j\} & \xrightarrow{\;\;t\;\;} & A_q \;\; \searrow^{t'} \\
\sigma \downarrow & & \sigma \downarrow \qquad \qquad \{\text{"meagre"}, \text{"fat"}\} \\
A_q \setminus \{j\} & \xrightarrow{\;\;u\;\;} & A_q \;\; \nearrow_{u'}
\end{array}
$$

Definition 15.3. An equivalence class of qth order labelled P-trees is called a *P-tree* of *order* q. The set of all P-trees of order q with a meagre root is denoted by TP_q^a, that with a fat root by TP_q^b. For a P-tree t we denote by $\varrho(t)$ the *order* of t, and by $\alpha(t)$ the number of elements in the equivalence class t.

Examples of P-trees together with the numbers $\varrho(t)$ and $\alpha(t)$ are given in Table 15.1 below. We first discuss a recursive representation of P-trees (extension of Definition 2.12), which is fundamental for the following theory.

Definition 15.4. Let $t_1, \ldots, t_m$ be P-trees. We then denote by

$$t = \,_a[t_1, \ldots, t_m] \tag{15.5}$$

the unique P-tree t such that the root is "meagre" and the P-trees $t_1, \ldots, t_m$ remain if the root and the adjacent branches are chopped off. Similarly, we denote by $_b[t_1, \ldots, t_m]$ the P-tree whose new root is "fat" (see Fig. 15.2). We further denote by τ_a and τ_b the meagre and fat P-trees of order one.

Our next aim is to make precise the connection between P-trees and the expressions of the formulas (15.4). For this we use the notation

$$w(t) = \begin{cases} a & \text{if the root of } t \text{ is meagre,} \\ b & \text{if the root of } t \text{ is fat.} \end{cases} \tag{15.6}$$

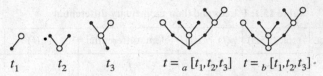

$$t_1 \qquad t_2 \qquad t_3 \qquad t = {}_a[t_1,t_2,t_3] \quad t = {}_b[t_1,t_2,t_3].$$

Fig. 15.2. Recursive definition of P-trees

Definition 15.5. The *elementary differentials,* corresponding to (15.2), are defined recursively by $(y = (y_a, y_b))$

$$F(\tau_a)(y) = f_a(y), \qquad F(\tau_b)(y) = f_b(y)$$

and

$$F(t)(y) = \frac{\partial^m f_{w(t)}(y)}{\partial y_{w(t_1)} \cdots \partial y_{w(t_m)}} \cdot \big(F(t_1)(y), \ldots, F(t_m)(y)\big)$$

for $t = {}_a[t_1, \ldots, t_m]$ or $t = {}_b[t_1, \ldots, t_m]$.

Elementary differentials for P-trees up to order 3 are given explicitly in Table 15.1.

We now return to the starting-point of this section and continue the differentiation of formulas (15.4). Using the notation of labelled P-trees, one sees that a differentiation of $F(t)(y_a, y_b)$ can be interpreted as an addition of a new branch with a meagre or fat vertex and a new summation letter to each vertex of the labelled P-tree t. In the same way as we proved Theorem 2.6 for non-partitioned differential equations, we arrive at

Theorem 15.6. *The derivatives of the exact solution of (15.2) satisfy*

$$y_a^{(q)} = \sum_{t \in LTP_q^a} F(t)(y_a, y_b) = \sum_{t \in TP_q^a} \alpha(t) F(t)(y_a, y_b) \qquad (15.4;q)$$

$$y_b^{(q)} = \sum_{t \in LTP_q^b} F(t)(y_a, y_b) = \sum_{t \in TP_q^b} \alpha(t) F(t)(y_a, y_b).$$

$\qquad\qquad\qquad\qquad\qquad\qquad\qquad\qquad\qquad\qquad\qquad\qquad\qquad\qquad\qquad\square$

Table 15.1. P-trees and their elementary differentials

P-tree	repr. (15.5)	$\varrho(t)$	$\alpha(t)$	elem. differential	$\Phi_j(t)$
•	τ_a	1	1	f_a	1
＼	$a[\tau_a]$	2	1	$\frac{\partial f_a}{\partial y_a} f_a$	$\sum_k a_{jk}$
✎	$a[\tau_b]$	2	1	$\frac{\partial f_a}{\partial y_b} f_b$	$\sum_k \widehat{a}_{jk}$
∨	$a[\tau_a,\tau_a]$	3	1	$\frac{\partial^2 f_a}{\partial y_a^2}(f_a,f_a)$	$\sum_{k,l} a_{jk}a_{jl}$
∨	$a[\tau_a,\tau_b]$	3	2	$\frac{\partial^2 f_a}{\partial y_a \partial y_b}(f_a,f_b)$	$\sum_{k,l} a_{jk}\widehat{a}_{jl}$
∨	$a[\tau_b,\tau_b]$	3	1	$\frac{\partial^2 f_a}{\partial y_b^2}(f_b,f_b)$	$\sum_{k,l} \widehat{a}_{jk}\widehat{a}_{jl}$
‹	$a[a[\tau_a]]$	3	1	$\frac{\partial f_a}{\partial y_a}\frac{\partial f_a}{\partial y_a} f_a$	$\sum_{k,l} a_{jk}a_{kl}$
‹	$a[a[\tau_b]]$	3	1	$\frac{\partial f_a}{\partial y_a}\frac{\partial f_a}{\partial y_b} f_b$	$\sum_{k,l} a_{jk}\widehat{a}_{kl}$
‹	$a[b[\tau_a]]$	3	1	$\frac{\partial f_a}{\partial y_b}\frac{\partial f_b}{\partial y_a} f_a$	$\sum_{k,l} \widehat{a}_{jk}a_{kl}$
‹	$a[b[\tau_b]]$	3	1	$\frac{\partial f_a}{\partial y_b}\frac{\partial f_b}{\partial y_b} f_b$	$\sum_{k,l} \widehat{a}_{jk}\widehat{a}_{kl}$
...	...	...	...	...	...
○	τ_b	1	1	f_b	1
ᴼ	$b[\tau_a]$	2	1	$\frac{\partial f_b}{\partial y_a} f_a$	$\sum_k a_{jk}$
ᴼ	$b[\tau_b]$	2	1	$\frac{\partial f_b}{\partial y_b} f_b$	$\sum_k \widehat{a}_{jk}$
...	...	...	...	...	...

P-Series

In Section II.12 we saw the importance of the key-lemma Corollary 12.7 for the derivation of the order conditions for Runge-Kutta methods. Therefore we extend this result also to partitioned ordinary differential equations.

It is convenient to introduce two new P-trees of order 0, namely $\emptyset_a$ and $\emptyset_b$. The corresponding elementary differentials are $F(\emptyset_a)(y) = y_a$ and $F(\emptyset_b)(y) = y_b$. We further set

$$TP^a = \{\emptyset_a\} \cup TP_1^a \cup TP_2^a \cup \ldots \qquad LTP^a = \{\emptyset_a\} \cup LTP_1^a \cup LTP_2^a \cup \ldots$$
$$TP^b = \{\emptyset_b\} \cup TP_1^b \cup TP_2^b \cup \ldots \qquad LTP^b = \{\emptyset_b\} \cup LTP_1^b \cup LTP_2^b \cup \ldots.$$

$$(15.7)$$

Definition 15.7. Let $\mathbf{c}(\emptyset_a)$, $\mathbf{c}(\emptyset_b)$, $\mathbf{c}(\tau_a)$, $\mathbf{c}(\tau_b)$, ... be real coefficients defined for all P-trees, i.e., $\mathbf{c} : TP^a \cup TP^b \to \mathbb{R}$. The series

$$P(\mathbf{c}, y) = \left(P_a(\mathbf{c}, y), P_b(\mathbf{c}, y) \right)^T$$

where

$$P_a(\mathbf{c}, y) = \sum_{t \in LTP^a} \frac{h^{\varrho(t)}}{\varrho(t)!} \mathbf{c}(t) F(t)(y), \qquad P_b(\mathbf{c}, y) = \sum_{t \in LTP^b} \frac{h^{\varrho(t)}}{\varrho(t)!} \mathbf{c}(t) F(t)(y)$$

is then called a *P-series*.

Theorem 15.6 simply states that the exact solution of (15.2) is a P-series

$$\left(y_a(x_0 + h), y_b(x_0 + h) \right)^T = P\left(\mathbf{y}, (y_a(x_0), y_b(x_0)) \right) \tag{15.8}$$

with $\mathbf{y}(t) = 1$ for all P-trees t.

Theorem 15.8. *Let* $\mathbf{c} : TP^a \cup TP^b \to \mathbb{R}$ *be a sequence of coefficients such that* $\mathbf{c}(\emptyset_a) = \mathbf{c}(\emptyset_b) = 1$. *Then*

$$h \left(\begin{array}{c} f_a\big(P(\mathbf{c}, (y_a, y_b))\big) \\ f_b\big(P(\mathbf{c}, (y_a, y_b))\big) \end{array} \right) = P(\mathbf{c}', (y_a, y_b)) \tag{15.9}$$

with

$$\mathbf{c}'(\emptyset_a) = \mathbf{c}'(\emptyset_b) = 0, \qquad \mathbf{c}'(\tau_a) = \mathbf{c}'(\tau_b) = 1 \tag{15.10}$$
$$\mathbf{c}'(t) = \varrho(t)\mathbf{c}(t_1)\ldots\mathbf{c}(t_m) \qquad if \quad t = {}_a[t_1,\ldots,t_m] \quad or \quad t = {}_b[t_1,\ldots,t_m].$$

The *proof* is related to that of Theorem 12.6. It is given with more details in Hairer (1981). □

Order Conditions for Partitioned Runge-Kutta Methods

With the help of Theorem 15.8 the order conditions for method (15.3) can readily be obtained. For this we denote the arguments in (15.3) by

$$g_i = y_{a0} + h \sum_{j=1}^{s} a_{ij} k_j, \qquad \widehat{g}_i = y_{b0} + h \sum_{j=1}^{s} \widehat{a}_{ij} \ell_j, \tag{15.11}$$

and we assume that $G_i = (g_i, \widehat{g}_i)^T$ and $K_i = h(k_i, \ell_i)^T$ are P-series with coefficients $\mathbf{G}_i(t)$ and $\mathbf{K}_i(t)$, respectively. The formulas (15.11) then yield $\mathbf{G}_i(\emptyset_a) = 1$, $\mathbf{G}_i(\emptyset_b) = 1$ and

$$\mathbf{G}_i(t) = \begin{cases} \sum_{j=1}^{s} a_{ij} \mathbf{K}_j(t) & \text{if the root of } t \text{ is meagre,} \\ \sum_{j=1}^{s} \widehat{a}_{ij} \mathbf{K}_j(t) & \text{if the root of } t \text{ is fat.} \end{cases} \tag{15.12}$$

Application of Theorem 15.8 to the relations $k_j = f_a(G_j)$, $\ell_j = f_b(G_j)$ shows that $\mathbf{K}_j(t) = \mathbf{G}'_j(t)$ which, together with (15.10) and (15.12), recursively defines the values $\mathbf{K}_j(t)$.

It is usual to write $\mathbf{K}_j(t) = \gamma(t)\Phi_j(t)$ where $\gamma(t)$ is the integer given in Definition 2.10 (see also (2.17)). The coefficient $\Phi_j(t)$ is then obtained in the same way as the corresponding value of standard Runge-Kutta methods (see Definition 2.9) with the exception that a factor a_{ik} has to be replaced by $\widehat{a}_{ik}$, if the vertex with label "k" is fat. A comparison of the P-series for the numerical solution $(y_{1a}, y_{1b})^T$ with that for the exact solution (15.8) yields the desired order conditions.

Theorem 15.9. *A partitioned Runge-Kutta method (15.3) is of order p iff*

$$\sum_{j=1}^{s} b_j \Phi_j(t) = \frac{1}{\gamma(t)} \qquad and \qquad \sum_{j=1}^{s} \widehat{b}_j \Phi_j(t) = \frac{1}{\gamma(t)} \qquad (15.13)$$

for all P-trees of order $\leq p$.   □

Example. A partitioned method (15.3) is of order 2, if and only if each of the two Runge-Kutta schemes has order 2 and if the coupling conditions

$$\sum_{i,j} b_i \widehat{a}_{ij} = \frac{1}{2}, \qquad \sum_{i,j} \widehat{b}_i a_{ij} = \frac{1}{2},$$

which correspond to trees $_a[\tau_b]$ and $_b[\tau_a]$ of Table 15.1 respectively, are satisfied. This happens if

$$c_i = \widehat{c}_i \quad \text{for all } i.$$

This last assumption simplifies the order conditions considerably (the "thickness" of terminating vertices then has no influence). The resulting conditions for order up to 4 have been tabulated by Griepentrog (1978).

Further Applications of P-Series

Runge-Kutta methods violating (1.9). For the non-autonomous differential equation $y' = f(x, y)$ we consider, as in Exercise 6 of Section II.1, the Runge-Kutta method

$$k_i = f\left(x_0 + \widehat{c}_i h, \ y_0 + h \sum_{j=1}^{s} a_{ij} k_j\right), \qquad y_1 = y_0 + h \sum_{i=1}^{s} b_i k_i, \qquad (15.14)$$

where $\widehat{c}_i$ is not necessarily equal to $c_i = \sum_j a_{ij}$. Therefore, the x and y components in

$$y' = f(x, y)$$
$$x' = 1.$$
(15.15)

are integrated differently. This system is of the form (15.2), if we put $y_a = y$, $y_b = x$, $f_a(y_a, y_b) = f(x, y)$ and $f_b(y_a, y_b) = 1$. Since f_b is constant, all elementary differentials that involve derivatives of f_b vanish identically. Thus, P-trees where at least one fat vertex is not an end-vertex need not be considered. It remains to treat the set

$$T_x = \{t \in TP_a; \text{ all fat vertices are end-vertices}\}.$$
(15.16)

Each tree of T_x gives rise to an order condition which is exactly that of Theorem 15.9. It is obtained in the usual way (Section II.2) with the exception that c_k has to be replaced by $\widehat{c}_k$, if the corresponding vertex is a fat one.

Fehlberg methods. The methods of Fehlberg, introduced in Section II.13, are equivalent to (15.14). However, it is known that the exact solution of the differential equation $y' = f(x, y)$ satisfies $y(x_0) = 0$, $y'(x_0) = 0, \ldots, y^{(m)}(x_0) = 0$ at the initial value $x = x_0$. As explained in II.13, this implies that the expressions $f, \partial f/\partial x, \ldots, \partial^{m-1} f/\partial x^{m-1}$ vanish at (x_0, y_0) and consequently also many of the elementary differentials disappear. The elements of T_x which remain to be considered are given in Fig. 15.3.

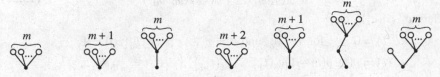

Fig. 15.3. P-trees for the methods of Fehlberg

Nyström methods. As a last application of Theorem 15.8 we present a new derivation of the order conditions for Nyström methods (Section II.14). The second order differential equation $y'' = f(y, y')$ can be written in partitioned form as

$$\binom{y}{y'}' = \binom{y'}{f(y, y')}.$$
(15.17)

In the notation of (15.2) we have $y_a = y$, $y_b = y'$, $f_a(y_a, y_b) = y_b$, $f_b(y_a, y_b) = f(y_a, y_b)$. The special structure of f_a implies that only P-trees which satisfy the condition (see Definition 14.2)

"meagre vertices have at most one son and this son has to be fat" (15.18)

have to be considered. The essential P-trees are thus
$$TN_q^a = \{t \in TP_q^a \; ; \; t \text{ satisfies } (15.18)\}$$
$$TN_q^b = \{t \in TP_q^b \; ; \; t \text{ satisfies } (15.18)\}.$$

It follows that each element of TN_{q+1}^a can be written as $t = {}_a[u]$ with $u \in TN_q^b$. This implies a one-to-one correspondence between TN_{q+1}^a and TN_q^b, leaving the elementary differentials invariant:

$$F({}_a[u])(y_a, y_b) = \frac{\partial y_b}{\partial y_b} \cdot F(u)(y_a, y_b) = F(u)(y_a, y_b).$$

From this property it follows that

$$hP_b(\mathbf{c}, (y_a, y_b)) = P_a(\mathbf{c}', (y_a, y_b)) \tag{15.19}$$

where $\mathbf{c}'(\emptyset_a) = 0$, $\mathbf{c}'(\tau_a) = \mathbf{c}(\emptyset_b)$ and

$$\mathbf{c}'(t) = \varrho(t)\mathbf{c}(u) \qquad \text{if } t = {}_a[u]. \tag{15.20}$$

This notation is in agreement with (15.10).

The order conditions of method (14.13) can now be derived as follows: assume g_i, g_i' to be P-series

$$g_i = P_a(\mathbf{c}_i, (y_0, y_0')), \qquad g_i' = P_b(\mathbf{c}_i, (y_0, y_0')).$$

Theorem 15.8 then implies that

$$hf(g_i, g_i') = P_b(\mathbf{c}_i', (y_0, y_0')). \tag{15.21}$$

Multiplying this relation by h it follows from (15.19) that

$$h^2 f(g_i, g_i') = P_a(\mathbf{c}_i'', (y_0, y_0')). \tag{15.22}$$

Here $\mathbf{c}_i'' = (\mathbf{c}_i')'$, i.e.,

$$\mathbf{c}_i''(t) = 0 \qquad \text{for } t = \emptyset_a \text{ and } t = \tau_a, \qquad \mathbf{c}_i''({}_a[\tau_b]) = 1,$$
$$\mathbf{c}_i''(t) = \varrho(t)(\varrho(t) - 1)\mathbf{c}_i(t_1) \ldots \mathbf{c}_i(t_m) \qquad \text{if } t = {}_a[{}_b[t_1, \ldots, t_m]].$$

The relations (15.21) and (15.22), when inserted into (14.13), yield

$$\mathbf{c}_i(\tau_a) = c_i,$$
$$\mathbf{c}_i(t) = \begin{cases} \sum_j \bar{a}_{ij} \mathbf{c}_j''(t) & \text{if the root of } t \text{ is meagre,} \\ \sum_j a_{ij} \mathbf{c}_j'(t) & \text{if the root of } t \text{ is fat.} \end{cases}$$

Finally, a comparison of the P-series for the exact and numerical solutions gives the order conditions (for order p)

$$\sum_i \bar{b}_i \mathbf{c}_i''(t) = 1 \qquad \text{for } t \in TN_q^a, \; q = 2, \ldots, p$$

$$\sum_i b_i \mathbf{c}_i'(t) = 1 \qquad \text{for } t \in TN_q^b, \; q = 1, \ldots, p. \tag{15.23}$$

Exercises

1. Denote the number of elements of TP_q^a (P-trees with meagre root of order q) by α_q (see Table 15.2). Prove that

$$\alpha_1 + \alpha_2 x + \alpha_3 x^2 + \ldots = (1-x)^{-2\alpha_1}(1-x^2)^{-2\alpha_2}(1-x^3)^{-2\alpha_3} \cdots .$$

   Compute the first α_q and compare them with the a_q of Table 2.1.

Table 15.2. Number of elements of TP_q^a

q	1	2	3	4	5	6	7	8	9	10
α_q	1	2	7	26	107	458	2058	9498	44947	216598

2. There is no explicit, 4-stage Runge-Kutta method of order 4, which does not satisfy condition (1.9).

   Hint. Use the techniques of the proof of Lemma 1.4.

3. Show that the order conditions (15.23) are the same as those given in Theorem 14.10.

4. Show that the partitioned method of Griepentrog (1978)

0		a_{ij}			0	0		$\widehat{a}_{ij}$
1/2	1/2				1/2	$-\beta/2$	$(1+\beta)/2$	
1	-1	2			1	$(3+5\beta)/2$	$-(1+3\beta)$	$(1+\beta)/2$
	1/6	2/3	1/6			1/6	2/3	1/6

   with $\beta = \sqrt{3}/3$ is of order 3 (the implicit method to the right is A-stable and is provided for the stiff part of the problem).

II.16 Symplectic Integration Methods

> It is natural to look forward to those discrete systems which preserve as much as possible the intrinsic properties of the continuous system.
>
> (Feng Kang 1985)

> Y.V. Rakitskii proposed ... a requirement of the most complete conformity between two dynamical systems: one resulting from the original differential equations and the other resulting from the difference equations of the computational method.
>
> (Y.B. Suris 1989)

Hamiltonian systems, given by

$$\dot{p}_i = -\frac{\partial H}{\partial q_i}(p,q), \qquad \dot{q}_i = \frac{\partial H}{\partial p_i}(p,q), \tag{16.1}$$

have been seen to possess two remarkable properties:

a) the solutions preserve the Hamiltonian $H(p,q)$ (Ex. 5 of Section I.6);

b) the corresponding flow is symplectic, i.e., preserves the differential 2-form

$$\omega^2 = \sum_{i=1}^{n} dp_i \wedge dq_i \tag{16.2}$$

(see Theorem I.14.12). In particular, the flow is volume preserving.
Both properties are usually destroyed by a numerical method applied to (16.1).

After some pioneering papers (de Vogelaere 1956, Ruth 1983, and Feng Kang (冯康 1985) an enormous avalanche of research started around 1988 on the characterization of existing numerical methods which preserve symplecticity or on the construction of new classes of symplectic methods. An excellent overview is presented by Sanz-Serna (1992).

Example 16.1. We consider the harmonic oscillator

$$H(p,q) = \frac{1}{2}(p^2 + k^2 q^2). \tag{16.3}$$

Here (16.1) becomes

$$\dot{p} = -k^2 q, \qquad \dot{q} = p \tag{16.4}$$

and we study the action of several steps of a numerical method on a well-known set of initial data (p_0, q_0) (see Fig. 16.1):

a) The explicit Euler method (I.7.3)

$$\begin{pmatrix} p_m \\ q_m \end{pmatrix} = \begin{pmatrix} 1 & -hk^2 \\ h & 1 \end{pmatrix} \begin{pmatrix} p_{m-1} \\ q_{m-1} \end{pmatrix}, \qquad h = \frac{\pi}{8k}, \ m = 1, \dots, 16; \tag{16.5a}$$

b) the implicit (or backward) Euler method (7.3)

$$\begin{pmatrix} p_m \\ q_m \end{pmatrix} = \frac{1}{1+h^2k^2} \begin{pmatrix} 1 & -hk^2 \\ h & 1 \end{pmatrix} \begin{pmatrix} p_{m-1} \\ q_{m-1} \end{pmatrix}, \quad h = \frac{\pi}{8k}, \; m = 1, \ldots, 16;$$
(16.5b)

c) Runge's method (1.4) of order 2

$$\begin{pmatrix} p_m \\ q_m \end{pmatrix} = \begin{pmatrix} 1 - \frac{h^2k^2}{2} & -hk^2 \\ h & 1 - \frac{h^2k^2}{2} \end{pmatrix} \begin{pmatrix} p_{m-1} \\ q_{m-1} \end{pmatrix}, \quad h = \frac{\pi}{4k}, \; m = 1, \ldots, 8;$$
(16.5c)

d) the implicit midpoint rule (7.4) of order 2

$$\begin{pmatrix} p_m \\ q_m \end{pmatrix} = \frac{1}{1+\frac{h^2k^2}{4}} \begin{pmatrix} 1 - \frac{h^2k^2}{4} & -hk^2 \\ h & 1 - \frac{h^2k^2}{4} \end{pmatrix} \begin{pmatrix} p_{m-1} \\ q_{m-1} \end{pmatrix}, \quad h = \frac{\pi}{4k}, \; m = 1, \ldots, 8.$$
(16.5d)

For the exact flow, the last of all these cats would precisely coincide with the first one and all cats would have the same area. Only the last method appears to be area preserving. It also preserves the Hamiltonian in this example.

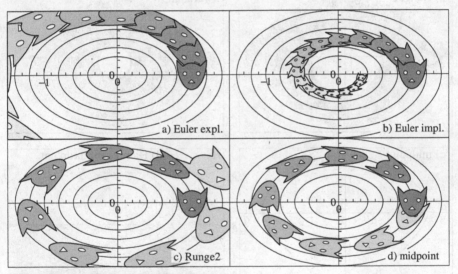

a) Euler expl.

b) Euler impl.

c) Runge2

d) midpoint

Fig. 16.1. Destruction of symplecticity of a Hamiltonian flow, $k = (\sqrt{5}+1)/2$

Example 16.2. For a nonlinear problem we choose

$$H(p, q) = \frac{p^2}{2} - \cos(q)\left(1 - \frac{p}{6}\right)$$
(16.6)

which is similar to the Hamiltonian of the pendulum (I.14.25), but with some of the pendulum's symmetry destroyed. Fig. 16.2 presents 12000 consecutive solution values (p_i, q_i) for

a) Runge's method of order 2 (see (1.4));

b) the implicit Radau method with $s = 2$ and order 3 (see Exercise 6 of Section II.7);

c) the implicit midpoint rule (7.4) of order 2.

The initial values are

$$p_0 = 0, \qquad q_0 = \begin{cases} \arccos(0.5) = \pi/3 & \text{for case (a)} \\ \arccos(-0.8) & \text{for cases (b) and (c).} \end{cases}$$

The computation is done with fixed step sizes

$$h = \begin{cases} 0.15 & \text{for case (a)} \\ 0.3 & \text{for cases (b) and (c).} \end{cases}$$

The solution of method (a) spirals out, that of method (b) spirals in and both by no means preserve the Hamiltonian. Method (c) behaves differently. Although the Hamiltonian is not precisely preserved (see picture (d)), its error remains bounded for long-scale computations.

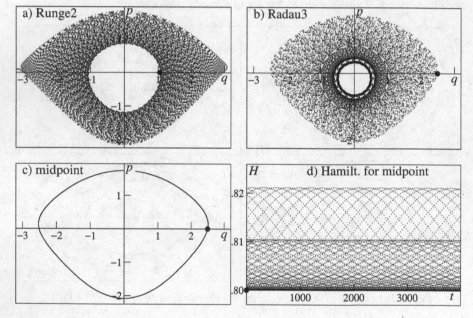

Fig. 16.2. A nonlinear pendulum and behaviour of H
($\bullet \ldots$ indicates the initial position)

Symplectic Runge-Kutta Methods

For a given Hamiltonian system (16.1), for a chosen one-step method (in particular a Runge-Kutta method) and a chosen step size h we denote by

$$\psi_h : \quad \mathbb{R}^{2n} \quad \longrightarrow \quad \mathbb{R}^{2n}$$
$$(p_0, q_0) \quad \longmapsto \quad (p_1, q_1) \tag{16.7}$$

the transformation defined by the method.

Remark. For implicit methods the numerical solution (p_1, q_1) need not exist for all h and all initial values (p_0, q_0) nor need it be uniquely determined (see Exercise 2). Therefore we usually will have to restrict the domain where ψ_h is defined and we will have to select a solution of the nonlinear system such that ψ_h is differentiable on this domain. The subsequent results hold for all possible choices of ψ_h.

Definition 16.4. A one-step method is called *symplectic* if for every smooth Hamiltonian H and for every step size h the mapping ψ_h is symplectic (see Definition I.14.11), i.e., preserves the differential 2-form ω^2 of (16.2).

We start with the easiest result.

Theorem 16.5. *The implicit s-stage Gauss methods of order $2s$ (Kuntzmann & Butcher methods of Section II.7) are symplectic for all s.*

Proof. We simplify the notation by putting $h = 1$ and $t_0 = 0$ and use the fact that the methods under consideration are collocation methods, i.e., the numerical solution after one step is defined by $(u(1), v(1))$ where $(u(t), v(t))$ are polynomials of degree s such that

$$u(0) = p_0, \quad u'(c_i) = -\frac{\partial H}{\partial q}\big(u(c_i), v(c_i)\big)$$
$$\qquad\qquad\qquad\qquad\qquad\qquad\qquad i = 1, \ldots, s. \tag{16.8}$$
$$v(0) = q_0, \quad v'(c_i) = \frac{\partial H}{\partial p}\big(u(c_i), v(c_i)\big)$$

The polynomials $u(t)$ and $v(t)$ are now considered as functions of the initial values. For arbitrary variations ξ_1^0 and ξ_2^0 of the initial point we denote the corresponding variations of u and v as

$$\xi_1^t = \frac{\partial(u(t), v(t))}{\partial(p_0, q_0)} \cdot \xi_1^0, \qquad \xi_2^t = \frac{\partial(u(t), v(t))}{\partial(p_0, q_0)} \cdot \xi_2^0.$$

Symplecticity of the method means that the expression

$$\omega^2(\xi_1^1, \xi_2^1) - \omega^2(\xi_1^0, \xi_2^0) = \int_0^1 \frac{d}{dt}\,\omega^2(\xi_1^t, \xi_2^t)\, dt \tag{16.9}$$

should vanish. Since ξ_1^t and ξ_2^t are polynomials in t of degree s, the expression $\frac{d}{dt}\omega^2(\xi_1^t, \xi_2^t)$ is a polynomial of degree $2s - 1$. We can thus exactly integrate (16.9) by the Gaussian quadrature formula and so obtain

$$\omega^2(\xi_1^1, \xi_2^1) - \omega^2(\xi_1^0, \xi_2^0) = \sum_{i=1}^{s} b_i \frac{d}{dt} \omega^2(\xi_1^t, \xi_2^t)\Big|_{t=c_i}. \tag{16.9'}$$

Differentiation of (16.8) with respect to (p_0, q_0) shows that (ξ_1^t, ξ_2^t) satisfies the variational equation (I.14.27) at the collocation points $t = c_i$, $i = 1, \ldots, s$. Therefore, the computations of the proof of Theorem I.14.12 imply that

$$\frac{d}{dt} \omega^2(\xi_1^t, \xi_2^t)\Big|_{t=c_i} = 0 \qquad \text{for} \quad i = 1, \ldots, s. \tag{16.10}$$

This, introduced into (16.9'), completes the proof of symplecticity.   □

The following theorem, discovered independently by at least three authors (F. Lasagni 1988, J.M. Sanz-Serna 1988, Y.B. Suris 1989) characterizes the class of all symplectic Runge-Kutta methods:

Theorem 16.6. *If the* $s \times s$ *matrix* M *with elements*

$$m_{ij} = b_i a_{ij} + b_j a_{ji} - b_i b_j, \qquad i, j = 1, \ldots, s \tag{16.11}$$

satisfies $M = 0$, *then the Runge-Kutta method (7.7) is symplectic.*

Proof. The matrix M has been known from nonlinear stability theory for many years (see Theorem IV.12.4). Both theorems have very similar proofs, the one works with the *inner* product, the other with the *exterior* product.

We write method (7.7) applied to problem (16.1) as

$$P_i = p_0 + h \sum_j a_{ij} k_j \qquad Q_i = q_0 + h \sum_j a_{ij} \ell_j \tag{16.12a}$$

$$p_1 = p_0 + h \sum_i b_i k_i \qquad q_1 = q_0 + h \sum_i b_i \ell_i \tag{16.12b}$$

$$k_i = -\frac{\partial H}{\partial q}(P_i, Q_i) \qquad \ell_i = \frac{\partial H}{\partial p}(P_i, Q_i), \tag{16.12c}$$

denote the Jth component of a vector by an upper index J and introduce the linear maps (one-forms)

$$dp_1^J : \mathbb{R}^{2n} \to \mathbb{R}, \qquad\qquad dP_i^J : \mathbb{R}^{2n} \to \mathbb{R},$$
$$\xi \mapsto \frac{\partial p_1^J}{\partial(p_0, q_0)} \xi \qquad\qquad \xi \mapsto \frac{\partial P_i^J}{\partial(p_0, q_0)} \xi \tag{16.13}$$

and similarly also dp_0^J, dk_i^J, dq_0^J, dq_1^J, dQ_i^J, $d\ell_i^J$ (the one-forms dp_0^J and dq_0^J correspond to dp_J and dq_J of Section I.14). Using the notation (16.13), symplecticity of the method is equivalent to

$$\sum_{J=1}^{n} dp_1^J \wedge dq_1^J = \sum_{J=1}^{n} dp_0^J \wedge dq_0^J. \tag{16.14}$$

To check this relation we differentiate (16.12) with respect to the initial values and obtain

$$dP_i^J = dp_0^J + h \sum_j a_{ij} dk_j^J \qquad dQ_i^J = dq_0^J + h \sum_j a_{ij} d\ell_j^J \tag{16.15a}$$

$$dp_1^J = dp_0^J + h \sum_i b_i dk_i^J \qquad dq_1^J = dq_0^J + h \sum_i b_i d\ell_i^J \tag{16.15b}$$

$$dk_i^J = -\sum_{L=1}^{n} \frac{\partial^2 H}{\partial q^J \partial p^L}(P_i, Q_i) \cdot dP_i^L - \sum_{L=1}^{n} \frac{\partial^2 H}{\partial q^J \partial q^L}(P_i, Q_i) \cdot dQ_i^L \tag{16.15c}$$

$$d\ell_i^J = \sum_{L=1}^{n} \frac{\partial^2 H}{\partial p^J \partial p^L}(P_i, Q_i) \cdot dP_i^L + \sum_{L=1}^{n} \frac{\partial^2 H}{\partial p^J \partial q^L}(P_i, Q_i) \cdot dQ_i^L. \tag{16.15d}$$

We now compute

$$dp_1^J \wedge dq_1^J - dp_0^J \wedge dq_0^J \tag{16.16}$$
$$= h \sum_i b_i\, dp_0^J \wedge d\ell_i^J + h \sum_i b_i\, dk_i^J \wedge dq_0^J + h^2 \sum_{i,j} b_i b_j\, dk_i^J \wedge d\ell_j^J$$

by using (16.15b) and the multilinearity of the wedge product. This formula corresponds precisely to (IV.12.6). Exactly as in the proof of Theorem IV.12.5, we now eliminate in (16.16) the quantities dp_0^J and dq_0^J with the help of (16.15a) to obtain

$$dp_1^J \wedge dq_1^J - dp_0^J \wedge dq_0^J \tag{16.17}$$
$$= h \sum_i b_i\, dP_i^J \wedge d\ell_i^J + h \sum_i b_i\, dk_i^J \wedge dQ_i^J - h^2 \sum_{i,j} m_{ij}\, dk_i^J \wedge d\ell_j^J,$$

the formula analogous to (IV.12.7). Equations (16.15c,d) are perfect analogues of the variational equation (I.14.27). Therefore the same computations as in (I.14.39) give

$$\sum_{J=1}^{n} dP_i^J \wedge d\ell_i^J + \sum_{J=1}^{n} dk_i^J \wedge dQ_i^J = 0 \tag{16.18}$$

and the first two terms in (16.17) disappear. The last term vanishes by hypothesis (16.11) and we obtain (16.14). ☐

Remark. F. Lasagni (1990) has proved in an unpublished manuscript that for *irreducible* methods (see Definitions IV.12.15 and IV.12.17) the condition $M = 0$ is also *necessary* for symplecticity. For a publication see Abia & Sanz-Serna (1993, Theorem 5.1), where this proof has been elaborated and adapted to a more general setting.

Remarks. a) Explicit Runge-Kutta methods are never symplectic (Ex. 1).

b) Equations (16.11) imply a substantial simplification of the order conditions (Sanz-Serna & Abia 1991). We shall return to this when treating partitioned methods (see (16.40)).

c) An important tool for the construction of symplectic methods is the W-transformation (see Section IV.5, especially Theorem IV.5.6). As can be seen from formula (IV.12.10), the method under consideration is symplectic if and only if the matrix X is skew-symmetric (with the exception of $x_{11} = 1/2$). Sun Geng (孙耿 1992) constructed several new classes of symplectic Runge-Kutta methods. One of his methods, based on Radau quadrature, is given in Table 16.1.

d) An inspection of Table IV.5.14 shows that all Radau IA, Radau IIA, Lobatto IIIA (in particular the trapezoidal rule), and Lobatto IIIC methods are not symplectic.

Table 16.1. 孙's symplectic Radau method of order 5

$\dfrac{4-\sqrt{6}}{10}$	$\dfrac{16-\sqrt{6}}{72}$	$\dfrac{328-167\sqrt{6}}{1800}$	$\dfrac{-2+3\sqrt{6}}{450}$
$\dfrac{4+\sqrt{6}}{10}$	$\dfrac{328+167\sqrt{6}}{1800}$	$\dfrac{16+\sqrt{6}}{72}$	$\dfrac{-2-3\sqrt{6}}{450}$
1	$\dfrac{85-10\sqrt{6}}{180}$	$\dfrac{85+10\sqrt{6}}{180}$	$\dfrac{1}{18}$
	$\dfrac{16-\sqrt{6}}{36}$	$\dfrac{16+\sqrt{6}}{36}$	$\dfrac{1}{9}$

Preservation of the Hamiltonian and of first integrals. In Exercise 5 of Section I.6 we have seen that the Hamiltonian $H(p,q)$ is a *first integral* of the system (16.1). This means that every solution $p(t), q(t)$ of (16.1) satisfies $H\big(p(t), q(t)\big) = Const$. The numerical solution of a symplectic integrator does not share this property in general (see Fig. 16.2). However, we will show that every *quadratic* first integral will be preserved.

Denote $y = (p,q)$ and let G be a symmetric $2n \times 2n$ matrix. We suppose that the quadratic functional

$$\langle y, y \rangle_G := y^T G y$$

is a first integral of the system (16.1). This means that

$$\langle y, J^{-1}\operatorname{grad} H(y)\rangle_G = 0 \qquad \text{with} \qquad J = \begin{pmatrix} 0 & I \\ -I & 0 \end{pmatrix} \qquad (16.19)$$

for all $y \in \mathbb{R}^{2n}$.

Theorem 16.7 (Sanz-Serna 1988).   *A symplectic Runge-Kutta method (i.e., a method satisfying (16.11)) leaves all quadratic first integrals of the system (16.1) invariant, i.e., the numerical solution* $y_n = (p_n, q_n)$ *satisfies*

$$\langle y_1, y_1 \rangle_G = \langle y_0, y_0 \rangle_G \qquad (16.20)$$

for all symmetric matrices G *satisfying (16.19).*

Proof (Cooper 1987). The Runge-Kutta method (7.7) applied to problem (16.1) is given by

$$y_1 = y_0 + \sum_i b_i k_i, \qquad Y_i = y_0 + \sum_j a_{ij} k_j,$$

$$k_i = J^{-1} \operatorname{grad} H(Y_i). \qquad (16.21)$$

As in the proof of Theorem 16.6 (see also Theorem IV.12.4) we obtain

$$\langle y_1, y_1 \rangle_G - \langle y_0, y_0 \rangle_G = 2h \sum_i b_i \langle Y_i, k_i \rangle_G - h^2 \sum_{i,j} m_{ij} \langle k_i, k_j \rangle_G.$$

The first term on the right-hand side vanishes by (16.19) and the second one by (16.11). □

An Example from Galactic Dynamics

> Always majestic, usually spectacularly beautiful, galaxies
> are . . .   (Binney & Tremaine 1987)

While the theoretical meaning of symplecticity of numerical methods is clear, its importance for practical computations is less easy to understand. Numerous numerical experiments have shown that symplectic methods, in a fixed step size mode, show an excellent behaviour for long-scale scientific computations of Hamiltonian systems. We shall demonstrate this on the following example chosen from galactic dynamics and give a theoretical justification later in this section. However, Calvo & Sanz-Serna (1992c) have made the interesting discovery that *variable step size* implementation can *destroy* the advantages of symplectic methods. In order to illustrate this phenomenon we shall include in our computations violent step changes; one with a random number generator and one with the step size changing in function of the solution position.

A galaxy is a set of N stars which are mutually attracted by Newton's law. A relatively easy way to study them is to perform a long-scale computation of the orbit of *one* of its stars in the potential formed by the $N-1$ remaining ones (see Binney & Tremaine 1987, Chapter 3); this potential is assumed to perform a uniform rotation with time, but not to change otherwise. The potential is determined

Fig. 16.3. Galactic orbit

by Poisson's differential equation $\Delta V = 4G\pi\varrho$, where ϱ is the density distribution of the galaxy, and real-life potential-density pairs are difficult to obtain (e.g., de Zeeuw & Pfenniger 1988). A popular issue is to choose a simple formula for V in such a way that the resulting ϱ corresponds to a reasonable galaxy, for example (Binney 1981, Binney & Tremaine 1987, p. 45f, Pfenniger 1990)

$$V = A \ln\Big(C + \frac{x^2}{a^2} + \frac{y^2}{b^2} + \frac{z^2}{c^2}\Big). \tag{16.22}$$

The Lagrangian for a coordinate system rotating with angular velocity Ω becomes

$$\mathcal{L} = \frac{1}{2}\Big((\dot{x} - \Omega y)^2 + (\dot{y} + \Omega x)^2 + \dot{z}^2\Big) - V(x, y, z). \tag{16.23}$$

This gives with the coordinates (see (I.6.23))

$$p_1 = \frac{\partial \mathcal{L}}{\partial \dot{x}} = \dot{x} - \Omega y, \qquad p_2 = \frac{\partial \mathcal{L}}{\partial \dot{y}} = \dot{y} + \Omega x, \qquad p_3 = \frac{\partial \mathcal{L}}{\partial \dot{z}} = \dot{z},$$

$$q_1 = x, \qquad\qquad q_2 = y, \qquad\qquad q_3 = z,$$

the Hamiltonian

$$H = p_1\dot{q}_1 + p_2\dot{q}_2 + p_3\dot{q}_3 - \mathcal{L} \tag{16.24}$$

$$= \frac{1}{2}(p_1^2 + p_2^2 + p_3^2) + \Omega(p_1 q_2 - p_2 q_1) + A\ln\Big(C + \frac{q_1^2}{a^2} + \frac{q_2^2}{b^2} + \frac{q_3^2}{c^2}\Big).$$

We choose the parameters and initial values as

$$a = 1.25, \quad b = 1, \quad c = 0.75, \quad A = 1, \quad C = 1, \quad \Omega = 0.25,$$
$$q_1(0) = 2.5, \quad q_2(0) = 0, \quad q_3(0) = 0, \quad p_1(0) = 0, \quad p_3(0) = 0.2, \tag{16.25}$$

and take for $p_2(0)$ the larger of the roots for which $H = 2$. Our star then sets out for its voyage through the galaxy, the orbit is represented in Fig. 16.3 for $0 \leq t \leq 15000$. We are interested in its Poincaré sections with the half-plane $q_2 = 0$, $q_1 > 0$, $\dot{q}_2 > 0$ for $0 \leq t \leq 1000000$. These consist, for the exact solution, in 47101 cut points which are presented in Fig. 16.6l. These points were computed with the (non-symplectic) code DOP853 with $Tol = 10^{-17}$ in quadruple precision on a VAX 8700 computer.

Fig. 16.4, Fig. 16.5, and Fig. 16.6 present the obtained numerical results for the methods and step sizes summarized in Table 16.2.

Table 16.2. Methods for numerical experiments

item	method	order	h	points $t < 1000000$	impl.	symplec.	symmet.
a)	Gauss	6	1/5	47093	yes	yes	yes
b)	"	"	2/5	46852	"	"	"
c)	Gauss	6	random	46717	yes	yes	yes
d)	Gauss	6	partially halved	46576	yes	yes	yes
e)	Radau	5	1/10	46597	yes	no	no
f)	"	"	1/5	46266	"	"	"
g)	RK44	4	1/40	47004	no	no	no
h)	"	"	1/10	46192	"	"	"
i)	Lobatto	6	1/5	47091	yes	no	yes
j)	"	"	2/5	46839	"	"	"
k)	Sun Geng	5	1/5	47092	yes	yes	no
l)	exact	–	–	47101	–	–	–

Remarks.

ad a): the Gauss6 method (Kuntzmann & Butcher method based on Gaussian quadrature with $s = 3$ and $p = 6$, see Table 7.4) for $h = 1/5$ is nearly identical to the exact solution;

ad b): Gauss6 for $h = 2/5$ is much better than Gauss6 with random or partially halved step sizes (see item (c) and (d)) where $h \leq 2/5$.

ad c): h was chosen at random uniformly distributed on $(0, 2/5)$;

ad d): h was chosen "partially halved" in the sense that

$$h = \begin{cases} 2/5 & \text{if } q_1 > 0, \\ 1/5 & \text{if } q_1 < 0. \end{cases}$$

This produced the worst result for the 6th order Gauss method. We thus

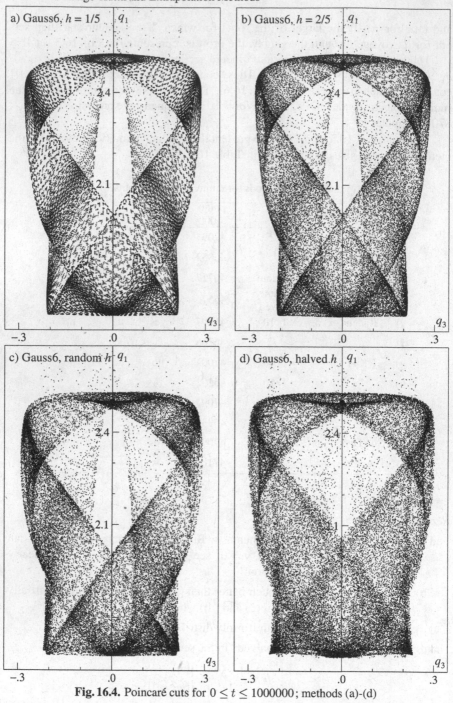

Fig. 16.4. Poincaré cuts for $0 \leq t \leq 1000000$; methods (a)-(d)

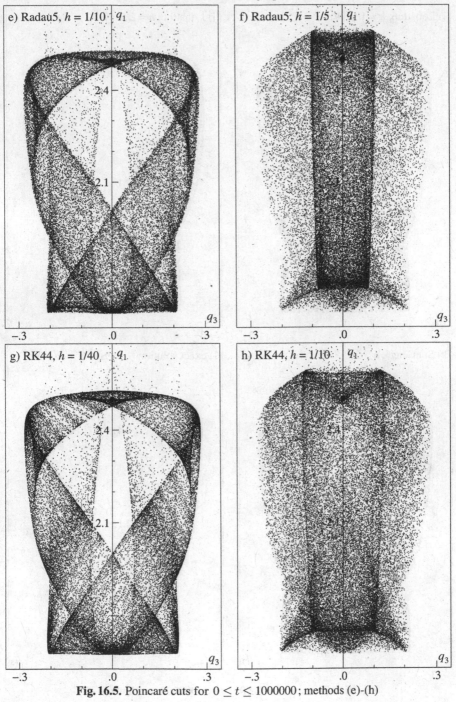

Fig. 16.5. Poincaré cuts for $0 \leq t \leq 1000000$; methods (e)-(h)

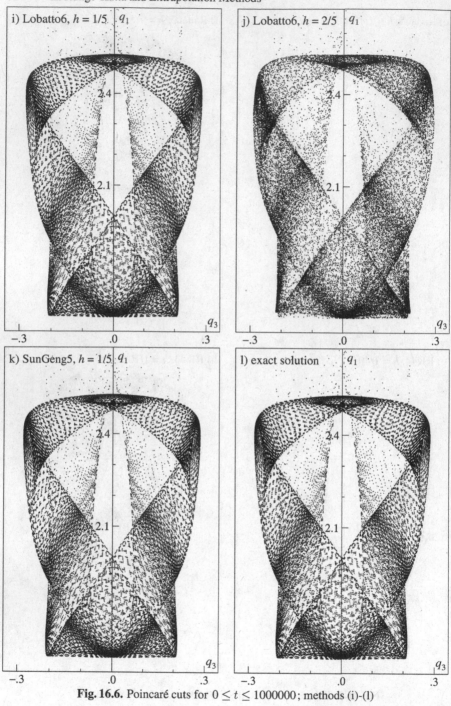

Fig. 16.6. Poincaré cuts for $0 \le t \le 1000000$; methods (i)-(l)

see that symplectic and symmetric methods compensate on the way back the errors committed on the outward journey.

ad e), f): Radau5 (method of Ehle based on Radau quadrature with $s = 3$ and $p = 5$, see Table 7.7) is here not at all satisfactory;

ad g): The explicit method RK44 (Runge-Kutta method with $s = p = 4$, see Table 1.2, left) is evidently much faster than the implicit methods, even with a smaller step size;

ad h): With increasing step size RK44 deteriorates drastically;

ad i): this is a non-symplectic but symmetric collocation method based on Lobatto quadrature with $s = 4$ of order 6 (see Table IV.5.8); its good performance on this nonlinear Hamiltonian problem is astonishing;

ad j): with increasing h Lobatto6 is less satisfactory (see also Fig. 16.7);

ad k): this is the symplectic non-symmetric method based on Radau quadrature of order 5 due to Sun Geng 孙耿 (Table 16.1).

The preservation of the Hamiltonian (correct value $H = 2$) during the computation for $0 \leq t \leq 1000000$ is shown in Fig. 16.7. While the errors for the symplectic and symmetric methods in constant step size mode remain bounded, random h (case c) results in a sort of Brownian motion, and the nonsymplectic methods as well as Gauss6 with partially halved step size result in permanent deterioration.

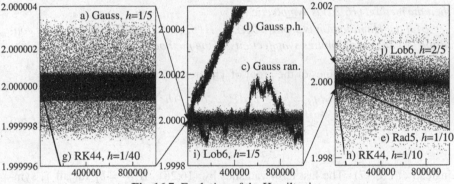

Fig. 16.7. Evolution of the Hamiltonian

Partitioned Runge-Kutta Methods

The fact that the system (16.1) possesses a natural partitioning suggests the use of partitioned Runge-Kutta methods as discussed in Section II.15. The main interest of such methods is for separable Hamiltonians where it is possible to obtain explicit symplectic methods.

A partitioned Runge-Kutta method for system (16.1) is defined by

$$P_i = p_0 + h \sum_j a_{ij} k_j \qquad Q_i = q_0 + h \sum_j \widehat{a}_{ij} \ell_j \tag{16.26a}$$

$$p_1 = p_0 + h \sum_i b_i k_i \qquad q_1 = q_0 + h \sum_i \widehat{b}_i \ell_i \tag{16.26b}$$

$$k_i = -\frac{\partial H}{\partial q}(P_i, Q_i) \qquad \ell_i = \frac{\partial H}{\partial p}(P_i, Q_i) \tag{16.26c}$$

where b_i, a_{ij} and $\widehat{b}_i, \widehat{a}_{ij}$ represent two different Runge-Kutta schemes.

Theorem 16.10 (Sanz-Serna 1992b, Suris 1990). *a) If the coefficients of (16.26) satisfy*

$$b_i = \widehat{b}_i, \qquad i = 1, \dots, s \tag{16.27}$$

$$b_i \widehat{a}_{ij} + \widehat{b}_j a_{ji} - b_i \widehat{b}_j = 0, \qquad i, j = 1, \dots, s \tag{16.28}$$

then the method (16.26) is symplectic.

b) If the Hamiltonian is separable (i.e., $H(p, q) = T(p) + U(q)$) then the condition (16.28) alone implies symplecticity of the method.

Proof. Following the lines of the proof of Theorem 16.6 we obtain

$$dp_1^J \wedge dq_1^J - dp_0^J \wedge dq_0^J = h \sum_i \widehat{b}_i \, dP_i^J \wedge d\ell_i^J + h \sum_i b_i \, dk_i^J \wedge dQ_i^J$$
$$- h^2 \sum_{i,j} (b_i \widehat{a}_{ij} + \widehat{b}_j a_{ji} - b_i \widehat{b}_j) \, dk_i^J \wedge d\ell_j^J, \tag{16.29}$$

instead of (16.17). The last term vanishes by (16.28). If $b_i = \widehat{b}_i$ for all i, symplecticity of the method follows from (16.18). If the Hamiltonian is separable (the mixed derivatives $\partial^2 H/\partial q^J \partial p^L$ and $\partial^2 H/\partial p^J \partial q^L$ are not present in (16.15c,d)) then each of the two terms in (16.18) vanishes separately and the method is symplectic without imposing (16.27). □

Remark. If (16.28) is satisfied and if the Hamiltonian is separable, it can be assumed without loss of generality that

$$b_i \neq 0, \qquad \widehat{b}_i \neq 0 \qquad \text{for all } i. \tag{16.30}$$

Indeed, the stage values P_i (for i with $\widehat{b}_i = 0$) and Q_j (for j with $b_j = 0$) don't influence the numerical solution (p_1, q_1) and can be removed from the scheme. Notice however that in the resulting scheme the number of stages P_i may be different from that of Q_j.

Explicit methods for separable Hamiltonians. Let the Hamiltonian be of the form $H(p, q) = T(p) + U(q)$ and consider a partitioned Runge-Kutta method satisfying

$$
\begin{aligned}
a_{ij} &= 0 \quad \text{for} \quad i < j \quad \text{(diagonally implicit)} \\
\widehat{a}_{ij} &= 0 \quad \text{for} \quad i \leq j \quad \text{(explicit)}.
\end{aligned}
\tag{16.31}
$$

Since $\partial H / \partial q$ depends only on q, the method (16.26) is explicit for such a choice of coefficients. Under the assumption (16.30), the symplecticity condition (16.28) then becomes

$$
a_{ij} = b_j \quad \text{for} \quad i \geq j, \qquad \widehat{a}_{ij} = \widehat{b}_j \quad \text{for} \quad i > j,
\tag{16.32}
$$

so that the method (16.26) is characterized by the two schemes

$$
\begin{array}{c|ccccc}
b_1 \\
b_1 & b_2 \\
b_1 & b_2 & b_3 \\
\vdots & \vdots & & \ddots & \ddots \\
b_1 & b_2 & \cdots & b_{s-1} & b_s \\
\hline
b_1 & b_2 & \cdots & b_{s-1} & b_s
\end{array}
\qquad
\begin{array}{c|ccccc}
0 \\
\widehat{b}_1 & 0 \\
\widehat{b}_1 & \widehat{b}_2 & 0 \\
\vdots & \vdots & & \ddots & \ddots \\
\widehat{b}_1 & \widehat{b}_2 & \cdots & \widehat{b}_{s-1} & 0 \\
\hline
\widehat{b}_1 & \widehat{b}_2 & \cdots & \widehat{b}_{s-1} & \widehat{b}_s
\end{array}
\tag{16.33}
$$

If we admit the cases $b_1 = 0$ and/or $\widehat{b}_s = 0$, it can be shown (Exercise 6) that this scheme already represents the most general method (16.26) which is symplectic and explicit. We denote this scheme by

$$
\begin{array}{cccccc}
b: & b_1 & b_2 & \cdots & b_s \\
\widehat{b}: & \widehat{b}_1 & \widehat{b}_2 & \cdots & \widehat{b}_s.
\end{array}
\tag{16.34}
$$

This method is particularly easy to implement:

$$
\begin{aligned}
&P_0 = p_0, \ Q_1 = q_0 \\
&\textbf{for } i := 1 \textbf{ to } s \textbf{ do} \\
&\quad P_i = P_{i-1} - h b_i \partial U / \partial q(Q_i) \\
&\quad Q_{i+1} = Q_i + h \widehat{b}_i \partial T / \partial p(P_i) \\
&p_1 = P_s, \ q_1 = Q_{s+1}
\end{aligned}
\tag{16.35}
$$

Special case $s = 1$. The combination of the implicit Euler method ($b_1 = 1$) with the explicit Euler method ($\widehat{b}_1 = 1$) gives the following symplectic method of order 1:

$$
p_1 = p_0 - h \frac{\partial U}{\partial q}(q_0), \qquad q_1 = q_0 + h \frac{\partial T}{\partial p}(p_1).
\tag{16.36a}
$$

By interchanging the roles of p and q we obtain the method

$$q_1 = q_0 + h\frac{\partial T}{\partial p}(p_0), \qquad p_1 = p_0 - h\frac{\partial U}{\partial q}(q_1) \tag{16.36b}$$

which is also symplectic. Methods (16.36a) and (16.36b) are mutually adjoint (see Section II.8).

Construction of higher order methods. The order conditions for general partitioned Runge-Kutta methods applied to general problems (15.2) are derived in Section II.15 (Theorem 15.9). Let us here discuss how these conditions simplify in our special situation.

A) We consider the system (16.1) with separable Hamiltonian. In the notation of Section II.15 this means that $f_a(y_a, y_b)$ depends only on y_b and $f_b(y_a, y_b)$ depends only on y_a. Therefore, many elementary differentials vanish and only P-trees whose meagre and fat vertices alternate in each branch have to be considered. This is a considerable reduction of the order conditions.

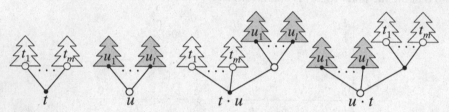

Fig. 16.8. Product of P-trees

B) As observed by Abia & Sanz-Serna (1993) the condition (16.28) acts as a simplifying assumption. Indeed, multiplying (16.28) by $\Phi_i(t) \cdot \Phi_j(u)$ (where $t = {}_a[t_1, \ldots, t_m] \in TP^a$, $u = {}_b[u_1, \ldots, u_l] \in TP^b$) and summing up over all i and j yields

$$\sum_i b_i \Phi_i(t \cdot u) + \sum_j \widehat{b}_j \Phi_j(u \cdot t) - \left(\sum_i b_i \Phi_i(t)\right)\left(\sum_j \widehat{b}_j \Phi_j(u)\right) = 0. \tag{16.37}$$

Here we have used the notation of Butcher (1987)

$$t \cdot u = {}_a[t_1, \ldots, t_m, u], \qquad u \cdot t = {}_b[u_1, \ldots, u_l, t], \tag{16.38}$$

illustrated in Fig. 16.8. Since

$$\frac{1}{\gamma(t \cdot u)} + \frac{1}{\gamma(u \cdot t)} - \frac{1}{\gamma(t)} \cdot \frac{1}{\gamma(u)} = 0 \tag{16.39}$$

(this relation follows from (16.37) by inserting the coefficients of a symplectic Runge-Kutta method of sufficiently high order, e.g., a Gauss method) we obtain the following fact:

let $\varrho(t) + \varrho(u) = p$ and assume that all order conditions for P-trees of order $< p$ are satisfied, then

$$\sum_i b_i \Phi_i(t \cdot u) = \frac{1}{\gamma(t \cdot u)} \qquad iff \qquad \sum_j \widehat{b}_j \Phi_j(u \cdot t) = \frac{1}{\gamma(u \cdot t)}. \quad (16.40)$$

From Fig. 16.8 we see that the P-trees $t \cdot u$ and $u \cdot t$ have the same geometrical structure. They differ only in the position of the root. Repeated application of this property implies that of all P-trees with identical geometrical structure only one has to be considered.

A *method of order* 3 (Ruth 1983). The above reductions leave five order conditions for a method of order 3 which, for $s = 3$, are the following:

$$b_1 + b_2 + b_3 = 1, \qquad \widehat{b}_1 + \widehat{b}_2 + \widehat{b}_3 = 1, \qquad b_2 \widehat{b}_1 + b_3(\widehat{b}_1 + \widehat{b}_2) = 1/2,$$
$$b_2 \widehat{b}_1^2 + b_3(\widehat{b}_1 + \widehat{b}_2)^2 = 1/3, \qquad \widehat{b}_1 b_1^2 + \widehat{b}_2(b_1 + b_2)^2 + \widehat{b}_3(b_1 + b_2 + b_3)^2 = 1/3.$$

This nonlinear system possesses many solutions. A particularly simple solution, proposed by Ruth (1983), is

$$
\begin{array}{cccc}
b: & 7/24 & 3/4 & -1/24 \\
\widehat{b}: & 2/3 & -2/3 & 1.
\end{array}
\qquad (16.41)
$$

Concatenation of a method with its adjoint. The adjoint method of (16.26) is obtained by replacing h by $-h$ and by exchanging the roles of p_0, q_0 and p_1, q_1 (see Section II.8). This results in a partitioned Runge-Kutta method with coefficients (compare Theorem 8.3)

$$a_{ij}^* = b_{s+1-j} - a_{s+1-i,s+1-j}, \qquad b_i^* = b_{s+1-i},$$
$$\widehat{a}_{ij}^* = \widehat{b}_{s+1-j} - \widehat{a}_{s+1-i,s+1-j}, \qquad \widehat{b}_i^* = \widehat{b}_{s+1-i}.$$

For the adjoint of (16.33) the first method is explicit and the second one is diagonally implicit, but otherwise it has the same structure. Adding dummy stages, it becomes of the form (16.33) with coefficients

$$
\begin{array}{cccccc}
b^*: & 0 & b_s & b_{s-1} & \dots & b_1 \\
\widehat{b}^*: & \widehat{b}_s & \widehat{b}_{s-1} & \dots & \widehat{b}_1 & 0.
\end{array}
\qquad (16.42)
$$

The following idea of Sanz-Serna (1992b) allows one to improve a method of odd order p: one considers the composition of method (16.33) (step size $h/2$) with its adjoint (again with step size $h/2$). The resulting method, which is represented by the coefficients

$$
\begin{array}{ccccccccc}
b_1/2 & b_2/2 & \dots & b_{s-1}/2 & b_s/2 & b_s/2 & b_{s-1}/2 & \dots & b_1/2 \\
\widehat{b}_1/2 & \widehat{b}_2/2 & \dots & \widehat{b}_{s-1}/2 & \widehat{b}_s & \widehat{b}_{s-1}/2 & & \dots & \widehat{b}_1/2 & 0,
\end{array}
$$

is symmetric and therefore has an even order which is $\geq p+1$. Concatenating

Ruth's method (16.41) with its adjoint yields the fourth order method

$$\begin{array}{lcccccc} b: & 7/48 & 3/8 & -1/48 & -1/48 & 3/8 & 7/48 \\ \widehat{b}: & 1/3 & -1/3 & 1 & -1/3 & 1/3 & 0. \end{array} \tag{16.43}$$

Symplectic Nyström Methods

A frequent special case of a separable Hamiltonian $H(p,q) = T(p) + U(q)$ is when $T(p)$ is a quadratic functional $T(p) = p^T M p/2$ (with M a constant symmetric matrix). In this situation the Hamiltonian system becomes

$$\dot{p} = -\frac{\partial U}{\partial q}(q), \qquad \dot{q} = Mp,$$

which is equivalent to the second order equation

$$\ddot{q} = -M\frac{\partial U}{\partial q}(q). \tag{16.44}$$

It is therefore natural to consider Nyström methods (Section II.14) which for the system (16.44) are given by

$$Q_i = q_0 + c_i h \dot{q}_0 + h^2 \sum_j \overline{a}_{ij} k'_j, \qquad k'_j = -M\frac{\partial U}{\partial q}(Q_j),$$

$$q_1 = q_0 + h\dot{q}_0 + h^2 \sum_i \overline{b}_i k'_i, \qquad \dot{q}_1 = \dot{q}_0 + h\sum_i b_i k'_i.$$

Replacing the variable $\dot{q}$ by Mp and k'_i by $M\ell_i$, this method reads

$$Q_i = q_0 + c_i h M p_0 + h^2 \sum_{j=1}^{s} \overline{a}_{ij} M\ell_j, \qquad \ell_j = -\frac{\partial U}{\partial q}(Q_j),$$

$$q_1 = q_0 + hMp_0 + h^2 \sum_{i=1}^{s} \overline{b}_i M\ell_i, \qquad p_1 = p_0 + h\sum_{i=1}^{s} b_i \ell_i. \tag{16.45}$$

Theorem 16.11 (Suris 1989). *Consider the system (16.44) where M is a symmetric matrix. Then, the s-stage Nyström method (16.45) is symplectic if the following two conditions are satisfied:*

$$\overline{b}_i = b_i(1 - c_i), \qquad i = 1, \dots, s \tag{16.46a}$$

$$b_i(\overline{b}_j - \overline{a}_{ij}) = b_j(\overline{b}_i - \overline{a}_{ji}), \qquad i, j = 1, \dots, s. \tag{16.46b}$$

Proof (Okunbor & Skeel 1992). As in the proof of Theorem 16.6 we differentiate the formulas (16.45) and compute

$$dp_1^J \wedge dq_1^J - dp_0^J \wedge dq_0^J$$
$$= h\sum_i b_i \, d\ell_i^J \wedge dq_0^J + h\sum_K M_{JK} \, dp_0^J \wedge dp_0^K \tag{16.47}$$

$$+ h^2 \sum_i b_i \sum_K M_{JK}\, d\ell_i^J \wedge dp_0^K + h^2 \sum_i \bar{b}_i \sum_K M_{JK}\, dp_0^J \wedge d\ell_i^K$$

$$+ h^3 \sum_{i,j} b_i \bar{b}_j \sum_K M_{JK}\, d\ell_i^J \wedge d\ell_j^K .$$

Next we eliminate dq_0^J with the help of the differentiated equation of Q_i, sum over all J and so obtain

$$\sum_{J=1}^n dp_1^J \wedge dq_1^J - \sum_{J=1}^n dp_0^J \wedge dq_0^J$$

$$= h \sum_i b_i \sum_J d\ell_i^J \wedge dQ_i^J + h \sum_{J,K} M_{JK}\, dp_0^J \wedge dp_0^K$$

$$+ h^2 \sum_i (b_i - \bar{b}_i - b_i c_i) \sum_{J,K} M_{JK}\, d\ell_i^J \wedge dp_0^K$$

$$+ h^3 \sum_{i<j} (b_i \bar{b}_j - b_j \bar{b}_i - b_i \bar{a}_{ij} + b_j \bar{a}_{ji}) \sum_{J,K} M_{JK}\, d\ell_i^J \wedge d\ell_j^K .$$

The last two terms disappear by (16.46) whereas the first two terms vanish due to the symmetry of M and of the second derivatives of $U(q)$.   □

We have already encountered condition (16.46a) in Lemma 14.13. There, it was used as a simplifying assumption. It implies that only the order conditions for $\dot{q}_1$ have to be considered.

For Nyström methods satisfying both conditions of (16.46), one can assume without loss of generality that

$$b_i \neq 0 \qquad \text{for} \quad i = 1, \dots, s. \tag{16.48}$$

Let $I = \{i \mid b_i = 0\}$, then $\bar{b}_i = 0$ for $i \in I$ and $\bar{a}_{ij} = 0$ for $i \notin I$, $j \in I$. Hence, the stage values Q_i $(i \in I)$ don't influence the numerical result (p_1, q_1) and can be removed from the scheme.

Explicit methods. Our main interest is in methods which satisfy

$$\bar{a}_{ij} = 0 \qquad \text{for} \quad i \leq j. \tag{16.49}$$

Under the assumption (16.48) the condition (16.46) then implies that the remaining coefficients are given by

$$\bar{a}_{ij} = b_j(c_i - c_j) \qquad \text{for} \quad i > j. \tag{16.50}$$

In this situation we may also suppose that

$$c_i \neq c_{i-1} \qquad \text{for} \quad i = 2, 3, \dots, s,$$

because equal consecutive c_i lead (via condition (16.50)) to equal stage values Q_i. Therefore the method is equivalent to one with a smaller number of stages.

The particular form of the coefficients $\bar{a}_{ij}$ allows the following simple implementation (Okunbor & Skeel 1992b)

$$Q_0 = q_0, \quad P_0 = p_0$$
for $i := 1$ **to** s **do**
$$Q_i = Q_{i-1} + h(c_i - c_{i-1})MP_{i-1} \qquad \text{(with } c_0 = 0) \qquad (16.51)$$
$$P_i = P_{i-1} - hb_i\partial U/\partial q(Q_i)$$
$$q_1 = Q_s + h(1 - c_s)MP_s, \quad p_1 = P_s.$$

Special case $s = 1$. Putting $b_1 = 1$ (c_1 is a free parameter) yields a symplectic, explicit Nyström method of order 1. For the choice $c_1 = 1/2$ it has order 2.

Special case $s = 3$. To obtain order 3, four order conditions have to be satisfied (see Table 14.3). The first three mean that (b_i, c_i) is a quadrature formula of order 3. They allow us to express b_1, b_2, b_3 in terms of c_1, c_2, c_3. The last condition then becomes (Okunbor & Skeel 1992b)

$$1 + 24\left(c_1 - \frac{1}{2}\right)\left(c_2 - \frac{1}{2}\right) + 24(c_2 - c_1)(c_3 - c_1)(c_3 - c_2) \qquad (16.52)$$

$$+ 144\left(c_1 - \frac{1}{2}\right)\left(c_2 - \frac{1}{2}\right)\left(c_3 - \frac{1}{2}\right)\left(c_1 + c_3 - c_2 - \frac{1}{2}\right) = 0.$$

We thus get a two-parameter family of third order methods. Okunbor & Skeel (1992b) suggest taking

$$c_2 = \frac{1}{2}, \qquad c_1 = 1 - c_3 = \frac{1}{6}\left(2 + \sqrt[3]{2} + \frac{1}{\sqrt[3]{2}}\right) \qquad (16.53)$$

(the real root of $12c_1(2c_1 - 1)^2 = 1$). This method is symmetric and thus of order 4. Another 3-stage method of order 4 has been found by Qin Meng-Zhao & Zhu Wen-jie (1991).

Higher order methods. For the construction of methods of order ≥ 4 it is worthwhile to investigate the effect of the condition (16.46b) on the order conditions. As for partitioned Runge-Kutta methods one can show that SN-trees with the same geometrical structure lead to equivalent order conditions. For details we refer to Calvo & Sanz-Serna (1992). With the notation of Table 14.3, the SN-trees t_6 and t_7 as well as the pairs t_9, t_{12} and t_{10}, t_{13} give rise to equivalent order conditions. Consequently, for order 5, one has to consider 10 conditions. Okunbor & Skeel (1992c) present explicit, symplectic Nyström methods of orders 5 and 6 with 5 and 7 stages, respectively. A 7th order method is given by Calvo & Sanz-Serna (1992b).

Conservation of the Hamiltonian; Backward Analysis

> The differential equation actually solved by the difference scheme will be called the modified equation.
>
> (Warming & Hyett 1974, p. 161)

> The *wrong* solution of the *right* equation; the *right* solution of the *wrong* equation.   (Feng Kang, Beijing Sept. 1, 1992)

We have observed above (Example 16.2 and Fig. 16.6) that for the numerical solution of symplectic methods the Hamiltonian H remained between fixed bounds over any long-term integration, i.e., so-called secular changes of H were absent. Following several authors (Yoshida 1993, Sanz-Serna 1992, Feng Kang 1991b) this phenomenon is explained by interpreting the numerical solution as the *exact* solution of a *perturbed Hamiltonian system*, which is obtained as the formal expansion (16.56) in powers of h. The *exact* conservation of the perturbed Hamiltonian $\widetilde{H}$ then involves the quasi-periodic behaviour of H along the computed points. This resembles Wilkinson's famous idea of backward error analysis in linear algebra and, in the case of differential equations, seems to go back to Warming & Hyett (1974). We demonstrate this idea for the symplectic Euler method (see (16.36b))

$$p_1 = p_0 - hH_q(p_0, q_1)$$
$$q_1 = q_0 + hH_p(p_0, q_1)$$
(16.54)

which, when expanded around the point (p_0, q_0), gives

$$p_1 = p_0 - hH_q - h^2 H_{qq}H_p - \frac{h^3}{2}H_{qqq}H_pH_p - h^3 H_{qq}H_{pq}H_p - \cdots \Big|_{p_0, q_0}$$
$$q_1 = q_0 + hH_p + h^2 H_{pq}H_p + \frac{h^3}{2}H_{pqq}H_pH_p + h^3 H_{pq}H_{pq}H_p + \cdots \Big|_{p_0, q_0}.$$
(16.54')

In the case of non-scalar equations the p's and q's must here be equipped with various summation indices. We suppress these in the sequel for the sake of simplicity and think of scalar systems only. The exact solution of a perturbed Hamiltonian

$$\dot{p} = -\widetilde{H}_q(p, q)$$
$$\dot{q} = \widetilde{H}_p(p, q)$$

has a Taylor expansion analogous to Theorem 2.6 as follows

$$p_1 = p_0 - h\widetilde{H}_q + \frac{h^2}{2}\left(\widetilde{H}_{qp}\widetilde{H}_q - \widetilde{H}_{qq}\widetilde{H}_p\right) + \cdots$$
$$q_1 = q_0 + h\widetilde{H}_p + \frac{h^2}{2}\left(-\widetilde{H}_{pp}\widetilde{H}_q + \widetilde{H}_{pq}\widetilde{H}_p\right) + \cdots .$$
(16.55)

We now set

$$\widetilde{H} = H + hH^{(1)} + h^2 H^{(2)} + h^3 H^{(3)} + \cdots$$
(16.56)

with unknown functions $H^{(1)}, H^{(2)}, \ldots$, insert this into (16.55) and compare the resulting formulas with (16.54'). Then the comparison of the h^2 terms gives

$$H_q^{(1)} = \frac{1}{2} H_{qq} H_p + \frac{1}{2} H_{qp} H_q, \qquad H_p^{(1)} = \frac{1}{2} H_{pp} H_q + \frac{1}{2} H_{pq} H_p$$

which by miracle (the "miracle" is in fact a consequence of the symplecticity of method (16.54)) allow the common primitive

$$H^{(1)} = \frac{1}{2} H_p H_q. \tag{16.56;1}$$

The h^3 terms lead to

$$H^{(2)} = \frac{1}{12} \left(H_{pp} H_q^2 + H_{qq} H_p^2 + 4 H_{pq} H_p H_q \right) \tag{16.56;2}$$

and so on.

Connection with the Campbell-Baker-Hausdorff formula. An elegant access to the expansion (16.56), which works for separable Hamiltonians $H(p, q) = T(p) + U(q)$, has been given by Yoshida (1993). We interpret method (16.54) as composition of the two symplectic maps

$$z_0 = \begin{pmatrix} p_0 \\ q_0 \end{pmatrix} \xrightarrow{S_T} z = \begin{pmatrix} p_0 \\ q_1 \end{pmatrix} \xrightarrow{S_U} z_1 = \begin{pmatrix} p_1 \\ q_1 \end{pmatrix} \tag{16.57}$$

which consist, respectively, in solving exactly the Hamiltonian systems

$$\begin{aligned} \dot{p} &= 0 \\ \dot{q} &= T_p(p) \end{aligned} \qquad \text{and} \qquad \begin{aligned} \dot{p} &= -U_q(q) \\ \dot{q} &= 0 \end{aligned} \tag{16.58}$$

and apply some Lie theory. If we introduce for these equations the differential operators given by (13.2')

$$D_T \Psi = \frac{\partial \Psi}{\partial q} T_p(p), \qquad D_U \Psi = -\frac{\partial \Psi}{\partial p} U_q(q), \tag{16.59}$$

the formulas (13.3) allow us to write the Taylor series of the map S_T as

$$z = \sum_{i=0}^{\infty} \frac{h^i}{i!} D_T^i z \bigg|_{z=z_0}. \tag{16.60}$$

If now $F(z)$ is an arbitrary function of the solution $z(t) = (p(t), q(t))$ (left equation of (16.58)), we find, as in (13.2), that

$$F(z)' = D_T F, \quad F(z)'' = D_T^2 F, \dots$$

and (16.60) extends to (Gröbner 1960)

$$F(z) = \sum_{i=0}^{\infty} \frac{h^i}{i!} D_T^i F(z) \bigg|_{z=z_0}. \tag{16.60'}$$

We now insert S_U for F and insert for S_U the formula analogous to (16.60) to obtain for the composition (16.57)

$$z_1 = (p_1, q_1) = \sum_{i=0}^{\infty} \frac{h^i}{i!} D_T^i \sum_{j=0}^{\infty} \frac{h^j}{j!} D_U^j z \Big|_{z=z_0} \tag{16.61}$$

$$= \exp(hD_T) \exp(hD_U)(p, q) \Big|_{p=p_0, q=q_0}.$$

But the product $\exp(hD_T) \exp(hD_U)$ is *not* $\exp(hD_T + hD_U)$, as we have all learned in school, because the operators D_T and D_U do not commute. This is precisely the content of the famous Campbell-Baker-Hausdorff Formula (claimed in 1898 by J.E. Campbell and proved independently by Baker (1905) and in the "kleine Untersuchung" of Hausdorff (1906)) which states, for our problem, that

$$\exp(hD_T) \exp(hD_U) = \exp(h\widetilde{D}) \tag{16.62}$$

where

$$\widetilde{D} = D_T + D_U + \frac{h}{2}[D_T, D_U] + \frac{h^2}{12}([D_T, [D_T, D_U]] + [D_U, [D_U, D_T]])$$

$$+ \frac{h^3}{24}[D_T, [D_U, [D_U, D_T]]] + \dots \tag{16.63}$$

and $[D_A, D_B] = D_A D_B - D_B D_A$ is the commutator. Equation (16.62) shows that the map (16.57) is the exact solution of the differential equation corresponding to the differential operator $\widetilde{D}$. A straightforward calculation now shows: If

$$D_A \Psi = -\frac{\partial \Psi}{\partial p} A_q + \frac{\partial \Psi}{\partial q} A_p \quad \text{and} \quad D_B \Psi = -\frac{\partial \Psi}{\partial p} B_q + \frac{\partial \Psi}{\partial q} B_p \tag{16.64}$$

are differential operators corresponding to Hamiltonians A and B respectively, then

$$[D_A, D_B]\Psi = D_C \Psi = -\frac{\partial \Psi}{\partial p} C_q + \frac{\partial \Psi}{\partial q} C_p$$

where

$$C = A_p B_q - A_q B_p. \tag{16.65}$$

A repeated application of (16.65) now allows us to obtain for all brackets in (16.63) a corresponding Hamiltonian which finally leads to

$$\widetilde{H} = T + U + \frac{h}{2} T_p U_q + \frac{h^2}{12}(T_{pp} U_q^2 + U_{qq} T_p^2) + \frac{h^3}{12} T_{pp} U_{qq} T_p U_q + \dots \tag{16.66}$$

which is the specialization of (16.56) to separable Hamiltonians.

Example 16.12 (Yoshida 1993). For the mathematical pendulum

$$H(p, q) = \frac{p^2}{2} - \cos q \tag{16.67}$$

series (16.66) becomes

$$\widetilde{H} = \frac{p^2}{2} - \cos q + \frac{h}{2} p \sin q + \frac{h^2}{12}(\sin^2 q + p^2 \cos q) + \frac{h^3}{12} p \cos q \sin q + \mathcal{O}(h^4). \tag{16.68}$$

Fig. 16.9 presents for various step sizes h and for various initial points ($p_0 = 0$, $q_0 = -1.5$; $p_0 = 0$, $q_0 = -2.5$; $p_0 = 1.5$, $q_0 = -\pi$; $p_0 = 2.5$, $q_0 = -\pi$) the numerically computed points for method (16.54) compared to the contour lines of $\widetilde{H} = Const$ given by the terms up to order h^3 in (16.68). The excellent agreement of the results with theory for $h \leq 0.6$ leaves nothing to be desired, while for h beyond 0.9 the dynamics of the numerical method turns rapidly into chaotic behaviour.

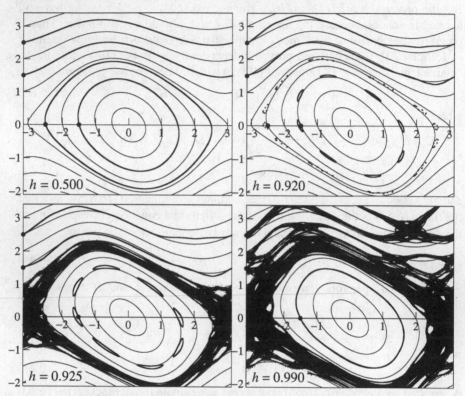

Fig. 16.9. Symplectic method compared to perturbed Hamiltonian ($\bullet \dots$ indicate the initial positions)

Remark. For much research, especially in the beginning of the "symplectic era", the central role for the construction of canonical difference schemes is played by the Hamilton-Jacobi theory and generating functions. For this, the reader may consult the papers Feng Kang (1986), Feng Kang, Wu Hua-mo, Qin Meng-zhao & Wang Dao-liu (1989), Channell & Scovel (1990) and Miesbach & Pesch (1992). Many additional numerical experiments can be found in Channell & Scovel (1990), Feng Kang (1991), and Pullin & Saffman (1991).

Exercises

1. Show that explicit Runge-Kutta methods are never symplectic.

   Hint. Compute the diagonal elements of M.

2. Study the existence and uniqueness of the numerical solution for the implicit mid-point rule when applied to the Hamiltonian system

$$\dot{p} = -q^2, \qquad \dot{q} = p.$$

   Show that the method possesses no solution at all for $h^2 q_0 + h^3 p_0/2 < -1$ and two solutions for $h^2 q_0 + h^3 p_0/2 > -1$ ($h \neq 0$). Only one of the solutions tends to (p_0, q_0) for $h \to 0$.

3. A Runge-Kutta method is called *linearly symplectic* if it is symplectic for all linear Hamiltonian systems

$$\dot{y} = J^{-1} C y$$

   (J is given in (16.19) and C is a symmetric matrix). Prove (Feng Kang 1985) that a Runge-Kutta method is linearly symplectic if and only if its stability function satisfies

$$R(-z)R(z) = 1 \qquad \text{for all} \quad z \in \mathbb{C}. \tag{16.69}$$

   Hint. For the definition of the stability function see Section IV.2 of Volume II. Then by Theorem I.14.14, linear symplecticity is equivalent to

$$R(hJ^{-1}C)^T J R(hJ^{-1}C) = J.$$

   Furthermore, the matrix $B := J^{-1}C$ is seen to verify $B^T J = -JB$ and hence also $(B^k)^T J = J(-B)^k$ for $k = 0, 1, 2, \ldots$. This implies that

$$R(hJ^{-1}C)^T J = JR(-hJ^{-1}C).$$

4. Prove that the stability function of a symmetric Runge-Kutta method satisfies (16.69).

5. Compute all quadratic first integrals of the Hamiltonian system (16.4).

6. For a separable Hamiltonian consider the method (16.26) where $a_{ij} = 0$ for $i < j$, $\widehat{a}_{ij} = 0$ for $i < j$ and for every i either $a_{ii} = 0$ or $\widehat{a}_{ii} = 0$. If the method satisfies (16.28) then it is equivalent to one given by scheme (16.33).

   Hint. Remove first all stages which don't influence the numerical result (see the remark after Theorem 16.10). Then deduce from (16.28) relations similar to (16.32). Finally, remove identical stages and add, if necessary, a dummy stage in order that both methods have the same number of stages.

7. (Lasagni 1990). Characterize symplecticity for multi-derivative Runge-Kutta methods. Show that the s-stage q-derivative method of Definition 13.1 is symplectic if its coefficients satisfy

$$b_i^{(r)} b_j^{(m)} - b_i^{(r)} a_{ij}^{(m)} - b_j^{(m)} a_{ji}^{(r)} = \begin{cases} b_i^{(r+m)} & \text{if } i = j \text{ and } r+m \leq q, \\ 0 & \text{otherwise.} \end{cases}$$
(16.70)

Hint. Denote $k^{(r)} = D_H^r p$, $\ell^{(r)} = D_H^r q$, where D_H is the differential operator as in (16.59) and (16.64), so that the exact solution of (16.1) is given by

$$p(x_0+h) = p_0 + \sum_{r \geq 1} \frac{h^r}{r!} k^{(r)}(p_0, q_0), \qquad q(x_0+h) = q_0 + \sum_{r \geq 1} \frac{h^r}{r!} \ell^{(r)}(p_0, q_0).$$

Then deduce from the symplecticity of the exact solution that

$$\frac{1}{\varrho!} \left(dp \wedge d\ell^{(\varrho)} + dk^{(\varrho)} \wedge dq \right) + \sum_{r+m=\varrho} \frac{1}{r!} \frac{1}{m!} dk^{(r)} \wedge d\ell^{(m)} = 0. \quad (16.71)$$

This, together with a modification of the proof of Theorem 16.6, allows us to obtain the desired result.

8. (Yoshida 1990, Qin Meng-Zhao & Zhu Wen-Jie 1992). Let $y_1 = \psi_h(y_0)$ denote a symmetric numerical scheme of order $p = 2k$. Prove that the composed method

$$\psi_{c_1 h} \circ \psi_{c_2 h} \circ \psi_{c_1 h}$$

is symmetric and has order $p+2$ if

$$2c_1 + c_2 = 1, \qquad 2c_1^{2k+1} + c_2^{2k+1} = 0. \quad (16.72)$$

Hence there exist, for separable Hamiltonians, explicit symplectic partitioned methods of arbitrarily high order.

Hint. Proceed as for (4.1)-(4.2) and use Theorem 8.10 (the order of a symmetric method is even).

9. The Hamiltonian function (16.24) for the galactic problem is *not* separable. Nevertheless, both methods (16.36a) and (16.36b) can be applied explicitly. Explain.

II.17 Delay Differential Equations

> Detailed studies of the real world impel us, albeit reluctantly, to take account of the fact that the rate of change of physical systems depends not only on their present state, but also on their past history.
> (Bellman & Cooke 1963)

Delay differential equations are equations with "retarded arguments" or "time lags" such as

$$y'(x) = f\big(x, y(x), y(x - \tau)\big) \tag{17.1}$$

or

$$y'(x) = f\big(x, y(x), y(x - \tau_1), y(x - \tau_2)\big) \tag{17.2}$$

or of even more general form. Here the derivative of the solutions depends also on its values at previous points.

Time lags are present in many models of applied mathematics. They can also be the source of interesting mathematical phenomena such as instabilities, limit cycles, periodic behaviour.

Existence

For equations of the type (17.1) or (17.2), where the delay values $x - \tau$ are bounded away from x by a positive constant, the question of existence is an easy matter: suppose that the solution is known, say

$$y(x) = \varphi(x) \qquad \text{for } x_0 - \tau \leq x \leq x_0.$$

Then $y(x - \tau)$ is a known function of x for $x_0 \leq x \leq x_0 + \tau$ and (17.1) becomes an ordinary differential equation, which can be treated by known existence theories. We then know $y(x)$ for $x_0 \leq x \leq x_0 + \tau$ and can compute the solution for $x_0 + \tau \leq x \leq x_0 + 2\tau$ and so on. This "method of steps" then yields existence and uniqueness results for all x. For more details we recommend the books of Bellman & Cooke (1963) and Driver (1977, especially Chapter V).

Example 1. We consider the equation

$$y'(x) = -y(x - 1), \qquad y(x) = 1 \quad \text{for } -1 \leq x \leq 0. \tag{17.3}$$

Proceeding as described above, we obtain

$$y(x) = 1 - x \qquad\qquad\qquad\qquad \text{for } 0 \le x \le 1,$$

$$y(x) = 1 - x + \frac{(x-1)^2}{2!} \qquad\qquad\quad \text{for } 1 \le x \le 2,$$

$$y(x) = 1 - x + \frac{(x-1)^2}{2!} - \frac{(x-2)^3}{3!} \quad \text{for } 2 \le x \le 3, \text{ etc.}$$

The solution is displayed in Fig. 17.1. We observe that despite the fact that the differential equation and the initial function are C^∞, the solution has discontinuities in its derivatives. This results from the fact that the initial function does not satisfy the differential equation. With every time step τ, however, these discontinuities are smoothed out more and more.

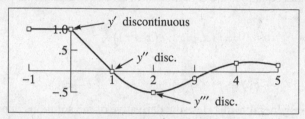

Fig. 17.1. Solution of (17.3)

Example 2. Our next example clearly illustrates the fact that the solutions of a delay equation depend on the entire history between $x_0 - \tau$ and x_0, and not only on the initial value:

$$y'(x) = -1.4 \cdot y(x-1) \tag{17.4}$$

a)	$\varphi(x) = 0.8$	for $-1 \le x \le 0,$	
b)	$\varphi(x) = 0.8 + x$	for $-1 \le x \le 0,$	
c)	$\varphi(x) = 0.8 + 2x$	for $-1 \le x \le 0.$	

The solutions are displayed in Fig. 17.2. An explanation for the oscillatory behaviour of the solutions will be given below.

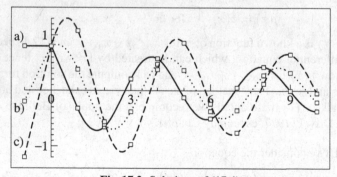

Fig. 17.2. Solutions of (17.4)

Constant Step Size Methods for Constant Delay

If we apply the Runge-Kutta method (1.8) (or (7.7)) to a delay equation (17.1) we obtain

$$g_i^{(n)} = y_n + h \sum_j a_{ij} f\big(x_n + c_j h, g_j^{(n)}, y(x_n + c_j h - \tau)\big)$$

$$y_{n+1} = y_n + h \sum_j b_j f\big(x_n + c_j h, g_j^{(n)}, y(x_n + c_j h - \tau)\big).$$

But which values should we give to $y(x_n + c_j h - \tau)$? If the delay is constant and satisfies $\tau = kh$ for some integer k, the most natural idea is to use the back-values of the old solution

$$g_i^{(n)} = y_n + h \sum_j a_{ij} f(x_n + c_j h, g_j^{(n)}, \gamma_j^{(n)}) \tag{17.5a}$$

$$y_{n+1} = y_n + h \sum_j b_j f(x_n + c_j h, g_j^{(n)}, \gamma_j^{(n)}) \tag{17.5b}$$

where

$$\gamma_j^{(n)} = \begin{cases} \varphi(x_n + c_j h - \tau) & \text{if } n < k \\ g_j^{(n-k)} & \text{if } n \geq k. \end{cases} \tag{17.5c}$$

This can be interpreted as solving successively

$$y'(x) = f\big(x, y(x), \varphi(x - \tau)\big) \tag{17.1a}$$

for the interval $[x_0, x_0 + \tau]$, then

$$y'(x) = f\big(x, y(x), z(x)\big)$$
$$z'(x) = f\big(x - \tau, z(x), \varphi(x - 2\tau)\big) \tag{17.1b}$$

for the interval $[x_0 + \tau, x_0 + 2\tau]$, then

$$y'(x) = f\big(x, y(x), z(x)\big)$$
$$z'(x) = f\big(x - \tau, z(x), v(x)\big)$$
$$v'(x) = f\big(x - 2\tau, v(x), \varphi(x - 3\tau)\big) \tag{17.1c}$$

for the interval $[x_0 + 2\tau, x_0 + 3\tau]$, and so on. This is the perfect numerical analog of the "method of steps" mentioned above.

Theorem 17.1. *If c_i, a_{ij}, b_j are the coefficients of a p-th order Runge-Kutta method, then (17.5) is convergent of order p.*

Proof. The sequence (17.1a), (17.1b),... are ordinary differential equations normally solved by a pth order Runge-Kutta method. Therefore the result follows immediately from Theorem 3.6. □

Remark. For the collocation method based on Gaussian quadrature formula, Theorem 17.1 yields superconvergence in spite of the use of the low order approximations $\gamma_j^{(n)}$ of (17.5c). Bellen (1984) generalizes this result to the situation where $\tau = \tau(x)$ and $\gamma_j^{(n)}$ is the value of the collocation polynomial at $x_n + c_j h - \tau(x_n + c_j h)$. He proves superconvergence if the grid-points are chosen such that every interval $[x_{n-1}, x_n]$ is mapped, by $x - \tau(x)$, into $[x_{j-1}, x_j]$ for some $j < n$.

Numerical Example. We have integrated the problem

$$y'(x) = \big(1.4 - y(x-1)\big) \cdot y(x)$$

(see (17.12) below) for $0 \le x \le 10$ with initial values $y(x) = 0$, $-1 \le x < 0$, $y(0) = 0.1$, and step sizes $h = 1, 1/2, 1/4, 1/8, \ldots, 1/128$ using Kutta's methods of order 4 (Table 1.2, left). The absolute value of the global errors (and the solution in grey) are presented in Fig. 17.3. The 4th order convergence can clearly be observed. The downward peaks are provoked by sign changes in the error.

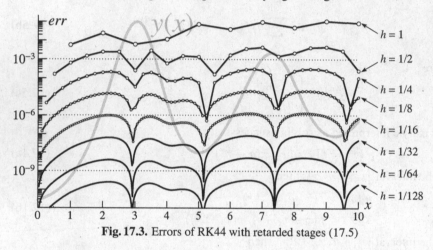

Fig. 17.3. Errors of RK44 with retarded stages (17.5)

Variable Step Size Methods

Although method (17.5) allows efficient and easy to code computations for simple problems with constant delays (such as all the examples of this section), it does not allow to change the step size arbitrarily, and an application to variable delay equations is not straightforward. If complete flexibility is desired, we need a *global* approximation to the solution. Such global approximations are furnished by multistep methods of Adams or BDF type (see Chapter III.1) or the modern Runge-Kutta methods which are constructed together with a dense output. The code RETARD of the appendix is a modification of the code DOPRI5 (method of Dormand &

Prince in Table 5.2 with Shampine's dense output; see (6.12), (6.13) and the subsequent discussion) in such a way that after every successful step of integration the coefficients of the continuous solution are written into memory. Back-values of the solution are then available by calling the function YLAG(I,X,PHI). For example, for problem (17.4) the subroutine FCN would read as

$$F(1) = -1.4D0 * YLAG(1, X - 1.D0, PHI).$$

As we have seen, the solutions possess discontinuities in the derivatives at several points, e.g. for (17.1) at $x_0 + \tau$, $x_0 + 2\tau$, $x_0 + 3\tau$, ... etc. Therefore the code RETARD provides a possibility to match given points of discontinuities exactly (specify IWORK(6) and WORK(11),...) which improves precision and computation time.

Earlier Runge-Kutta codes for delay equations have been written by Oppelstrup (1976), Oberle & Pesch (1981) and Bellen & Zennaro (1985). Bock & Schlöder (1981) exploited the natural dense output of multistep methods.

Stability

It can be observed from Fig. 17.1 and Fig. 17.2 that the solutions, after the initial phase, seem to tend to something like $e^{\alpha x} \cos \beta(x - \delta)$. We now try to determine α and β. We study the equation

$$y'(x) = \lambda y(x) + \mu y(x - 1). \tag{17.6}$$

There is no loss of generality in supposing the delay $\tau = 1$, since any delay $\tau \neq 1$ can be reduced to $\tau = 1$ by a coordinate change.

We search for a solution of the form

$$y(x) = e^{\gamma x} \qquad \text{where} \quad \gamma = \alpha + i\beta. \tag{17.7}$$

Introducing this into (17.6) we obtain the following "characteristic equation" for γ

$$\gamma - \lambda - \mu e^{-\gamma} = 0, \tag{17.8}$$

which, for $\mu \neq 0$, possesses an infinity of solutions: in fact, if $|\gamma|$ becomes large, we obtain from (17.8), since λ is fixed, that $\mu e^{-\gamma}$ must be large too and

$$\gamma \approx \mu e^{-\gamma}. \tag{17.8'}$$

This implies that $\gamma = \alpha + i\beta$ is close to the imaginary axis. Hence $|\gamma| \approx |\beta|$ and from (17.8')

$$|\beta| \approx |\mu| e^{-\alpha}.$$

Therefore the roots of (17.8) lie asymptotically on the curves $-\alpha = \log |\beta| - \log |\mu|$. Again from (17.8'), we have a root whenever the argument of $\mu e^{-i\beta}$ is close to $\pi/2$ (for $\beta > 0$), i.e. if

$$\beta \approx \arg \mu - \frac{\pi}{2} + 2k\pi \qquad k = 1, 2, \dots$$

There are thus two sequences of characteristic values which tend to infinity on logarithmic curves left of the imaginary axis, with 2π as asymptotic distance between two consecutive values.

The "general solution" of (17.6) is thus a Fourier-like superposition of solutions of type (17.7) (Wright 1946, see also Bellman & Cooke 1963, Chapter 4). The larger $-\operatorname{Re}\gamma$ is, the faster these solutions "die out" as $x \to \infty$. The dominant solutions are thus (provided that the corresponding coefficients are not zero) those which correspond to the largest real part, i.e., those closest to the origin. For equations (17.3) and (17.4) the characteristic equations are $\gamma + e^{-\gamma} = 0$ and $\gamma + 1.4e^{-\gamma} = 0$ with solutions $\gamma = -0.31813 \pm 1.33724i$ and $\gamma = -0.08170 \pm 1.51699i$ respectively, which explains nicely the behaviour of the asymptotic solutions of Fig. 17.1 and Fig. 17.2.

Remark. For the case of *matrix equations*

$$y'(x) = Ay(x) + By(x-1)$$

where A and B are not simultaneously diagonalizable, we set $y(x) = ve^{\gamma x}$ where $v \neq 0$ is a given vector. The equation now leads to

$$\gamma v = Av + Be^{-\gamma}v,$$

which has a nontrivial solution if

$$\det(\gamma I - A - Be^{-\gamma}) = 0, \tag{17.8"}$$

the characteristic equation for the more general case. The shape of the solutions of (17.8") is similar to those of (17.8), there are just $r = \operatorname{rank}(B)$ points in each strip of width 2π instead of one.

All solutions of (17.6) remain *stable* for $x \to \infty$ if all characteristic roots of (17.8) remain in the negative half plane. This result follows either from the above expansion theorem or from the theory of Laplace transforms (e.g., Bellmann & Cooke (1963), Chapter 1), which, in fact, is closely related.

In order to study the boundary of the stability domain, we search for (λ, μ) values for which the first solution γ crosses the imaginary axis, i.e. $\gamma = i\theta$ for θ real. If we insert this into (17.8), we obtain

$$\lambda = -\mu \qquad \text{for } \theta = 0 \ (\gamma \text{ real})$$
$$\lambda = i\theta - \mu e^{-i\theta} \qquad \text{for } \theta \neq 0$$

or, by separating real and imaginary parts,

$$\lambda = \frac{\cos\theta \cdot \theta}{\sin\theta}, \qquad \mu = -\frac{\theta}{\sin\theta}$$

valid for real λ and μ. These paths are sketched in Fig. 17.4 and separate in the (λ, μ)-plane the domains of stability and instability for the solutions of (17.6) (a result of Hayes 1950).

If we put $\theta = \pi/2$, we find that the solutions of $y'(x) = \mu y(x-1)$ remain *stable* for

$$-\frac{\pi}{2} \le \mu \le 0 \tag{17.9a}$$

and are *unstable* for

$$\mu < -\frac{\pi}{2} \qquad \text{as well as} \qquad \mu > 0. \tag{17.9b}$$

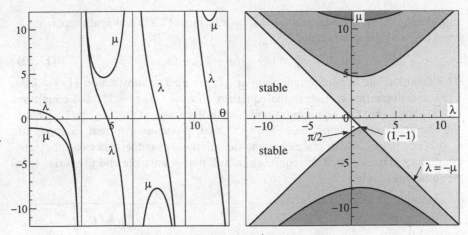

Fig. 17.4. Domain of stability for $y'(x) = \lambda y(x) + \mu y(x-1)$

An Example from Population Dynamics

> Lord Cherwell drew my attention to an equation, equivalent to (8) (here: (17.12)) with $a = \log 2$, which he had encountered in his application of probability methods to the problem of distribution of primes. My thanks are due to him for thus introducing me to an interesting problem.   (E.M. Wright 1945)

We now demonstrate the phenomena discussed above and the power of our programs on a couple of examples drawn from applications. For supplementary applications of delay equations to all sorts of sciences, consult the impressive list in Driver (1977, p. 239-240).

Let $y(x)$ represent the population of a certain species, whose development as a function of time is to be studied. The simple model of infinite exponential growth $y' = \lambda y$ was soon replaced by the hypothesis that the growth rate λ will decrease with increasing population y due to illness and lack of food and space. One then arrives at the model (Verhulst 1845, Pearl & Reed 1922)

$$y'(x) = k \cdot \big(a - y(x)\big) \cdot y(x). \tag{17.10}$$

"Nous donnerons le nom *logistique* à la courbe caractérisée par l'équation précédente" (Verhulst). It can be solved by elementary functions (Exercise 1). All solutions with initial value $y_0 > 0$ tend asymptotically to a as $x \to \infty$. If we assume the growth rate to depend on the population of the *preceding* generation, (17.10) becomes a delay equation (Cunningham 1954, Wright 1955, Kakutani & Markus 1958)

$$y'(x) = k \cdot \big(a - y(x - \tau)\big) \cdot y(x). \tag{17.11}$$

Introducing the new function $z(x) = k\tau y(\tau x)$ into (17.11) and again replacing z by y and $ka\tau$ by a we obtain

$$y'(x) = \big(a - y(x - 1)\big) \cdot y(x). \tag{17.12}$$

This equation has an equilibrium point at $y(x) = a$. The substitution $y(x) = a + z(x)$ and linearization leads to the equation $z'(x) = -az(x - 1)$, and condition (17.9) shows that this equilibrium point is locally stable if $0 < a \le \pi/2$. Hence the characteristic equation, here $\gamma + ae^{-\gamma} = 0$, possesses two real solutions iff $a < 1/e = 0.368$, which makes monotonic solutions possible; otherwise they are oscillatory. For $a > \pi/2$ the equilibrium solution is unstable and gives rise to a periodic limit cycle.

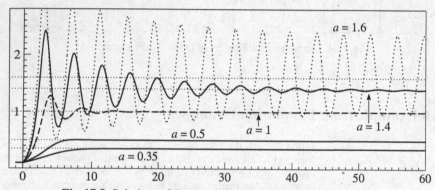

Fig. 17.5. Solutions of the population dynamics problem (17.12)

The solutions in Fig. 17.5 have been computed by the code RETARD of the appendix with subroutine FCN as

F(1) = (A − YLAG(1, X − 1.D0, PHI)) ∗ Y(1),   A = 0.35, 0.5, 1., 1.4, and 1.6.

Infectious Disease Modelling

> De tous ceux qui ont traité cette matière, c'est sans contredit M.
> *de la Condamine* qui l'a fait avec plus de succès. Il est déjà venu
> à bout de persuader la meilleure partie du monde raisonnable de
> la grande utilité de l'inoculation: quant aux autres, il serait inutile
> de vouloir employer la raison avec eux: puisqu'ils n'agissent pas
> par principes. Il faut les conduire comme des enfants vers leur
> mieux ...   (Daniel Bernoulli 1760)

Daniel Bernoulli ("Docteur en medecine, Professeur de Physique en l'Université de Bâle, Associé étranger de l'Academie des Sciences") was the first to use differential calculus to model infectious diseases in his 1760 paper on smallpox vaccination. At the beginning of our century, mathematical modelling of epidemics gained new interest. This finally led to the classical model of Kermack & McKendrick (1927): let $y_1(x)$ measure the *susceptible* portion of the population, $y_2(x)$ the *infected,* and $y_3(x)$ the *removed* (e.g. immunized) one. It is then natural to assume that the number of newly infected people per time unit is proportional to the product $y_1(x)y_2(x)$, just as in bimolecular chemical reactions (see Section I.16). If we finally assume the number of newly removed persons to be proportional to the infected ones, we arrive at the model

$$y_1' = -y_1 y_2, \qquad y_2' = y_1 y_2 - y_2, \qquad y_3' = y_2 \qquad (17.13)$$

where we have taken for simplicity all rate constants equal to one. This system can be integrated by elementary methods (divide the first two equations and solve $dy_2/dy_1 = -1 + 1/y_1$). The numerical solution with initial values $y_1(0) = 5$, $y_2(0) = 0.1$, $y_3(0) = 1$ is painted in gray color in Fig. 17.6: an epidemic breaks out, everybody finally becomes "removed" and nothing further happens.

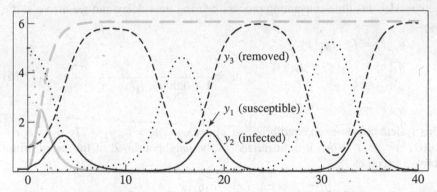

Fig. 17.6. Periodic outbreak of disease, model (17.14)
(in gray: Solution of Kermack - McKendrick model (17.13))

We arrive at a periodic outbreak of the disease, if we assume that immunized people become susceptible again, say after a fixed time τ $(\tau = 10)$. If we also introduce an incubation period of, say, $\tau_2 = 1$, we arrive at the model

$$
\begin{aligned}
y_1'(x) &= -y_1(x)y_2(x-1) + y_2(x-10) \\
y_2'(x) &= y_1(x)y_2(x-1) - y_2(x) \\
y_3'(x) &= y_2(x) - y_2(x-10)
\end{aligned}
\tag{17.14}
$$

instead of (17.13). The solutions of (17.14), for the initial phases $y_1(x) = 5$, $y_2(x) = 0.1$, $y_3(x) = 1$ for $x \leq 0$, are shown in Fig. 17.6 and illustrate the periodic outbreak of the disease.

An Example from Enzyme Kinetics

Our next example, more complicated than the preceding ones, is from enzyme kinetics (Okamoto & Hayashi 1984). Consider the following consecutive reactions

$$
I \longrightarrow Y_1 \xrightarrow{z} Y_2 \xrightarrow{k_2} Y_3 \xrightarrow{k_3} Y_4 \xrightarrow{k_4}
\tag{17.15}
$$

where I is an exogenous substrate supply which is maintained constant and n molecules of the end product Y_4 inhibit co-operatively the reaction step of $Y_1 \to Y_2$ as

$$
z = \frac{k_1}{1 + \alpha(y_4(x))^n}.
$$

It is generally expected that the inhibitor molecule must be moved to the position of the regulatory enzyme by forces such as diffusion or active transport. Thus, we consider this time consuming process causing time-delay and we arrive at the model

$$
\begin{aligned}
y_1'(x) &= I - zy_1(x) \\
y_2'(x) &= zy_1(x) - y_2(x) \\
y_3'(x) &= y_2(x) - y_3(x) \\
y_4'(x) &= y_3(x) - 0.5y_4(x)
\end{aligned}
\qquad z = \frac{1}{1 + 0.0005(y_4(x-4))^3}.
\tag{17.16}
$$

This system possesses an equilibrium at $zy_1 = y_2 = y_3 = I$, $y_4 = 2I$, $y_1 = I(1 + 0.004I^3) =: c_1$. When it is linearized in the neighbourhood of this equilibrium point, it becomes

$$
\begin{aligned}
y_1'(x) &= -c_1 y_1(x) + c_2 y_4(x-4) \\
y_2'(x) &= c_1 y_1(x) - y_2(x) - c_2 y_4(x-4) \\
y_3'(x) &= y_2(x) - y_3(x) \\
y_4'(x) &= y_3(x) - 0.5y_4(x)
\end{aligned}
\tag{17.17}
$$

where $c_2 = c_1 \cdot I^3 \cdot 0.006$. By setting $y(x) = v \cdot e^{\gamma x}$ we arrive at the characteristic equation (see (17.8")), which becomes after some simplifications

$$(c_1 + \gamma)(1 + \gamma)^2(0.5 + \gamma) + c_2 \gamma e^{-4\gamma} = 0. \tag{17.18}$$

As in the paper of Okamoto & Hayashi, we put $I = 10.5$. Then (17.18) possesses one pair of complex solutions in $\mathbb{C}^+$, namely

$$\gamma = 0.04246 \pm 0.47666i$$

and the equilibrium solution is unstable (see Fig. 17.7). The period of the solution of the linearized equation is thus $T = 2\pi/0.47666 = 13.18$. The solutions then tend to a limit cycle of approximately the same period.

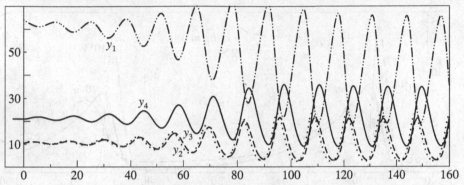

Fig. 17.7. Solutions of the enzyme kinetics problem (17.16), $I = 10.5$.
Initial values close to equilibrium position

A Mathematical Model in Immunology

We conclude our series of examples with Marchuk's model (Marchuk 1975) for the struggle of viruses $V(t)$, antibodies $F(t)$ and plasma cells $C(t)$ in the organism of a person infected by a viral disease. The equations are

$$\frac{dV}{dt} = (h_1 - h_2 F)V$$

$$\frac{dC}{dt} = \xi(m)h_3 F(t - \tau)V(t - \tau) - h_5(C - 1) \tag{17.19}$$

$$\frac{dF}{dt} = h_4(C - F) - h_8 FV.$$

The first is a Volterra - Lotka like predator-prey equation. The second equation describes the creation of new plasma cells with time lag due to infection, in the absence of which the second term creates an equilibrium at $C = 1$. The third equation models the creation of antibodies from plasma cells $(h_4 C)$ and their

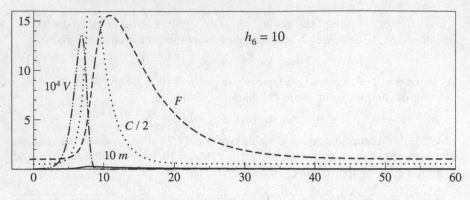

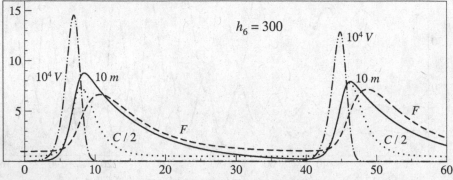

Fig. 17.8. Solutions of the Marchuk immunology model

decrease due to aging $(-h_4 F)$ and binding with antigens $(-h_8 FV)$. The term $\xi(m)$, finally, is defined by

$$\xi(m) = \begin{cases} 1 & \text{if } m \le 0.1 \\ (1-m)\dfrac{10}{9} & \text{if } 0.1 \le m \le 1 \end{cases}$$

and expresses the fact that the creation of plasma cells slows down when the organism is damaged by the viral infection. The relative characteristic $m(t)$ of damaging is given by a fourth equation

$$\frac{dm}{dt} = h_6 V - h_7 m$$

where the first term expresses the damaging and the second recuperation.

This model allows us, by changing the coefficients $h_1, h_2, \ldots, h_8$, to model all sorts of behaviour of stable health, unstable health, acute form of a disease, chronic form etc. See Chapter 2 of Marchuk (1983). In Fig. 17.8 we plot the solutions of this model for $\tau = 0.5$, $h_1 = 2$, $h_2 = 0.8$, $h_3 = 10^4$, $h_4 = 0.17$, $h_5 = 0.5$, $h_7 = 0.12$, $h_8 = 8$ and initial values $V(t) = \max(0, 10^{-6} + t)$ if $t \le 0$, $C(0) = 1$, $F(t) = 1$ if $t \le 0$, $m(0) = 0$. In dependence of the value of h_6 ($h_6 = 10$

or $h_6 = 300$), we then observe either complete recovery (defined by $V(t) < 10^{-16}$), or periodic outbreak of the disease due to damaging ($m(t)$ becomes nearly 1).

Integro-Differential Equations

Often the hypothesis that a system depends on the time lagged solution at a specified fixed value $x - \tau$ is not very realistic, and one should rather suppose this dependence to be stretched out over a longer period of time. Then, instead of (17.1), we would have for example

$$y'(x) = f\left(x, y(x), \int_{x-\tau}^{x} K\big(x, \xi, y(\xi)\big) \, d\xi\right). \tag{17.20}$$

The numerical treatment of these problems becomes much more expensive (see Brunner & van der Houwen (1986) for a study of various discretization methods). If $K(x, \xi, y)$ is zero in the neighbourhood of the diagonal $x = \xi$, one can eventually use RETARD and call a quadrature routine for each function evaluation.

Fortunately, many integro-differential equations can be reduced to ordinary or delay differential equations by introducing new variables for the integral function.

Example (Volterra 1934). Consider the equation

$$y'(x) = \left(\varepsilon - \alpha y(x) - \int_0^x k(x - \xi) y(\xi) \, d\xi\right) \cdot y(x) \tag{17.21}$$

for population dynamics, where the integral term represents a decrease of the reproduction rate due to pollution. If now for example $k(x) = c$, we put

$$\int_0^x y(\xi) \, d\xi = v(x), \qquad y(x) = v'(x)$$

and obtain

$$v''(x) = \big(\varepsilon - \alpha v'(x) - c v(x)\big) \cdot v'(x),$$

an ordinary differential equation.

The same method is possible for equations (17.20) with "degenerate kernel"; i.e., where

$$K(x, \xi, y) = \sum_{i=1}^{m} a_i(x) b_i(\xi, y). \tag{17.22}$$

If we insert this into (17.20) and put

$$v_i(x) = \int_{x-\tau}^{x} b_i\big(\xi, y(\xi)\big) \, d\xi, \tag{17.23}$$

we obtain

$$y'(x) = f\left(x, y(x), \sum_{i=1}^{m} a_i(x)v_i(x)\right)$$

$$v_i'(x) = b_i(x, y(x)) - b_i(x - \tau, y(x - \tau)) \qquad i = 1, \ldots, m,$$

(17.20')

a system of delay differential equations.

Exercises

1. Compute the solution of the Verhulst & Pearl equation (17.10).

2. Compute the equilibrium points of Marchuk's equation (17.19) and study their stability.

3. Assume that the kernel $k(x)$ in Volterra's equation (17.21) is given by

$$k(x) = p(x)e^{-\beta x}$$

where $p(x)$ is some polynomial. Show that this problem can be transformed into an ordinary differential equation.

4. Consider the integro-differential equation

$$y'(x) = f\left(x, y(x), \int_0^x K(x, \xi, y(\xi)) \, d\xi\right).$$

(17.24)

a) For the degenerate kernel (17.22) problem (17.24) becomes equivalent to the ordinary differential equation

$$y'(x) = f\left(x, y(x), \sum_{j=1}^{m} a_j(x)v_j(x)\right)$$

$$v_j'(x) = b_j(x, y(x)).$$

(17.25)

b) Show that an application of an explicit (pth order) Runge-Kutta method to (17.25) yields the formulas (Pouzet 1963)

$$y_{n+1} = y_n + h\sum_{i=1}^{s} b_i f(x_n + c_i h, g_i^{(n)}, u_i^{(n)})$$

$$g_i^{(n)} = y_n + h\sum_{j=1}^{i-1} a_{ij} f(x_n + c_j h, g_j^{(n)}, u_j^{(n)})$$

$$u_i^{(n)} = F_n(x_n + c_i h) + h\sum_{j=1}^{i-1} a_{ij} K(x_n + c_i h, x_n + c_j h, g_j^{(n)})$$

(17.26)

where

$$F_0(x) = 0, \qquad F_{n+1}(x) = F_n(x) + h \sum_{i=1}^{s} b_i K(x, x_n + c_i h, g_i^{(n)}).$$

c) If we apply method (17.26) to problem (17.24), where the kernel does not necessarily satisfy (17.22), we nevertheless have convergence of order p.

Hint. Approximate the kernel by a degenerate one.

5. (Zennaro 1986). For the delay equation (17.1) consider the method (17.5) where (17.5c) is replaced by

$$\gamma_j^{(n)} = \begin{cases} \varphi(x_n + c_j h - \tau) & \text{if } n < k \\ q_{n-k}(c_j) & \text{if } n \geq k. \end{cases} \qquad (17.5\text{c'})$$

Here $q_n(\theta)$ is the polynomial given by a continuous Runge-Kutta method (Section II.6)

$$q_n(\theta) = y_n + h \sum_{j=1}^{s} b_j(\theta) f(x_n + c_j h, g_j^{(n)}, \gamma_j^{(n)}).$$

a) Prove that the orthogonality conditions

$$\int_0^1 \theta^{q-1} \left(\gamma(t) \sum_{j=1}^{s} b_j(\theta) \Phi_j(t) - \theta^{\varrho(t)} \right) d\theta = 0 \qquad \text{for } q + \varrho(t) \leq p$$

$$(17.27)$$

imply convergence of order p, if the underlying Runge-Kutta method is of order p for ordinary differential equations.

Hint. Use the theory of B-series and the Gröbner - Alekseev formula (I.14.18) of Section I.14.

b) If for a given Runge-Kutta method the polynomials $b_j(\theta)$ of degree $\leq [(p+1)/2]$ are such that $b_j(0) = 0$, $b_j(1) = b_j$ and

$$\int_0^1 \theta^{q-1} b_j(\theta) \, d\theta = \frac{1}{q} b_j(1 - c_j^q), \qquad q = 1, \ldots, [(p-1)/2], \quad (17.28)$$

then (17.27) is satisfied. In addition one has the order conditions

$$\sum_{j=1}^{s} b_j(\theta) \Phi_j(t) = \frac{\theta^{\varrho(t)}}{\gamma(t)} \qquad \text{for } \varrho(t) \leq [(p+1)/2].$$

c) Show that the conditions (17.28) admit unique polynomials $b_j(\theta)$ of degree $[(p+1)/2]$.

6. Solve Volterra's equation (17.21) with $k(x) = c$ and compare the solution with the "pollution free" problem (17.10). Which population lives better, that *with* pollution, or that *without*?

Chapter III. Multistep Methods and General Linear Methods

This chapter is devoted to the study of multistep and general multivalue methods. After retracing their historical development (Adams, Nyström, Milne, BDF) we study in the subsequent sections the order, stability and convergence properties of these methods. Convergence is most elegantly set in the framework of one-step methods in higher dimensions. Sections III.5 and III.6 are devoted to variable step size and Nordsieck methods. We then discuss the various available codes and compare them on the numerical examples of Section II.10 as well as on some equations of high dimension. Before closing the chapter with a section on special methods for second order equations, we discuss two highly theoretical subjects: one on general linear methods, including Runge-Kutta methods as well as multistep methods and many generalizations, and the other on the asymptotic expansion of the global error of such methods.

III.1 Classical Linear Multistep Formulas

> ..., and my undertaking must have ended here, if I had depended
> upon my own resources. But at this point Professor J.C. Adams
> furnished me with a perfectly satisfactory method of calculating
> by quadratures the exact theoretical forms of drops of fluids from
> the Differential Equation of Laplace,...   (F. Bashforth 1883)

Another improvement of Euler's method was considered even earlier than Runge-Kutta methods — the methods of Adams. These were devised by John Couch Adams in order to solve a problem of F. Bashforth, which occurred in an investigation of capillary action. Both the problem and the numerical integration schemes are published in Bashforth (1883). The actual origin of these methods must date back to at least 1855, since in that year F. Bashforth made an application to the Royal Society for assistance from the Government grant. There he wrote: "..., but I am indebted to Mr Adams for a method of treating the differential equation

$$\frac{\frac{ddz}{du^2}}{\left(1+\frac{dz^2}{du^2}\right)^{3/2}} + \frac{\frac{1}{u}\frac{dz}{du}}{\left(1+\frac{dz^2}{du^2}\right)^{1/2}} - 2\alpha z = \frac{2}{b},$$

when put under the form

$$\frac{b}{\varrho} + \frac{b}{x}\sin\varphi = 2 + 2\alpha b^2\frac{z}{b} = 2 + \beta\frac{z}{b},$$

which gives the theoretical form of the drop with an accuracy exceeding that of the most refined measurements."

In contrast to one-step methods, where the numerical solution is obtained solely from the differential equation and the initial value, the algorithm of Adams consists of two parts: firstly, a *starting procedure* which provides $y_1, \ldots, y_{k-1}$ (approximations to the exact solution at the points $x_0 + h, \ldots, x_0 + (k-1)h$) and, secondly, a *multistep formula* to obtain an approximation to the exact solution $y(x_0 + kh)$. This is then applied recursively, based on the numerical approximations of k successive steps, to compute $y(x_0 + (k+1)h)$, etc.

There are several possibilities for obtaining the missing starting values. J.C. Adams actually computed them using the Taylor series expansion of the exact solution (as described in Section I.8, see also Exercise 2). Another possibility is the use of any one-step method, e.g., a Runge-Kutta method (see Chapter II). It is also usual to start with low-order Adams methods and very small step sizes.

Explicit Adams Methods

We now derive, following Adams, the first explicit multistep formulas. We introduce the notation $x_i = x_0 + ih$ for the grid points and suppose we know the numerical approximations $y_n, y_{n-1}, \ldots, y_{n-k+1}$ to the exact solution $y(x_n), \ldots, y(x_{n-k+1})$ of the differential equation

$$y' = f(x, y), \quad y(x_0) = y_0. \tag{1.1}$$

Adams considers (1.1) in integrated form,

$$y(x_{n+1}) = y(x_n) + \int_{x_n}^{x_{n+1}} f(t, y(t))\, dt. \tag{1.2}$$

On the right hand side of (1.2) there appears the unknown solution $y(x)$. But since the approximations $y_{n-k+1}, \ldots, y_n$ are known, the values

$$f_i = f(x_i, y_i) \qquad \text{for} \quad i = n-k+1, \ldots, n \tag{1.3}$$

are also available and it is natural to replace the function $f(t, y(t))$ in (1.2) by the interpolation polynomial through the points $\{(x_i, f_i) \mid i = n-k+1, \ldots, n\}$ (see Fig. 1.1).

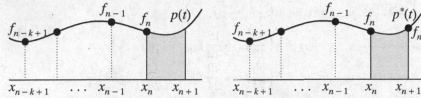

Fig. 1.1. Explicit Adams methods   **Fig. 1.2.** Implicit Adams methods

This polynomial can be expressed in terms of backward differences

$$\nabla^0 f_n = f_n, \qquad \nabla^{j+1} f_n = \nabla^j f_n - \nabla^j f_{n-1}$$

as follows:

$$p(t) = p(x_n + sh) = \sum_{j=0}^{k-1} (-1)^j \binom{-s}{j} \nabla^j f_n \tag{1.4}$$

(Newton's interpolation formula of 1676, published in Newton (1711), see e.g. Henrici (1962), p. 190). The numerical analogue to (1.2) is then given by

$$y_{n+1} = y_n + \int_{x_n}^{x_{n+1}} p(t)\, dt$$

or after insertion of (1.4) by

$$y_{n+1} = y_n + h \sum_{j=0}^{k-1} \gamma_j \nabla^j f_n \tag{1.5}$$

where the coefficients γ_j satisfy

$$\gamma_j = (-1)^j \int_0^1 \binom{-s}{j} ds \qquad (1.6)$$

(see Table 1.1 for their numerical values). A simple recurrence relation for these coefficients will be derived below (formula (1.7)).

Table 1.1. Coefficients for the explicit Adams methods

j	0	1	2	3	4	5	6	7	8
γ_j	1	$\dfrac{1}{2}$	$\dfrac{5}{12}$	$\dfrac{3}{8}$	$\dfrac{251}{720}$	$\dfrac{95}{288}$	$\dfrac{19087}{60480}$	$\dfrac{5257}{17280}$	$\dfrac{1070017}{3628800}$

Special cases of (1.5). For $k = 1, 2, 3, 4$, after expressing the backward differences in terms of f_{n-j}, one obtains the formulas

$$k = 1: \quad y_{n+1} = y_n + h f_n \qquad \text{(explicit Euler method)}$$

$$k = 2: \quad y_{n+1} = y_n + h\left(\frac{3}{2}f_n - \frac{1}{2}f_{n-1}\right)$$

$$k = 3: \quad y_{n+1} = y_n + h\left(\frac{23}{12}f_n - \frac{16}{12}f_{n-1} + \frac{5}{12}f_{n-2}\right) \qquad (1.5')$$

$$k = 4: \quad y_{n+1} = y_n + h\left(\frac{55}{24}f_n - \frac{59}{24}f_{n-1} + \frac{37}{24}f_{n-2} - \frac{9}{24}f_{n-3}\right).$$

Recurrence relation for the coefficients. Using Euler's method of *generating functions* we can deduce a simple recurrence relation for γ_i (see e.g. Henrici 1962). Denote by $G(t)$ the series

$$G(t) = \sum_{j=0}^{\infty} \gamma_j t^j.$$

With the definition of γ_j and the binomial theorem one obtains

$$G(t) = \sum_{j=0}^{\infty} (-t)^j \int_0^1 \binom{-s}{j} ds = \int_0^1 \sum_{j=0}^{\infty} (-t)^j \binom{-s}{j} ds$$

$$= \int_0^1 (1-t)^{-s} ds = -\frac{t}{(1-t)\log(1-t)}.$$

This can be written as

$$-\frac{\log(1-t)}{t} G(t) = \frac{1}{1-t}$$

or as

$$\left(1 + \frac{1}{2}t + \frac{1}{3}t^2 + \ldots\right)\left(\gamma_0 + \gamma_1 t + \gamma_2 t^2 + \ldots\right) = (1 + t + t^2 + \ldots).$$

Comparing the coefficients of t^m we get the desired recurrence relation

$$\gamma_m + \frac{1}{2}\gamma_{m-1} + \frac{1}{3}\gamma_{m-2} + \ldots + \frac{1}{m+1}\gamma_0 = 1. \tag{1.7}$$

Implicit Adams Methods

The formulas (1.5) are obtained by integrating the interpolation polynomial (1.4) from x_n to x_{n+1}, i.e., outside the interpolation interval (x_{n-k+1}, x_n). It is well known that an interpolation polynomial is usually a rather poor approximation outside this interval. Adams therefore also investigated methods where (1.4) is replaced by the interpolation polynomial which uses in addition the point (x_{n+1}, f_{n+1}), i.e.,

$$p^*(t) = p^*(x_n + sh) = \sum_{j=0}^{k} (-1)^j \binom{-s+1}{j} \nabla^j f_{n+1} \tag{1.8}$$

(see Fig. 1.2). Inserting this into (1.2) we obtain the following implicit method

$$y_{n+1} = y_n + h \sum_{j=0}^{k} \gamma_j^* \nabla^j f_{n+1} \tag{1.9}$$

where the coefficients γ_j^* satisfy

$$\gamma_j^* = (-1)^j \int_0^1 \binom{-s+1}{j} ds \tag{1.10}$$

and are given in Table 1.2 for $j \le 8$. Again, a simple recurrence relation can be derived for these coefficients (Exercise 3).

Table 1.2. Coefficients for the implicit Adams methods

j	0	1	2	3	4	5	6	7	8
γ_j^*	1	$-\frac{1}{2}$	$-\frac{1}{12}$	$-\frac{1}{24}$	$-\frac{19}{720}$	$-\frac{3}{160}$	$-\frac{863}{60480}$	$-\frac{275}{24192}$	$-\frac{33953}{3628800}$

The formulas thus obtained are generally of the form

$$y_{n+1} = y_n + h\big(\beta_k f_{n+1} + \ldots + \beta_0 f_{n-k+1}\big). \tag{1.9'}$$

The first examples are as follows

$$
\begin{aligned}
k=0: \quad & y_{n+1} = y_n + h f_{n+1} = y_n + h f(x_{n+1}, y_{n+1}) \\
k=1: \quad & y_{n+1} = y_n + h\left(\frac{1}{2} f_{n+1} + \frac{1}{2} f_n\right) \\
k=2: \quad & y_{n+1} = y_n + h\left(\frac{5}{12} f_{n+1} + \frac{8}{12} f_n - \frac{1}{12} f_{n-1}\right) \\
k=3: \quad & y_{n+1} = y_n + h\left(\frac{9}{24} f_{n+1} + \frac{19}{24} f_n - \frac{5}{24} f_{n-1} + \frac{1}{24} f_{n-2}\right).
\end{aligned} \tag{1.9''}
$$

The special cases $k=0$ and $k=1$ are the implicit Euler method and the trapezoidal rule, respectively. They are actually one-step methods and have already been considered in Chapter II.7.

The methods (1.9) give in general more accurate approximations to the exact solution than (1.5). This will be discussed in detail when the concepts of order and error constant are introduced (Section III.2). The price for this higher accuracy is that y_{n+1} is only defined implicitly by formula (1.9). Therefore, in general a nonlinear equation has to be solved at each step.

Predictor-corrector methods. One possibility for solving this nonlinear equation is to apply fixed point iteration. In practice one proceeds as follows:

P: compute the predictor $\widehat{y}_{n+1} = y_n + h \sum_{j=0}^{k-1} \gamma_j \nabla^j f_n$ by the explicit Adams method (1.5); this already yields a reasonable approximation to $y(x_{n+1})$;

E: evaluate the function at this approximation: $\widehat{f}_{n+1} = f(x_{n+1}, \widehat{y}_{n+1})$;

C: apply the corrector formula

$$
y_{n+1} = y_n + h\left(\beta_k \widehat{f}_{n+1} + \beta_{k-1} f_n + \ldots + \beta_0 f_{n-k+1}\right) \tag{1.11}
$$

to obtain y_{n+1}.

E: evaluate the function anew, i.e., compute $f_{n+1} = f(x_{n+1}, y_{n+1})$.

This is the most common procedure, denoted by PECE. Other possibilities are: PECECE (two fixed point iterations per step) or PEC (one uses $\widehat{f}_{n+1}$ instead of f_{n+1} in the subsequent steps).

This predictor-corrector technique has been used by F.R. Moulton (1926) as well as by W.E. Milne (1926). J.C. Adams actually solved the implicit equation (1.9) by Newton's method, in the same way as is now usual for stiff equations (see Volume II).

Remark. Formula (1.5) is often attributed to Adams-Bashforth. Similarly, the multistep formula (1.9) is usually attributed to Adams-Moulton (Moulton 1926). In fact, both formulas are due to Adams.

Numerical Experiment

We consider the Van der Pol equation (I.16.2) with $\varepsilon = 1$, take as initial values $y_1(0) = A$, $y_2(0) = 0$ on the limit cycle and integrate over one period T (for the values of A and T see Exercise I.16.1). This is exactly the same problem as the one used for the comparison of Runge-Kutta methods (Fig. II.1.1). We have applied the above explicit and implicit Adams methods with several fixed step sizes. The missing starting values were computed with high accuracy by an explicit Runge-Kutta method. Fig. 1.3 shows the errors of both components in dependence of the number of function evaluations. Since we have implemented the implicit method (1.9) in PECE mode it requires 2 function evaluations per step, whereas the explicit method (1.5) needs only one.

This experiment shows that, for the same value of k, the implicit methods usually give a better result (the strange behaviour in the error of the y_2-component for $k \geq 3$ is due to a sign change). Since we have used double logarithmic scales, it is possible to read the "numerical order" from the slope of the corresponding lines. We observe that the global error of the explicit Adams methods behaves like $\mathcal{O}(h^k)$ and that of the implicit methods like $\mathcal{O}(h^{k+1})$. This will be proved in the following sections.

We also remark that the scales used in Fig. 1.3 are exactly the same as those of Fig. II.1.1. This allows a comparison with the Runge-Kutta methods of Section II.1.

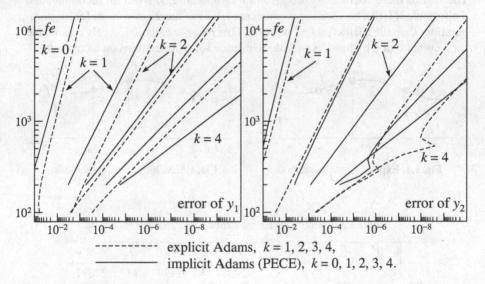

------------ explicit Adams, $k = 1, 2, 3, 4$,

——————— implicit Adams (PECE), $k = 0, 1, 2, 3, 4$.

Fig. 1.3. Global errors versus number of function evaluations

Explicit Nyström Methods

> Die angenäherte Integration hat, besonders in der letzten Zeit, ein ausgedehntes Anwendungsgebiet innerhalb der exakten Wissenschaften und der Technik gefunden.   (E.J. Nyström 1925)

In his review article on the numerical integration of differential equations (which we have already encountered in Section II.14), Nyström (1925) also presents a new class of multistep methods. He considers instead of (1.2) the integral equation

$$y(x_{n+1}) = y(x_{n-1}) + \int_{x_{n-1}}^{x_{n+1}} f\big(t, y(t)\big)\, dt. \tag{1.12}$$

In the same way as above he replaces the unknown function $f(t, y(t))$ by the polynomial $p(t)$ of (1.4) and so obtains the formula (see Fig. 1.4)

$$y_{n+1} = y_{n-1} + h \sum_{j=0}^{k-1} \kappa_j \nabla^j f_n \tag{1.13}$$

with the coefficients

$$\kappa_j = (-1)^j \int_{-1}^{1} \binom{-s}{j}\, ds. \tag{1.14}$$

The first of these coefficients are given in Table 1.3. E.J. Nyström recommended the formulas (1.13), because the coefficients κ_j were more convenient for his computations than the coefficients γ_j of (1.6). This recommendation, surely reasonable for a computation by hand, is of little relevance for computations on a computer.

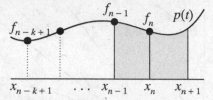

Fig. 1.4. Explicit Nyström methods

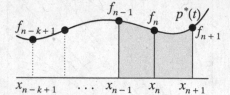

Fig. 1.5. Milne-Simpson methods

Table 1.3. Coefficients for the explicit Nyström methods

j	0	1	2	3	4	5	6	7	8
κ_j	2	0	$\dfrac{1}{3}$	$\dfrac{1}{3}$	$\dfrac{29}{90}$	$\dfrac{14}{45}$	$\dfrac{1139}{3780}$	$\dfrac{41}{140}$	$\dfrac{32377}{113400}$

Special cases. For $k = 1$ the formula

$$y_{n+1} = y_{n-1} + 2hf_n \tag{1.13'}$$

is obtained. It is called the *mid-point rule* and is the simplest two-step method. Its symmetry was extremely useful in the extrapolation schemes of Section II.9. The case $k = 2$ yields nothing new, because $\kappa_1 = 0$. For $k = 3$ one gets

$$y_{n+1} = y_{n-1} + h\left(\frac{7}{3}f_n - \frac{2}{3}f_{n-1} + \frac{1}{3}f_{n-2}\right). \tag{1.13"}$$

Milne–Simpson Methods

We consider again the integral equation (1.12). But now we replace the integrand by the polynomial $p^*(t)$ of (1.8), which in addition to $f_n, \ldots, f_{n-k+1}$ also interpolates the value f_{n+1} (see Fig. 1.5). Proceeding as usual, we get the implicit formulas

$$y_{n+1} = y_{n-1} + h\sum_{j=0}^{k} \kappa_j^* \nabla^j f_{n+1}. \tag{1.15}$$

The coefficients κ_j^* are defined by

$$\kappa_j^* = (-1)^j \int_{-1}^{1} \binom{-s+1}{j} ds, \tag{1.16}$$

and the first of these are given in Table 1.4.

Table 1.4. Coefficients for the Milne-Simpson methods

j	0	1	2	3	4	5	6	7	8
κ_j^*	2	-2	$\dfrac{1}{3}$	0	$-\dfrac{1}{90}$	$-\dfrac{1}{90}$	$-\dfrac{37}{3780}$	$-\dfrac{8}{945}$	$-\dfrac{119}{16200}$

If the backward differences in (1.15) are expressed in terms of f_{n-j}, one obtains the following methods for special values of k:

$$k = 0: \quad y_{n+1} = y_{n-1} + 2hf_{n+1},$$
$$k = 1: \quad y_{n+1} = y_{n-1} + 2hf_n, \tag{1.15'}$$
$$k = 2: \quad y_{n+1} = y_{n-1} + h\left(\frac{1}{3}f_{n+1} + \frac{4}{3}f_n + \frac{1}{3}f_{n-1}\right),$$
$$k = 4: \quad y_{n+1} = y_{n-1} + h\left(\frac{29}{90}f_{n+1} + \frac{124}{90}f_n + \frac{24}{90}f_{n-1} + \frac{4}{90}f_{n-2} - \frac{1}{90}f_{n-3}\right).$$

The special case $k = 0$ is just Euler's implicit method applied with step size $2h$. For $k = 1$ one obtains the previously derived mid-point rule. The particular case

$k = 2$ is an interesting method, known as the *Milne method* (Milne 1926, 1970, p. 66). It is a direct generalization of Simpson's rule.

Many other similar methods have been investigated. They are all based on an integral equation of the form

$$y(x_{n+1}) = y(x_{n-\ell}) + \int_{x_{n-\ell}}^{x_{n+1}} f(t, y(t))\, dt, \qquad (1.17)$$

where $f(t, y(t))$ is replaced either by the interpolating polynomial $p(t)$ (formula (1.4)) or by $p^*(t)$ (formula (1.8)). E.g., for $\ell = 3$ one obtains

$$y_{n+1} = y_{n-3} + h\left(\frac{8}{3}f_n - \frac{4}{3}f_{n-1} + \frac{8}{3}f_{n-2}\right). \qquad (1.18)$$

This particular method has been used by Milne (1926) as a "predictor" for his method: in order to solve the implicit equation (1.15'), Milne uses one or two fixed-point iterations with the numerical value of (1.18) as starting point.

Methods Based on Differentiation (BDF)

> "My name is Gear." — "pardon?"
> "Gear, dshii, ii, ay, are." — "Mr. Jiea?"
> (In a hotel of Paris)

The multistep formulas considered until now are all based on numerical integration, i.e., the integral in (1.17) is approximated numerically using some quadrature formula. The underlying idea of the following multistep formulas is totally different as they are based on the numerical differentiation of a given function.

Assume that the approximations $y_{n-k+1}, \ldots, y_n$ to the exact solution of (1.1) are known. In order to derive a formula for y_{n+1} we consider the polynomial $q(x)$ which interpolates the values $\{(x_i, y_i) \mid i = n-k+1, \ldots, n+1\}$. As in (1.8) this polynomial can be expressed in terms of backward differences, namely

$$q(x) = q(x_n + sh) = \sum_{j=0}^{k} (-1)^j \binom{-s+1}{j} \nabla^j y_{n+1}. \qquad (1.19)$$

The unknown value y_{n+1} will now be determined in such a way that the polynomial $q(x)$ satisfies the differential equation at at least one grid-point, i.e.,

$$q'(x_{n+1-r}) = f(x_{n+1-r}, y_{n+1-r}). \qquad (1.20)$$

For $r = 1$ we obtain *explicit* formulas. For $k = 1$ and $k = 2$, these are equivalent to the explicit Euler method and the mid-point rule, respectively. The case $k = 3$ yields

$$\frac{1}{3}y_{n+1} + \frac{1}{2}y_n - y_{n-1} + \frac{1}{6}y_{n-2} = hf_n. \qquad (1.21)$$

This formula, however, as well as those for $k > 3$, is unstable (see Section III.3) and therefore useless.

Much more interesting are the formulas one obtains when (1.20) is taken for $r = 0$ (see Fig. 1.6).

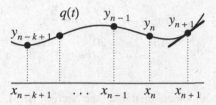

Fig. 1.6. Definition of BDF

In this case one gets the *implicit* formulas

$$\sum_{j=0}^{k} \delta_j^* \nabla^j y_{n+1} = h f_{n+1} \tag{1.22}$$

with the coefficients

$$\delta_j^* = (-1)^j \frac{d}{ds} \binom{-s+1}{j} \bigg|_{s=1}.$$

Using the definition of the binomial coefficient

$$(-1)^j \binom{-s+1}{j} = \frac{1}{j!}(s-1)s(s+1)\dots(s+j-2)$$

the coefficients δ_j^* are obtained by direct differentiation:

$$\delta_0^* = 0, \qquad \delta_j^* = \frac{1}{j} \quad \text{for } j \geq 1. \tag{1.23}$$

Formula (1.22) therefore becomes

$$\sum_{j=1}^{k} \frac{1}{j} \nabla^j y_{n+1} = h f_{n+1}. \tag{1.22'}$$

These multistep formulas, known as *backward differentiation formulas* (or *BDF-methods)*, are, since the work of Gear (1971), widely used for the integration of stiff differential equations (see Volume II). They were introduced by Curtiss & Hirschfelder (1952); Mitchell & Craggs (1953) call them "standard step-by-step methods".

For the sake of completeness we give these formulas also in the form which expresses the backward differences in terms of the y_{n-j}.

$$k = 1: \quad y_{n+1} - y_n = h f_{n+1},$$

$$k = 2: \quad \frac{3}{2} y_{n+1} - 2 y_n + \frac{1}{2} y_{n-1} = h f_{n+1}, \tag{1.22''}$$

$$k = 3: \quad \frac{11}{6}y_{n+1} - 3y_n + \frac{3}{2}y_{n-1} - \frac{1}{3}y_{n-2} = hf_{n+1},$$

$$k = 4: \quad \frac{25}{12}y_{n+1} - 4y_n + 3y_{n-1} - \frac{4}{3}y_{n-2} + \frac{1}{4}y_{n-3} = hf_{n+1},$$

$$k = 5: \quad \frac{137}{60}y_{n+1} - 5y_n + 5y_{n-1} - \frac{10}{3}y_{n-2} + \frac{5}{4}y_{n-3} - \frac{1}{5}y_{n-4} = hf_{n+1},$$

$$k = 6: \quad \frac{147}{60}y_{n+1} - 6y_n + \frac{15}{2}y_{n-1} - \frac{20}{3}y_{n-2} + \frac{15}{4}y_{n-3} - \frac{6}{5}y_{n-4} + \frac{1}{6}y_{n-5}$$
$$= hf_{n+1}.$$

For $k > 6$ the BDF-methods are unstable (see Section III.3).

Exercises

1. Let the differential equation $y' = y^2$, $y(0) = 1$ and the exact starting values $y_i = 1/(1 - x_i)$ for $i = 0, 1, \ldots, k-1$ be given. Apply the methods of Adams and study the expression $y(x_k) - y_k$ for small step sizes.

2. Consider the differential equation at the beginning of this section. It describes the form of a drop and can be written as (F. Bashforth 1883, page 26; the same problem as Exercise 2 of Section II.1 in a different coordinate system)

$$\frac{dx}{d\varphi} = \varrho \cos \varphi, \qquad \frac{dz}{d\varphi} = \varrho \sin \varphi \tag{1.24}$$

where

$$\frac{1}{\varrho} + \frac{\sin \varphi}{x} = 2 + \beta z. \tag{1.25}$$

ϱ may be considered as a function of the coordinates x and z. It can be interpreted as the radius of curvature and φ denotes the angle between the normal to the curve and the z-axis (see Fig. 1.7 for $\beta = 3$). The initial values are given by $x(0) = 0$, $z(0) = 0$, $\varrho(0) = 1$.

Solve the above differential equation along the lines of J.C. Adams:

a) Assuming

$$\varrho = 1 + b_2\varphi^2 + b_4\varphi^4 + \ldots$$

and inserting this expression into (1.24) we obtain after integration the truncated Taylor series of $x(\varphi)$ and $z(\varphi)$ in terms of $b_2, b_4, \ldots$. These parameters can then be calculated from (1.25) by comparing the coefficients of φ^m. In this way one obtains the solution for small values of φ (starting values).

b) Use one of the proposed multistep formulas and calculate the solution for fixed β (say $\beta = 3$) over the interval $[0, \pi]$.

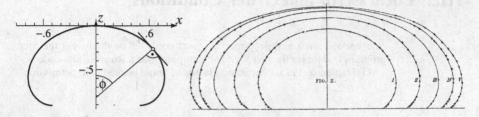

Fig. 1.7. Solution of the differential equation (1.24)
and an illustration from the book of Bashforth

3. Prove that the coefficients γ_j^*, defined by (1.10), satisfy $\gamma_0^* = 1$ and

$$\gamma_m^* + \frac{1}{2}\gamma_{m-1}^* + \frac{1}{3}\gamma_{m-2}^* + \ldots + \frac{1}{m+1}\gamma_0^* = 0 \qquad \text{for} \quad m \geq 1.$$

4. Let $\kappa_j, \kappa_j^*, \gamma_j, \gamma_j^*$ be the coefficients defined by (1.14), (1.16), (1.6), (1.10), respectively. Show that (with $\gamma_{-1} = \gamma_{-1}^* = 0$)

$$\kappa_j = 2\gamma_j - \gamma_{j-1}, \qquad \kappa_j^* = 2\gamma_j^* - \gamma_{j-1}^* \qquad \text{for} \quad j \geq 0.$$

Hint. By splitting the integral in (1.14) one gets $\kappa_j = \gamma_j + \gamma_j^*$. The relation $\gamma_j^* = \gamma_j - \gamma_{j-1}$ is obtained by using a well-known identity for binomial coefficients.

III.2 Local Error and Order Conditions

> You know, I am a multistep man ... and don't tell anybody, but the first
> program I wrote for the first Swedish computer was a Runge-Kutta code ...
> (G. Dahlquist, 1982, after some glasses of wine; printed with permission)

A general theory of multistep methods was started by the work of Dahlquist
(1956, 1959), and became famous through the classical book of Henrici (1962).
All multistep formulas considered in the previous section have this in common that
the numerical approximations y_i as well as the values f_i appear linearly. We thus
consider the general difference equation

$$\alpha_k y_{n+k} + \alpha_{k-1} y_{n+k-1} + \ldots + \alpha_0 y_n = h(\beta_k f_{n+k} + \ldots + \beta_0 f_n) \tag{2.1}$$

which includes all considered methods as special cases. In this formula the α_i and
β_i are real parameters, h denotes the step size and

$$f_i = f(x_i, y_i), \qquad x_i = x_0 + ih.$$

Throughout this chapter we shall assume that

$$\alpha_k \neq 0, \qquad |\alpha_0| + |\beta_0| > 0. \tag{2.2}$$

The first assumption expresses the fact that the implicit equation (2.1) can be solved
with respect to y_{n+k} at least for sufficiently small h. The second relation in (2.2)
can always be achieved by reducing the index k, if necessary.

Formula (2.1) will be called a *linear multistep method* or more precisely a
linear k-step method. We also distinguish between *explicit* $(\beta_k = 0)$ and *implicit*
$(\beta_k \neq 0)$ multistep methods.

Local Error of a Multistep Method

As the numerical solution of a multistep method does not depend only on the initial
value problem (1.1) but also on the choice of the starting values, the definition of
the local error is not as straightforward as for one-step methods (compare Sections
II.2 and II.3).

Definition 2.1. The *local error* of the multistep method (2.1) is defined by

$$y(x_k) - y_k$$

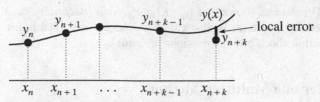

Fig. 2.1. Illustration of the local error

where $y(x)$ is the exact solution of $y' = f(x, y)$, $y(x_0) = y_0$, and y_k is the numerical solution obtained from (2.1) by using the exact starting values $y_i = y(x_i)$ for $i = 0, 1, \ldots, k-1$ (see Fig. 2.1).

In the case $k = 1$ this definition coincides with the definition of the local error for one-step methods. In order to show the connection with other possible definitions of the local error, we associate with (2.1) the linear difference operator L defined by

$$L(y, x, h) = \sum_{i=0}^{k} \Big(\alpha_i y(x + ih) - h\beta_i y'(x + ih) \Big). \tag{2.3}$$

Here $y(x)$ is some differentiable function defined on an interval that contains the values $x + ih$ for $i = 0, 1, \ldots, k$.

Lemma 2.2. *Consider the differential equation (1.1) with $f(x, y)$ continuously differentiable and let $y(x)$ be its solution. For the local error one has*

$$y(x_k) - y_k = \Big(\alpha_k I - h\beta_k \frac{\partial f}{\partial y}(x_k, \eta) \Big)^{-1} L(y, x_0, h).$$

Here η is some value between $y(x_k)$ and y_k, if f is a scalar function. In the case of a vector valued function f, the matrix $\frac{\partial f}{\partial y}(x_k, \eta)$ is the Jacobian whose rows are evaluated at possibly different values lying on the segment joining $y(x_k)$ and y_k.

Proof. By Definition 2.1, y_k is determined implicitly by the equation

$$\sum_{i=0}^{k-1} \Big(\alpha_i y(x_i) - h\beta_i f(x_i, y(x_i)) \Big) + \alpha_k y_k - h\beta_k f(x_k, y_k) = 0.$$

Inserting (2.3) we obtain

$$L(y, x_0, h) = \alpha_k \big(y(x_k) - y_k \big) - h\beta_k \big(f(x_k, y(x_k)) - f(x_k, y_k) \big)$$

and the statement follows from the mean value theorem.   □

This lemma shows that $\alpha_k^{-1} L(y, x_0, h)$ is essentially equal to the local error. Sometimes this term is also called *the* local error (Dahlquist 1956, 1959). For explicit methods both expressions are equal.

Order of a Multistep Method

Once the local error of a multistep method is defined, one can introduce the concept of order in the same way as for one-step methods.

Definition 2.3. The multistep method (2.1) is said to be of *order p*, if one of the following two conditions is satisfied:

   i) for all sufficiently regular functions $y(x)$ we have $L(y, x, h) = \mathcal{O}(h^{p+1})$;

   ii) the local error of (2.1) is $\mathcal{O}(h^{p+1})$ for all sufficiently regular differential equations (1.1).

Observe that by Lemma 2.2 the above conditions (i) and (ii) are equivalent. Our next aim is to characterize the order of a multistep method in terms of the free parameters α_i and β_i. Dahlquist (1956) was the first to observe the fundamental role of the polynomials

$$
\begin{aligned}
\varrho(\zeta) &= \alpha_k \zeta^k + \alpha_{k-1} \zeta^{k-1} + \ldots + \alpha_0 \\
\sigma(\zeta) &= \beta_k \zeta^k + \beta_{k-1} \zeta^{k-1} + \ldots + \beta_0.
\end{aligned}
\tag{2.4}
$$

They will be called the *generating polynomials* of the multistep method (2.1).

Theorem 2.4. *The multistep method (2.1) is of order p, if and only if one of the following equivalent conditions is satisfied:*

   i) $\displaystyle\sum_{i=0}^{k} \alpha_i = 0 \quad and \quad \sum_{i=0}^{k} \alpha_i i^q = q \sum_{i=0}^{k} \beta_i i^{q-1} \quad for \quad q = 1, \ldots, p;$

   ii) $\varrho(e^h) - h\sigma(e^h) = \mathcal{O}(h^{p+1}) \quad for \quad h \to 0;$

   iii) $\dfrac{\varrho(\zeta)}{\log \zeta} - \sigma(\zeta) = \mathcal{O}((\zeta-1)^p) \quad for \quad \zeta \to 1.$

Proof. Expanding $y(x+ih)$ and $y'(x+ih)$ into a Taylor series and inserting these series (truncated if necessary) into (2.3) yields

$$
\begin{aligned}
L(y, x, h) &= \sum_{i=0}^{k} \left(\alpha_i \sum_{q \geq 0} \frac{i^q}{q!} h^q y^{(q)}(x) - h\beta_i \sum_{r \geq 0} \frac{i^r}{r!} h^r y^{(r+1)}(x) \right) \\
&= y(x) \sum_{i=0}^{k} \alpha_i + \sum_{q \geq 1} \frac{h^q}{q!} y^{(q)}(x) \left(\sum_{i=0}^{k} \alpha_i i^q - q \sum_{i=0}^{k} \beta_i i^{q-1} \right).
\end{aligned}
\tag{2.5}
$$

This implies the equivalence of condition (i) with $L(y, x, h) = \mathcal{O}(h^{p+1})$ for all sufficiently regular functions $y(x)$.

It remains to prove that the three conditions of Theorem 2.4 are equivalent. The identity

$$L(\exp, 0, h) = \varrho(e^h) - h\sigma(e^h)$$

where exp denotes the exponential function, together with

$$L(\exp, 0, h) = \sum_{i=0}^{k} \alpha_i + \sum_{q \geq 1} \frac{h^q}{q!} \left(\sum_{i=0}^{k} \alpha_i i^q - q \sum_{i=0}^{k} \beta_i i^{q-1} \right),$$

which follows from (2.5), shows the equivalence of the conditions (i) and (ii).

By use of the transformation $\zeta = e^h$ (or $h = \log \zeta$) condition (ii) can be written in the form

$$\varrho(\zeta) - \log \zeta \cdot \sigma(\zeta) = \mathcal{O}\big((\log \zeta)^{p+1}\big) \qquad \text{for } \zeta \to 1.$$

But this condition is equivalent to (iii), since

$$\log \zeta = (\zeta - 1) + \mathcal{O}\big((\zeta - 1)^2\big) \qquad \text{for } \zeta \to 1. \qquad \square$$

Remark. The conditions for a multistep method to be of order 1, which are usually called *consistency* conditions, can also be written in the form

$$\varrho(1) = 0, \qquad \varrho'(1) = \sigma(1). \tag{2.6}$$

Once the proofs of the above order conditions have been understood, it is not difficult to treat the more general situation of non-equidistant grids (see Section III.5 and the book of Stetter (1973), p. 191).

Example 2.5. *Order of the explicit Adams methods.* Let us first investigate for which differential equations the explicit Adams methods give theoretically the exact solution. This is the case if the polynomial $p(t)$ of (1.4) is equal to $f(t, y(t))$. Suppose now that $f(t, y) = f(t)$ does not depend on y and is a polynomial of degree less than k. Then the explicit Adams methods integrate the differential equations

$$y' = qx^{q-1}, \qquad \text{for } q = 0, 1, \ldots, k$$

exactly. This means that the local error is zero and hence, by Lemma 2.2,

$$0 = L(x^q, 0, h) = h^q \left(\sum_{i=0}^{k} \alpha_i i^q - q \sum_{i=0}^{k} \beta_i i^{q-1} \right) \qquad \text{for } q = 0, \ldots, k.$$

This is just condition (i) of Theorem 2.4 with $p = k$ so that the order of the explicit Adams methods is at least k. In fact it will be shown that the order of these methods is not greater than k (Example 2.7).

Example 2.6. For *implicit Adams methods* the polynomial $p^*(t)$ of (1.8) has degree one higher than that of $p(t)$. Thus the same considerations as in Example 2.5 show that these methods have order at least $k+1$.

All methods of Section III.1 can be treated analogously (see Exercise 3 and Table 2.1).

Table 2.1. Order and error constant of multistep methods

method	formula	order	error constant
explicitAdams	(1.5)	k	γ_k
implicitAdams	(1.9)	$k+1$	γ_{k+1}^*
midpoint rule	(1.13')	2	$1/6$
Nyström, $k>2$	(1.13)	k	$\kappa_k/2$
Milne, $k=2$	(1.15')	4	$-1/180$
Milne-Simpson, $k>3$	(1.15)	$k+1$	$\kappa_{k+1}^*/2$
BDF	(1.22')	k	$-1/(k+1)$

Error Constant

The order of a multistep method indicates how fast the error tends to zero if $h \to 0$. Different methods of the *same* order, however, can have different errors; they are distinguished by the *error constant*. Formula (2.5) shows that the difference operator L, associated with a pth order multistep method, is such that for all sufficiently regular functions $y(x)$

$$L(y, x, h) = C_{p+1} h^{p+1} y^{(p+1)}(x) + \mathcal{O}(h^{p+2}) \tag{2.7}$$

where the constant C_{p+1} is given by

$$C_{p+1} = \frac{1}{(p+1)!} \left(\sum_{i=0}^{k} \alpha_i i^{p+1} - (p+1) \sum_{i=0}^{k} \beta_i i^p \right). \tag{2.8}$$

This constant is not suitable as a measure of accuracy, since multiplication of formula (2.1) by a constant can give any value for C_{p+1}, whereas the numerical solution $\{y_n\}$ remains unchanged. A better choice would be the constant $\alpha_k^{-1} C_{p+1}$, since the local error of a multistep method is given by (Lemma 2.2 and formula (2.7))

$$y(x_k) - y_k = \alpha_k^{-1} C_{p+1} h^{p+1} y^{(p+1)}(x_0) + \mathcal{O}(h^{p+2}). \tag{2.9}$$

For several reasons, however, this is not yet a satisfactory definition, as we shall see from the following motivation: let

$$e_n = \frac{y(x_n) - y_n}{h^p}$$

be the global error scaled by h^p, and assume for this motivation that $e_n = \mathcal{O}(1)$. Subtracting (2.1) from (2.3) and using (2.7) we have

$$\sum_{i=0}^{k} \alpha_i e_{n+i} = h^{1-p} \sum_{i=0}^{k} \beta_i \Big(f\big(x_{n+i}, y(x_{n+i})\big) - f(x_{n+i}, y_{n+i}) \Big) \tag{2.10}$$
$$+ C_{p+1} h y^{(p+1)}(x_n) + \mathcal{O}(h^2).$$

The point is now to use

$$y^{(p+1)}(x_n) = \frac{1}{\sigma(1)} \sum_{i=0}^{k} \beta_i y^{(p+1)}(x_{n+i}) + \mathcal{O}(h) \tag{2.11}$$

which brings the error term in (2.10) inside the sum with the β_i. We linearize

$$f\big(x_{n+i}, y(x_{n+i})\big) - f(x_{n+i}, y_{n+i}) = \frac{\partial f}{\partial y} \big(x_{n+i}, y(x_{n+i})\big) h^p e_{n+i} + \mathcal{O}(h^{2p})$$

and insert this together with (2.11) into (2.10). Neglecting the $\mathcal{O}(h^2)$ and $\mathcal{O}(h^{2p})$ terms, we can interpret the obtained formula as the multistep method applied to

$$e'(x) = \frac{\partial f}{\partial y} \big(x, y(x)\big) e(x) + C y^{(p+1)}(x), \qquad e(x_0) = 0, \tag{2.12}$$

where

$$C = \frac{C_{p+1}}{\sigma(1)} \tag{2.13}$$

is seen to be a natural measure for the global error and is therefore called *the error constant*.

Another derivation of Definition (2.13) will be given in the section on global convergence (see Exercise 2 of Section III.4). Further, the solution of (2.12) gives the first term of the asymptotic expansion of the global error (see Section III.9).

Example 2.7. *Error constant of the explicit Adams methods.* Consider the differential equation $y' = f(x)$ with $f(x) = (k+1)x^k$, the exact solution of which is $y(x) = x^{k+1}$. As this differential equation is integrated exactly by the $(k+1)$-step explicit Adams method (see Example 2.5), we have

$$y(x_k) - y(x_{k-1}) = h \sum_{j=0}^{k} \gamma_j \nabla^j f_{k-1}.$$

The local error of the k-step explicit Adams method (1.5) is therefore given by

$$y(x_k) - y_k = h \gamma_k \nabla^k f_{k-1} = h^{k+1} \gamma_k f^{(k)}(x_0) = h^{k+1} \gamma_k y^{(k+1)}(x_0).$$

As $\gamma_k \neq 0$, this formula shows that the order of the k-step method is not greater than k (compare Example 2.5). Furthermore, since $\alpha_k = 1$, a comparison with formula (2.9) yields $C_{k+1} = \gamma_k$. Finally, for Adams methods we have $\varrho(\zeta) = \zeta^k - \zeta^{k-1}$ and $\varrho'(1) = 1$, so that by the use of (2.6) the error constant is given by $C = \gamma_k$.

The error constants of all other previously considered multistep methods are summarized in Table 2.1 (observe that $\sigma(1) = 2$ for explicit Nyström and Milne-Simpson methods).

Irreducible Methods

Let $\varrho(\zeta)$ and $\sigma(\zeta)$ of formula (2.4) be the generating polynomials of (2.1) and suppose that they have a common factor $\varphi(\zeta)$. Then the polynomials

$$\varrho^*(\zeta) = \frac{\varrho(\zeta)}{\varphi(\zeta)}, \qquad \sigma^*(\zeta) = \frac{\sigma(\zeta)}{\varphi(\zeta)},$$

are the generating polynomials of a new and simpler multistep method. Using the shift operator E, defined by

$$Ey_n = y_{n+1} \qquad \text{or} \qquad Ey(x) = y(x+h),$$

this multistep method can be written in compact form as

$$\varrho^*(E)y_n = h\sigma^*(E)f_n.$$

Multiplication by $\varphi(E)$ shows that any solution $\{y_n\}$ of this method is also a solution of $\varrho(E)y_n = h\sigma(E)f_n$. The two methods are thus essentially equal. Denote by L^* the difference operator associated with the new reduced method, and by C^*_{p+1} the constant given by (2.7). As

$$L(y, x, h) = \varphi(E)L^*(y, x, h) = C^*_{p+1}h^{p+1}\varphi(E)y^{(p+1)}(x) + \mathcal{O}(h^{p+2})$$
$$= C^*_{p+1}\varphi(1)h^{p+1}y^{(p+1)}(x) + \mathcal{O}(h^{p+2})$$

one immediately obtains $C_{p+1} = \varphi(1)C^*_{p+1}$ and therefore also the relation

$$C_{p+1}/\sigma(1) = C^*_{p+1}/\sigma^*(1)$$

holds. Both methods thus have the same error constant.

The above analysis has shown that multistep methods whose generating polynomials have a common factor are not interesting. We therefore usually assume that

$$\varrho(\zeta) \text{ and } \sigma(\zeta) \text{ have no common factor.} \tag{2.14}$$

Multistep methods satisfying this property are called *irreducible*.

The Peano Kernel of a Multistep Method

The order and the error constant above do not yet give a complete description of the error, since the subsequent terms of the series for the error may be much larger than C_{p+1}. Several attempts have therefore been made, originally for the error of a quadrature formula, to obtain a complete description of the error. The following discussion is an extension of the ideas of Peano (1913).

Theorem 2.8. *Let the multistep method (2.1) be of order p and let q $(1 \leq q \leq p)$ be an integer. For any $(q+1)$-times continuously differentiable function $y(x)$ we then have*

$$L(y, x, h) = h^{q+1} \int_0^k K_q(s) y^{(q+1)}(x + sh) \, ds, \qquad (2.15)$$

where

$$K_q(s) = \frac{1}{q!} \sum_{i=0}^{k} \alpha_i (i - s)_+^q - \frac{1}{(q-1)!} \sum_{i=0}^{k} \beta_i (i - s)_+^{q-1} \qquad (2.16a)$$

with

$$(i - s)_+^r = \begin{cases} (i - s)^r & \text{for } i - s > 0 \\ 0 & \text{for } i - s \leq 0. \end{cases}$$

$K_q(s)$ *is called the qth Peano kernel of the multistep method (2.1).*

Remark. We see from (2.16a) that $K_q(s)$ is a piecewise polynomial and satisfies

$$K_q(s) = \frac{1}{q!} \sum_{i=j}^{k} \alpha_i (i - s)^q - \frac{1}{(q-1)!} \sum_{i=j}^{k} \beta_i (i - s)^{q-1} \quad \text{for } s \in [j - 1, j).$$

$$(2.16b)$$

Proof. Taylor's theorem with the integral representation of the remainder yields

$$y(x + ih) = \sum_{r=0}^{q} \frac{i^r}{r!} h^r y^{(r)}(x) + h^{q+1} \int_0^i \frac{(i - s)^q}{q!} y^{(q+1)}(x + sh) \, ds,$$

$$hy'(x + ih) = \sum_{r=1}^{q} \frac{i^{r-1}}{(r-1)!} h^r y^{(r)}(x) + h^{q+1} \int_0^i \frac{(i - s)^{q-1}}{(q-1)!} y^{(q+1)}(x + sh) \, ds.$$

Inserting these two expressions into (2.3), the same considerations as in the proof of Theorem 2.4 show that for $q \leq p$ the polynomials before the integral cancel. The statement then follows from

$$\int_0^i \frac{(i - s)^q}{q!} y^{(q+1)}(x + sh) \, ds = \int_0^k \frac{(i - s)_+^q}{q!} y^{(q+1)}(x + sh) \, ds. \qquad \square$$

Besides the representation (2.16), the Peano kernel $K_q(s)$ has the following properties:

$$K_q(s) = 0 \text{ for } s \in (-\infty, 0) \cup [k, \infty) \text{ and } q = 1, \ldots, p; \tag{2.17}$$

$K_q(s)$ is $(q-2)$-times continuously differentiable and
$$K_q'(s) = -K_{q-1}(s) \text{ for } q = 2, \ldots, p \text{ (for } q = 2 \text{ piecewise)}; \tag{2.18}$$

$K_1(s)$ is a piecewise linear function with discontinuities at $0, 1, \ldots, k$. It has a jump of size β_j at the point j and its slope over the interval $(j-1, j)$ is given by $-(\alpha_j + \alpha_{j+1} + \ldots + \alpha_k)$;
$$\tag{2.19}$$

For the constant C_{p+1} of (2.8) we have $C_{p+1} = \int_0^k K_p(s)ds$. $\tag{2.20}$

The proofs of Statements (2.17) to (2.20) are as follows: it is an immediate consequence of the definition of the Peano kernel that $K_q(s) = 0$ for $s \geq k$ and $q \leq p$. In order to prove that $K_q(s) = 0$ also for $s < 0$ we consider the polynomial $y(x) = (x-s)^q$ with s as a parameter. Theorem 2.8 then shows that

$$L(y, 0, 1) = \sum_{i=0}^{k} \alpha_i (i-s)^q - q \sum_{i=0}^{k} \beta_i (i-s)^{q-1} \equiv 0 \quad \text{for } q \leq p$$

and hence $K_q(s) = 0$ for $s < 0$. This gives (2.17). The relation (2.18) is seen by partial integration of (2.15). As an example, the Peano kernels for the 3-step Nyström method (1.13") are plotted in Fig. 2.2.

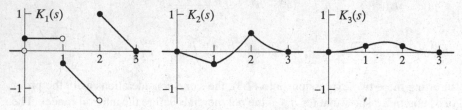

Fig. 2.2. Peano kernels of the 3-step Nyström method

Exercises

1. Construction of multistep methods. Let $\varrho(\zeta)$ be a kth degree polynomial satisfying $\varrho(1) = 0$.

   a) There exists exactly one polynomial $\sigma(\zeta)$ of degree $\leq k$, such that the order of the corresponding multistep method is at least $k+1$.

   b) There exists exactly one polynomial $\sigma(\zeta)$ of degree $< k$, such that the corresponding multistep method, which is then explicit, has order at least k.

   Hint. Use condition (iii) of Theorem 2.4.

2. Find the multistep method of the form
$$y_{n+2} + \alpha_1 y_{n+1} + \alpha_0 y_n = h(\beta_1 f_{n+1} + \beta_0 f_n)$$
   of the highest possible order. Apply this formula to the example $y' = y$, $y(0) = 1$, $h = 0.1$.

3. Verify that the order and the error constant of the BDF-formulas are those of Table 2.1.

4. Show that the Peano kernel $K_p(s)$ does not change sign for the explicit and implicit Adams methods, nor for the BDF-formulas. Deduce from this property that
$$L(y, x, h) = h^{p+1} C_{p+1} y^{(p+1)}(\zeta) \quad \text{with } \zeta \in (x, x + kh)$$
   where the constant C_{p+1} is given by (2.8).

5. Let $y(x)$ be an exact solution of $y' = f(x, y)$ and let $y_i = y(x_i)$, $i = 0, 1, \ldots, k - 1$. Assume that f is continuous and satisfies a Lipschitz condition with respect to y (f not necessarily differentiable). Prove that for consistent multistep methods (i.e., methods with (2.6)) the local error satisfies
$$\|y(x_k) - y_k\| \leq h\omega(h)$$
   where $\omega(h) \to 0$ for $h \to 0$.

III.3 Stability and the First Dahlquist Barrier

... hat der Verfasser seither öfters Verfahren zur numerischen Integration von Differentialgleichungen beobachtet, die, obschon zwar mit bestechend kleinem Abbruchfehler behaftet, doch die grosse Gefahr der numerischen Instabilität in sich bergen.

(H. Rutishauser 1952)

Rutishauser observed in his famous paper that high order and a small local error are not sufficient for a useful multistep method. The numerical solution can be "unstable", even though the step size h is taken very small. The same observation was made by Todd (1950), who applied certain difference methods to second order differential equations. Our presentation will mainly follow the lines of Dahlquist (1956), where this effect has been studied systematically. An interesting presentation of the historical development of numerical stability concepts can be found in Dahlquist (1985) "33 years of numerical instability, Part I".

Let us start with an example, taken from Dahlquist (1956). Among all explicit 2-step methods we consider the formula with the highest order (see Exercise 2 of Section III.2). A short calculation using Theorem 2.4 shows that this method of order 3 is given by

$$y_{n+2} + 4y_{n+1} - 5y_n = h(4f_{n+1} + 2f_n). \tag{3.1}$$

Application to the differential equation

$$y' = y, \qquad\qquad y(0) = 1 \tag{3.2}$$

yields the linear difference relation

$$y_{n+2} + 4(1-h)y_{n+1} - (5+2h)y_n = 0. \tag{3.3}$$

As starting values we take $y_0 = 1$ and $y_1 = \exp(h)$, the values on the exact solution. The numerical solution together with the exact solution $\exp(x)$ is plotted in Fig. 3.1 for the step sizes $h = 1/10$, $h = 1/20$, $h = 1/40$, etc. In spite of the small local error, the results are very bad and become even worse as the step size decreases.

An explanation for this effect can easily be given. As usual for linear difference equations (Dan. Bernoulli 1728, Lagrange 1775), we insert $y_j = \zeta^j$ into (3.3). This leads to the characteristic equation

$$\zeta^2 + 4(1-h)\zeta - (5+2h) = 0. \tag{3.4}$$

The general solution of (3.3) is then given by

$$y_n = A\zeta_1^n(h) + B\zeta_2^n(h) \tag{3.5}$$

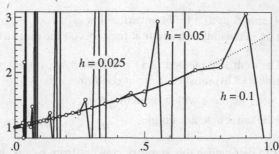

Fig. 3.1. Numerical solution of the unstable method (3.1)

where

$$\zeta_1(h) = 1 + h + \mathcal{O}(h^2), \qquad \zeta_2(h) = -5 + \mathcal{O}(h)$$

are the roots of (3.4) and the coefficients A and B are determined by the starting values y_0 and y_1. Since $\zeta_1(h)$ approximates $\exp(h)$, the first term in (3.5) approximates the exact solution $\exp(x)$ at the point $x = nh$. The second term in (3.5), often called a *parasitic solution,* is the one which causes trouble in our method: since for $h \to 0$ the absolute value of $\zeta_2(h)$ is larger than one, this parasitic solution becomes very large and dominates the solution y_n for increasing n.

We now turn to the stability discussion of the general method (2.1). The essential part is the behaviour of the solution as $n \to \infty$ (or $h \to 0$) with nh fixed. We see from (3.3) that for $h \to 0$ we obtain

$$\alpha_k y_{n+k} + \alpha_{k-1} y_{n+k-1} + \ldots + \alpha_0 y_n = 0. \tag{3.6}$$

This can be interpreted as the numerical solution of the method (2.1) for the differential equation

$$y' = 0. \tag{3.7}$$

We put $y_j = \zeta^j$ in (3.6), divide by ζ^n, and find that ζ must be a root of

$$\varrho(\zeta) = \alpha_k \zeta^k + \alpha_{k-1} \zeta^{k-1} + \ldots + \alpha_0 = 0. \tag{3.8}$$

As in Section I.13, we again have some difficulty when (3.8) possesses a root of *multiplicity* $m > 1$. In this case (Lagrange 1792, see Exercise 1 below) $y_n = n^{j-1} \zeta^n$ $(j = 1, \ldots, m)$ are solutions of (3.6) and we obtain by superposition:

Lemma 3.1. *Let $\zeta_1, \ldots, \zeta_l$ be the roots of $\varrho(\zeta)$, of respective multiplicity $m_1, \ldots, m_l$. Then the general solution of (3.6) is given by*

$$y_n = p_1(n)\zeta_1^n + \ldots + p_l(n)\zeta_l^n \tag{3.9}$$

where the $p_j(n)$ are polynomials of degree $m_j - 1$. □

Formula (3.9) shows us that for boundedness of y_n, as $n \to \infty$, we need that the roots of (3.8) lie in the unit disc and that the roots on the unit circle be simple.

Definition 3.2. The multistep method (2.1) is called *stable*, if the generating polynomial $\varrho(\zeta)$ (formula (3.8)) satisfies the *root condition*, i.e.,

i) The roots of $\varrho(\zeta)$ lie on or within the unit circle;

ii) The roots on the unit circle are simple.

Remark. In order to distinguish this stability concept from others, it is sometimes called *zero-stability* or, in honour of Dahlquist, also *D-stability*.

Examples. For the explicit and implicit *Adams methods*, $\varrho(\zeta) = \zeta^k - \zeta^{k-1}$. Besides the simple root 1, there is a $(k-1)$-fold root at 0. The Adams methods are therefore stable.

The same is true for the explicit *Nyström* and the *Milne-Simpson methods*, where $\varrho(\zeta) = \zeta^k - \zeta^{k-2}$. Note that here we have a simple root at -1. This root can be dangerous for certain differential equations (see Section III.9 and Section V.1 of Volume II).

Stability of the BDF-Formulas

The investigation of the stability of the BDF-formulas is more difficult. As the characteristic polynomial of $\nabla^j y_{k+n} = 0$ is given by $\zeta^{k-j}(\zeta - 1)^j = 0$ it follows from the representation (1.22') that the generating polynomial $\varrho(\zeta)$ of the BDF-formulas has the form

$$\varrho(\zeta) = \sum_{j=1}^{k} \frac{1}{j} \zeta^{k-j} (\zeta - 1)^j. \tag{3.10}$$

In order to study the zeros of (3.10) it is more convenient to consider the polynomial

$$p(z) = (1-z)^k \varrho\left(\frac{1}{1-z}\right) = \sum_{j=1}^{k} \frac{z^j}{j} \tag{3.11}$$

via the transformation $\zeta = 1/(1-z)$. This polynomial is just the kth partial sum of $-\log(1-z)$. As the roots of $p(z)$ and $\varrho(\zeta)$ are related by the above transformation, we have:

Lemma 3.3. *The k-step BDF-formula (1.22') is stable iff all roots of the polynomial (3.11) are outside the disc $\{z; |z - 1| \le 1\}$, with simple roots allowed on the boundary.* □

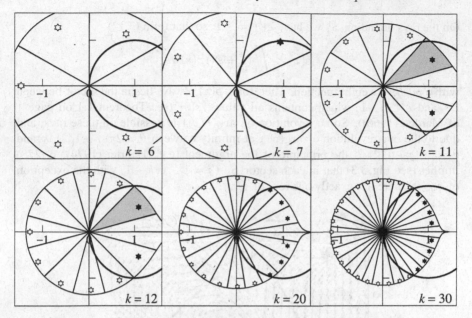

Fig. 3.2. Roots of the polynomial $p(z)$ of (3.11)

The roots of (3.11) are displayed in Fig. 3.2 for different values of k.

Theorem 3.4. *The k-step BDF-formula (1.22') is stable for $k \le 6$, and unstable for $k \ge 7$.*

Proof. The first assertion can be verified simply by a finite number of numerical calculations (see Fig. 3.2). This was first observed by Mitchell & Craggs (1953). The second statement, however, contains an infinity of cases and is more difficult. The first complete proof was given by Cryer (1971) in a technical report, a condensed version of which is published in Cryer (1972). A second proof is given in Creedon & Miller (1975) (see also Grigorieff (1977), p. 135), based on the Schur-Cohn criterion. This proof is outlined in Exercise 4 below. The following proof, which is given in Hairer & Wanner (1983), is based on the representation

$$p(z) = \int_0^z \sum_{j=1}^k \zeta^{j-1} d\zeta = \int_0^z \frac{1-\zeta^k}{1-\zeta}\, d\zeta = \int_0^r \left(1 - e^{ik\theta} s^k\right)\varphi(s)\, ds \qquad (3.12)$$

with

$$\zeta = s e^{i\theta}, \qquad z = r e^{i\theta}, \qquad \varphi(s) = \frac{e^{i\theta}}{1 - s e^{i\theta}}.$$

We cut the complex plane into k sectors

$$S_j = \left\{ z \;;\; \frac{2\pi}{k}\left(j - \frac{1}{2}\right) < \arg(z) < \frac{2\pi}{k}\left(j + \frac{1}{2}\right) \right\}, \qquad j = 0, 1, \dots, k-1.$$

On the rays bounding S_j we have $e^{ik\theta} = -1$, so that from (3.12)

$$p(z) = \int_0^r (1 + s^k)\varphi(s)\, ds$$

with a *positive* weight function. Therefore, $p(z)$ always lies in the sector between $e^{i\theta}$ and $e^{i\pi} = -1$, which contains all values $\varphi(s)$ (see Theorem 1.1 on page 1 of Marden (1966)). So no revolution of $\arg(p(z))$ is possible on these rays, and due to the one revolution of $\arg(z^k)$ at infinity between $\theta = 2\pi(j - 1/2)/k$ and $\theta = 2\pi(j + 1/2)/k$ the principle of the argument (e.g., Henrici (1974), p. 278) implies (see Fig. 3.3) that in each sector S_j $(j = 1, \ldots, k-1$, with the exception of $j = 0$) there lies exactly one root of $p(z)$.

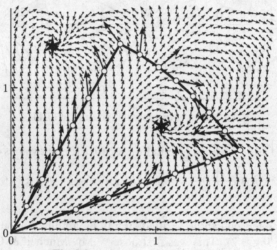

Fig. 3.3. Argument of $p(z)$ of (3.11)

In order to complete the proof, we still have to bound the zeros of $p(z)$ from above: we observe that in (3.12) the term s^k becomes large for $s > 1$. We therefore partition (3.12) into two integrals $p(z) = I_1 - I_2$, where

$$I_1 = \int_0^r \varphi(s)\, ds - \int_0^1 e^{ik\theta} s^k \varphi(s)\, ds, \qquad I_2 = e^{ik\theta} \int_1^r s^k \varphi(s)\, ds.$$

Since $|\varphi(s)| \le B(\theta)$ where

$$B(\theta) = \begin{cases} |\sin\theta|^{-1} & \text{if } 0 < \theta \le \pi/2 \text{ or } 3\pi/2 \le \theta < 2\pi, \\ 1 & \text{otherwise,} \end{cases}$$

we obtain

$$|I_1| \le \left(r + \frac{1}{k+1}\right) B(\theta) < r B(\theta) \frac{k+2}{k+1}, \qquad (r > 1). \tag{3.13}$$

Secondly, since s^k is positive,

$$I_2 = e^{ik\theta} \Phi \int_1^r s^k \, ds \quad \text{with} \quad \Phi \in \text{ convex hull of } \{\varphi(s); \, 1 \le s \le r\}.$$

Any element of the above convex hull can be written in the form

$$\Phi = \alpha\varphi(s_1) + (1 - \alpha)\varphi(s_2) = \frac{\varphi(s_1)\varphi(s_2)}{\varphi(\widehat{s})}$$

with $\widehat{s} = \alpha s_2 + (1-\alpha)s_1$, $0 \le \alpha \le 1$, $1 \le s_1, s_2 \le r$. Since $|\varphi(s)|$ decreases monotonically for $s \ge 1$, we have $|\Phi| \ge |\varphi(r)|$. Some elementary geometry then leads to $|\Phi| \ge 1/2r$ and we get

$$|I_2| \ge \frac{r^{k+1} - 1}{2r(k+1)} > \frac{r(r^{k-1} - 1)}{2k+2}, \quad (r > 1). \tag{3.14}$$

From (3.13) and (3.14) we see that

$$r \ge R(\theta) = \left((2k+4)B(\theta) + 1\right)^{1/(k-1)} \tag{3.15}$$

implies $|I_2| > |I_1|$, so that $p(z)$ cannot be zero. The curve $R(\theta)$ is also plotted in Fig. 3.2 and cuts from the sectors S_j what we call Madame Imhof's cheese pie, each slice of which (with $j \ne 0$) must contain precisely one zero of $p(z)$. A simple analysis shows that for $k = 12$ the cheese pie, cut from S_1, is small enough to ensure the presence of zeros of $p(z)$ inside the disc $\{z; |z - 1| \le 1\}$. As $R(\theta)$, for fixed θ, as well as $R(\pi/k)$ are monotonically decreasing in k, the same is true for all $k \ge 12$.

For $6 < k < 12$ numerical calculations show that the method is unstable (see Fig. 3.2 or Exercise 4). $\qquad\qquad\qquad\qquad\qquad\qquad\qquad\qquad\qquad\qquad\qquad\quad\square$

Highest Attainable Order of Stable Multistep Methods

It is a natural task to investigate the stability of the multistep methods with highest possible order. This has been performed by Dahlquist (1956), resulting in the famous "first Dahlquist-barrier".

Counting the order conditions (Theorem 2.4) shows that for order p the parameters of a linear multistep method have to satisfy $p+1$ linear equations. As $2k+1$ free parameters are involved (without loss of generality one can assume $\alpha_k = 1$), this suggests that $2k$ is the highest attainable order. Indeed, this can be verified (see Exercise 5). However, these methods are of no practical significance, because we shall prove

Theorem 3.5 (The first Dahlquist-barrier). *The order p of a stable linear k-step method satisfies*

$$
\begin{aligned}
&p \leq k+2 \quad \textit{if } k \textit{ is even,}\\
&p \leq k+1 \quad \textit{if } k \textit{ is odd,}\\
&p \leq k \qquad \textit{if } \beta_k/\alpha_k \leq 0 \textit{ (in particular if the method is explicit).}
\end{aligned}
$$

We postpone the verification of this theorem and give some notations and lemmas, which will be useful for the proof. First of all we introduce the "Greek-Roman transformation"

$$
\zeta = \frac{z+1}{z-1} \qquad \text{or} \qquad z = \frac{\zeta+1}{\zeta-1}. \tag{3.16}
$$

This transformation maps the disk $|\zeta| < 1$ onto the half-plane $\mathrm{Re}\, z < 0$, the upper half-plane $\mathrm{Im}\, z > 0$ onto the lower half-plane, the circle $|\zeta| = 1$ to the imaginary axis, the point $\zeta = 1$ to $z = \infty$ and the point $\zeta = -1$ to $z = 0$. We then consider the polynomials

$$
R(z) = \left(\frac{z-1}{2}\right)^k \varrho(\zeta) = \sum_{j=0}^{k} a_j z^j,
$$

$$
\tag{3.17}
$$

$$
S(z) = \left(\frac{z-1}{2}\right)^k \sigma(\zeta) = \sum_{j=0}^{k} b_j z^j.
$$

Since the zeros of $R(z)$ and of $\varrho(\zeta)$ are connected via the transformation (3.16), the stability condition of a multistep method can be formulated in terms of $R(z)$ as follows: all zeros of $R(z)$ lie in the negative half-plane $\mathrm{Re}\, z \leq 0$ and no multiple zero of $R(z)$ lies on the imaginary axis.

Lemma 3.6. *Suppose the multistep method to be stable and of order at least 0. We then have*

i) $a_k = 0$ *and* $a_{k-1} = 2^{1-k}\varrho'(1) \neq 0$;

ii) *All non-vanishing coefficients of $R(z)$ have the same sign.*

Proof. Dividing formula (3.17) by z^k and putting $z = \infty$, one sees that $a_k = 2^{-k}\varrho(1)$. This expression must vanish, because the method is of order 0. In the same way one gets $a_{k-1} = 2^{1-k}\varrho'(1)$, which is different from zero, since by stability 1 cannot be a multiple root of $\varrho(\zeta)$. The second statement follows from the factorization

$$
R(z) = a_{k-1} \prod (z+x_j) \prod ((z+u_j)^2 + v_j^2).
$$

where $-x_j$ are the real roots and $-u_j \pm iv_j$ are the conjugate pairs of complex roots. By stability $x_j \geq 0$ and $u_j \geq 0$, implying that all coefficients of $R(z)$ have the same sign. $\qquad\square$

We next express the order conditions of Theorem 2.4 in terms of the polynomials $R(z)$ and $S(z)$.

Lemma 3.7. *The multistep method is of order p if and only if*

$$R(z)\left(\log\frac{z+1}{z-1}\right)^{-1} - S(z) = C_{p+1}\left(\frac{2}{z}\right)^{p-k} + \mathcal{O}\left(\left(\frac{2}{z}\right)^{p-k+1}\right) \quad \textit{for } z \to \infty \tag{3.18}$$

Proof. First, observe that the $\mathcal{O}((\zeta-1)^p)$ term in condition (iii) of Theorem 2.4 is equal to $C_{p+1}(\zeta-1)^p + \mathcal{O}((\zeta-1)^{p+1})$ by formula (2.7). Application of the transformation (3.16) then yields (3.18), because $(\zeta-1) = 2/(z-1) = 2/z + \mathcal{O}((2/z)^2)$ for $z \to \infty$. □

Lemma 3.8. *The coefficients of the Laurent series*

$$\left(\log\frac{z+1}{z-1}\right)^{-1} = \frac{z}{2} - \mu_1 z^{-1} - \mu_3 z^{-3} - \mu_5 z^{-5} - \dots \tag{3.19}$$

satisfy $\mu_{2j+1} > 0$ for all $j \geq 0$.

Proof. We consider the branch of $\log\zeta$ which is analytic in the complex ζ-plane cut along the negative real axis and satisfies $\log 1 = 0$. The transformation (3.16) maps this cut onto the segment from -1 to $+1$ on the real axis. The function $\log((z+1)/(z-1))$ is thus analytic on the complex z-plane cut along this segment (see Fig. 3.4). From the formula

$$\log\frac{z+1}{z-1} = \frac{2}{z}\left(1 + \frac{z^{-2}}{3} + \frac{z^{-4}}{5} + \frac{z^{-6}}{7} + \dots\right), \tag{3.20}$$

the existence of (3.19) becomes clear. In order to prove the positivity of the coefficients, we use Cauchy's formula for the coefficients of the function $f(z) = \sum_{n\in\mathbb{Z}} a_n (z - z_0)^n$,

$$a_n = \frac{1}{2\pi i}\int_\gamma \frac{f(z)}{(z - z_0)^{n+1}}\, dz,$$

i.e., in our situation

$$\mu_{2j+1} = -\frac{1}{2\pi i}\int_\gamma z^{2j}\left(\log\frac{z+1}{z-1}\right)^{-1} dz$$

(Cauchy 1831; see also Behnke & Sommer 1962). Here γ is an arbitrary curve enclosing the segment $(-1, 1)$, e.g., the curve plotted in Fig. 3.4.

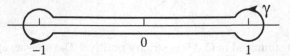

Fig. 3.4. Cut z-plane with curve γ

Observing that $\log((z+1)/(z-1)) = \log((1+x)/(1-x)) - i\pi$ when z approaches the real value $x \in (-1, 1)$ from above, and that $\log((z+1)/(z-1)) = \log((1+x)/(1-x)) + i\pi$ when z approaches x from below, we obtain

$$\mu_{2j+1} = -\frac{1}{2\pi i} \int_{-1}^{1} x^{2j} \left[\left(\log \frac{1+x}{1-x} + i\pi \right)^{-1} - \left(\log \frac{1+x}{1-x} - i\pi \right)^{-1} \right] dx$$

$$= \int_{-1}^{1} x^{2j} \left[\left(\log \frac{1+x}{1-x} \right)^2 + \pi^2 \right]^{-1} dx > 0. \qquad \square$$

For another proof of this lemma, which avoids complex analysis, see Exercise 10.

Proof of Theorem 3.5. We insert the series (3.19) into (3.18) and obtain

$$R(z) \left(\log \frac{z+1}{z-1} \right)^{-1} - S(z) = \text{polynomial}(z) + d_1 z^{-1} + d_2 z^{-2} + \mathcal{O}(z^{-3}) \quad (3.21)$$

where

$$\begin{aligned} d_1 &= -\mu_1 a_0 - \mu_3 a_2 - \mu_5 a_4 - \dots \\ d_2 &= -\mu_3 a_1 - \mu_5 a_3 - \mu_7 a_5 - \dots. \end{aligned} \quad (3.22)$$

Lemma 3.6 together with the positivity of the μ_j (Lemma 3.8) implies that all summands in the above formulas for d_1 and d_2 have the same sign. Since $a_{k-1} \neq 0$ we therefore have $d_2 \neq 0$ for k even and $d_1 \neq 0$ for k odd. The first two bounds of Theorem 3.5 are now an immediate consequence of formula (3.18).

Finally, we prove that $p \leq k$ for $\beta_k / \alpha_k \leq 0$: assume, by contradiction, that the order is greater than k. Then by formula (3.18), $S(z)$ is equal to the principal part of $R(z)(\log((z+1)/(z-1)))^{-1}$, and we may write (putting $\mu_j = 0$ for even j)

$$S(z) = R(z) \left(\frac{z}{2} - \sum_{j=1}^{k-1} \mu_j z^{-j} \right) + \sum_{j=1}^{k-1} \left(\sum_{s=j}^{k-1} \mu_s a_{s-j} \right) z^{-j}.$$

Setting $z = 1$ we obtain

$$\frac{S(1)}{R(1)} = \left(\frac{1}{2} - \sum_{j=1}^{k-1} \mu_j \right) + \sum_{j=1}^{k-1} \left(\sum_{s=j}^{k-1} \mu_s a_{s-j} \right) \frac{1}{R(1)}. \quad (3.23)$$

Since by formula (3.17), $S(1) = \beta_k$ and $R(1) = \alpha_k$, it is sufficient to prove $S(1)/R(1) > 0$. Formula (3.19), for $z \to 1$, gives

$$\sum_{j=1}^{\infty} \mu_j = \frac{1}{2},$$

so that the first summand in (3.23) is strictly positive. The non-negativeness of the second summand is seen from Lemmas 3.6 and 3.8. $\qquad \square$

The stable multistep methods which attain the highest possible order $k+2$ have a very special structure.

Theorem 3.9. *Stable multistep methods of order $k+2$ are symmetric, i.e.,*

$$\alpha_j = -\alpha_{k-j}, \qquad \beta_j = \beta_{k-j} \qquad \text{for all } j. \tag{3.24}$$

Remark. For symmetric multistep methods we have $\varrho(\zeta) = -\zeta^k \varrho(1/\zeta)$ by definition. Since with ζ_i also $1/\zeta_i$ is a zero of $\varrho(\zeta)$, all roots of stable symmetric multistep methods lie on the unit circle and are simple.

Proof. A comparison of the formulas (3.18) and (3.21) shows that $d_1 = 0$ is necessary for order $k+2$. Since the method is assumed to be stable, Lemma 3.6 implies that all even coefficients of $R(z)$ vanish. Hence, k is even and $R(z)$ satisfies the relation $R(z) = -R(-z)$. By definition of $R(z)$ this relation is equivalent to $\varrho(\zeta) = -\zeta^k \varrho(1/\zeta)$, which implies the first condition of (3.24). Using the above relation for $R(z)$ one obtains from formula (3.18) that $S(z) - S(-z) = \mathcal{O}((2/z)^2)$, implying $S(z) = S(-z)$. If this relation is transformed into an equivalent one for $\sigma(\zeta)$, one gets the second condition of (3.24). $\qquad\square$

Exercises

1. Consider the linear difference equation (3.6) with

$$\varrho(\zeta) = \alpha_k \zeta^k + \alpha_{k-1}\zeta^{k-1} + \ldots + \alpha_0$$

   as characteristic polynomial. Let $\zeta_1, \ldots, \zeta_l$ be the different roots of $\varrho(\zeta)$ and let $m_j \geq 1$ be the multiplicity of the root ζ_j. Show that for $1 \leq j \leq l$ and $0 \leq i \leq m_j - 1$ the sequences

$$\left\{ \binom{n}{i} \zeta_j^{n-i} \right\}_{n \geq 0}$$

   form a system of k linearly independent solutions of (3.6).

2. Show that all roots of the polynomial $p(z)$ of formula (3.11) except the simple root 0 lie in the annulus

$$\frac{k}{k-1} \leq |z| \leq 2.$$

   Hint. Use the following lemma, which can be found in Marden (1966), p.137: if all coefficients of the polynomial $a_k z^k + a_{k-1}z^{k-1} + \ldots + a_0$ are real and positive, then its roots lie in the annulus $\varrho_1 \leq |z| \leq \varrho_2$ with $\varrho_1 = \min(a_j/a_{j+1})$ and $\varrho_2 = \max(a_j/a_{j+1})$.

3. Apply the lemma of the above exercise to $\varrho(\zeta)/(\zeta - 1)$ and show that the BDF-formulas are stable for $k = 1, 2, 3, 4$.

4. Give a different proof of Theorem 3.4 by applying the Schur-Cohn criterion to the polynomial

$$f(z) = z^k \varrho\left(\frac{1}{z}\right) = \sum_{j=1}^{k} \frac{1}{j}(1-z)^j. \tag{3.25}$$

Schur-Cohn criterion (see e.g., Marden (1966), Chapter X). For a given polynomial with real coefficients

$$f(z) = a_0 + a_1 z + \ldots + a_k z^k$$

we consider the coefficients $a_i^{(j)}$ where

$$
\begin{aligned}
a_i^{(0)} &= a_i & i &= 0, 1, \ldots, k \\
a_i^{(j+1)} &= a_0^{(j)} a_i^{(j)} - a_{k-j}^{(j)} a_{k-j-i}^{(j)} & i &= 0, 1, \ldots, k-j-1
\end{aligned}
\tag{3.26}
$$

and also the products

$$P_1 = a_0^{(1)}, \qquad P_{j+1} = P_j a_0^{(j+1)} \quad \text{for } j = 1, \ldots, k-1. \tag{3.27}$$

We further denote by n the number of negative elements among the values $P_1, \ldots, P_k$ and by p the number of positive elements. Then $f(z)$ has at least n zeros inside the unit disk and at least p zeros outside it.

a) Prove the following formulas for the coefficients of (3.25):

$$a_0 = \sum_{i=1}^{k} \frac{1}{i}, \qquad\qquad a_1 = -k, \qquad\qquad a_2 = \frac{k(k-1)}{4},$$

$$a_{k-2} = (-1)^k \frac{k(k-1)}{2(k-2)}, \quad a_{k-1} = (-1)^{k-1} \frac{k}{k-1}, \quad a_k = (-1)^k \frac{1}{k}. \tag{3.28}$$

b) Verify that the coefficients $a_0^{(j)}$ of (3.26) have the sign structure of Table 3.1. For $k < 13$ these tedious calculations can be performed on a computer. The verification of $a_0^{(1)} > 0$ and $a_0^{(2)} > 0$ is easy for all $k > 2$. In order to verify $a_0^{(3)} = (a_0^{(2)})^2 - (a_{k-2}^{(2)})^2 < 0$ for $k \geq 13$ consider the expression

$$
\begin{aligned}
a_0^{(2)} - (-1)^k a_{k-2}^{(2)} = {}& a_0^{(1)}\left(a_0^2 - a_k^2 - a_0|a_{k-2}| + a_2|a_k|\right) \\
& - |a_{k-1}^{(1)}| \cdot (a_0 + |a_k|)(|a_{k-1}| + a_1)
\end{aligned}
\tag{3.29}
$$

Table 3.1. Signs of $a_0^{(j)}$.

k	2	3	4	5	6	7	8	9	10	11	12	13	> 13
$j=1$	+	+	+	+	+	+	+	+	+	+	+	+	+
$j=2$	0	+	+	+	+	+	+	+	+	+	+	+	+
$j=3$		0	+	+	+	+	+	+	+	+	+	−	−
$j=4$			0	+	+	+	−	−	−	−	−	−	−
$j=5$				0	+	−							

which can be written in the form $(a_0 + |a_k|)\varphi(k)$ with

$$\varphi(k) = (a_0 - |a_k|)\left(a_0^2 - a_k^2 - a_0|a_{k-2}| + a_2|a_k|\right) - |a_{k-1}^{(1)}|(a_1 + |a_{k-1}|)$$

$$= a_0^3 - a_0^2\left(\frac{k}{2} + \frac{1}{2} + \frac{1}{k-2} + \frac{1}{k}\right)$$

$$+ a_0\left(\frac{5k}{4} + \frac{1}{4} + \frac{1}{2k-4} - \frac{1}{k-1} - \frac{1}{(k-1)^2} - \frac{1}{k^2}\right)$$

$$- \left(k - \frac{3}{4} - \frac{1}{k-1} - \frac{1}{4k} - \frac{1}{k^3}\right).$$

Show that $\varphi(13) < 0$ and that φ is monotonically decreasing for $k \geq 13$ (observe that $a_0 = a_0(k)$ actually depends on k and that $a_0(k+1) = a_0(k) + 1/(k+1)$). Finally, deduce from the negativeness of (3.29) that $a_0^{(3)} < 0$ for $k \geq 13$.

c) Use Table 3.1 and the Schur-Cohn criterion for the verification of Theorem 3.4.

5. (Multistep methods of maximal order). Verify the following statements:

  a) there is no k-step method of order $2k+1$,

  b) there is a unique (implicit) k-step method of order $2k$,

  c) there is a unique explicit k-step method of order $2k-1$.

6. Prove that symmetric multistep methods are always of even order. More precisely, if a symmetric multistep method is of order $2s-1$ then it is also of order $2s$.

7. Show that all stable 4-step methods of order 6 are given by

$$\varrho(\zeta) = (\zeta^2 - 1)(\zeta^2 + 2\mu\zeta + 1), \qquad |\mu| < 1,$$

$$\sigma(\zeta) = \frac{1}{45}(14 - \mu)(\zeta^4 + 1) + \frac{1}{45}(64 + 34\mu)\zeta(\zeta^2 + 1) + \frac{1}{15}(8 + 38\mu)\zeta^2.$$

Compute the error constant and observe that it cannot become arbitrarily small.

Result. $C = -(16 - 5\mu)/(7560(1 + \mu))$.

8. Prove the following bounds for the error constant:

   a) For stable methods of order $k + 2$
   $$C \le -2^{-1-k} \mu_{k+1}.$$

   b) For stable methods of order $k + 1$ with odd k we have
   $$C \le '-2^{-k} \mu_k.$$

   c) For stable explicit methods of order k we have ($\mu_j = 0$ for even j)
   $$C \ge 2^{1-k} \left(\frac{1}{2} - \sum_{j=1}^{k-1} \mu_j \right).$$

   Show that all these bounds are optimal.

   Hint. Compare the formulas (3.18) and (3.21) and use the relation $\sigma(1) = 2^{k-1} a_{k-1}$ of Lemma 3.6.

9. The coefficients μ_j of formula (3.19) satisfy the recurrence relation
   $$\mu_{2j+1} + \frac{1}{3} \mu_{2j-1} + \ldots + \frac{1}{2j+1} \mu_1 = \frac{1}{4j+6}. \tag{3.30}$$

   The first of these coefficients are given by
   $$\mu_1 = \frac{1}{6}, \quad \mu_3 = \frac{2}{45}, \quad \mu_5 = \frac{22}{945}, \quad \mu_7 = \frac{214}{14175}.$$

10. Another proof of Lemma 3.8: multiplying (3.30) by $2j + 3$ and subtracting from it the same formula with j replaced by $j - 1$ yields
    $$(2j+3) \mu_{2j+1} + \sum_{i=0}^{j-1} \mu_{2i+1} \left(\frac{2j+3}{2j-2i+1} - \frac{2j+1}{2j-2i-1} \right) = 0.$$

    Show that the expression in brackets is negative and deduce the result of Lemma 3.8 by a simple induction argument.

III.4 Convergence of Multistep Methods

..., ist das Adams'sche Verfahren jedem andern bedeutend überlegen. Wenn es gleichwohl nicht genügend allgemein angewandt wird und, besonders in Deutschland, gegenüber den von Runge, Heun und Kutta entwickelten Methoden zurücktritt, so mag dies daran liegen, dass bisher eine brauchbare Untersuchung der Genauigkeit der Adams'schen Integration gefehlt hat. Diese Lücke soll hier ausgefüllt, werden, ...

(R. v. Mises 1930)

The convergence of Adams methods was investigated in the influential article of von Mises (1930), which was followed by an avalanche of papers improving the error bounds and applying the ideas to other special multistep methods, e.g., Tollmien (1938), Fricke (1949), Weissinger (1950), Vietoris (1953). A general convergence proof for the method (2.1), however, was first given by Dahlquist (1956), who gave necessary and sufficient conditions for convergence. Great elegance was introduced in the proofs by the ideas of Butcher (1966), where multistep formulas are written as one-step formulas in a higher dimensional space. Furthermore, the resulting presentation can easily be extended to a more general class of integration methods (see Section III.8).

We cannot expect reasonable convergence of numerical methods, if the differential equation problem

$$y' = f(x, y), \qquad y(x_0) = y_0 \tag{4.1}$$

does not possess a unique solution. We therefore make the following assumptions, which were seen in Sections I.7 and I.9 to be natural for our purpose:

$$f \text{ is continuous on } D = \{(x, y) \, ; \, x \in [x_0, \widehat{x}], \|y(x) - y\| \le b\} \tag{4.2a}$$

where $y(x)$ denotes the exact solution of (4.1) and b is some positive number. We further assume that f satisfies a Lipschitz condition, i.e.,

$$\|f(x, y) - f(x, z)\| \le L\|y - z\| \quad \text{for } (x, y), (x, z) \in D. \tag{4.2b}$$

If we apply the multistep method (2.1) with step size h to the problem (4.1) we obtain a sequence $\{y_i\}$. For given x and h such that $(x - x_0)/h = n$ is an integer, we introduce the following notation for the numerical solution:

$$y_h(x) = y_n \quad \text{if } x - x_0 = nh. \tag{4.3}$$

Definition 4.1 (Convergence). i) The linear multistep method (2.1) is called *convergent*, if for all initial value problems (4.1) satisfying (4.2),

$$y(x) - y_h(x) \to 0 \qquad \text{for } h \to 0, \ x \in [x_0, \widehat{x}]$$

whenever the starting values satisfy

$$y(x_0 + ih) - y_h(x_0 + ih) \to 0 \qquad \text{for } h \to 0, \ i = 0, 1, \ldots, k-1.$$

ii) Method (2.1) is *convergent of order p*, if to any problem (4.1) with f sufficiently differentiable, there exists a positive h_0 such that

$$\|y(x) - y_h(x)\| \le C h^p \qquad \text{for } h \le h_0$$

whenever the starting values satisfy

$$\|y(x_0 + ih) - y_h(x_0 + ih)\| \le C_0 h^p \qquad \text{for } h \le h_0, \ i = 0, 1, \ldots, k-1.$$

In this definition we clearly assume that a solution of (4.1) exists on $[x_0, \widehat{x}]$.

The aim of this section is to prove that stability together with consistency are necessary and sufficient for the convergence of a multistep method. This is expressed in the famous slogan

$$\text{convergence} = \text{stability} + \text{consistency}$$

(compare also Lax & Richtmyer 1956). We begin with the study of necessary conditions for convergence.

Theorem 4.2. *If the multistep method (2.1) is convergent, then it is necessarily*

> i)   *stable and*
> ii) *consistent (i.e. of order 1: $\varrho(1) = 0$, $\varrho'(1) = \sigma(1)$).*

Proof. Application of the multistep method (2.1) to the differential equation $y' = 0$, $y(0) = 0$ yields the difference equation (3.6). Suppose, by contradiction, that $\varrho(\zeta)$ has a root ζ_1 with $|\zeta_1| > 1$, or a root ζ_2 on the unit circle whose multiplicity exceeds 1. ζ_1^n and $n\zeta_2^n$ are then divergent solutions of (3.6). Multiplying by $\sqrt{h}$ we achieve that the starting values converge to $y_0 = 0$ for $h \to 0$. Since $y_h(x) = \sqrt{h}\,\zeta_1^{x/h}$ and $y_h(x) = (x/\sqrt{h})\,\zeta_2^{x/h}$ remain divergent for every fixed x, we have a contradiction to the assumption of convergence. The method (2.1) must therefore be stable.

We next consider the initial value problem $y' = 0$, $y(0) = 1$ with exact solution $y(x) = 1$. The corresponding difference equation is again that of (3.6), which, in the new notation, can be written as

$$\alpha_k y_h(x + kh) + \alpha_{k-1} y_h(x + (k-1)h) + \ldots + \alpha_0 y_h(x) = 0.$$

Letting $h \to 0$, convergence immediately implies that $\varrho(1) = 0$.

Finally we apply method (2.1) to the problem $y' = 1$, $y(0) = 0$. The exact solution is $y(x) = x$. Since we already know that $\varrho(1) = 0$, it is easy to verify that a particular numerical solution is given by $y_n = nhK$ or $y_h(x) = xK$ where $K = \sigma(1)/\varrho'(1)$. By convergence, $K = 1$ is necessary. $\qquad\square$

Although the statement of Theorem 4.2 was derived from a consideration of almost trivial differential equations, it is remarkable that conditions (i) and (ii) turn out to be not only necessary but also sufficient for convergence.

Formulation as One-Step Method

We are now at the point where it is useful to rewrite a multistep method as a one-step method in a higher dimensional space (see Butcher 1966, Skeel 1976). For this let $\psi = \psi(x_i, y_i, ..., y_{i+k-1}, h)$ be defined implicitly by

$$\psi = \sum_{j=0}^{k-1} \beta_j' f\left(x_i + jh, y_{i+j}\right) + \beta_k' f\left(x_i + kh, h\psi - \sum_{j=0}^{k-1} \alpha_j' y_{i+j}\right) \tag{4.4}$$

where $\alpha_j' = \alpha_j/\alpha_k$ and $\beta_j' = \beta_j/\alpha_k$. Multistep formula (2.1) can then be written as

$$y_{i+k} = -\sum_{j=0}^{k-1} \alpha_j' y_{i+j} + h\psi. \tag{4.5}$$

Introducing the $m \cdot k$-dimensional vectors (m is the dimension of the differential equation)

$$Y_i = (y_{i+k-1}, y_{i+k-2}, \ldots, y_i)^T, \qquad i \geq 0 \tag{4.6}$$

and

$$A = \begin{pmatrix} -\alpha_{k-1}' & -\alpha_{k-2}' & \cdots & \cdot & -\alpha_0' \\ 1 & 0 & \cdots & \cdot & 0 \\ & 1 & & \cdot & 0 \\ & & \ddots & \vdots & \vdots \\ & & & 1 & 0 \end{pmatrix}, \qquad e_1 = \begin{pmatrix} 1 \\ 0 \\ 0 \\ \vdots \\ 0 \end{pmatrix}, \tag{4.7}$$

the multistep method (4.5) can be written — after adding some trivial identities — in compact form as

$$Y_{i+1} = (A \otimes I)Y_i + h\Phi(x_i, Y_i, h), \qquad i \geq 0 \tag{4.8}$$

with

$$\Phi(x_i, Y_i, h) = (e_1 \otimes I)\psi(x_i, Y_i, h). \tag{4.8a}$$

Here, $A \otimes I$ denotes the Kronecker tensor product, i.e. the $m \cdot k$-dimensional block matrix with (m, m)-blocks $a_{ij}I$. Readers unfamiliar with the notation and properties of this product may assume for simplicity that (4.1) is a scalar equation ($m = 1$) and $A \otimes I = A$.

The following lemmas express the concepts of order and stability in this new notation.

Lemma 4.3. *Let $y(x)$ be the exact solution of (4.1). For $i = 0, 1, 2, \ldots$ we define the vector $\widehat{Y}_{i+1}$ as the numerical solution of one step*

$$\widehat{Y}_{i+1} = (A \otimes I) Y(x_i) + h \Phi(x_i, Y(x_i), h)$$

with correct starting values

$$Y(x_i) = \big(y(x_{i+k-1}), y(x_{i+k-2}), \ldots, y(x_i)\big)^T.$$

i) *If the multistep method (2.1) is of order 1 and if f satisfies (4.2), then an $h_0 > 0$ exists such that for $h \le h_0$,*

$$\|Y(x_{i+1}) - \widehat{Y}_{i+1}\| \le h\omega(h), \qquad 0 \le i \le \widehat{x}/h - k$$

where $\omega(h) \to 0$ for $h \to 0$.

ii) *If the multistep method (2.1) is of order p and if f is sufficiently differentiable then a constant M exists such that for h small enough,*

$$\|Y(x_{i+1}) - \widehat{Y}_{i+1}\| \le Mh^{p+1}, \qquad 0 \le i \le \widehat{x}/h - k.$$

Proof. The first component of $Y(x_{i+1}) - \widehat{Y}_{i+1}$ is the local error as given by Definition 2.1. Since the remaining components all vanish, Exercise 5 of Section III.2 and Definition 2.3 yield the result.   □

Lemma 4.4. *Suppose that the multistep method (2.1) is stable. Then there exists a vector norm (on $\mathbb{R}^{mk}$) such that the matrix A of (4.7) satisfies*

$$\|A \otimes I\| \le 1$$

in the subordinate matrix norm.

Proof. If λ is a root of $\varrho(\zeta)$, then the vector $(\lambda^{k-1}, \lambda^{k-2}, \ldots, 1)$ is an eigenvector of the matrix A with eigenvalue λ. Therefore the eigenvalues of A (which are the roots of $\varrho(\zeta)$) satisfy the root condition by Definition 3.2. A transformation to Jordan canonical form therefore yields (see Section I.12)

$$T^{-1}AT = J = \text{diag}\left\{\lambda_1, \ldots, \lambda_l, \begin{pmatrix} \lambda_{l+1} & \varepsilon_{l+1} & \\ & \ddots & \varepsilon_{k-1} \\ & & \lambda_k \end{pmatrix}\right\} \tag{4.9}$$

where $\lambda_1, \ldots, \lambda_l$ are the eigenvalues of modulus 1, which must be simple, each ε_j is either 0 or 1. We further find by a suitable multiplication of the columns of T that $|\varepsilon_j| < 1 - |\lambda_j|$ for $j = l+1, \ldots, k-1$. Because of (9.11') of Chapter I we then have $\|J \otimes I\|_\infty \le 1$. Using the transformation T of (4.9) we define the norm

$$\|x\| := \|(T^{-1} \otimes I)x\|_\infty.$$

This yields

$$\|(A \otimes I)x\| = \|(T^{-1} \otimes I)(A \otimes I)x\|_\infty = \|(J \otimes I)(T^{-1} \otimes I)x\|_\infty$$
$$\leq \|(T^{-1} \otimes I)x\|_\infty = \|x\|$$

and hence also $\|A \otimes I\| \leq 1$. $\qquad\square$

Proof of Convergence

The convergence theorem for multistep methods can now be established.

Theorem 4.5. *If the multistep method (2.1) is stable and of order 1 then it is convergent. If method (2.1) is stable and of order p then it is convergent of order p.*

Proof. As in the convergence theorem for one-step methods (Section II.3) we may assume without loss of generality that $f(x, y)$ is defined for all $y \in \mathbb{R}^m$, $x \in [x_0, \widehat{x}]$ and satisfies there a (global) Lipschitz condition. This implies that for sufficiently small h the functions $\psi(x_i, Y_i, h)$ and $\Phi(x_i, Y_i, h)$ satisfy a Lipschitz condition with respect to the second argument (with Lipschitz constant L^*). For the function G, defined by formula (4.8), which maps the vector Y_i onto Y_{i+1} we thus obtain from Lemma 4.4

$$\|G(Y_i) - G(Z_i)\| \leq (1 + hL^*)\|Y_i - Z_i\|. \tag{4.10}$$

The rest of the proof now proceeds in the same way as for one-step methods and is illustrated in Fig. 4.1.

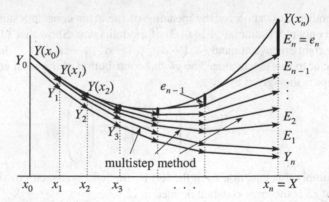

Fig. 4.1. Lady Windermere's Fan for multistep methods

The arrows in Fig. 4.1 indicate the application of G. From Lemma 4.3 we know that $\|Y(x_{i+1}) - G(Y(x_i))\| \leq h\omega(h)$. This together with (4.10) shows that the local error $Y(x_{i+1}) - G(Y(x_i))$ at stage $i+1$ causes an error at stage n, which is at most $h\omega(h)(1+hL^*)^{n-i+1}$. Thus we have

$$\|Y(x_n) - Y_n\| \le \|Y(x_0) - Y_0\|(1+hL^*)^n$$
$$+ h\omega(h)\Big((1+hL^*)^{n-1} + (1+hL^*)^{n-2} + \ldots + 1\Big) \tag{4.11}$$
$$\le \|Y(x_0) - Y_0\| \exp(nhL^*) + \frac{\omega(h)}{L^*}\big(\exp(nhL^*) - 1\big).$$

Convergence of method (2.1) is now an immediate consequence of formula (4.11). If the multistep method is of order p, the same proof with $\omega(h)$ replaced by Mh^p yields convergence of order p.   □

Exercises

1. Consider the function (for $x \ge 0$)

$$f(x,y) = \begin{cases} 2x & \text{for } y \le 0, \\ 2x - \dfrac{4y}{x} & \text{for } 0 < y < x^2, \\ -2x & \text{for } y \ge x^2. \end{cases}$$

a) Show that $y(x) = x^2/3$ is the unique solution of $y' = f(x,y)$, $y(0) = 0$, although f does not satisfy a Lipschitz condition near the origin.

b) Apply the mid-point rule (1.13') with starting values $y_0 = 0$, $y_1 = -h^2$ to the above problem and verify that the numerical solution at $x = nh$ is given by $y_h(x) = (-1)^n x^2$ (Taubert 1976, see also Grigorieff 1977).

2. Another motivation for the meaning of the error constant: suppose that 1 is the only eigenvalue of A in (4.7) of modulus one. Show that $(1,1,\ldots,1)^T$ is the right eigenvector and $(1, 1+\alpha'_{k-1}, 1+\alpha'_{k-1}+\alpha'_{k-2}, \ldots)$ is the left eigenvector to this eigenvalue. The *global* contribution of the *local* error after many steps is then given by

$$A^\infty \begin{pmatrix} C_{p+1} \\ 0 \\ \vdots \\ 0 \end{pmatrix} = C \begin{pmatrix} 1 \\ 1 \\ \vdots \\ 1 \end{pmatrix}. \tag{4.12}$$

Multiply this equation from the left by the left eigenvector to show with (2.6) that C is the error constant defined in (2.13).

Remark. For multistep methods with several eigenvalues of modulus 1, formula (4.12) remains valid if A^∞ is replaced by E (see Section III.8).

III.5 Variable Step Size Multistep Methods

Des war a harter Brockn, des ...   (Tyrolean dialect)

It is clear from the considerations of Section II.4 that an efficient integrator must be able to change the step size. However, changing the step size with multistep methods is difficult since the formulas of the preceding sections require the numerical approximations at equidistant points. There are in principle two possibilities for overcoming this difficulty:

  i) use polynomial interpolation to reproduce the starting values at the new (equidistant) grid;

 ii) construct methods which are adjusted to variable grid points.

This section is devoted to the second approach. We investigate consistency, stability and convergence. The actual implementation (order and step size strategies) will be considered in Section III.7.

Variable Step Size Adams Methods

F. Ceschino (1961) was apparently the first person to propose a "smooth" transition from a step size h to a new step size ωh. C.V.D. Forrington (1961) and later on F.T. Krogh (1969) extended his ideas: we consider an arbitrary grid (x_n) and denote the step sizes by $h_n = x_{n+1} - x_n$. We assume that approximations y_j to $y(x_j)$ are known for $j = n - k + 1, \ldots, n$ and we put $f_j = f(x_j, y_j)$. In the same way as in Section III.1 we denote by $p(t)$ the polynomial which interpolates the values (x_j, f_j) for $j = n - k + 1, \ldots, n$. Using Newton's interpolation formula we have

$$p(t) = \sum_{j=0}^{k-1} \prod_{i=0}^{j-1} (t - x_{n-i})\, \delta^j f[x_n, x_{n-1}, \ldots, x_{n-j}] \tag{5.1}$$

where the divided differences $\delta^j f[x_n, \ldots, x_{n-j}]$ are defined recursively by

$$\delta^0 f[x_n] = f_n$$

$$\delta^j f[x_n, \ldots, x_{n-j}] = \frac{\delta^{j-1} f[x_n, \ldots, x_{n-j+1}] - \delta^{j-1} f[x_{n-1}, \ldots, x_{n-j}]}{x_n - x_{n-j}}. \tag{5.2}$$

For actual computations (see Krogh 1969) it is practical to rewrite (5.1) as

$$p(t) = \sum_{j=0}^{k-1} \prod_{i=0}^{j-1} \frac{t - x_{n-i}}{x_{n+1} - x_{n-i}} \cdot \Phi_j^*(n) \tag{5.1'}$$

where

$$\Phi_j^*(n) = \prod_{i=0}^{j-1} (x_{n+1} - x_{n-i}) \cdot \delta^j f[x_n, \ldots, x_{n-j}]. \tag{5.3}$$

We now define the approximation to $y(x_{n+1})$ by

$$y_{n+1} = y_n + \int_{x_n}^{x_{n+1}} p(t)\, dt. \tag{5.4}$$

Inserting formula (5.1') into (5.4) we obtain

$$y_{n+1} = y_n + h_n \sum_{j=0}^{k-1} g_j(n) \Phi_j^*(n) \tag{5.5}$$

with

$$g_j(n) = \frac{1}{h_n} \int_{x_n}^{x_{n+1}} \prod_{i=0}^{j-1} \frac{t - x_{n-i}}{x_{n+1} - x_{n-i}}\, dt. \tag{5.6}$$

Formula (5.5) is the extension of the explicit Adams method (1.5) to variable step sizes. Observe that for constant step sizes the above expressions reduce to (Exercise 1)

$$g_j(n) = \gamma_j, \qquad \Phi_j^*(n) = \nabla^j f_n.$$

The variable step size *implicit* Adams methods can be deduced similarly. In analogy to Section III.1 we let $p^*(t)$ be the polynomial of degree k that interpolates (x_j, f_j) for $j = n-k+1, \ldots, n, n+1$ (the value $f_{n+1} = f(x_{n+1}, y_{n+1})$ contains the unknown solution y_{n+1}). Again, using Newton's interpolation formula we obtain

$$p^*(t) = p(t) + \prod_{i=0}^{k-1} (t - x_{n-i}) \cdot \delta^k f[x_{n+1}, x_n, \ldots, x_{n-k+1}].$$

The numerical solution, defined by

$$y_{n+1} = y_n + \int_{x_n}^{x_{n+1}} p^*(t)\, dt,$$

is now given by

$$y_{n+1} = p_{n+1} + h_n g_k(n) \Phi_k(n+1), \tag{5.7}$$

where p_{n+1} is the numerical approximation obtained by the explicit Adams method

$$p_{n+1} = y_n + h_n \sum_{j=0}^{k-1} g_j(n) \Phi_j^*(n)$$

and where

$$\Phi_k(n+1) = \prod_{i=0}^{k-1}(x_{n+1} - x_{n-i}) \cdot \delta^k f[x_{n+1}, x_n, \ldots, x_{n-k+1}]. \qquad (5.8)$$

Recurrence Relations for $g_j(n)$, $\Phi_j(n)$ and $\Phi_j^*(n)$

> The cost of computing integration coefficients is the biggest disadvantage to permitting arbitrary variations in the step size.
>
> (F.T. Krogh 1973)

The values $\Phi_j^*(n)$ $(j = 0, \ldots, k-1)$ and $\Phi_k(n+1)$ can be computed efficiently with the recurrence relations

$$\Phi_0(n) = \Phi_0^*(n) = f_n$$
$$\Phi_{j+1}(n) = \Phi_j(n) - \Phi_j^*(n-1) \qquad (5.9)$$
$$\Phi_j^*(n) = \beta_j(n)\Phi_j(n),$$

which are an immediate consequence of Definitions (5.3) and (5.8). The coefficients

$$\beta_j(n) = \prod_{i=0}^{j-1} \frac{x_{n+1} - x_{n-i}}{x_n - x_{n-i-1}}$$

can be calculated by

$$\beta_0(n) = 1, \qquad \beta_j(n) = \beta_{j-1}(n)\frac{x_{n+1} - x_{n-j+1}}{x_n - x_{n-j}}.$$

The calculation of the coefficients $g_j(n)$ is trickier (F.T. Krogh 1974). We introduce the q-fold integral

$$c_{jq}(x) = \frac{(q-1)!}{h_n^q} \int_{x_n}^x \int_{x_n}^{\xi_{q-1}} \cdots \int_{x_n}^{\xi_1} \prod_{i=0}^{j-1} \frac{\xi_0 - x_{n-i}}{x_{n+1} - x_{n-i}} \, d\xi_0 \ldots d\xi_{q-1} \qquad (5.10)$$

and observe that

$$g_j(n) = c_{j1}(x_{n+1}).$$

Lemma 5.1. *We have*

$$c_{0q}(x_{n+1}) = \frac{1}{q}, \qquad c_{1q}(x_{n+1}) = \frac{1}{q(q+1)},$$

$$c_{jq}(x_{n+1}) = c_{j-1,q}(x_{n+1}) - c_{j-1,q+1}(x_{n+1})\frac{h_n}{x_{n+1} - x_{n-j+1}}.$$

Proof. The first two relations follow immediately from (5.10). In order to prove the recurrence relation we denote by $d(x)$ the difference

$$d(x) = c_{jq}(x) - c_{j-1,q}(x)\frac{x - x_{n-j+1}}{x_{n+1} - x_{n-j+1}} + c_{j-1,q+1}(x)\frac{h_n}{x_{n+1} - x_{n-j+1}}.$$

Clearly, $d^{(i)}(x_n) = 0$ for $i = 0, 1, \ldots, q-1$. Moreover, the q-th derivative of $d(x)$ vanishes, since by the Leibniz rule

$$\frac{d^q}{dx^q}\left(c_{j-1,q}(x) \cdot \frac{x - x_{n-j+1}}{x_{n+1} - x_{n-j+1}}\right)$$

$$= c_{j-1,q}^{(q)}(x)\frac{x - x_{n-j+1}}{x_{n+1} - x_{n-j+1}} + qc_{j-1,q}^{(q-1)}(x)\frac{1}{x_{n+1} - x_{n-j+1}}$$

$$= c_{j,q}^{(q)}(x) + c_{j-1,q+1}^{(q)}(x)\frac{h_n}{x_{n+1} - x_{n-j+1}}.$$

Therefore we have $d(x) \equiv 0$ and the statement follows by putting $x = x_{n+1}$.

$\square$

Using the above recurrence relation one can successively compute $c_{2q}(x_{n+1})$ for $q = 1, \ldots, k-1$; $c_{3q}(x_{n+1})$ for $q = 1, \ldots, k-2$; $\ldots$; $c_{kq}(x_{n+1})$ for $q = 1$. This procedure yields in an efficient way the coefficients $g_j(n) = c_{j1}(x_{n+1})$ of the Adams methods.

Variable Step Size BDF

The BDF-formulas (1.22) can also be extended in a natural way to variable step size. Denote by $q(t)$ the polynomial of degree k that interpolates (x_i, y_i) for $i = n+1, n, \ldots, n-k+1$. It can be expressed, using divided differences, by

$$q(t) = \sum_{j=0}^{k} \prod_{i=0}^{j-1} (t - x_{n+1-i}) \cdot \delta^j y[x_{n+1}, x_n, \ldots, x_{n-j+1}]. \tag{5.11}$$

The requirement

$$q'(x_{n+1}) = f(x_{n+1}, y_{n+1})$$

immediately leads to the variable step size BDF-formulas

$$\sum_{j=1}^{k} h_n \prod_{i=1}^{j-1} (x_{n+1} - x_{n+1-i}) \cdot \delta^j y[x_{n+1}, \ldots, x_{n-j+1}] = h_n f(x_{n+1}, y_{n+1}). \tag{5.12}$$

The computation of the coefficients is much easier here than for the Adams methods.

General Variable Step Size Methods and Their Orders

For theoretical investigations it is convenient to write the methods in a form where the y_j and f_j values appear linearly. For example, the implicit Adams method (5.7) becomes ($k = 2$)

$$y_{n+1} = y_n + \frac{h_n}{6(1+\omega_n)}\left((3+2\omega_n)f_{n+1} + (3+\omega_n)(1+\omega_n)f_n - \omega_n^2 f_{n-1}\right),$$

$$(5.13)$$

where we have introduced the notation $\omega_n = h_n/h_{n-1}$ for the step size ratio. Or, the 2-step BDF-formula (5.12) can be written as

$$y_{n+1} - \frac{(1+\omega_n)^2}{1+2\omega_n}y_n + \frac{\omega_n^2}{1+2\omega_n}y_{n-1} = h_n\frac{1+\omega_n}{1+2\omega_n}f_{n+1}. \qquad (5.14)$$

In order to give a unified theory for all these variable step size multistep methods we consider formulas of the form

$$y_{n+k} + \sum_{j=0}^{k-1}\alpha_{jn}y_{n+j} = h_{n+k-1}\sum_{j=0}^{k}\beta_{jn}f_{n+j}. \qquad (5.15)$$

The coefficients α_{jn} and β_{jn} actually depend on the ratios $\omega_i = h_i/h_{i-1}$ for $i = n+1,\ldots,n+k-1$. In analogy to the constant step size case we give

Definition 5.2. Method (5.15) is *consistent of order* p, if

$$q(x_{n+k}) + \sum_{j=0}^{k-1}\alpha_{jn}q(x_{n+j}) = h_{n+k-1}\sum_{j=0}^{k}\beta_{jn}q'(x_{n+j})$$

holds for all polynomials $q(x)$ of degree $\leq p$ and for all grids (x_j).

By definition, the explicit Adams method (5.5) is of order k, the implicit Adams method (5.7) is of order $k+1$, and the BDF-formula (5.12) is of order k.

The notion of consistency certainly has to be related to the local error. Indeed, if the method is of order p, if the ratios h_j/h_n are bounded for $j = n+1,\ldots,n+k-1$ and if the coefficients satisfy

$$\alpha_{jn}, \beta_{jn} \text{ are bounded}, \qquad (5.16)$$

then a Taylor expansion argument implies that

$$y(x_{n+k}) + \sum_{j=0}^{k-1}\alpha_{jn}y(x_{n+j}) - h_{n+k-1}\sum_{j=0}^{k}\beta_{jn}y'(x_{n+j}) = \mathcal{O}(h_n^{p+1}) \qquad (5.17)$$

for sufficiently smooth $y(x)$. Interpreting $y(x)$ as the solution of the differential equation, a trivial extension of Lemma 2.2 to variable step sizes shows that the local error at x_{n+k} (cf. Definition 2.1) is also $\mathcal{O}(h_n^{p+1})$.

This motivates the investigation of condition (5.16). The methods (5.13) and (5.14) are seen to satisfy (5.16) whenever the step size ratio h_n/h_{n-1} is bounded from above. In general we have

Lemma 5.3. *For the explicit and implicit Adams methods as well as for the BDF-formulas the coefficients α_{jn} and β_{jn} are bounded whenever for some Ω*

$$h_n/h_{n-1} \le \Omega.$$

Proof. We prove the statement for the explicit Adams methods only. The proof for the other methods is similar and thus omitted. We see from formula (5.5) that the coefficients α_{jn} do not depend on n and hence are bounded. The β_{jn} are composed of products of $g_j(n)$ with the coefficients of $\Phi_j^*(n)$, when written as a linear combination of $f_n, \ldots, f_{n-j}$. From formula (5.6) we see that $|g_j(n)| \le 1$. It follows from $(x_{n+1} - x_{n-j+1}) \le \max(1, \Omega^j)(x_n - x_{n-j})$ and from an induction argument that the coefficients of $\Phi_j^*(n)$ are also bounded. Hence the β_{jn} are bounded, which proves the lemma.   □

The condition $h_n/h_{n-1} \le \Omega$ is a reasonable assumption which can easily be satisfied by a code.

Stability

> So geht das einfach ...   (R.D. Grigorieff, Halle 1983)

The study of stability for variable step size methods was begun in the articles of Gear & Tu (1974) and Gear & Watanabe (1974). Further investigations are due to Grigorieff (1983) and Crouzeix & Lisbona (1984).

We have seen in Section III.3 that for equidistant grids stability is equivalent to the boundedness of the numerical solution, when applied to the scalar differential equation $y' = 0$. Let us do the same here for the general case. Method (5.15), applied to $y' = 0$, gives the difference equation with variable coefficients

$$y_{n+k} + \sum_{j=0}^{k-1} \alpha_{jn} y_{n+j} = 0.$$

If we introduce the vector $Y_n = (y_{n+k-1}, \ldots, y_n)^T$, this difference equation is equivalent to

$$Y_{n+1} = A_n Y_n$$

with

$$
A_n = \begin{pmatrix} -\alpha_{k-1,n} & \cdots & \cdots & -\alpha_{1,n} & -\alpha_{0,n} \\ 1 & 0 & \cdots & 0 & 0 \\ & \ddots & \ddots & \vdots & \vdots \\ & & 1 & 0 & 0 \\ & & & 1 & 0 \end{pmatrix}, \tag{5.18}
$$

the companion matrix.

Definition 5.4. Method (5.15) is called *stable*, if

$$
\| A_{n+l} A_{n+l-1} \cdots A_{n+1} A_n \| \leq M \tag{5.19}
$$

for all n and $l \geq 0$.

Observe that in general A_n depends on the step ratios $\omega_{n+1}, \ldots, \omega_{n+k-1}$. Therefore, condition (5.19) will usually lead to a restriction on these values. For the Adams methods (5.5) and (5.7) the coefficients α_{jn} do not depend on n and hence are stable for any step size sequence.

In the following three theorems we present stability results for general variable step size methods. The first one, taken from Crouzeix & Lisbona (1984), is a sort of perturbation result: the variable step size method is considered as a perturbation of a strongly stable fixed step size method.

Theorem 5.5. *Let the method (5.15) satisfy the following properties:*

a) *it is of order $p \geq 0$, i.e.,* $\quad 1 + \displaystyle\sum_{j=0}^{k-1} \alpha_{jn} = 0$;

b) *the coefficients $\alpha_{jn} = \alpha_j(\omega_{n+1}, \ldots, \omega_{n+k-1})$ are continuous in a neighbourhood of $(1, \ldots, 1)$;*

c) *the underlying constant step size formula is strongly stable, i.e., all roots of*

$$
\zeta^k + \sum_{j=0}^{k-1} \alpha_j(1, \ldots, 1)\zeta^j = 0
$$

lie in the open unit disc $|\zeta| < 1$, with the exception of $\zeta_1 = 1$.
Then there exist real numbers ω, Ω ($\omega < 1 < \Omega$) such that the method is stable if

$$
\omega \leq h_n / h_{n-1} \leq \Omega \qquad \text{for all } n. \tag{5.20}
$$

Proof. Let A be the companion matrix of the constant step size formula. As in the proof of Lemma 4.4 we transform A to Jordan canonical form and obtain

$$
T^{-1}AT = \begin{pmatrix} \widehat{A} & \begin{matrix} 0 \\ \vdots \\ 0 \\ 1 \end{matrix} \end{pmatrix}
$$

where, by assumption (c), $\|\widehat{A}\|_1 < 1$. Observe that the last column of T, the eigenvector of A corresponding to 1, is given by $t_k = (1, \ldots, 1)^T$. Assumption (a) implies that this vector t_k is also an eigenvector for each A_n. Therefore we have

$$T^{-1} A_n T = \begin{pmatrix} & & 0 \\ \widehat{A}_n & & \vdots \\ & & 0 \\ & & 1 \end{pmatrix}$$

and, by continuity, $\|\widehat{A}_n\|_1 \le 1$, if $\omega_{n+1}, \ldots, \omega_{n+k-1}$ are sufficiently close to 1. Stability now follows from the fact that

$$\|T^{-1} A_n T\|_1 = \max(\|\widehat{A}_n\|_1, 1) = 1,$$

which implies that

$$\|A_{n+l} \cdots A_{n+1} A_n\| \le \|T\| \cdot \|T^{-1}\|. \qquad \square$$

The next result (Grigorieff 1983) is based on a reduction of the dimension of the matrices A_n by one. The idea is to use the transformation

$$T = \begin{pmatrix} 1 & 1 & 1 & .. & 1 \\ & 1 & 1 & .. & 1 \\ & & 1 & .. & 1 \\ 0 & & & \ddots & \vdots \\ & & & & 1 \end{pmatrix}, \qquad T^{-1} = \begin{pmatrix} 1 & -1 & & 0 \\ & 1 & -1 & \\ & & 1 & \ddots \\ 0 & & & \ddots & -1 \\ & & & & 1 \end{pmatrix}.$$

Observe that the last column of T is just t_k of the above proof. A simple calculation shows that

$$T^{-1} A_n T = \begin{pmatrix} A_n^* & 0 \\ e_{k-1}^T & 1 \end{pmatrix}$$

where $e_{k-1}^T = (0, \ldots, 0, 1)$ and

$$A_n^* = \begin{pmatrix} -\alpha_{k-2,n}^* & -\alpha_{k-3,n}^* & .. & -\alpha_{1n}^* & -\alpha_{0n}^* \\ 1 & 0 & .. & . & 0 \\ & 1 & .. & . & 0 \\ & & \ddots & \vdots & \vdots \\ & & & 1 & 0 \end{pmatrix} \qquad (5.21)$$

with

$$\alpha_{k-2,n}^* = 1 + \alpha_{k-1,n}, \qquad \alpha_{0n}^* = -\alpha_{0n},$$

$$\alpha_{k-j-1,n}^* - \alpha_{k-j,n}^* = \alpha_{k-j,n} \quad \text{for } j = 2, \ldots, k-1.$$

We remark that the coefficients $\alpha_{j,n}^*$ are just the coefficients of the polynomial defined by

$$(\zeta^k + \alpha_{k-1,n}\zeta^{k-1} + \ldots + \alpha_{1,n}\zeta + \alpha_{0,n})$$
$$= (\zeta - 1)(\zeta^{k-1} + \alpha_{k-2,n}^*\zeta^{k-2} + \ldots + \alpha_{1,n}^*\zeta + \alpha_{0,n}^*).$$

Theorem 5.6. *Let the method (5.15) be of order $p \geq 0$. Then the method is stable if and only if for all n and $l \geq 0$,*

$$a) \qquad \left\| A^*_{n+l} \cdots A^*_{n+1} A^*_n \right\| \leq M_1$$

$$b) \qquad \left\| e^T_{k-1} \sum_{j=n}^{n+l} \prod_{i=n}^{j-1} A^*_i \right\| \leq M_2.$$

Proof. A simple induction argument shows that

$$T^{-1} A_{n+l} \cdots A_n T = \begin{pmatrix} A^*_{n+l} \cdots A^*_n & 0 \\ b^T_{n,l} & 1 \end{pmatrix}$$

with

$$b^T_{n,l} = e^T_{k-1} \sum_{j=n}^{n+l} \prod_{i=n}^{j-1} A^*_i.$$

□

Since in this theorem the dimension of the matrices under consideration is reduced by one, it is especially useful for the stability investigation of two-step methods.

Example. Consider the two-step BDF-method (5.14). Here

$$\alpha_{0n} = \frac{\omega^2_{n+1}}{1 + 2\omega_{n+1}}, \qquad \alpha_{1n} = -1 - \alpha_{0n}.$$

The matrix (5.21) becomes in this case

$$A^*_n = (-\alpha^*_{0n}), \qquad -\alpha^*_{0n} = \frac{\omega^2_{n+1}}{1 + 2\omega_{n+1}}.$$

If $|\alpha^*_{0n}| \leq q < 1$ the conditions of Theorem 5.6 are satisfied and imply stability. This is the case, if

$$0 < h_{n+1}/h_n \leq \Omega < 1 + \sqrt{2}.$$

An interesting consequence of the theorem above is the *instability* of the two-step BDF-formula if the step sizes increase at least like $h_{n+1}/h_n \geq 1 + \sqrt{2}$.

The investigation of stability for k-step $(k \geq 3)$ methods becomes much more difficult, because several step size ratios $\omega_{n+1}, \omega_{n+2}, \ldots$ are involved. Grigorieff (1983) calculated the bounds (5.20) given in Table 5.1 for the higher order BDF-methods which *ensure* stability. These bounds are surely unrealistic, since all pathological step size variations are admitted.

A less pessimistic result is obtained if the step sizes are supposed to vary more smoothly (Gear & Tu 1974): the local error is known to be of the form $d(x_n)h_n^{p+1} + \mathcal{O}(h_n^{p+2})$, where $d(x)$ is the principal error function. This local error

Table 5.1. Bounds (5.20) for k-step BDF formulas

k	2	3	4	5
ω	0	0.836	0.979	0.997
Ω	2.414	1.127	1.019	1.003

is, by the step size control, kept equal to *Tol*. Hence, if $d(x)$ is bounded away from zero we have

$$h_n = |Tol/d(x_n)|^{1/(p+1)} + \mathcal{O}(h_n)$$

which implies (if $h_{n+1}/h_n \leq \Omega$) that

$$h_{n+1}/h_n = |d(x_n)/d(x_{n+1})|^{1/(p+1)} + \mathcal{O}(h_n).$$

If $d(x)$ is differentiable, we obtain

$$|h_{n+1}/h_n - 1| \leq C h_n. \tag{5.22}$$

Several stability results of Gear & Tu are based on this hypothesis ("Consequently, we can expect either method to be stable if the fixed step method is stable. ...."). Adding up (5.22) we obtain

$$\sum_{j=n}^{n+l} |h_{j+1}/h_j - 1| \leq C(\hat{x} - x_0),$$

a condition which contains only step size ratios. This motivates the following theorem:

Theorem 5.7. *Let the coefficients α_{jn} of method (5.15) be continuously differentiable functions of $\omega_{n+1}, \ldots, \omega_{n+k-1}$ in a neighbourhood of the set*

$$\{(\omega_{n+1}, \ldots, \omega_{n+k-1}) \, ; \, \omega \leq \omega_j \leq \Omega\}$$

and assume that the method is stable for constant step sizes (i.e., for $\omega_j = 1$). Then the condition

$$\sum_{j=n}^{n+l} |h_{j+1}/h_j - 1| \leq C \qquad \text{for all } n \text{ and } l \geq 0, \tag{5.23}$$

together with $\omega \leq h_{j+1}/h_j \leq \Omega$, imply the stability condition (5.19).

Proof. As in the proof of Theorem 5.5 we denote by A the companion matrix of the constant step size formula and by T a suitable transformation such that $\|T^{-1}AT\| = 1$. The mean value theorem, applied to $\alpha_j(\omega_{n+1}, \ldots, \omega_{n+k-1}) - \alpha_j(1, \ldots, 1)$, implies that

$$\|T^{-1}A_n T - T^{-1}AT\| \leq K \sum_{j=n+1}^{n+k-1} |\omega_j - 1|.$$

Hence

$$\|T^{-1}A_nT\| \le 1 + K \sum_{j=n+1}^{n+k-1} |\omega_j - 1| \le \exp\Big(K \sum_{j=n+1}^{n+k-1} |\omega_j - 1|\Big).$$

From this inequality we deduce that

$$\|A_{n+l}\cdots A_{n+1}A_n\| \le \|T\| \cdot \|T^{-1}\| \cdot \exp\big(K \cdot (k-1)C\big). \qquad \square$$

Convergence

Convergence for variable step size Adams methods was first studied by Piotrowski (1969). In order to prove convergence for the general case we introduce the vector $Y_n = (y_{n+k-1}, \ldots, y_{n+1}, y_n)^T$. In analogy to (4.8) the method (5.15) then becomes equivalent to

$$Y_{n+1} = (A_n \otimes I)Y_n + h_{n+k-1}\Phi_n(x_n, Y_n, h_n) \qquad (5.24)$$

where A_n is given by (5.18) and

$$\Phi_n(x_n, Y_n, h_n) = (e_1 \otimes I)\Psi_n(x_n, Y_n, h_n).$$

The value $\Psi = \Psi_n(x_n, Y_n, h_n)$ is defined implicitly by

$$\Psi = \sum_{j=0}^{k-1} \beta_{jn} f(x_{n+j}, y_{n+j}) + \beta_{kn} f\Big(x_{n+k}, h\Psi - \sum_{j=0}^{k-1} \alpha_{jn}y_{n+j}\Big).$$

Let us further denote by

$$Y(x_n) = \big(y(x_{n+k-1}), \ldots, y(x_{n+1}), y(x_n)\big)^T$$

the exact values to be approximated by Y_n. The convergence theorem can now be formulated as follows:

Theorem 5.8. *Assume that*

a) *the method (5.15) is stable, of order p, and has bounded coefficients α_{jn} and β_{jn};*

b) *the starting values satisfy $\|Y(x_0) - Y_0\| = \mathcal{O}(h_0^p)$;*

c) *the step size ratios are bounded ($h_n/h_{n-1} \le \Omega$).*

Then the method is convergent of order p, i.e., for each differential equation $y' = f(x, y)$, $y(x_0) = y_0$ with f sufficiently differentiable the global error satisfies

$$\|y(x_n) - y_n\| \le Ch^p \qquad for\ x_n \le \widehat{x},$$

where $h = \max h_j$.

Proof. Since the method is of order p and the coefficients and step size ratios are bounded, formula (5.17) shows that the local error

$$\delta_{n+1} = Y(x_{n+1}) - (A_n \otimes I)Y(x_n) - h_{n+k-1}\Phi_n(x_n, Y(x_n), h_n) \qquad (5.25)$$

satisfies

$$\delta_{n+1} = \mathcal{O}(h_n^{p+1}). \qquad (5.26)$$

Subtracting (5.24) from (5.25) we obtain

$$Y(x_{n+1}) - Y_{n+1} = (A_n \otimes I)(Y(x_n) - Y_n)$$
$$+ h_{n+k-1}\big(\Phi_n(x_n, Y(x_n), h_n) - \Phi_n(x_n, Y_n, h_n)\big) + \delta_{n+1}$$

and by induction it follows that

$$Y(x_{n+1}) - Y_{n+1} = \big((A_n \dots A_0) \otimes I\big)(Y(x_0) - Y_0)$$
$$+ \sum_{j=0}^{n} h_{j+k-1}\big((A_n \dots A_{j+1}) \otimes I\big)\big(\Phi_j(x_j, Y(x_j), h_j) - \Phi_j(x_j, Y_j, h_j)\big)$$
$$+ \sum_{j=0}^{n} \big((A_n \dots A_{j+1}) \otimes I\big)\delta_{j+1}.$$

As in the proof of Theorem 4.5 we deduce that the Φ_n satisfy a uniform Lipschitz condition with respect to Y_n. This, together with stability and (5.26), implies that

$$\|Y(x_{n+1}) - Y_{n+1}\| \leq \sum_{j=0}^{n} h_{j+k-1}L\|Y(x_j) - Y_j\| + C_1 h^p.$$

In order to solve this inequality we introduce the sequence $\{\varepsilon_n\}$ defined by

$$\varepsilon_0 = \|Y(x_0) - Y_0\|, \qquad \varepsilon_{n+1} = \sum_{j=0}^{n} h_{j+k-1}L\varepsilon_j + C_1 h^p. \qquad (5.27)$$

A simple induction argument shows that

$$\|Y(x_n) - Y_n\| \leq \varepsilon_n. \qquad (5.28)$$

From (5.27) we obtain for $n \geq 1$

$$\varepsilon_{n+1} = \varepsilon_n + h_{n+k-1}L\varepsilon_n \leq \exp(h_{n+k-1}L)\varepsilon_n$$

so that also

$$\varepsilon_n \leq \exp((\widehat{x} - x_0)L)\varepsilon_1 = \exp((\widehat{x} - x_0)L) \cdot (h_{k-1}L\|Y(x_0) - Y_0\| + C_1 h^p).$$

This inequality together with (5.28) completes the proof of Theorem 5.8. $\qquad \square$

Exercises

1. Prove that for constant step sizes the expressions $g_j(n)$ and $\Phi_j^*(n)$ (formulas (5.3) and (5.6)) reduce to

$$g_j(n) = \gamma_j, \qquad \Phi_j^*(n) = \nabla^j f_n,$$

where γ_j is given by (1.6).

2. (Grigorieff 1983). For the k-step BDF-methods consider grids with constant mesh ratio ω, i.e., $h_n = \omega h_{n-1}$ for all n. In this case the elements of A_n^* (see (5.21)) are independent of n. Show numerically that all eigenvalues of A_n^* are of absolute value less than one for $0 < \omega < R_k$ where

k	2	3	4	5	6
R_k	2.414	1.618	1.280	1.127	1.044

III.6 Nordsieck Methods

> While [the method] is primarily designed to optimize the effi-
> ciency of large-scale calculations on automatic computers, its es-
> sential procedures also lend themselves well to hand computation.
>
> (A. Nordsieck 1962)

> Two further problems must be dealt with in order to implement the
> automatic choice and revision of the elementary interval, namely,
> choosing which quantities to remember in such a way that the
> interval may be changed rapidly and conveniently . . .
>
> (A. Nordsieck 1962)

In an important paper Nordsieck (1962) considered a class of methods for ordinary differential equations which allow a convenient way of changing the step size
(see Section III.7). He already remarked that his methods are equivalent to the implicit Adams methods, in a certain sense. Let us begin with his derivation of these
methods and then investigate their relation to linear multistep methods.

Nordsieck (1962) remarked ". . . that all methods of numerical integration are
equivalent to finding an approximating polynomial for $y(x)$ . . .". His idea was to
represent such a polynomial by the 0th to kth derivatives, i.e., by a vector ("the
Nordsieck vector")

$$z_n = \left(y_n, hy'_n, \frac{h^2}{2!}y''_n, \ldots, \frac{h^k}{k!}y_n^{(k)} \right)^T. \tag{6.1}$$

The $y_n^{(j)}$ are meant to be approximations to $y^{(j)}(x_n)$, where $y(x)$ is the exact
solution of the differential equation

$$y' = f(x, y). \tag{6.2}$$

In order to define the integration procedure we have to give a rule for determining
z_{n+1} when z_n and the differential equation (6.2) are given. By Taylor's expansion,
such a rule is (e.g., for $k = 3$)

$$
\begin{aligned}
y_{n+1} &= y_n + hy'_n + \tfrac{h^2}{2!}y''_n + \tfrac{h^3}{3!}y'''_n + \tfrac{h^4}{4!}e \\
hy'_{n+1} &= \phantom{y_n + {}} hy'_n + 2\tfrac{h^2}{2!}y''_n + 3\tfrac{h^3}{3!}y'''_n + 4\tfrac{h^4}{4!}e \\
\tfrac{h^2}{2!}y''_{n+1} &= \tfrac{h^2}{2!}y''_n + 3\tfrac{h^3}{3!}y'''_n + 6\tfrac{h^4}{4!}e \\
\tfrac{h^3}{3!}y'''_{n+1} &= \phantom{\tfrac{h^2}{2!}y''_n + 3} \tfrac{h^3}{3!}y'''_n + 4\tfrac{h^4}{4!}e,
\end{aligned}
\tag{6.3}
$$

where the value e is determined in such a way that

$$y'_{n+1} = f(x_{n+1}, y_{n+1}). \tag{6.4}$$

Inserting (6.4) into the second relation of (6.3) yields

$$4\frac{h^4}{4!}e = h\left(f(x_{n+1}, y_{n+1}) - f_n^p \right) \tag{6.5}$$

with

$$hf_n^p = hy_n' + 2\frac{h^2}{2!}y_n'' + 3\frac{h^3}{3!}y_n'''.$$

With this relation for e the above method becomes

$$
\begin{aligned}
y_{n+1} &= y_n + hy_n' + \tfrac{h^2}{2!}y_n'' + \tfrac{h^3}{3!}y_n''' + \tfrac{1}{4}h\Big(f(x_{n+1},y_{n+1}) - f_n^p\Big)\\
hy_{n+1}' &= \quad\;\; hy_n' + 2\tfrac{h^2}{2!}y_n'' + 3\tfrac{h^3}{3!}y_n''' + h\Big(f(x_{n+1},y_{n+1}) - f_n^p\Big)\\
\tfrac{h^2}{2!}y_{n+1}'' &= \qquad\qquad \tfrac{h^2}{2!}y_n'' + 3\tfrac{h^3}{3!}y_n''' + \tfrac{3}{2}h\Big(f(x_{n+1},y_{n+1}) - f_n^p\Big)\\
\tfrac{h^3}{3!}y_{n+1}''' &= \qquad\qquad\qquad\qquad \tfrac{h^3}{3!}y_n''' + h\Big(f(x_{n+1},y_{n+1}) - f_n^p\Big)
\end{aligned}
\tag{6.6}
$$

The first equation constitutes an implicit formula for y_{n+1}, the others are explicit. Observe that for sufficiently accurate approximations $y_n^{(j)}$ to $y^{(j)}(x_n)$ the value e (formula (6.5)) is an approximation to $y^{(4)}(x_n)$. This seems to be a desirable property from the point of view of accuracy. Unfortunately, method (6.6) is unstable. To see this, we put $f(x,y) = 0$ in (6.6). In this case the method becomes the linear transformation

$$z_{n+1} = Mz_n \tag{6.7}$$

where

$$M = \begin{pmatrix} 1 & 1 & 1 & 1\\ 0 & 1 & 2 & 3\\ 0 & 0 & 1 & 3\\ 0 & 0 & 0 & 1 \end{pmatrix} - \begin{pmatrix} 1/4\\ 1\\ 3/2\\ 1 \end{pmatrix} \begin{pmatrix} 0 & 1 & 2 & 3 \end{pmatrix}.$$

The eigenvalues of M are seen to be $1, 0, -(2+\sqrt{3})$ and $-1/(2+\sqrt{3})$, implying that (6.6) is unstable and therefore of no use. The phenomenon that highly accurate methods are often unstable is, after our experiences in Section III.3, no longer astonishing.

To overcome this difficulty Nordsieck proposed to replace the constants $1/4$, 1, $3/2$, 1 which appear in front of the brackets in (6.6) by arbitrary values (l_0, l_1, l_2, l_3), and to use this extra freedom to achieve stability. In compact form this modification can be written as

$$z_{n+1} = (P \otimes I)z_n + (l \otimes I)\Big(hf(x_{n+1},y_{n+1}) - (e_1^T P \otimes I)z_n\Big). \tag{6.8}$$

Here z_n is given by (6.1), P is the Pascal triangle matrix defined by

$$p_{ij} = \begin{cases} \displaystyle\binom{j}{i} & \text{for } 0 \le i \le j \le k,\\[2mm] 0 & \text{else}, \end{cases}$$

$l = (l_0, l_1, \ldots, l_k)^T$ and $e_1 = (0, 1, 0, \ldots, 0)^T$. Observe that the indices of vectors and matrices start from zero.

For notational simplicity in the following theorems, we consider from now on scalar differential equations only, so that method (6.8) becomes

$$z_{n+1} = Pz_n + l(hf_{n+1} - e_1^T Pz_n). \tag{6.8'}$$

All results, of course, remain valid for systems of equations. Condition (6.4), which relates the method to the differential equation, fixes the value of l_1 as

$$l_1 = 1. \tag{6.9}$$

The above stability analysis applied to the general method (6.8) leads to the difference equation (6.7) with

$$M = P - le_1^T P. \tag{6.10}$$

For instance, for $k = 3$ this matrix is given by

$$M = \begin{pmatrix} 1 & 1-l_0 & 1-2l_0 & 1-3l_0 \\ 0 & 0 & 0 & 0 \\ 0 & -l_2 & 1-2l_2 & 3-3l_2 \\ 0 & -l_3 & -2l_3 & 1-3l_3 \end{pmatrix}.$$

One observes that 1 and 0 are two eigenvalues of M and that its characteristic polynomial is independent of l_0. Nordsieck determined $l_2, \ldots, l_k$ in such a way that the remaining eigenvalues of M are zero. For $k = 3$ this yields $l_2 = 3/4$ and $l_3 = 1/6$. The coefficient l_0 can be chosen such that the error constant of the method (see Theorem 6.2 below) vanishes. In our situation one gets $l_0 = 3/8$, so that the resulting method is given by

$$l = (3/8, 1, 3/4, 1/6)^T.$$

It is interesting to note that this method is equivalent to the implicit 3-step Adams method. Indeed, an elimination of the terms $(h^3/3!)y_n'''$ and $(h^2/2!)y_n''$ by using formula (6.8) with reduced indices leads to (cf. formula (1.9''))

$$y_{n+1} = y_n + \frac{h}{24}\left(9y_{n+1}' + 19y_n' - 5y_{n-1}' + y_{n-2}'\right). \tag{6.11}$$

Equivalence with Multistep Methods

More insight into the connection between Nordsieck methods and multistep methods is due to Descloux (1963), Osborne (1966), and Skeel (1979). The following two theorems show that every Nordsieck method is equivalent to a multistep formula and that the order of this method is at least k.

Theorem 6.1. *Consider the Nordsieck method (6.8) where $l_1 = 1$. The first two components of z_n then satisfy the linear multistep formula (for $n \geq 0$)*

$$\sum_{i=0}^{k} \alpha_i y_{n+i} = h \sum_{i=0}^{k} \beta_i f_{n+i} \tag{6.12}$$

where the generating polynomials are given by

$$\varrho(\zeta) = \det(\zeta I - P) \cdot e_1^T (\zeta I - P)^{-1} l$$
$$\sigma(\zeta) = \det(\zeta I - P) \cdot e_0^T (\zeta I - P)^{-1} l. \tag{6.13}$$

Proof. The proof of the original papers simplifies considerably, if we work with the generating functions (discrete Laplace transformation)

$$Z(\zeta) = \sum_{n \geq 0} z_n \zeta^n, \quad Y(\zeta) = \sum_{n \geq 0} y_n \zeta^n, \quad F(\zeta) = \sum_{n \geq 0} f_n \zeta^n, \quad \dots .$$

Multiplying formula (6.8') by ζ^{n+1} and adding up we obtain

$$Z(\zeta) = \zeta P Z(\zeta) + l\left(h F(\zeta) - e_1^T P \zeta Z(\zeta)\right) + (z_0 - l h f_0). \tag{6.14}$$

Similarly, the linear multistep method (6.12) can be written as

$$\widehat{\varrho}(\zeta) Y(\zeta) = h \widehat{\sigma}(\zeta) F(\zeta) + p_{k-1}(\zeta), \tag{6.15}$$

where

$$\widehat{\varrho}(\zeta) = \zeta^k \varrho(1/\zeta), \qquad \widehat{\sigma}(\zeta) = \zeta^k \sigma(1/\zeta) \tag{6.16}$$

and p_{k-1} is a polynomial of degree $k-1$ depending on the starting values. In order to prove the theorem we have to show that the first two components of $Z(\zeta)$ satisfy a relation of the form (6.15). We first rewrite equation (6.14) in the form

$$Z(\zeta) = (I - \zeta P)^{-1} l\left(h F(\zeta) - e_1^T P \zeta Z(\zeta)\right) + (I - \zeta P)^{-1}(z_0 - l h f_0)$$

so that its first two components become

$$Y(\zeta) = e_0^T (I - \zeta P)^{-1} l\left(h F(\zeta) - e_1^T P \zeta Z(\zeta)\right) + e_0^T (I - \zeta P)^{-1}(z_0 - l h f_0)$$
$$h F(\zeta) = e_1^T (I - \zeta P)^{-1} l\left(h F(\zeta) - e_1^T P \zeta Z(\zeta)\right) + e_1^T (I - \zeta P)^{-1}(z_0 - l h f_0).$$

Eliminating the term in brackets and multiplying by $\det(I - \zeta P)$ we arrive at formula (6.15) with

$$\widehat{\varrho}(\zeta) = \det(I - \zeta P) \cdot e_1^T (I - \zeta P)^{-1} l$$
$$\widehat{\sigma}(\zeta) = \det(I - \zeta P) \cdot e_0^T (I - \zeta P)^{-1} l$$
$$p_{k-1}(\zeta) = \det(I - \zeta P)\left(e_1^T (I - \zeta P)^{-1} l e_0^T (I - \zeta P)^{-1} \right. \tag{6.17}$$
$$\left. - e_0^T (I - \zeta P)^{-1} l e_1^T (I - \zeta P)^{-1}\right) z_0.$$

With the help of (6.16) we immediately get formulas (6.13). Therefore, it remains to show that p_{k-1}, given by (6.17), is a polynomial of degree $k-1$. Since the dimension of P is $(k+1)$, p_{k-1} behaves like ζ^{k-1} for $|\zeta| \to \infty$. Finally, the relation (6.15) implies that the Laurent series of p_{k-1} cannot contain negative powers. □

Putting $(\zeta I - P)^{-1} l = u$ in (6.13) and applying Cramer's rule to the linear system $(\zeta I - P)u = l$ we obtain from (6.13) the elegant expressions

$$\varrho(\zeta) = \det \begin{pmatrix} \zeta - 1 & l_0 & -1 & .. & -1 \\ 0 & l_1 & -2 & .. & -k \\ 0 & l_2 & \zeta - 1 & .. & . \\ \vdots & \vdots & \vdots & & \vdots \\ 0 & l_k & 0 & .. & \zeta - 1 \end{pmatrix} \tag{6.13a}$$

$$\sigma(\zeta) = \det \begin{pmatrix} l_0 & -1 & -1 & .. & -1 \\ l_1 & \zeta - 1 & -2 & .. & -k \\ l_2 & 0 & \zeta - 1 & .. & . \\ \vdots & \vdots & \vdots & & \vdots \\ l_k & 0 & 0 & .. & \zeta - 1 \end{pmatrix}. \tag{6.13b}$$

We observe that $\varrho(\zeta)$ does not depend on l_0. Further, $\zeta_0 = 1$ is a simple root of $\varrho(\zeta)$ if and only if $l_k \neq 0$. We have

$$\varrho'(1) = \sigma(1) = k! \, l_k. \tag{6.18}$$

Condition (6.9) is equivalent to $\alpha_k = 1$.

Theorem 6.2. *Assume that $l_k \neq 0$. The multistep method defined by (6.13) is of order at least k and its error constant (see (2.13)) is given by*

$$C = -\frac{b^T l}{k! \, l_k}.$$

Here the components of

$$b^T = (B_0, B_1, \ldots, B_k) = \left(1, -\frac{1}{2}, \frac{1}{6}, 0, -\frac{1}{30}, 0, \frac{1}{42}, \ldots\right)$$

are the Bernoulli numbers.

Proof. By Theorem 2.4 we have order k iff

$$\varrho(\zeta) - \log \zeta \cdot \sigma(\zeta) = C_{k+1}(\zeta - 1)^{k+1} + \mathcal{O}\big((\zeta - 1)^{k+2}\big).$$

Since $\det(\zeta I - P) = (\zeta - 1)^{k+1}$ this is equivalent to

$$e_1^T (\zeta I - P)^{-1} l - \log \zeta \cdot e_0^T (\zeta I - P)^{-1} l = C_{k+1} + \mathcal{O}\big((\zeta - 1)\big)$$

and, by (6.18), it suffices to show that

$$(\log \zeta \cdot e_0^T - e_1^T)(\zeta I - P)^{-1} = b^T + \mathcal{O}((\zeta - 1)). \tag{6.19}$$

Denoting the left-hand side of (6.19) by $b^T(\zeta)$ we obtain

$$(\zeta I - P)^T b(\zeta) = (\log \zeta \cdot e_0 - e_1). \tag{6.20}$$

The qth component $(q \geq 2)$ of this equation

$$\zeta b_q(\zeta) - \sum_{j=0}^{q} \binom{q}{j} b_j(\zeta) = 0$$

is equivalent to

$$\frac{\zeta b_q(\zeta)}{q!} - \sum_{j=0}^{q} \frac{b_j(\zeta)}{j!} \frac{1}{(q-j)!} = 0,$$

which is seen to be a Cauchy product. Hence, formula (6.20) becomes

$$\zeta \sum_{q \geq 0} \frac{t^q}{q!} b_q(\zeta) - e^t \sum_{q \geq 0} \frac{t^q}{q!} b_q(\zeta) = \log \zeta - t$$

which yields

$$\sum_{q \geq 0} \frac{t^q}{q!} b_q(\zeta) = \frac{t - \log \zeta}{e^t - \zeta}.$$

If we set $\zeta = 1$ in this formula we obtain

$$\sum_{q \geq 0} \frac{t^q}{q!} b_q(1) = \frac{t}{e^t - 1},$$

therefore $b_q(1) = B_q$, the qth Bernoulli number (see Abramowitz & Stegun, Chapter 23). $\qquad \square$

We have thus shown that to each Nordsieck method (6.8) there corresponds a linear multistep method of order at least k. Our next aim is to establish a correspondence in the opposite direction.

Theorem 6.3. *Let (ϱ, σ) be the generating polynomials of a k-step method (6.12) of order at least k and assume $\alpha_k = 1$. Then we have:*

a) *There exists a unique vector l such that ϱ and σ are given by (6.13).*

b) *If, in addition, the multistep method is irreducible, then there exists a non-singular transformation T such that the solution of (6.8') is related to that of (6.12) by*

$$z_n = T^{-1} u_n \tag{6.21}$$

where the jth component of u_n is given by

$$u_j^{(n)} = \begin{cases} \sum_{i=0}^{j}(\alpha_{k-j+i}y_{n+i} - h\beta_{k-j+i}f_{n+i}) & \text{for } 0 \leq j \leq k-1, \\ hf_n & \text{for } j = k. \end{cases}$$

$$(6.22)$$

Proof. a) For every kth order multistep method the polynomial $\varrho(\zeta)$ is uniquely determined by $\sigma(\zeta)$ (see Theorem 2.4). Expanding the determinant in (6.13b) with respect to the first column we see that

$$\sigma(\zeta) = l_0(\zeta - 1)^k + l_1(\zeta - 1)^{k-1}r_1(\zeta) + \ldots + l_k r_k(\zeta),$$

where $r_j(\zeta)$ is a polynomial of degree j satisfying $r_j(1) \neq 0$. Hence, l can be computed from $\sigma(\zeta)$.

b) Let $y_0, \ldots, y_{k-1}$ and $f_0, \ldots, f_{k-1}$ be given. Then the polynomial $p_{k-1}(\zeta)$ in (6.15) satisfies

$$p_{k-1}(\zeta) = u_0^{(0)} + u_1^{(0)}\zeta + \ldots + u_{k-1}^{(0)}\zeta^{k-1}.$$

On the other hand, if the starting vector z_0 for the Nordsieck method defined by l of (a) is known, then $p_{k-1}(\zeta)$ is given by (6.17). Equating both expressions we obtain

$$\sum_{j=0}^{k-1} u_j^{(0)}\zeta^j = (\widehat{\varrho}(\zeta)e_0^T - \widehat{\sigma}(\zeta)e_1^T)(I - \zeta P)^{-1}z_0. \tag{6.23}$$

We now denote by t_j^T ($j = 0, \ldots, k-1$) the coefficients of the vector polynomial

$$(\widehat{\varrho}(\zeta)e_0^T - \widehat{\sigma}(\zeta)e_1^T)(I - \zeta P)^{-1} = \sum_{j=0}^{k-1} t_j^T \zeta^j \tag{6.24}$$

and set $t_k^T = e_1^T$. Then let T be the square matrix whose jth row is t_j^T so that $u_0 = Tz_0$ is a consequence of (6.23) and $hf_n = hy'_n$. The same argument applied to $y_n, \ldots, y_{n+k-1}$ and $f_n, \ldots, f_{n+k-1}$ instead of $y_0, \ldots, y_{k-1}$ and $f_0, \ldots, f_{k-1}$ yields $u_n = Tz_n$ for all n.

To complete the proof it remains to verify the non-singularity of T. Let $v = (v_0, v_1, \ldots, v_k)^T$ be a non-zero vector satisfying $Tv = 0$. By definition of t_k^T we have $v_1 = 0$ and from (6.24) it follows (using the transformation (6.16)) that

$$\varrho(\zeta)\tau_0(\zeta) = \sigma(\zeta)\tau_1(\zeta), \tag{6.25}$$

where $\tau_i(\zeta) = \det(\zeta I - P)e_i^T(\zeta I - P)^{-1}v$ are polynomials of degree at most k. Moreover, Cramer's rule shows that the degree of $\tau_1(\zeta)$ is at most $k-1$, since $v_1 = 0$. Hence from (6.25) at least one of the roots of $\varrho(\zeta)$ must be a root of $\sigma(\zeta)$. This is in contradiction with the assumption that the method is irreducible.   □

Table 6.1. Coefficients l_j of the k-step implicit Adams methods

	l_0	l_1	l_2	l_3	l_4	l_5	l_6
$k=1$	1/2	1					
$k=2$	5/12	1	1/2				
$k=3$	3/8	1	3/4	1/6			
$k=4$	251/720	1	11/12	1/3	1/24		
$k=5$	95/288	1	25/24	35/72	5/48	1/120	
$k=6$	19087/60480	1	137/120	5/8	17/96	1/40	1/720

Table 6.2. Coefficients l_j of the k-step BDF-methods

	l_0	l_1	l_2	l_3	l_4	l_5	l_6
$k=1$	1	1					
$k=2$	2/3	1	1/3				
$k=3$	6/11	1	6/11	1/11			
$k=4$	12/25	1	7/10	1/5	1/50		
$k=5$	60/137	1	225/274	85/274	15/274	1/274	
$k=6$	20/49	1	58/63	5/12	25/252	1/84	1/1764

The vectors l which correspond to the implicit Adams methods and to the BDF-methods are given in Tables 6.1 and 6.2. For these two classes of methods we shall investigate the equivalence in some more detail.

Implicit Adams Methods

The following results are due to Byrne & Hindmarsh (1975). Since their "efficient package" EPISODE and the successor VODE are based on the Nordsieck representation of variable step size methods, we extend our considerations to this case. The Adams methods define in a natural way a polynomial which approximates the unknown solution of (6.2). Namely, if y_n and $f_n, \ldots, f_{n-k+1}$ are given, then the k-step Adams method is equivalent to the construction of a polynomial $p_{n+1}(x)$ of degree $k+1$ which satisfies

$$p_{n+1}(x_n) = y_n, \qquad p_{n+1}(x_{n+1}) = y_{n+1},$$
$$p'_{n+1}(x_j) = f_j \qquad \text{for } j = n-k+1, \ldots, n+1. \tag{6.26}$$

Condition (6.26) defines y_{n+1} implicitly. We observe that the difference of two consecutive polynomials, $p_{n+1}(x) - p_n(x)$, vanishes at x_n and that its derivative is zero at $x_{n-k+1}, \ldots, x_n$. Therefore, if we let $e_{n+1} = y_{n+1} - p_n(x_{n+1})$, this difference can be written as

$$p_{n+1}(x) - p_n(x) = \Lambda\left(\frac{x - x_{n+1}}{x_{n+1} - x_n}\right)e_{n+1} \tag{6.27}$$

where Λ is the unique polynomial of degree $(k+1)$ defined by

$$\Lambda(0) = 1, \qquad \Lambda(-1) = 0$$
$$\Lambda'\left(\frac{x_j - x_{n+1}}{x_{n+1} - x_n}\right) = 0 \qquad \text{for } j = n-k+1, \ldots, n. \tag{6.28}$$

The derivative of (6.27) taken at $x = x_{n+1}$ shows that with $h_n = x_{n+1} - x_n$,

$$h_n f_{n+1} - h_n p_n'(x_{n+1}) = \Lambda'(0)e_{n+1}.$$

If we introduce the Nordsieck vector

$$\widetilde{z}_n = \left(p_n(x_n), h_n p_n'(x_n), \ldots, \frac{h_n^{k+1}}{(k+1)!}p_n^{(k+1)}(x_n)\right)^T$$

and the coefficients $\widetilde{l}_j$ by

$$\Lambda(t) = \sum_{j=0}^{k+1} \widetilde{l}_j t^j, \tag{6.29}$$

then (6.27) becomes equivalent to

$$\widetilde{z}_{n+1} = P\widetilde{z}_n + \widetilde{l}\,\widetilde{l}_1^{-1}\left(hf_{n+1} - e_1^T P\widetilde{z}_n\right) \tag{6.30}$$

with $\widetilde{l} = (\widetilde{l}_0, \widetilde{l}_1, \ldots, \widetilde{l}_{k+1})^T$. This method is of the form (6.8'). However, it is of dimension $k+2$ and not, as expected by Theorem 6.3, of dimension $k+1$. The reason is the following: let $\widetilde{\varrho}(\zeta)$ and $\widetilde{\sigma}(\zeta)$ be the generating polynomials of the multistep method which corresponds to (6.30). Then the conditions $\Lambda(-1) = 0$ and $\Lambda'(-1) = 0$ imply that $\widetilde{\sigma}(0) = \widetilde{\varrho}(0) = 0$, so that this method is reducible. Nevertheless, method (6.30) is useful, since the last component of $\widetilde{z}_n$ can be used for step size control.

Remark. For $k \geq 2$ the coefficients $\widetilde{l}_j$, defined by (6.29), depend on the step size ratios h_j/h_{j-1} for $j = n-k+2, \ldots, n$. They can be computed from the formula

$$\Lambda(t) = \frac{\int_{-1}^t \prod_{j=1}^k (s - t_j)\, ds}{\int_{-1}^0 \prod_{j=1}^k (s - t_j)\, ds} \tag{6.31}$$

where $t_j = (x_{n-j+1} - x_{n+1})/(x_{n+1} - x_n)$ (see also Exercise 1).

BDF-Methods

One step of the k-step BDF method consists in constructing a polynomial $q_{n+1}(x)$ of degree k which satisfies

$$q_{n+1}(x_j) = y_j \qquad \text{for } j = n-k+1, \ldots, n+1$$
$$q'_{n+1}(x_{n+1}) = f_{n+1}$$

(6.32)

and in computing a value y_{n+1} which makes this possible. As for the Adams methods we have

$$q_{n+1}(x) - q_n(x) = \Lambda\Big(\frac{x - x_{n+1}}{x_{n+1} - x_n}\Big) \cdot (y_{n+1} - q_n(x_{n+1})),$$

(6.33)

where $\Lambda(t)$ is the polynomial of degree k defined by

$$\Lambda\Big(\frac{x_j - x_{n+1}}{x_{n+1} - x_n}\Big) = 0 \qquad \text{for } j = n-k+1, \ldots, n,$$
$$\Lambda(0) = 1.$$

With the vector

$$\widetilde{z}_n = \Big(q_n(x_n), \, h_n q'_n(x_n), \ldots, \frac{h_n^k}{k!} q_n^{(k)}(x_n)\Big)^T$$

and the coefficients $\widetilde{l}_j$ given by

$$\Lambda(t) = \sum_{j=0}^{k} \widetilde{l}_j t^j,$$

equation (6.33) becomes

$$\widetilde{z}_{n+1} = P\widetilde{z}_n + \widetilde{l}\,\widetilde{l}_1^{-1}\big(hf_{n+1} - e_1^T P\widetilde{z}_n\big).$$

(6.34)

The vector $\widetilde{l} = (\widetilde{l}_0, \widetilde{l}_1, \ldots, \widetilde{l}_k)^T$ can be computed from the formula

$$\Lambda(t) = \prod_{j=1}^{k}\Big(1 + \frac{t}{t_j}\Big)$$

where $t_j = (x_{n-j+1} - x_{n+1})/(x_{n+1} - x_n)$. For constant step sizes formula (6.34) corresponds to that of Theorem 6.3 and the coefficients $l_j = \widetilde{l}_j/\widetilde{l}_1$ coincide with those of Table 6.2.

Exercises

1. Let $l_j^{(k)} (j = 0, \ldots, k)$ be the Nordsieck coefficients of the k-step implicit Adams methods (defined by Theorem 6.3 and given in Table 6.1). Further, denote by $\widetilde{l}_j^{(k)} \ (j = 0, \ldots, k+1)$ the coefficients given by (6.29) and (6.31) for the case of constant step sizes. Show that

$$\frac{\widetilde{l}_j^{(k)}}{\widetilde{l}_1^{(k)}} = \begin{cases} l_j^{(k)} & \text{for } j = 0 \\ l_j^{(k+1)} & \text{for } j = 1, \ldots, k+1. \end{cases}$$

Use these relations to verify Table 6.1.

2. a) Calculate the matrix T of Theorem 6.3 for the 3-step implicit Adams method.

   Result.

   $$T^{-1} = \begin{pmatrix} 1 & 0 & 0 & 3/8 \\ 0 & 0 & 0 & 1 \\ 0 & 6 & 6 & 3/4 \\ 0 & 4 & 12 & 1/6 \end{pmatrix}.$$

   Show that the Nordsieck vector z_n is given by

   $$z_n = \Big(y_n, \ hf_n, \ (3hf_n - 4hf_{n-1} + hf_{n-2})/4, \ (hf_n - 2hf_{n-1} + hf_{n-2})/6 \Big)^T.$$

   b) The vector $\widetilde{z}_n$ for the 2-step implicit Adams method (6.30) (constant step sizes) also satisfies

   $$\widetilde{z}_n = \Big(y_n, \ hf_n, \ (3hf_n - 4hf_{n-1} + hf_{n-2})/4, \ (hf_n - 2hf_{n-1} + hf_{n-2})/6 \Big)^T,$$

   but this time y_n is a less accurate approximation to $y(x_n)$.

III.7 Implementation and Numerical Comparisons

There is a great deal of freedom in the implementation of multistep methods (even if we restrict our considerations to the Adams methods). One can either directly use the *variable step size methods* of Section III.5 or one can take a fixed step size method and determine the necessary offgrid values, which are needed for a change of step size, by *interpolation*. Further, it is possible to choose between the *divided difference* formulation (5.7) and the *Nordsieck* representation (6.30).

The historical approach was the use of formula (1.9) together with interpolation (J.C. Adams (1883): "We may, of course, change the value of ω (the step size) whenever the more or less rapid rate of diminution of the successive differences shews that it is expedient to increase or diminish the interval. It is only necessary, by selection from or interpolation between the values already calculated, to find the coordinates for a few values of φ separated from each other by the newly chosen interval."). It is theoretically more satisfactory and more elegant to work with the variable step size method (5.7). For both of these approaches the change of step size is rather expensive whereas the change of order is very simple — one just has to add a further term to the expansion (1.9). If the Nordsieck representation (6.30) is implemented, the situation is the opposite. There, the change of order is not as direct as above, but the step size can be changed simply by multiplying the Nordsieck-vector (6.1) by the diagonal matrix with entries $(1, \omega, \omega^2, \ldots)$ where $\omega = h_{\text{new}}/h_{\text{old}}$ is the step size ratio. Indeed, this was the main reason for introducing this representation.

Step Size and Order Selection

Much was made of the starting of multistep computations and the need for Runge-Kutta methods in the literature of the 60ies (see e.g., Ralston 1962). Nowadays, codes for multistep methods simply start with order one and very small step sizes and are therefore self-starting. The following step size and order selection is closely related to the description of Shampine & Gordon (1975).

Suppose that the numerical integration has proceeded successfully until x_n and that a further step with step size h_n and order $k+1$ is taken, which yields the approximation y_{n+1} to $y(x_{n+1})$. To decide whether y_{n+1} will be accepted or not, we need an estimate of the local truncation error. Such an estimate is e.g. given by

$$le_{k+1}(n+1) = y^*_{n+1} - y_{n+1}$$

where y^*_{n+1} is the result of the $(k+2)$nd order implicit Adams formula. Subtracting formula (5.7) from the same formula with k replaced by $k+1$, we obtain

$$le_{k+1}(n+1) = h_n \big(g_{k+1}(n) - g_k(n)\big)\Phi_{k+1}(n+1). \qquad (7.1)$$

Without changing the leading term in this expression we can replace the expression $\Phi_{k+1}(n+1)$ by

$$\Phi^p_{k+1}(n+1) = \prod_{i=0}^{k}(x_{n+1} - x_{n-i})\, \delta^{k+1} f^p[x_{n+1}, x_n, \ldots, x_{n-k}]. \qquad (7.2)$$

The superscript p of f indicates that $f_{n+1} = f(x_{n+1}, y_{n+1})$ is replaced by $f(x_{n+1}, p_{n+1})$ when forming the divided differences. If the implicit equation (5.7) is solved iteratively with p_{n+1} as predictor, then $\Phi^p_{k+1}(n+1)$ has to be calculated anyway. Therefore, the only cost for computing the estimate

$$LE_{k+1}(n+1) = h_n\big(g_{k+1}(n) - g_k(n)\big)\Phi^p_{k+1}(n+1) \qquad (7.3)$$

is the computation of $g_{k+1}(n)$. After the expression (7.3) has been calculated, we require (in the norm (4.11) of Section II.4)

$$\|LE_{k+1}(n+1)\| \leq 1 \qquad (7.4)$$

for the step to be successful.

If the Nordsieck representation (6.30) is considered instead of (5.7), then the estimate of the local error is not as simple, since the $\widetilde{l}$-vectors in (6.30) are totally different for different orders. For a possible error-estimate we refer to the article of Byrne & Hindmarsh (1975).

Suppose now that y_{n+1} is accepted. We next have to choose a new step size and a new order. The idea of the *step size selection* is to find the largest h_{n+1} for which the predicted local error is acceptable, i.e., for which

$$h_{n+1} \cdot \big|g_{k+1}(n+1) - g_k(n+1)\big| \cdot \|\Phi^p_{k+1}(n+2)\| \leq 1.$$

However, this procedure is of no practical use, since the expressions $g_j(n+1)$ and $\Phi^p_{k+1}(n+2)$ depend in a complicated manner on the unknown step size h_{n+1}. Also, the coefficients $g_{k+1}(n+1)$ and $g_k(n+1)$ are too expensive to calculate. To overcome this difficulty we assume the grid to be equidistant (this is a doubtful assumption, but leads to a simple formula for the new step size). In this case the local error (for the method of order $k+1$) is of the form $C(x_{n+2})h^{k+2} + \mathcal{O}(h^{k+3})$ with C depending smoothly on x. The local error at x_{n+2} can thus be approximated by that at x_{n+1} and in the same way as for one-step methods (cf. Section II.4

formula (4.12)) we obtain

$$h_{\text{opt}}^{(k+1)} = h_n \cdot \left(\frac{1}{\|LE_{k+1}(n+1)\|} \right)^{1/(k+2)} \tag{7.5}$$

as optimal step size. The local error $LE_{k+1}(n+1)$ is given by (7.3) or, again under the assumption of an equidistant grid, by

$$LE_{k+1}(n+1) = h_n \gamma_{k+1}^* \Phi_{k+1}^p(n+1) \tag{7.6}$$

with γ_{k+1}^* from Table 1.2 (see Exercise 1 of Section III.5 and Exercise 4 of Section III.1).

We next describe how an *optimal order* can be determined. Since the number of necessary function evaluations is the same for all orders, there are essentially two strategies for selecting the new order. One can choose the order $k+1$ either such that the local error estimate is minimal, or such that the new optimal step size is maximal. Because of the exponent $1/(k+2)$ in formula (7.5), the two strategies are not always equivalent. For more details see the description of the code DEABM below. It should be mentioned that each implementation of the Adams methods — and there are many — contains refinements of the above description and has in addition several ad-hoc devices. One of them is to keep the step size constant if $h_{\text{new}}/h_{\text{old}}$ is near to 1. In this way the computation of the coefficients $g_j(n)$ is simplified.

Some Available Codes

We have chosen the three codes DEABM, VODE and LSODE to illustrate the order- and step size strategies for multistep methods.

DEABM is a modification of the code DE/STEP/INTRP described in the book of Shampine & Gordon (1975). It belongs to the package DEPAC, designed by Shampine & Watts (1979). Our numerical tests use the revised version from February 1984. For European users it is available from the "Rechenzentrum der RWTH Aachen, Seffenter Weg 23, D-5100 Aachen, Germany".

This code implements the variable step size, divided difference representation (5.7) of the Adams formulas. In order to solve the nonlinear equation (5.7) for y_{n+1} the value p_{n+1} is taken as predictor (P), then $f_{n+1}^p = f(x_{n+1}, p_{n+1})$ is calculated (E) and *one* corrector iteration (C) is performed, to obtain y_{n+1}. Finally, in the case of a successful step, $f_{n+1} = f(x_{n+1}, y_{n+1})$ is evaluated (E) for the next step. This PECE implementation needs two function evaluations for each successful step. Let us also outline the order strategy of this code: after performing a step with order $k+1$, one computes $LE_{k-1}(n+1)$, $LE_k(n+1)$ and $LE_{k+1}(n+1)$ using a slight modification of (7.6). Then the order is reduced by one, if

$$\max\Big(\|LE_{k-1}(n+1)\|, \|LE_k(n+1)\| \Big) \leq \|LE_{k+1}(n+1)\|. \tag{7.7}$$

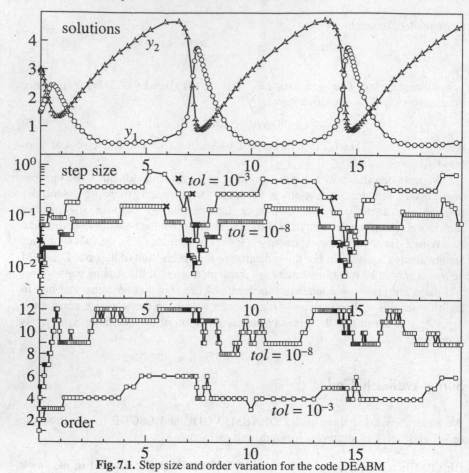

Fig. 7.1. Step size and order variation for the code DEABM

An increase in the order is considered only if the step is successful, (7.7) is violated and a constant step size is used. In this case one computes the estimate

$$LE_{k+2}(n+1) = h_n \gamma^*_{k+2} \Phi_{k+2}(n+1)$$

using the new value $f_{n+1} = f(x_{n+1}, y_{n+1})$ and increases the order by one if

$$\|LE_{k+2}(n+1)\| < \|LE_{k+1}(n+1)\|.$$

In Fig. 7.1 we demonstrate the variation of the step size and order on the example of Section II.4 (see Fig. 4.1 and also Fig. 9.5 of Section II.9). We plot the solution obtained with $Rtol = Atol = 10^{-3}$, the step size and order for the tolerances 10^{-3} and 10^{-8}. We observe that the step size — and not the order — drops significantly at passages where the solution varies more rapidly. Furthermore, constant step sizes are taken over long intervals, and the order is changed rather often (especially for $Tol = 10^{-8}$). This is in agreement with the observation of Shampine &

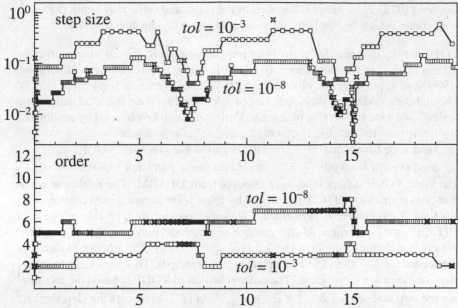

Fig. 7.2. Step size and order variation for the code VODE

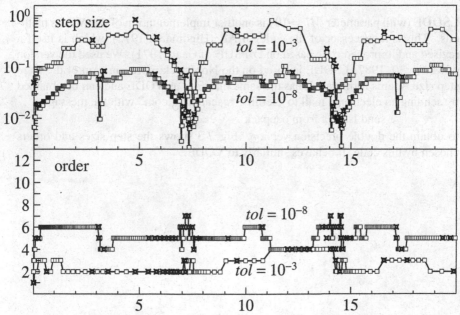

Fig. 7.3. Step size and order variation for the code LSODE

Gordon (1975): "... small reductions in the estimated error may cause the order to fluctuate, which in turn helps the code continue with constant step size."

VODE with parameter $MF = 10$ is an implementation of the variable-coefficient Adams method in Nordsieck form (6.30). It is due to Brown, Byrne & Hindmarsh (1989) and supersedes the older code EPISODE of Byrne & Hindmarsh (1975). The authors recommend their code "for problems with widely different active time scales". We used the version of August 31, 1992. It can be obtained by sending an electronic mail to "netlib@research.att.com" with the message

send vode.f from ode   to obtain double precision VODE,

send svode.f from ode   to obtain single precision VODE.

The code VODE differs in several respects from DEABM. The nonlinear equation (first component of (6.30)) is solved by fixed-point iteration until convergence. No final f-evaluation is performed. This method can thus be interpreted as a $P(EC)^M$-method, where M, the number of iterations, may be different from step to step. E.g., in the example of Fig. 7.2 ($Tol = 10^{-8}$) only 930 function evaluations are needed for 535 steps (519 accepted and 16 rejected). This shows that for many steps one iteration is sufficient. The order selection in VODE is based on maximizing the step size among $h_{\text{opt}}^{(k)}$, $h_{\text{opt}}^{(k+1)}$, $h_{\text{opt}}^{(k+2)}$. Fig. 7.2 presents the step size and order variation for VODE for the same example as above: compared to DEABM we observe that much lower orders are taken. Further, the order is constant over long intervals. This is reasonable, since a change in the order is not natural for the Nordsieck representation.

LSODE (with parameter $MF = 10$) is another implementation of the Adams methods. This is a successor of the code GEAR (Hindmarsh 1972), which is itself a revised and improved code based on DIFSUB of Gear (1971). We used the version of March 30, 1987. LSODE is based on the Nordsieck representation of the fixed step size Adams formulas. It has the same interface as VODE and can be obtained by sending an electronic mail to "netlib@research.att.com" with the message

send lsode.f from odepack

to obtain the double precision version. Fig. 7.3 shows the step sizes and orders chosen by this code. It behaves similarly to VODE.

Numerical Comparisons

> Of the three families of methods, the fixed order Runge-Kutta is
> the simplest, in several respects the best understood, and the least
> efficient.   (Shampine & Gordon 1975)

It is, of course, interesting to study the numerical performance of the above implementations of the Adams methods:

DEABM — symbol $\boxtimes$
VODE — symbol $\bigcirc$
LSODE — symbol $\triangle$

In order to compare the results with those of a typical one-step Runge-Kutta method
we include the results of the code

DOP853 — symbol $\stackrel{\wedge}{\vee}$

described in Section II.5.

With all these methods we have computed the numerical solution for the
six problems EULR, AREN, LRNZ, PLEI, ROPE, BRUS of Section II.10 us-
ing many different tolerances between 10^{-3} and 10^{-14} (the "integer" tolerances
10^{-3}, 10^{-4},... are distinguished by enlarged symbols). Fig. 7.4 gives the number
of *function evaluations* plotted against the achieved accuracy in double logarithmic scale. Some general tendencies can be distinguished in the crowds of numerical results. LSODE and DEABM require, for equal obtained accuracy, usually
less function evaluations, with DEABM becoming champion for higher precision
($Tol \leq 10^{-6}$).

The situation changes dramatically in favour of the Runge-Kutta code DOP853
if *computing time* is measured instead of function evaluations (see Fig. 7.5; the CPU
time is that of a Sun Workstation, SunBlade 100). We observe that for problems
with cheap function evaluations (EULR, AREN, LRNZ) the Runge-Kutta code
needs much less CPU time than the multistep codes, although more function evaluations are necessary in general. For the problems PLEI and ROPE, where the right
hand side is rather expensive to evaluate, the discrepancy is not as large. For the
last problem (BRUS) the dimension is very high, but the individual components are
not too complicated. In this situation, the CPU time of DOP853 is also significantly
less than for the multistep codes; this indicates that their overhead also increases
with the dimension of the problem.

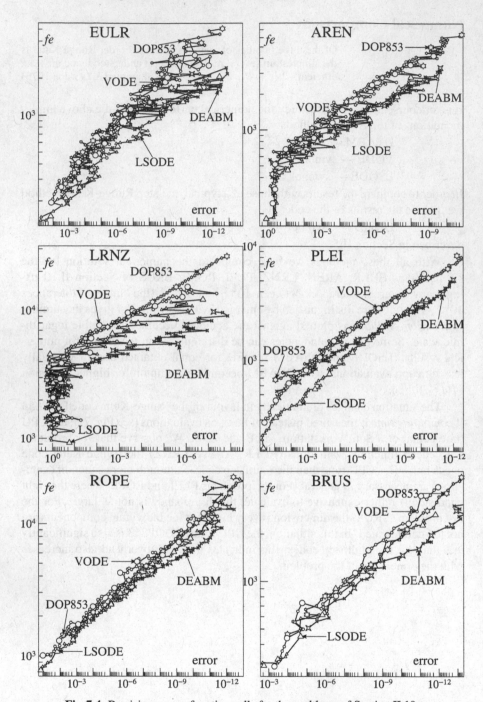

Fig. 7.4. Precision versus function calls for the problems of Section II.10

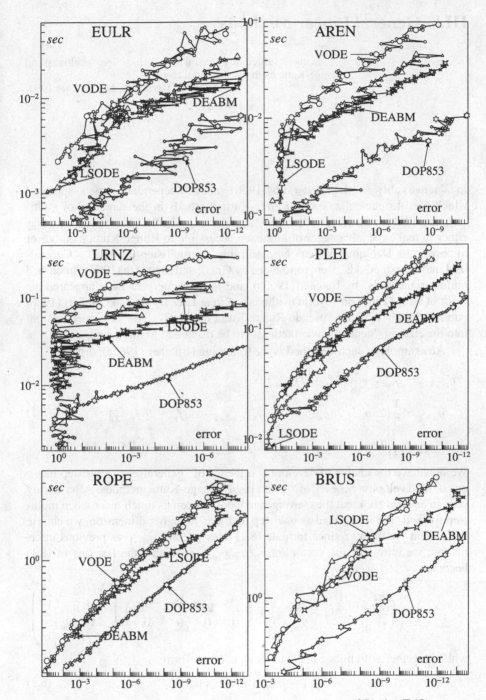

Fig. 7.5. Precision versus computing time for the problems of Section II.10

III.8 General Linear Methods

... methods sufficiently general as to include linear multistep and Runge-Kutta methods as special cases ...

(K. Burrage & J.C. Butcher 1980)

In a remarkably short period (1964-1966) many independent papers appeared which tried to generalize either Runge-Kutta methods in the direction of multistep or multistep methods in the direction of Runge-Kutta. The motivation was either to make the advantages of multistep accessible to Runge-Kutta methods or to "break the Dahlquist barrier" by modifying the multistep formulas. "Generalized multistep methods" were introduced by Gragg and Stetter in (1964), "modified multistep methods" by Butcher (1965a), and in the same year there appeared the work of Gear (1965) on "hybrid methods". A year later Byrne and Lambert (1966) published their work on "pseudo Runge-Kutta methods". All these methods fall into the class of "general linear methods" to be discussed in this section.

An example of such a method is the following (Butcher (1965a), order 5)

$$\widehat{y}_{n+1/2} = y_{n-1} + \frac{h}{8}\left(9f_n + 3f_{n-1}\right)$$

$$\widehat{y}_{n+1} = \frac{1}{5}\left(28y_n - 23y_{n-1}\right) + \frac{h}{5}\left(32\widehat{f}_{n+1/2} - 60f_n - 26f_{n-1}\right) \tag{8.1}$$

$$y_{n+1} = \frac{1}{31}\left(32y_n - y_{n-1}\right) + \frac{h}{93}\left(64\widehat{f}_{n+1/2} + 15\widehat{f}_{n+1} + 12f_n - f_{n-1}\right).$$

We now have the choice of developing a theory of "generalized" multistep methods or of developing a theory of "generalized" Runge-Kutta methods. After having seen in Section III.4 that the convergence theory becomes much nicer when multistep methods are interpreted as one-step methods in higher dimension, we choose the second possibility: since formula (8.1) uses y_n and y_{n-1} as previous information, we introduce the vector $u_n = (y_n, y_{n-1})^T$ so that the last line of (8.1) becomes

$$\begin{pmatrix} y_{n+1} \\ y_n \end{pmatrix} = \begin{pmatrix} \frac{32}{31} & -\frac{1}{31} \\ 1 & 0 \end{pmatrix} \begin{pmatrix} y_n \\ y_{n-1} \end{pmatrix} + \begin{pmatrix} \frac{64}{93} & \frac{15}{93} & \frac{12}{93} & -\frac{1}{93} \\ 0 & 0 & 0 & 0 \end{pmatrix} \begin{pmatrix} hf(\widehat{y}_{n+1/2}) \\ hf(\widehat{y}_{n+1}) \\ hf(y_n) \\ hf(y_{n-1}) \end{pmatrix}$$

which, together with lines 1 and 2 of (8.1), is of the form

$$u_{n+1} = Su_n + h\Phi(x_n, u_n, h). \tag{8.2}$$

Properties of such general methods have been investigated by Butcher (1966),

Hairer & Wanner (1973), Skeel (1976), Cooper (1978), Albrecht (1978, 1985) and others. Clearly, nothing prevents us from letting S and Φ be arbitrary, or from allowing also other interpretations of u_n.

A General Integration Procedure

We consider the system

$$y' = f(x, y), \qquad y(x_0) = y_0 \tag{8.3}$$

where f satisfies the regularity condition (4.2). Let m be the dimension of the differential equation (8.3), $q \geq m$ be the dimension of the difference equation (8.2) and $x_n = x_0 + nh$ be the subdivision points of an equidistant grid. The methods under consideration consist of three parts:

i) a *forward step procedure*, i.e., a formula (8.2), where the square matrix S is independent of (8.3).

ii) a *correct value function* $z(x, h)$, which gives an interpretation of the values u_n; $z_n = z(x_n, h)$ is to be approximated by u_n, so that the global error is given by $u_n - z_n$. It is assumed that the exact solution $y(x)$ of (8.3) can be recovered from $z(x, h)$.

iii) a *starting procedure* $\varphi(h)$, which specifies the starting value $u_0 = \varphi(h)$. $\varphi(h)$ approximates $z_0 = z(x_0, h)$.

The discrete problem corresponding to (8.3) is thus given by

$$u_0 = \varphi(h), \tag{8.4a}$$

$$u_{n+1} = S u_n + h \Phi(x_n, u_n, h), \qquad n = 0, 1, 2, \ldots, \tag{8.4b}$$

which yields the numerical solution $u_0, u_1, u_2, \ldots$. We remark that the increment function $\Phi(x, u, h)$, the starting procedure $\varphi(h)$ and the correct value function $z(x, h)$ depend on the differential equation (8.3), although this is not stated explicitly.

Example 8.1. The most simple cases are *one-step methods*. A characteristic feature of these is that the dimensions of the differential and difference equation are equal (i.e., $m = q$) and that S is the identity matrix. Furthermore, $\varphi(h) = y_0$ and $z(x, h) = y(x)$. They have been investigated in Chapter II.

Example 8.2. We have seen in Section III.4 that linear *multistep methods* also fall into the class (8.4). For k-step methods the dimension of the difference equation is $q = km$ and the forward step procedure is given by formula (4.8). A starting procedure yields the vector $\varphi(h) = (y_{k-1}, \ldots, y_1, y_0)^T$ and, finally, the correct value function is given by

$$z(x, h) = \big(y(x + (k-1)h), \ldots, y(x+h), y(x) \big)^T.$$

The most common way of implementing an implicit multistep method is a *predictor-corrector* process (compare (1.11) and Section III.7): an approximation $y_{n+k}^{(0)}$ to y_{n+k} is "predicted" by an explicit multistep method, say

$$\alpha_k^p y_{n+k}^{(0)} + \alpha_{k-1}^p y_{n+k-1} + \ldots + \alpha_0^p y_n = h(\beta_{k-1}^p f_{n+k-1} + \ldots + \beta_0^p f_n) \quad (8.5;\text{P})$$

and is then "corrected" (usually once or twice) by

$$f_{n+k}^{(l-1)} := f(x_{n+k}, y_{n+k}^{(l-1)}) \quad\quad\quad (8.5;\text{E})$$

$$\alpha_k y_{n+k}^{(l)} + \alpha_{k-1} y_{n+k-1} + \ldots + \alpha_0 y_n = h(\beta_k f_{n+k}^{(l-1)} + \beta_{k-1} f_{n+k-1} + \ldots + \beta_0 f_n).$$
$$(8.5;\text{C})$$

If the iteration (8.5) is carried out until convergence, the process is identical to that of Example 8.2. In practice, however, only a fixed number, say M, of iterations are carried out and the method is theoretically no longer a "pure" multistep method. We distinguish two predictor-corrector (PC) methods, depending on whether it ends with a correction (8.5;C) or not. The first algorithm is symbolized as $P(EC)^M$ and the second possibility, where f_{n+k} is once more updated by (8.5;E) for further use in the subsequent steps, as $P(EC)^M E$. We shall now see how these two procedures can be interpreted as methods of type (8.4).

Example 8.2a. $P(EC)^M E$-methods. The starting procedure and the correct value function are the same as for multistep methods and also $q = km$. Furthermore we have $S = A \otimes I$, where A is given by (4.7) and I is the m-dimensional identity matrix. Observe that S depends only on the corrector-formula and not on the predictor-formula. Here, the increment function is given by

$$\Phi(x, u, h) = (e_1 \otimes I)\psi(x, u, h)$$

with $e_1 = (1, 0, \ldots, 0)^T$. For $u = (u^1, \ldots, u^k)^T$ with $u^j \in \mathbb{R}^m$ the function $\psi(x, u, h)$ is defined by

$$\psi(x, u, h) = \alpha_k^{-1}\Big(\beta_k f(x + kh, y^{(M)})$$

$$+ \beta_{k-1} f(x + (k-1)h, u^1) + \ldots + \beta_0 f(x, u^k)\Big)$$

where the value $y^{(M)}$ is calculated from

$$\alpha_k^p y^{(0)} + \alpha_{k-1}^p u^1 + \ldots + \alpha_0^p u^k$$
$$= h\Big(\beta_{k-1}^p f(x + (k-1)h, u^1) + \ldots + \beta_0^p f(x, u^k)\Big)$$
$$\alpha_k y^{(l)} + \alpha_{k-1} u^1 + \ldots + \alpha_0 u^k$$
$$= h\Big(\beta_k f(x+kh, y^{(l-1)}) + \beta_{k-1} f(x+(k-1)h, u^1) + \ldots + \beta_0 f(x, u^k)\Big)$$

(for $l = 1, \ldots, M$).

Example 8.2b. For P(EC)^M-methods, the formulation as a method of type (8.4) becomes more complicated, since the information to be carried over to the next step is determined not only by $y_{n+k-1}, \ldots, y_n$, but also depends on the values $hf_{n+k-1}, \ldots, hf_n$, where $hf_{n+j} = hf(x_{n+j}, y_{n+j}^{(M-1)})$. Therefore the dimension of the difference equation becomes $q = 2km$. A usual starting procedure (as for multistep methods) yields

$$\varphi(h) = \Big(y_{k-1}, \ldots, y_0, hf(x_{k-1}, y_{k-1}), \ldots, hf(x_0, y_0)\Big)^T.$$

If we define the correct value function by

$$z(x, h) = \Big(y\big(x + (k-1)h\big), \ldots, y(x), hy'\big(x + (k-1)h\big), \ldots, hy'(x)\Big)^T,$$

the forward step procedure is given by

$$S = \begin{pmatrix} A & B \\ 0 & N \end{pmatrix}, \qquad \Phi(x, u, h) = \begin{pmatrix} \beta'_k e_1 \\ e_1 \end{pmatrix} \Psi(x, u, h).$$

Here A is the matrix given by (4.7), $\beta'_j = \beta_j/\alpha_k$ and

$$N = \begin{pmatrix} 0 & 0 & \ldots & 0 & 0 \\ 1 & 0 & \ldots & 0 & 0 \\ \vdots & \vdots & & \vdots & \vdots \\ 0 & 0 & \ldots & 1 & 0 \end{pmatrix}, \qquad B = \begin{pmatrix} \beta'_{k-1} & \ldots & \beta'_0 \\ 0 & \ldots & 0 \\ \vdots & & \vdots \\ 0 & \ldots & 0 \end{pmatrix}, \qquad e_1 = \begin{pmatrix} 1 \\ 0 \\ \vdots \\ 0 \end{pmatrix}.$$

For $u = (u^1, \ldots, u^k, hv^1, \ldots, hv^k)$ the function $\psi(x, u, h) \in \mathbb{R}^q$ is defined by

$$\psi(x, u, h) = f(x + kh, y^{(M-1)})$$

where $y^{(M-1)}$ is given by

$$\alpha_k^p y^{(0)} + \alpha_{k-1}^p u^1 + \ldots + \alpha_0^p u^k = h(\beta_{k-1}^p v^1 + \ldots + \beta_0^p v^k)$$

$$\alpha_k y^{(l)} + \alpha_{k-1} u^1 + \ldots + \alpha_0 u^k = h\big(\beta_k f(x + kh, y^{(l-1)}) + \beta_{k-1} v^1 + \ldots + \beta_0 v^k\big).$$

Again we observe that S depends only on the corrector-formula.

Example 8.3. *Nordsieck methods* are also of the form (8.4). This follows immediately from the representation (6.8). In this case the correct value function

$$z(x, h) = \Big(y(x), hy'(x), \frac{h^2}{2!} y''(x), \ldots, \frac{h^k}{k!} y^{(k)}(x)\Big)^T$$

is composed not only of values of the exact solution, but also contains their derivatives.

Example 8.4. *Cyclic multistep methods.* Donelson & Hansen (1971) have investigated the possibility of basing a discretization scheme on several different k-step methods which are used cyclically. Let S_j and Φ_j represent the forward step procedure of the jth multistep method; then the numerical solution $u_0, u_1, \ldots$ is defined by

$$u_0 = \varphi(h)$$
$$u_{n+1} = S_j u_n + h\Phi_j(x_n, u_n, h) \qquad \text{if } n \equiv (j-1) \bmod m.$$

In order to get a method (8.4) with S independent of the step number, we consider one cycle of the method as one step of a new method

$$u_0^* = \varphi(\frac{h^*}{m})$$
$$u_{n+1}^* = S u_n^* + h^* \Phi(x_n^*, u_n^*, h^*) \tag{8.6}$$

with step size $h^* = mh$. Here $x_n^* = x_0 + nh^*$, $S = S_m \ldots S_2 S_1$ and Φ has to be chosen suitably. E.g., in the case $m = 2$ we have

$$\Phi(x^*, u^*, h^*) = \frac{1}{2} S_2 \Phi_1\left(x^*, u^*, \frac{h^*}{2}\right)$$
$$+ \frac{1}{2} \Phi_2\left(x^* + \frac{h^*}{2}, S_1 u^* + \frac{h^*}{2}\Phi_1(x^*, u^*, \frac{h^*}{2}), \frac{h^*}{2}\right).$$

It is interesting to note that cyclically used k-step methods can lead to convergent methods of order $2k - 1$ (or even $2k$). The "first Dahlquist barrier" (Theorem 3.5) can be broken in this way. For more details see Stetter (1973), Albrecht (1979) and Exercise 2.

Example 8.5. *General linear methods.*

> Following the advice of Aristotle ... (the original Greek can be found in Butcher's paper) ... we look for the greatest good as a mean between extremes. (J.C. Butcher 1985a)

Introduced by Burrage & Butcher (1980), these methods are general enough to include all previous examples as special cases, but at the same time the increment function is given explicitly in terms of the differential equation and several free parameters. They are defined by

$$v_i^{(n)} = \sum_{j=1}^{k} \widetilde{a}_{ij} u_j^{(n)} + h \sum_{j=1}^{s} \widetilde{b}_{ij} f(x_n + c_j h, v_j^{(n)}) \quad i = 1, \ldots, s, \tag{8.7a}$$

$$u_i^{(n+1)} = \sum_{j=1}^{k} a_{ij} u_j^{(n)} + h \sum_{j=1}^{s} b_{ij} f(x_n + c_j h, v_j^{(n)}) \quad i = 1, \ldots, k. \tag{8.7b}$$

The stages $v_i^{(n)}$ $(i = 1, \ldots, s)$ are the *internal stages* and do not leave the "black box" of the current step. The stages $u_i^{(n)}$ $(i = 1, \ldots, k)$ are called the *external stages* since they contain all the necessary information from the previous step used in carrying out the current step. The coefficients a_{ij} in (8.7b) form the matrix S of (8.4b). Very often, some internal stages are identical to external ones, as for example in method (8.1), where

$$v_n = (\widehat{y}_{n+1/2}, \widehat{y}_{n+1}, y_n, y_{n-1})^T.$$

One-step Runge-Kutta methods are characterized by $k = 1$. At the end of this section we shall discuss the algebraic conditions for general linear methods to be of order p.

Example 8.6. In order to illustrate the fact that the analysis of this section is not only applicable to numerical methods that discretize first order differential equations, we consider the second order initial value problem

$$y'' = g(x, y), \quad y(x_0) = y_0, \quad y'(x_0) = y_0' \tag{8.8}$$

Replacing $y''(x)$ by a central difference yields

$$y_{n+1} - 2y_n + y_{n-1} = h^2 g(x_n, y_n),$$

and with the additional variables

$$hy_n' = y_{n+1} - y_n$$

this method can be written as

$$\begin{pmatrix} y_{n+1} \\ y_{n+1}' \end{pmatrix} = \begin{pmatrix} 1 & 0 \\ 0 & 1 \end{pmatrix} \begin{pmatrix} y_n \\ y_n' \end{pmatrix} + h \begin{pmatrix} y_n' \\ g(x_{n+1}, y_n + hy_n') \end{pmatrix}.$$

It now has the form of a method (8.4) with the correct value function $z(x, h) = \left(y(x), (y(x+h) - y(x))/h \right)^T$. Here $y(x)$ denotes the exact solution of (8.8).

Clearly, all Nyström methods (Section II.14) fit into this framework, as do multistep methods for second order differential equations. They will be investigated in more detail in Section III.10.

Example 8.7. *Multi-step multi-stage multi-derivative* methods seem to be the most general class of explicitly given linear methods and generalize the methods of Section II.13. In the notation of that section, we can write

$$v_i^{(n)} = \sum_{j=1}^{k} \widetilde{a}_{ij} u_j^{(n)} + \sum_{r=1}^{q} \frac{h^r}{r!} \sum_{j=1}^{s} \widetilde{b}_{ij}^{(r)} D^r y(x_n + c_j h, v_j^{(n)}) \quad i = 1, \ldots, s,$$

$$u_i^{(n+1)} = \sum_{j=1}^{k} a_{ij} u_j^{(n)} + \sum_{r=1}^{q} \frac{h^r}{r!} \sum_{j=1}^{s} b_{ij}^{(r)} D^r y(x_n + c_j h, v_j^{(n)}) \quad i = 1, \ldots, k.$$

Such methods have been studied in Hairer & Wanner (1973).

Stability and Order

The following study of stability, order and convergence follows mainly the lines of Skeel (1976). Stability of a numerical scheme just requires that for $h \to 0$ the numerical solution remain bounded. This motivates the following definition.

Definition 8.8. Method (8.4) is called *stable* if $\|S^n\|$ is uniformly bounded for all $n \geq 0$.

The local error of method (8.4) is defined in exactly the same way as for one-step methods (Section II.3) and multistep methods (Section III.2).

Definition 8.9. Let $z(x, h)$ be the correct value function for the method (8.4) and let $z_n = z(x_n, h)$. The *local error* is then given by (see Fig. 8.1)

$$
\begin{aligned}
d_0 &= z_0 - \varphi(h) \\
d_{n+1} &= z_{n+1} - Sz_n - h\Phi(x_n, z_n, h), \qquad n = 0, 1, \ldots
\end{aligned}
\tag{8.9}
$$

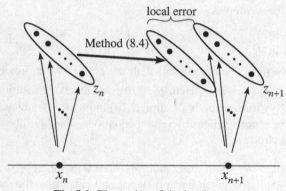

Fig. 8.1. Illustration of the local error

The definition of order is not as straightforward. The requirement that the local error be $\mathcal{O}(h^{p+1})$ (cf. one-step and multistep methods) will turn out to be sufficient but in general not necessary for convergence of order p. For an appropriate definition we need the *spectral decomposition* of the matrix S.

First observe that, whenever the local error (8.9) tends to zero for $h \to 0$ ($nh = x - x_0$ fixed), we get

$$
0 = z(x, 0) - Sz(x, 0), \tag{8.10}
$$

so that 1 is an eigenvalue of S and $z(x, 0)$ a corresponding eigenvector. Furthermore, by stability, no eigenvalue of S can lie outside the unit disc and the eigenvalues of modulus one can not give rise to Jordan chains. Denoting the eigenvalues of modulus one by $\zeta_1 (= 1), \zeta_2, \ldots, \zeta_l$, the Jordan canonical form of S (see

(I.12.14)) is therefore the block diagonal matrix

$$S = T \operatorname{diag} \left\{ \begin{pmatrix} 1 & & \\ & \ddots & \\ & & 1 \end{pmatrix}, \begin{pmatrix} \zeta_2 & & \\ & \ddots & \\ & & \zeta_2 \end{pmatrix}, \dots, \begin{pmatrix} \zeta_l & & \\ & \ddots & \\ & & \zeta_l \end{pmatrix}, \tilde{J} \right\} T^{-1}.$$

If we decompose this matrix into the terms which correspond to the single eigenvalues we obtain

$$S = E + \zeta_2 E_2 + \dots + \zeta_l E_l + \tilde{E} \tag{8.11}$$

where

$$E = T \operatorname{diag} \left\{ I, 0, 0, \dots \right\} T^{-1}, \tag{8.12}$$

$$E_2 = T \operatorname{diag} \left\{ 0, I, 0, \dots \right\} T^{-1}, \dots, \quad E_l = T \operatorname{diag} \left\{ 0, \dots, 0, I, 0 \right\} T^{-1},$$

$$\tilde{E} = T \operatorname{diag} \left\{ 0, 0, 0, \dots, \tilde{J} \right\} T^{-1}.$$

We are now prepared to give

Definition 8.10. The method (8.4) is of *order p (consistent of order p)*, if for all problems (8.3) with p times continuously differentiable f, the local error satisfies

$$
\begin{aligned}
d_0 &= \mathcal{O}(h^p) \\
E(d_0 + d_1 + \dots + d_n) + d_{n+1} &= \mathcal{O}(h^p) \qquad \text{for } 0 \le nh \le Const.
\end{aligned}
\tag{8.13}
$$

Remark. This property is called *quasi-consistency of order p* by Skeel (1976).

If the right-hand side of the differential equation (8.3) is p-times continuously differentiable then, in general, $\varphi(h)$, $\Phi(x, u, h)$ and $z(x, h)$ are also smooth, so that the local error (8.9) can be expanded into a Taylor series in h:

$$
\begin{aligned}
d_0 &= \gamma_0 + \gamma_1 h + \dots + \gamma_{p-1} h^{p-1} + \mathcal{O}(h^p) \\
d_{n+1} &= \delta_0(x_n) + \delta_1(x_n) h + \dots + \delta_p(x_n) h^p + \mathcal{O}(h^{p+1}).
\end{aligned}
\tag{8.14}
$$

The function $\delta_j(x)$ is then $(p - j + 1)$-times continuously differentiable. The following lemma gives a more practical characterization of the order of the methods (8.4).

Lemma 8.11. *Assume that the local error of method (8.4) satisfies (8.14) with continuous $\delta_j(x)$. The method is then of order p, if and only if*

$$d_n = \mathcal{O}(h^p) \quad \text{for } 0 \le nh \le Const, \quad \text{and} \quad E\delta_p(x) = 0. \tag{8.15}$$

Proof. The condition (8.15) is equivalent to

$$d_n = \mathcal{O}(h^p), \qquad E d_{n+1} = \mathcal{O}(h^{p+1}) \qquad \text{for } 0 \le nh \le Const, \tag{8.16}$$

which is clearly sufficient for order p. We now show that (8.15) is also necessary. Since $E^2 = E$ (see (8.12)) order p implies

$$d_n = \mathcal{O}(h^p), \qquad E(d_1 + \ldots + d_n) = \mathcal{O}(h^p) \qquad \text{for } 0 \leq nh \leq \textit{Const.} \quad (8.17)$$

This is best seen by multiplying (8.13) by E. Consider now pairs (n, h) such that $nh = x - x_0$ for some fixed x. We insert (8.14) (observe that $d_n = \mathcal{O}(h^p)$) into $E(d_1 + \ldots + d_n)$ and approximate the resulting sum by the corresponding Riemann integral

$$E(d_1 + \ldots + d_n) = h^p E \sum_{j=1}^{n} \delta_p(x_{j-1}) + \mathcal{O}(h^p) = h^{p-1} E \int_{x_0}^{x} \delta_p(s)\, ds + \mathcal{O}(h^p).$$

It follows from (8.17) that $E \int_{x_0}^{x} \delta_p(s)\, ds = 0$ and by differentiation that $E\delta_p(x) = 0$. $\qquad \square$

Convergence

In addition to the numerical solution given by (8.4) we consider a perturbed numerical solution $(\widehat{u}_n)$ defined by

$$\begin{aligned}
\widehat{u}_0 &= \varphi(h) + r_0 \\
\widehat{u}_{n+1} &= S\widehat{u}_n + h\Phi(x_n, \widehat{u}_n, h) + r_{n+1}, \quad n = 0, 1, \ldots, N-1
\end{aligned} \qquad (8.18)$$

for some perturbation $R = (r_0, r_1, \ldots, r_N)$. For example, the exact solution $z_n = z(x_n, h)$ can be interpreted as a perturbed solution, where the perturbation is just the local error. The following lemma gives the best possible qualitative bound on the difference $u_n - \widehat{u}_n$ in terms of the perturbation R. We have to assume that the increment function $\Phi(x, u, h)$ satisfies a Lipschitz condition with respect to u (on a compact neighbourhood of the solution). This is the case for all reasonable methods.

Lemma 8.12. *Let the method (8.4) be stable and assume the sequences (u_n) and $(\widehat{u}_n)$ be given by (8.4) and (8.18), respectively. Then there exist positive constants c and C such that for any perturbation R and for $hN \leq Const$*

$$c\|R\|_S \leq \max_{0 \leq n \leq N} \|u_n - \widehat{u}_n\| \leq C\|R\|_S$$

with

$$\|R\|_S = \max_{0 \leq n \leq N} \Big\| \sum_{j=0}^{n} S^{n-j} r_j \Big\|.$$

Remark. $\|R\|_S$ is a norm on $\mathbb{R}^{(N+1)q}$. Its positivity is seen as follows: if $\|R\|_S = 0$ then for $n = 0, 1, 2, \ldots$ one obtains $r_0 = 0, r_1 = 0, \ldots$ recursively.

Proof. Set $\Delta u_n = \widehat{u}_n - u_n$ and $\Delta \Phi_n = \Phi(x_n, \widehat{u}_n, h) - \Phi(x_n, u_n, h)$. Then we have

$$\Delta u_{n+1} = S \Delta u_n + h \Delta \Phi_n + r_{n+1}. \tag{8.19}$$

By assumption there exists a constant L such that $\|\Delta \Phi_n\| \le L \|\Delta u_n\|$. Solving the difference equation (8.19) gives $\Delta u_0 = r_0$ and

$$\Delta u_{n+1} = \sum_{j=0}^{n} S^{n-j} h \Delta \Phi_j + \sum_{j=0}^{n+1} S^{n+1-j} r_j. \tag{8.20}$$

By stability there exists a constant B such that

$$\|S^n\| L \le B \qquad \text{for all } n \ge 0. \tag{8.21}$$

Thus (8.20) becomes

$$\|\Delta u_{n+1}\| \le hB \sum_{j=0}^{n} \|\Delta u_j\| + \|R\|_S.$$

By induction on n it follows that

$$\|\Delta u_n\| \le (1 + hB)^n \|R\|_S \le \exp(Const \cdot B) \cdot \|R\|_S,$$

which proves the second inequality in the lemma. From (8.20) and (8.21)

$$\Big\| \sum_{j=0}^{n} S^{n-j} r_j \Big\| \le (1 + nhB) \max_{0 \le n \le N} \|\Delta u_n\|,$$

and we thus obtain for $Nh \le Const$

$$\|R\|_S \le (1 + Const \cdot B) \cdot \max_{0 \le n \le N} \|\widehat{u}_n - u_n\|. \qquad \square$$

Remark. Two-sided error bounds, such as in Lemma 8.12, were first studied, in the case of multi-step methods, by Spijker (1971). This theory has become prominent through the treatment of Stetter (1973, pp. 81-84). Extensions to general linear methods are due to Skeel (1976) and Albrecht (1978).

Using the lemma above we can prove

Theorem 8.13. *Consider a stable method (8.4) and assume that the local error satisfies (8.14) with $\delta_p(x)$ continuously differentiable. The method is then convergent of order p, i.e., the global error $u_n - z_n$ satisfies*

$$u_n - z_n = \mathcal{O}(h^p) \qquad \text{for } 0 \le nh \le Const,$$

if and only if it is consistent of order p.

Proof. The identity

$$E(d_0 + \ldots + d_n) + d_{n+1} = \sum_{j=0}^{n+1} S^{n+1-j} d_j - (S - E) \sum_{j=0}^{n} S^{n-j} d_j,$$

which is a consequence of $ES = E$ (see (8.11) and (8.12)), implies that for $n \leq N - 1$ and $D = (d_0, \ldots, d_N)$,

$$\|E(d_0 + \ldots + d_n) + d_{n+1}\| \leq (1 + \|S - E\|) \cdot \|D\|_S. \tag{8.22}$$

The lower bound of Lemma 8.12, with r_n and $\widehat{u}_n$ replaced by d_n and z_n respectively, yields the "only if" part of the theorem.

For the "if" part we use the upper bound of Lemma 8.12. We have to show that consistency of order p implies

$$\max_{0 \leq n \leq N} \Big\| \sum_{j=0}^{n} S^{n-j} d_j \Big\| = \mathcal{O}(h^p). \tag{8.23}$$

By (8.11) and (8.12) we have

$$S^{n-j} = E + \zeta_2^{n-j} E_2 + \ldots + \zeta_l^{n-j} E_l + \widetilde{E}^{n-j}.$$

This identity together with Lemma 8.11 implies

$$\sum_{j=0}^{n} S^{n-j} d_j = h^p E_2 \sum_{j=1}^{n} \zeta_2^{n-j} \delta_p(x_{j-1}) + \ldots$$

$$+ h^p E_l \sum_{j=1}^{n} \zeta_l^{n-j} \delta_p(x_{j-1}) + \sum_{j=0}^{n} \widetilde{E}^{n-j} d_j + \mathcal{O}(h^p).$$

The last term in this expression is $\mathcal{O}(h^p)$ since in a suitable norm $\|\widetilde{E}\| < 1$ and therefore

$$\Big\| \sum_{j=0}^{n} \widetilde{E}^{n-j} d_j \Big\| \leq \sum_{j=0}^{n} \|\widetilde{E}\|^{n-j} \|d_j\| \leq \frac{1}{1 - \|\widetilde{E}\|} \cdot \max_{0 \leq n \leq N} \|d_n\|.$$

For the rest we use partial summation (Abel 1826)

$$\sum_{j=1}^{n} \zeta^{n-j} \delta(x_{j-1}) = \frac{1 - \zeta^n}{1 - \zeta} \cdot \delta(x_0) + \sum_{j=1}^{n} \frac{1 - \zeta^{n-j}}{1 - \zeta} \cdot \Big(\delta(x_j) - \delta(x_{j-1}) \Big) = \mathcal{O}(1),$$

whenever $|\zeta| = 1$, $\zeta \neq 1$ and δ is of bounded variation. $\qquad \square$

Order Conditions for General Linear Methods

For the construction of a pth order general linear method (8.7) the conditions (8.15) are still not very practical. One would like to have instead algebraic conditions in the free parameters, as is the case for Runge-Kutta methods. We shall demonstrate how this can be achieved using the theory of B-series of Section II.12 (see also Burrage & Moss 1980). In order to avoid tensor products we assume in what follows that the differential equation under consideration is a scalar one. All results, however, are also valid for systems. We further assume the differential equation to be autonomous, so that the theory of Section II.12 is directly applicable. This will be justified in Remark 8.17 below.

Suppose now that the components of the correct value function $z(x, h) = (z_1(x, h), \ldots, z_k(x, h))^T$ possess an expansion as a B-series

$$z_i(x, h) = B(\mathbf{z}_i, y(x))$$

so that with $\mathbf{z}(t) = (\mathbf{z}_1(t), \ldots, \mathbf{z}_k(t))^T$,

$$z(x, h) = \mathbf{z}(\emptyset) y(x) + h \mathbf{z}(\tau) f(y(x)) + \ldots. \tag{8.24}$$

Before deriving the order conditions we observe that (8.7a) makes sense only if $v_j^{(n)} \to y(x_n)$ for $h \to 0$. Otherwise $f(v_j^{(n)})$ need not be defined. Since $u_j^{(n)}$ is an approximation of $z_j(x_n, h)$, this leads to the condition $\sum \widetilde{a}_{ij} z_j(\emptyset) = 1$. This together with (8.10) are the so-called *preconsistency conditions:*

$$A\mathbf{z}(\emptyset) = \mathbf{z}(\emptyset), \qquad \widetilde{A} z(\emptyset) = \mathbb{1}. \tag{8.25}$$

A and $\widetilde{A}$ are the matrices with entries a_{ij} and $\widetilde{a}_{ij}$, respectively, and $\mathbb{1}$ is the column vector $(1, \ldots, 1)^T$. Recall that the local error (8.9) for the general linear method (8.7) is given by

$$d_i^{(n+1)} = z_i(x_n + h, h) - \sum_{j=1}^{k} a_{ij} z_j(x_n, h) - \sum_{j=1}^{s} b_{ij} h f(v_j) \tag{8.26a}$$

where

$$v_i = \sum_{j=1}^{k} \widetilde{a}_{ij} z_j(x_n, h) + \sum_{j=1}^{s} \widetilde{b}_{ij} h f(v_j). \tag{8.26b}$$

For the derivation of the order conditions we write v_i and $d_i^{(n+1)}$ as B-series

$$v_i = B(\mathbf{v}_i, y(x_n)), \qquad d_i^{(n+1)} = B(\mathbf{d}_i, y(x_n)).$$

By the composition theorem for B-series and by formula (12.10) of Section II.12 we have

$$z_i(x_n + h, h) = B(\mathbf{z}_i, y(x_n + h)) = B(\mathbf{z}_i, B(\mathbf{p}, y(x_n))) = B(\mathbf{p}\mathbf{z}_i, y(x_n)).$$

Inserting all these series into (8.26) and comparing the coefficients we arrive at

$$\mathbf{d}_i(t) = (\mathbf{p}\mathbf{z}_i)(t) - \sum_{j=1}^{k} a_{ij}\mathbf{z}_j(t) - \sum_{j=1}^{s} b_{ij}\mathbf{v}'_j(t)$$

$$\mathbf{v}_i(t) = \sum_{j=1}^{k} \widetilde{a}_{ij}\mathbf{z}_j(t) + \sum_{j=1}^{s} \widetilde{b}_{ij}\mathbf{v}'_j(t).$$
(8.27)

An application of Lemma 8.11 now yields

Theorem 8.14. *Let* $\mathbf{d}(t) = \big(\mathbf{d}_1(t), \ldots, \mathbf{d}_k(t)\big)^T$ *with* $\mathbf{d}_i(t)$ *be given by (8.27). The general linear method (8.7) is of order p, iff*

$$\mathbf{d}(t) = 0 \qquad for \ t \in T, \ \varrho(t) \leq p - 1,$$
$$E\mathbf{d}(t) = 0 \qquad for \ t \in T, \ \varrho(t) = p,$$
(8.28)

where the matrix E is defined in (8.12).   □

Corollary 8.15. *Sufficient conditions for the general linear method to be of order p are*

$$\mathbf{d}(t) = 0 \qquad for \ t \in T, \ \varrho(t) \leq p.$$
(8.29)

                                                                                    □

Remark 8.16. The expression $(\mathbf{p}\mathbf{z}_i)(t)$ in (8.27) can be computed using formula (12.8) of Section II.12. Since $\mathbf{p}(t) = 1$ for all trees t, we have

$$(\mathbf{p}\mathbf{z}_i)(t) = \sum_{j=0}^{\varrho(t)} \binom{\varrho(t)}{j} \frac{1}{\alpha(t)} \sum_{\text{all labellings}} \mathbf{z}_i(s_j(t)).$$
(8.30)

This rather complicated formula simplifies considerably if we assume that the coefficients $\mathbf{z}_i(t)$ of the correct value function depend only on the order of t, i.e., that

$$\mathbf{z}_i(t) = \mathbf{z}_i(u) \qquad \text{whenever} \ \varrho(t) = \varrho(u) \ .$$
(8.31)

In this case formula (8.30) becomes

$$(\mathbf{p}\mathbf{z}_i)(t) = \sum_{j=0}^{\varrho(t)} \binom{\varrho(t)}{j} \mathbf{z}_i(\tau^j).$$
(8.32)

Here τ^j represents any tree of order j, e.g.,

$$\tau^j = [\underbrace{\tau, \ldots, \tau}_{j-1}], \quad \tau^1 = \tau, \quad \tau^0 = \emptyset.$$
(8.33)

Usually the components of $z(x, h)$ are composed of

$$y(x), \ y(x + jh), \ hy'(x), \ h^2 y''(x), \ldots,$$

in which case assumption (8.31) is satisfied.

Remark 8.17. Non-autonomous systems. For the differential equation $x' = 1$, formula (8.7a) becomes

$$v_n = \widetilde{A} u_n + h \widetilde{B} \mathbb{1}.$$

Assuming that $x' = 1$ is integrated exactly, i.e., $u_n = z(\emptyset) x_n + h z(\tau)$ we obtain $v_n = x_n \mathbb{1} + hc$, where $c = (c_1, \ldots, c_s)^T$ is given by

$$c = \widetilde{A} z(\tau) + \widetilde{B} e. \tag{8.34}$$

This definition of the c_i implies that the numerical results for $y' = f(x, y)$ and for the augmented autonomous differential equation are the same and the above results are also valid in the general case.

Table 8.1 presents the order conditions up to order 3 in addition to the preconsistency conditions (8.25). We assume that (8.31) is satisfied and that c is given by (8.34). Furthermore, c^j denotes the vector $(c_1^j, \ldots, c_s^j)^T$.

Table 8.1. Order conditions for general linear methods

t	$\varrho(t)$	order condition
τ	1	$Az(\tau) + B \mathbb{1} = z(\tau) + z(\emptyset)$
τ^2	2	$Az(\tau^2) + 2Bc = z(\tau^2) + 2z(\tau) + z(\emptyset)$
τ^3	3	$Az(\tau^3) + 3Bc^2 = z(\tau^3) + 3z(\tau^2) + 3z(\tau) + z(\emptyset)$
$[\tau^2]$	3	$Az(\tau^3) + 3Bv(\tau^2) = z(\tau^3) + 3z(\tau^2) + 3z(\tau) + z(\emptyset)$
		with $v(\tau^2) = \widetilde{A} z(\tau^2) + 2 \widetilde{B} c$

Construction of General Linear Methods

Let us demonstrate on an example how low order methods can be constructed: we set $k = s = 2$ and fix the correct value function as

$$z(x, h) = \big(y(x), \ y(x - h)\big)^T.$$

This choice satisfies (8.24) and (8.31) with

$$z(\emptyset) = \begin{pmatrix} 1 \\ 1 \end{pmatrix}, \qquad z(\tau) = \begin{pmatrix} 0 \\ -1 \end{pmatrix}, \qquad z(\tau^2) = \begin{pmatrix} 0 \\ 1 \end{pmatrix}, \ldots.$$

Since the second component of $z(x+h,h)$ is equal to the first component of $z(x,h)$, it is natural to look for methods with

$$A = \begin{pmatrix} a_{11} & a_{12} \\ 1 & 0 \end{pmatrix}, \qquad B = \begin{pmatrix} b_{11} & b_{12} \\ 0 & 0 \end{pmatrix}.$$

We further impose

$$\widetilde{B} = \begin{pmatrix} 0 & 0 \\ \widetilde{b}_{21} & 0 \end{pmatrix}$$

so that the resulting method is explicit.

The preconsistency condition (8.25), formula (8.34) and the order conditions of Table 8.1 yield the following equations to be solved:

$$a_{11} + a_{12} = 1 \tag{8.35a}$$

$$\widetilde{a}_{11} + \widetilde{a}_{12} = 1, \qquad \widetilde{a}_{21} + \widetilde{a}_{22} = 1 \tag{8.35b}$$

$$c_1 = -\widetilde{a}_{12}, \qquad c_2 = \widetilde{b}_{21} - \widetilde{a}_{22} \tag{8.35c}$$

$$-a_{12} + b_{11} + b_{12} = 1 \tag{8.35d}$$

$$a_{12} + 2(b_{11}c_1 + b_{12}c_2) = 1 \tag{8.35e}$$

$$-a_{12} + 3(b_{11}c_1^2 + b_{12}c_2^2) = 1 \tag{8.35f}$$

$$-a_{12} + 3\left(b_{11}\widetilde{a}_{12} + b_{12}(\widetilde{a}_{22} + 2\widetilde{b}_{21}c_1)\right) = 1. \tag{8.35g}$$

These are 9 equations in 11 unknowns. Letting c_1 and c_2 be free parameters, we obtain the solution in the following way: compute a_{12}, b_{11} and b_{12} from the linear system (8.35d,e,f), then $\widetilde{a}_{12}, \widetilde{a}_{22}$ and $\widetilde{b}_{21}$ from (8.35c,g) and finally $a_{11}, \widetilde{a}_{11}$ and $\widetilde{a}_{21}$ from (8.35a,b). A particular solution for $c_1 = 1/2$, $c_2 = -2/5$ is:

$$A = \begin{pmatrix} 16/11 & -5/11 \\ 1 & 0 \end{pmatrix}, \quad B = \begin{pmatrix} 104/99 & -50/99 \\ 0 & 0 \end{pmatrix},$$

$$\widetilde{A} = \begin{pmatrix} 3/2 & -1/2 \\ 3/2 & -1/2 \end{pmatrix}, \quad \widetilde{B} = \begin{pmatrix} 0 & 0 \\ -9/10 & 0 \end{pmatrix}. \tag{8.36}$$

This method, which represents a stable explicit 2-step, 2-stage method of order 3, is due to Butcher (1984).

The construction of higher order methods soon becomes very complicated, and the use of *simplifying assumptions* will be very helpful:

Theorem 8.18 (Burrage & Moss 1980). *Assume that the correct value function satisfies (8.31). The simplifying assumptions*

$$\widetilde{A}\mathbf{z}(\tau^j) + j\widetilde{B}c^{j-1} = c^j \qquad j = 1, \ldots, p-1 \tag{8.37}$$

together with the preconsistency relations (8.25) and the order conditions for the "bushy trees"

$$\mathbf{d}(\tau^j) = 0 \qquad j = 1, \ldots, p$$

imply that the method (8.7) is of order p.

Proof. An induction argument based on (8.27) implies that

$$\mathbf{v}(t) = \mathbf{v}(\tau^j) \qquad \text{for } \varrho(t) = j, \ j = 1, \ldots, p-1$$

and consequently also that

$$\mathbf{d}(t) = \mathbf{d}(\tau^j) \qquad \text{for } \varrho(t) = j, \ j = 1, \ldots, p. \qquad \square$$

The simplifying assumptions (8.37) allow an interesting interpretation: they are equivalent to the fact that the internal stages $v_1^{(n)}$ approximate the exact solution at $x_n + c_i h$ up to order $p-1$, i.e., that

$$v_i^{(n)} - y(x_n + c_i h) = \mathcal{O}(h^p).$$

In the case of Runge-Kutta methods (8.37) reduces to the conditions $C(p-1)$ of Section II.7.

For further examples of general linear methods satisfying (8.37) we refer to Burrage & Moss (1980) and Butcher (1981). See also Burrage (1985) and Butcher (1985a).

Exercises

1. Consider the composition of (cf. Example 8.5)

   a) explicit and implicit Euler method;

   b) implicit and explicit Euler method.

   To which methods are they equivalent? What is the order of the composite methods?

2. a) Suppose that each of the m multistep methods (ϱ_i, σ_i) $i = 1, \ldots, m$ is of order p. Prove that the corresponding cyclic method is of order at least p.

   b) Construct a stable, 2-cyclic, 3-step linear multistep method of order 5: find first a one-parameter family of linear 3-step methods of order 5 (which are necessarily unstable).

   Result.

   $$\varrho_c(\zeta) = c\zeta^3 + \left(\frac{19}{30} - c\right)\zeta^2 - \left(\frac{8}{30} + c\right)\zeta + \left(c - \frac{11}{30}\right)$$

   $$\sigma_c(\zeta) = \left(\frac{1}{9} - \frac{c}{3}\right)\zeta^3 + \left(c + \frac{8}{30}\right)\zeta^2 + \left(\frac{19}{30} - c\right)\zeta + \left(\frac{c}{3} - \frac{1}{90}\right).$$

   Then determine c_1 and c_2, such that the eigenvalues of the matrix S for the composite method become $1, 0, 0$.

3. Prove that the composition of two different general linear methods (with the same correct value function) again gives a general linear method. As a consequence, the cyclic methods of Example 8.4 are general linear methods.

4. Suppose that all eigenvalues of S (except $\zeta_1 = 1$) lie inside the unit circle. Then

$$\|R\|_E = \max_{0 \leq n \leq N} \left\| r_n + E \sum_{j=0}^{n-1} r_j \right\|$$

is a minimal stability functional.

5. Verify for linear multistep methods that the consistency conditions (2.6) are equivalent to consistency of order 1 in the sense of Lemma 8.11.

6. Write method (8.1) as general linear method (8.7) and determine its order (answer: $p = 5$).

7. Interpret the method of Caira, Costabile & Costabile (1990)

$$k_i^n = hf\left(x_n + c_i h, \; y_n + \sum_{j=1}^{s} \bar{a}_{ij} k_j^{n-1} + \sum_{j=1}^{i-1} a_{ij} k_j^n\right)$$

$$y_{n+1} = y_n + \sum_{i=1}^{s} b_i k_i^n$$

as general linear method. Show that, if

$$\|k_i^{-1} - hy'(x_0 + (c_i - 1)h)\| \leq C \cdot h^p,$$

$$\sum_{i=1}^{s} b_i c_i^{q-1} = \frac{1}{q}, \qquad q = 1, \ldots, p,$$

$$\sum_{j=1}^{s} \bar{a}_{ij}(c_j - 1)^{q-1} + \sum_{j=1}^{i-1} a_{ij} c_j^{q-1} = \frac{c_i^q}{q}, \qquad q = 1, \ldots, p-1,$$

then the method is of order at least p. Find parallels of these conditions with those of Theorem 8.18.

8. Jackiewicz & Zennaro (1992) propose the following two-step Runge-Kutta method

$$Y_i^{n-1} = y_{n-1} + h_{n-1} \sum_{j=1}^{i-1} a_{ij} f(Y_j^{n-1}), \qquad Y_i^n = y_n + h_{n-1} \xi \sum_{j=1}^{i-1} a_{ij} f(Y_j^n),$$

$$y_{n+1} = y_n + h_{n-1} \sum_{i=1}^{s} v_i f(Y_i^{n-1}) + h_{n-1} \xi \sum_{i=1}^{s} w_i f(Y_i^n), \qquad (8.38)$$

where $\xi = h_n/h_{n-1}$. The coefficients v_i, w_i may depend on ξ, but the a_{ij} do not. Hence, this method requires s function evaluations per step.

a) Show that the order of method (8.38) is p (according to Definition 8.10) if and only if for all trees t with $1 \leq \varrho(t) \leq p$

$$\xi^{\varrho(t)} = \sum_{i=1}^{s} v_i (\mathbf{y}^{-1} \mathbf{g}'_i)(t) + \xi^{\varrho(t)} \sum_{i=1}^{s} w_i \mathbf{g}'_i(t), \tag{8.39}$$

where, as for Runge-Kutta methods, $\mathbf{g}_i(t) = \sum_{j=1}^{i-1} a_{ij} \mathbf{g}'_j(t)$. The coefficients $\mathbf{y}^{-1}(t) = (-1)^{\varrho(t)}$ are those of $y(x_n - h) = B(\mathbf{y}^{-1}, y(x_n))$.

b) Under the assumption

$$v_i + \xi^p w_i = 0 \qquad \text{for} \quad i = 2, \dots, s \tag{8.40}$$

the order conditions (8.39) are equivalent to

$$\xi = \sum_{i=1}^{s} v_i + \xi \sum_{i=1}^{s} w_i, \tag{8.41a}$$

$$\xi^r = \sum_{j=1}^{r-1} j \binom{r}{j} (-1)^{r-j} \sum_{i=1}^{s} v_i c_i^{j-1} + (1 - \xi^{r-p}) r \sum_{i=1}^{s} v_i c_i^{r-1}, \quad r = 2, \dots, p, \tag{8.41b}$$

$$\sum_{i=1}^{s} v_i \left(\mathbf{g}'_i(u) - \varrho(u) c_i^{\varrho(u)-1} \right) = 0 \qquad \text{for } \varrho(u) \leq p - 1. \tag{8.41c}$$

c) The conditions (8.41a,b) uniquely define $\sum_i w_i$, $\sum_i v_i c_i^{j-1}$ as functions of $\xi > 0$ (for $j = 1, \dots, p-1$).

d) For each continuous Runge-Kutta method of order $p - 1 \geq 2$ there exists a method (8.38) of order p with the same coefficient matrix (a_{ij}).

Hints. To obtain (8.41c) subtract equation (8.40) from the same equation where t is replaced by the bushy tree of order $\varrho(t)$. Then proceed by induction. The conditions $\sum_i v_i c_i^{j-1} = f_j^p(\xi)$, $j = 1, \dots, p-1$, obtained from (c), together with (8.41c) have the same structure as the order conditions (order $p-1$) of a continuous Runge-Kutta method (Theorem II.6.1).

III.9 Asymptotic Expansion of the Global Error

The asymptotic expansion of the global error of multistep methods was studied in the famous thesis of Gragg (1964). His proof is very technical and can also be found in a modified version in the book of Stetter (1973), pp. 234-245. The existence of asymptotic expansions for general linear methods was conjectured by Skeel (1976). The proof given below (Hairer & Lubich 1984) is based on the ideas of Section II.8.

An Instructive Example

Let us start with an example in order to understand which kind of asymptotic expansion may be expected. We consider the simple differential equation

$$y' = -y, \qquad y(0) = 1,$$

take a constant step size h and apply the 3-step BDF-formula (1.22') with one of the following three starting procedures:

$$y_0 = 1, \qquad y_1 = \exp(-h), \qquad\qquad y_2 = \exp(-2h) \quad \text{(exact values)} \quad (9.1a)$$

$$y_0 = 1, \qquad y_1 = 1 - h + \frac{h^2}{2} - \frac{h^3}{6}, \qquad y_2 = 1 - 2h + 2h^2 - \frac{4h^3}{3}, \qquad (9.1b)$$

$$y_0 = 1, \qquad y_1 = 1 - h + \frac{h^2}{2}, \qquad\qquad y_2 = 1 - 2h + 2h^2. \qquad (9.1c)$$

The three pictures on the left of Fig. 9.1 (they correspond to the three starting procedures in the same order) show the global error divided by h^3 for the five step sizes $h = 1/5, 1/10, 1/20, 1/40, 1/80$.

For the first two starting procedures we observe uniform convergence to the function $e_3(x) = xe^{-x}/4$ (cf. formula (2.12)), so that

$$y_n - y(x_n) = e_3(x_n)h^3 + \mathcal{O}(h^4), \qquad (9.2)$$

valid uniformly for $0 \le nh \le Const$. In the third case we have convergence to $e_3(x) = (9+x)e^{-x}/4$ (Exercise 2), but this time the convergence is no longer uniform. Therefore (9.2) only holds for x_n bounded away from x_0, i.e., for $0 < \alpha \le nh \le Const$. In the three pictures on the right of Fig. 9.1 the functions

$$\left(y_n - y(x_n) - e_3(x_n)h^3\right)/h^4 \qquad (9.3)$$

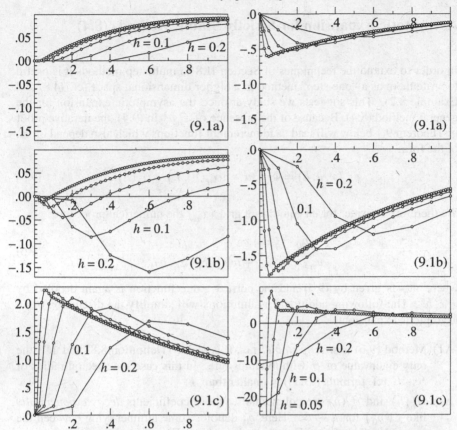

Fig. 9.1. The values $(y_n - y(x_n))/h^3$ (left), $(y_n - y(x_n) - e_3(x_n)h^3)/h^4$ (right) for the 3-step BDF method and for three different starting procedures

are plotted. Convergence to functions $e_4(x)$ is observed in all cases. Clearly, since $e_3(x_0) \neq 0$ for the starting procedure (9.1c), the sequence (9.3) diverges at x_0 like $\mathcal{O}(1/h)$ in this case.

We conclude from this example that for linear multistep methods there is in general no asymptotic expansion of the form

$$y_n - y(x_n) = e_p(x_n)h^p + e_{p+1}(x_n)h^{p+1} + \dots$$

which holds uniformly for $0 \leq nh \leq Const$. It will be necessary to add perturbation terms

$$y_n - y(x_n) = \left(e_p(x_n) + \varepsilon_n^p\right)h^p + \left(e_{p+1}(x_n) + \varepsilon_n^{p+1}\right)h^{p+1} + \dots \qquad (9.4)$$

which compensate the irregularity near x_0. If the perturbations ε_n^j decay exponentially (for $n \to \infty$), then they have no influence on the asymptotic expansion for x_n bounded away from x_0.

Asymptotic Expansion for Strictly Stable Methods (8.4)

In order to extend the techniques of Section II.8 to multistep methods it is useful to write them as a "one-step" method in a higher dimensional space (cf. (4.8) and Example 8.2). This suggests we study at once the asymptotic expansion for the general method (8.4). Because of the presence of $\varepsilon_n^j h^j$ in (9.4), the iterative proof of Theorem 9.1 below will lead us to increment functions which also depend on n, of the form

$$\Phi_n(x, u, h) = \Phi(x, u + h\alpha_n(h), h) + \beta_n(h). \tag{9.5}$$

We therefore consider for an equidistant grid (x_n) the numerical procedure

$$\begin{aligned} u_0 &= \varphi(h) \\ u_{n+1} &= Su_n + h\Phi_n(x_n, u_n, h), \end{aligned} \tag{9.6}$$

where Φ_n is given by (9.5) and the correct value function is again denoted by $z(x, h)$. The following additional assumptions will simplify the discussion of an asymptotic expansion:

A1) Method (9.6) is *strictly stable;* i.e., it is stable (Definition 8.8) and 1 is the only eigenvalue of S with modulus one. In this case the spectral radius of $S - E$ (cf. formula (8.11)) is smaller than 1;

A2) $\alpha_n(h)$ and $\beta_n(h)$ are polynomials, whose coefficients *decay exponentially* like $\mathcal{O}(\varrho_0^n)$ for $n \to \infty$. Here ϱ_0 denotes some number lying between the spectral radius of $S - E$ and one; i.e. $\varrho(S - E) < \varrho_0 < 1$;

A3) the functions φ, z and Φ are sufficiently differentiable.

Assumption A3 allows us to expand the local error, defined by (8.9), into a Taylor series:

$$\begin{aligned} d_{n+1} &\doteq z(x_n + h, h) - Sz(x_n, h) - h\Phi(x_n, z(x_n, h) + h\alpha_n(h), h) - h\beta_n(h) \\ &= d_0(x_n) + d_1(x_n)h + \ldots + d_{N+1}(x_n)h^{N+1} \\ &\quad - h^2 \frac{\partial \Phi}{\partial u}(x_n, z(x_n, 0), 0)\alpha_n(h) - \ldots - h\beta_n(h) + \mathcal{O}(h^{N+1}). \end{aligned}$$

The expressions involving $\alpha_n(h)$ can be simplified further. Indeed, for a smooth function $G(x)$ we have

$$G(x_n)\alpha_n(h) = G(x_0)\alpha_n(h) + hG'(x_0)n\alpha_n(h) + \ldots + h^{N+1}R(n, h).$$

We observe that $n^j \alpha_n(h)$ is again a polynomial in h and that its coefficients decay like $\mathcal{O}(\varrho^n)$ where ϱ satisfies $\varrho_0 < \varrho < 1$. The same argument shows the boundedness of the remainder $R(n, h)$ for $0 \leq nh \leq Const$. As a consequence we can write the local error in the form

$$d_0 = \gamma_0 + \gamma_1 h + \ldots + \gamma_N h^N + \mathcal{O}(h^{N+1})$$

$$d_{n+1} = \left(d_0(x_n) + \delta_n^0\right) + \ldots + \left(d_{N+1}(x_n) + \delta_n^{N+1}\right)h^{N+1} + \mathcal{O}(h^{N+2}) \quad (9.7)$$

$$\text{for } 0 \le nh \le Const.$$

The functions $d_j(x)$ are smooth and the perturbations δ_n^j satisfy $\delta_n^j = \mathcal{O}(\varrho^n)$. The expansion (9.7) is unique, because $\delta_n^j \to 0$ for $n \to \infty$.

Method (9.6) is called *consistent of order* p, if the local error (9.7) satisfies (Lemma 8.11)

$$d_n = \mathcal{O}(h^p) \qquad \text{for } 0 \le nh \le Const, \qquad \text{and} \qquad E d_p(x) = 0. \qquad (9.8)$$

Observe that by this definition the perturbations δ_n^j have to vanish for $j = 0, \ldots,$ $p - 1$, but no condition is imposed on δ_n^p. The exponential decay of these terms implies that we still have

$$d_{n+1} + E(d_n + \ldots + d_0) = \mathcal{O}(h^p) \qquad \text{for } 0 \le nh \le Const,$$

in agreement with Definition 8.10. One can now easily verify that Lemma 8.12 (Φ_n satisfies a Lipschitz condition with the same constant as Φ) and the Convergence Theorem 8.13 remain valid for method (9.6). In the following theorem we use, as for one-step methods, the notation $u_h(x) = u_n$ when $x = x_n$.

Theorem 9.1 (Hairer & Lubich 1984). *Let the method (9.6) satisfy* A1-A3 *and be consistent of order* $p \ge 1$. *Then the global error has an asymptotic expansion of the form*

$$u_h(x) - z(x, h) = e_p(x)h^p + \ldots + e_N(x)h^N + E(x, h)h^{N+1} \qquad (9.9)$$

where the $e_j(x)$ *are given in the proof (cf. formula (9.18)) and* $E(x, h)$ *is bounded uniformly in* $h \in [0, h_0]$ *and for* x *in compact intervals not containing* x_0. *More precisely than (9.9), there is an expansion*

$$u_n - z_n = \left(e_p(x_n) + \varepsilon_n^p\right)h^p + \ldots + \left(e_N(x_n) + \varepsilon_n^N\right)h^N + \widetilde{E}(n, h)h^{N+1} \qquad (9.10)$$

where $\varepsilon_n^j = \mathcal{O}(\varrho^n)$ *with* $\varrho(S - E) < \varrho < 1$ *and* $\widetilde{E}(n, h)$ *is bounded for* $0 \le nh \le Const.$

Remark. We obtain from (9.10) and (9.9)

$$E(x_n, h) = \widetilde{E}(n, h) + h^{-1}\varepsilon_n^N + h^{-2}\varepsilon_n^{N-1} + \ldots + h^{p-N-1}\varepsilon_n^p,$$

so that the remainder term $E(x, h)$ is in general not uniformly bounded in h for x varying in an interval $[x_0, \bar{x}]$. However, if x is bounded away from x_0, say $x \ge x_0 + \delta$ ($\delta > 0$ fixed), the sequence ε_n^j goes to zero faster than any power of $\delta/n \le h$.

Proof. a) As for one-step methods (cf. proof of Theorem 8.1, Chapter II) we construct a new method, which has as numerical solution

$$\widehat{u}_n = u_n - \left(e(x_n) + \varepsilon_n\right)h^p \tag{9.11}$$

for a given smooth function $e(x)$ and a given sequence ε_n satisfying $\varepsilon_n = \mathcal{O}(\varrho^n)$. Such a method is given by

$$\widehat{u}_0 = \widehat{\varphi}(h)$$
$$\widehat{u}_{n+1} = S\widehat{u}_n + h\widehat{\Phi}_n(x_n, \widehat{u}_n, h) \tag{9.12}$$

where $\widehat{\varphi}(h) = \varphi(h) - \left(e(x_0) + \varepsilon_0\right)h^p$ and

$$\widehat{\Phi}_n(x, u, h) = \Phi_n\left(x, u + (e(x) + \varepsilon_n)h^p, h\right)$$
$$- \left(e(x+h) - Se(x)\right)h^{p-1} - (\varepsilon_{n+1} - S\varepsilon_n)h^{p-1}.$$

Since Φ_n is of the form (9.5), $\widehat{\Phi}_n$ is also of this form, so that its local error has an expansion (9.7). We shall now determine $e(x)$ and ε_n in such a way that the method (9.12) is consistent of order $p+1$.

b) The local error $\widehat{d}_n$ of (9.12) can be expanded as

$$\widehat{d}_0 = z_0 - \widehat{u}_0 = \left(\gamma_p + e(x_0) + \varepsilon_0\right)h^p + \mathcal{O}(h^{p+1})$$
$$\widehat{d}_{n+1} = z_{n+1} - Sz_n - h\widehat{\Phi}_n(x_n, z_n, h)$$
$$= d_{n+1} + \left((I - S)e(x_n) + (\varepsilon_{n+1} - S\varepsilon_n)\right)h^p$$
$$+ \left(-G(x_n)(e(x_n) + \varepsilon_n) + e'(x_n)\right)h^{p+1} + \mathcal{O}(h^{p+2}).$$

Here

$$G(x) = \frac{\partial \Phi_n}{\partial u}\left(x, z(x, 0), 0\right)$$

which is independent of n by (9.5). The method (9.12) is consistent of order $p+1$, if (see (9.8))

    i) $\varepsilon_0 = -\gamma_p - e(x_0)$,

    ii) $d_p(x) + (I - S)e(x) + \delta_n^p + \varepsilon_{n+1} - S\varepsilon_n = 0$ for $x = x_n$,

    iii) $Ee'(x) = EG(x)e(x) - Ed_{p+1}(x)$.

We assume for the moment that the system (i)-(iii) can be solved for $e(x)$ and ε_n. This will actually be demonstrated in part (d) of the proof. By the Convergence Theorem 8.13 the method (9.12) is convergent of order $p+1$. Hence

$$\widehat{u}_n - z_n = \mathcal{O}(h^{p+1}) \qquad \text{uniformly for } 0 \le nh \le Const,$$

which yields the statement (9.10) for $N = p$.

c) The method (9.12) satisfies the assumptions of the theorem with p replaced by $p+1$ and ϱ_0 by ϱ. As in Theorem 8.1 (Section II.8) an induction argument yields the result.

d) It remains to find a solution of the system (i)-(iii). Condition (ii) is satisfied if

$$\text{(iia)} \quad d_p(x) = (S - I)(e(x) + c)$$
$$\text{(iib)} \quad \varepsilon_{n+1} - c = S(\varepsilon_n - c) - \delta_n^p$$

hold for some constant c. Using $(I - S + E)^{-1}(I - S) = (I - E)$, which is a consequence of $SE = E^2 = E$ (see (8.11)), formula (iia) is equivalent to

$$(I - S + E)^{-1}d_p(x) = -(I - E)(e(x) + c). \tag{9.13}$$

From (i) we obtain $\varepsilon_0 - c = -\gamma_p - (e(x_0) + c)$, so that by (9.13)

$$(I - E)(\varepsilon_0 - c) = -(I - E)\gamma_p + (I - S + E)^{-1}d_p(x_0).$$

Since $Ed_p(x_0) = 0$, this relation is satisfied in particular if

$$\varepsilon_0 - c = -(I - E)\gamma_p + (I - S + E)^{-1}d_p(x_0). \tag{9.14}$$

The numbers $\varepsilon_n - c$ are now determined by the recurrence relation (iib)

$$\varepsilon_n - c = S^n(\varepsilon_0 - c) - \sum_{j=1}^{n} S^{n-j}\delta_{j-1}^p$$

$$= E(\varepsilon_0 - c) + (S - E)^n(\varepsilon_0 - c) - E\sum_{j=0}^{\infty}\delta_j^p + E\sum_{j=n}^{\infty}\delta_j^p - \sum_{j=1}^{n}(S - E)^{n-j}\delta_{j-1}^p,$$

where we have used $S^n = E + (S - E)^n$. If we put

$$c = E\sum_{j=0}^{\infty}\delta_j^p \tag{9.15}$$

the sequence $\{\varepsilon_n\}$ defined above satisfies $\varepsilon_n = \mathcal{O}(\varrho^n)$, since $E(\varepsilon_0 - c) = 0$ by (9.14) and since $\delta_n^p = \mathcal{O}(\varrho^n)$.

In order to find $e(x)$ we define

$$v(x) = Ee(x).$$

With the help of formulas (9.15) and (9.13) we can recover $e(x)$ from $v(x)$ by

$$e(x) = v(x) - (I - S + E)^{-1}d_p(x). \tag{9.16}$$

Equation (iii) can now be rewritten as the differential equation

$$v'(x) = EG(x)\Big(v(x) - (I - S + E)^{-1}d_p(x)\Big) - Ed_{p+1}(x), \tag{9.17}$$

and condition (i) yields the starting value $v(x_0) = -E(\gamma_p + \varepsilon_0)$. This initial value problem can be solved for $v(x)$ and we obtain $e(x)$ by (9.16). This function and the ε_n defined above represent a solution of (i)-(iii). $\square$

Remarks. a) It follows from (9.15)-(9.17) that the principal error term satisfies

$$e'_p(x) = EG(x)e_p(x) - Ed_{p+1}(x) - (I - S + E)^{-1}d'_p(x)$$

$$e_p(x_0) = -E\gamma_p - E\sum_{j=0}^{\infty} \delta_j^p - (I - S + E)^{-1}d_p(x_0). \tag{9.18}$$

b) Since $e_{p+1}(x)$ is just the principal error term of method (9.12), it satisfies the differential equation (9.18) with d_j replaced by $\widehat{d}_{j+1}$. By an induction argument we therefore have for $j \geq p$

$$e'_j(x) = EG(x)e_j(x) + \text{inhomogeneity}(x).$$

Weakly Stable Methods

We next study the asymptotic expansion for stable methods, which are not strictly stable. For example, the explicit mid-point rule (1.13'), treated in connection with the GBS-algorithm (Section II.9), is of this type. As at the beginning of this section, we apply the mid-point rule to the problem $y' = -y$, $y(0) \doteq 1$ and consider the following three starting procedures

$$y_0 = 1, \qquad y_1 = \exp(-h) \tag{9.19a}$$

$$y_0 = 1, \qquad y_1 = 1 - h + \frac{h^2}{2} \tag{9.19b}$$

$$y_0 = 1, \qquad y_1 = 1 - h. \tag{9.19c}$$

The three pictures on the left of Fig. 9.2 show the global error divided by h^2. For the first two starting procedures we have convergence to the function $xe^{-x}/6$, while for (9.19c) the divided error $(y_n - y(x_n))/h^2$ converges to

$$e^{-x}\left(\frac{2x-3}{12}\right) + \frac{e^x}{4} \qquad \text{for } n \text{ even,}$$

$$e^{-x}\left(\frac{2x-3}{12}\right) - \frac{e^x}{4} \qquad \text{for } n \text{ odd.}$$

We then subtract the h^2-term from the global error and divide by h^3 in the case (9.19a) and by h^4 for (b) and (c). The result is plotted in the pictures on the right of Fig. 9.2.

This example nicely illustrates the fact that we no longer have an asymptotic expansion of the form (9.9) or (9.10) but that there exists one expansion for x_n with n even, and a different expansion for x_n with n odd (see also Exercise 2 of Section II.9). Similar results for more general methods will be obtained here.

We say that a method of the form (8.4) is *weakly stable,* if it is stable, but if the matrix S has, besides $\zeta_1 = 1$, further eigenvalues of modulus 1, say $\zeta_2, \ldots, \zeta_l$.

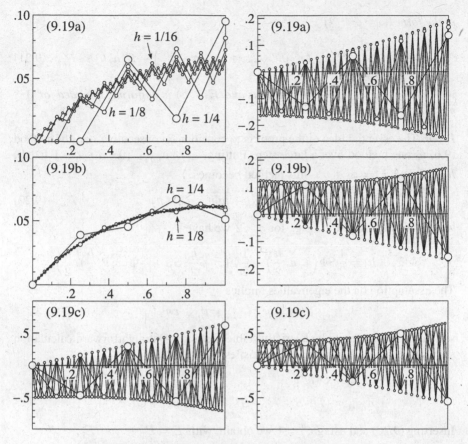

Fig. 9.2. Asymptotic expansion of the mid-point rule
(three different starting procedures)

The matrix S therefore has the representation (cf. (8.11))

$$S = \zeta_1 E_1 + \zeta_2 E_2 + \ldots + \zeta_l E_l + R \tag{9.20}$$

where the E_j are the projectors (corresponding to ζ_j) and the spectral radius of R satisfies $\varrho(R) < 1$.

In what follows we restrict ourselves to the case where all ζ_j $(j = 1, \ldots, l)$ are roots of unity. This allows a simple proof for the existence of an asymptotic expansion and is at the same time by far the most important special case. For the general situation we refer to Hairer & Lubich (1984).

Theorem 9.2. *Let the method (9.6) with Φ_n independent of n be stable, consistent of order p and satisfy A3. If all eigenvalues (of S) of modulus 1 satisfy $\zeta_j^q = 1$ $(j = 1, \ldots, l)$ for some positive integer q, then we have an asymptotic expansion* *of the form* $(\omega = e^{2\pi i/q})$

$$u_n - z_n = \sum_{s=0}^{q-1} \omega^{ns} \Big(e_{ps}(x_n)h^p + \ldots + e_{Ns}(x_n)h^N \Big) + E(n,h)h^{N+1} \quad (9.21)$$

where the $e_{js}(x)$ *are smooth functions and* $E(n,h)$ *is uniformly bounded for* $0 < \delta \le nh \le Const.$

Proof. The essential idea of the proof is to consider q consecutive steps of method (9.6) as one method over a large step. Putting $\widetilde{u}_n = u_{nq+i}$ $(0 \le i \le q-1$ fixed), $\widetilde{h} = qh$ and $\widetilde{x}_n = x_i + n\widetilde{h}$, this method becomes

$$\widetilde{u}_{n+1} = S^q \widetilde{u}_n + \widetilde{h}\widetilde{\Phi}(\widetilde{x}_n, \widetilde{u}_n, \widetilde{h}) \quad (9.22)$$

with a suitably chosen $\widetilde{\Phi}$. E.g., for $q=2$ we have

$$\widetilde{\Phi}(\widetilde{x}, \widetilde{u}, \widetilde{h}) = \frac{1}{2}S\Phi\Big(\widetilde{x}, \widetilde{u}, \frac{\widetilde{h}}{2}\Big) + \frac{1}{2}\Phi\Big(\widetilde{x} + \frac{\widetilde{h}}{2}, S\widetilde{u} + \frac{\widetilde{h}}{2}\Phi(\widetilde{x}, \widetilde{u}, \frac{\widetilde{h}}{2}), \frac{\widetilde{h}}{2}\Big).$$

The assumption on the eigenvalues implies

$$S^q = E_1 + \ldots + E_l + R^q$$

so that (9.22) is seen to be a strictly stable method. A straightforward calculation shows that the local error of (9.22) satisfies

$$\widetilde{d}_0 = \mathcal{O}(h^p)$$
$$\widetilde{d}_{n+1} = (I + S + \ldots + S^{q-1})d_p(\widetilde{x}_n)h^p + \mathcal{O}(h^{p+1}).$$

Inserting (9.20) and using $\zeta_j^q = 1$ we obtain, with $\widetilde{E} = E_1 + \ldots + E_l$,

$$\widetilde{E}(I + S + \ldots + S^{q-1})d_p(x)$$
$$= \widetilde{E}\Big(I - \widetilde{E} + qE_1 + \sum_{j=2}^{l}\frac{1-\zeta_j^q}{1-\zeta_j}E_j + \sum_{j=1}^{q-1}R^j\Big)d_p(x) = qE_1 d_p(x),$$

which vanishes by (8.15). Hence, also method (9.22) is consistent of order p. All the assumptions of Theorem 9.1 are thus verified for method (9.22). We therefore obtain

$$u_{nq+i} - z_{nq+i} = \widetilde{e}_{pi}(x_{nq+i})h^p + \ldots + \widetilde{e}_{Ni}(x_{nq+i})h^N + E_i(n,h)h^{N+1}$$

where $E_i(n,h)$ has the desired boundedness properties. If we define $e_{js}(x)$ as a solution of the Vandermonde-type system

$$\sum_{s=0}^{q-1} \omega^{is} e_{js}(x) = \widetilde{e}_{ji}(x)$$

we obtain (9.21). $\square$

The Adjoint Method

For a method (8.4) the correct value function $z(x, h)$, the starting procedure $\varphi(h)$ and the increment function $\Phi(x, u, h)$ are usually also defined for negative h (see the examples of Section III.8). As for one-step methods (Section II.8) we shall give here a precise meaning to the numerical solution $u_h(x)$ for negative h. This then leads in a natural way to the study of asymptotic expansions in even powers of h.

With the notation $u_h(x) = u_n$ for $x = x_0 + nh$ $(h > 0)$ the method (8.4) becomes

$$u_h(x_0) = \varphi(h)$$
$$u_h(x + h) = Su_h(x) + h\Phi\big(x, u_h(x), h\big) \qquad \text{for } x = x_0 + nh. \tag{9.23}$$

We first replace h by $-h$ in (9.23) to obtain

$$u_{-h}(x_0) = \varphi(-h)$$
$$u_{-h}(x - h) = Su_{-h}(x) - h\Phi\big(x, u_{-h}(x), -h\big)$$

and then x by $x + h$ which gives

$$u_{-h}(x_0) = \varphi(-h)$$
$$u_{-h}(x) = Su_{-h}(x + h) - h\Phi\big(x + h, u_{-h}(x + h), -h\big).$$

For sufficiently small h this equation can be solved for $u_{-h}(x + h)$ (Implicit Function Theorem) and we obtain

$$u_{-h}(x_0) = \varphi(-h),$$
$$u_{-h}(x + h) = S^{-1}u_{-h}(x) + h\Phi^*\big(x, u_{-h}(x), h\big). \tag{9.24}$$

The method (9.24), which is again of the form (8.4), is called the *adjoint method* of (9.23). Its correct value function is $z^*(x, h) = z(x, -h)$. Observe that for given S and Φ the new increment function Φ^* is just defined by the pair of formulas

$$v = Su - h\Phi(x + h, u, -h)$$
$$u = S^{-1}v + h\Phi^*(x, v, h). \tag{9.25}$$

Example 9.3. Consider a linear multistep method with generating functions

$$\varrho(\zeta) = \sum_{j=0}^{k} \alpha_j \zeta^j, \qquad \sigma(\zeta) = \sum_{j=0}^{k} \beta_j \zeta^j.$$

Then we have

$$S = \begin{pmatrix} -\alpha_{k-1}/\alpha_k & -\alpha_{k-2}/\alpha_k & \cdots & -\alpha_0/\alpha_k \\ 1 & 0 & \cdots & 0 \\ & 1 & & 0 \\ & & \ddots & \vdots \\ & & 1 & 0 \end{pmatrix}, \qquad \Phi(x, u, h) = \begin{pmatrix} 1 \\ 0 \\ \vdots \\ 0 \end{pmatrix} \psi(x, u, h)$$

where $\psi = \psi(x, u, h)$ is the solution of $(u = (u_{k-1}, \ldots, u_0)^T)$

$$\alpha_k \psi = \sum_{j=0}^{k-1} \beta_j f(x + jh, u_j) + \beta_k f\left(x + kh, h\psi - \sum_{j=0}^{k-1} \frac{\alpha_j}{\alpha_k} u_j\right).$$

A straightforward use of the formulas (9.25) shows that

$$S^{-1} = \begin{pmatrix} 0 & 1 & & \\ 0 & 0 & & \\ \vdots & \vdots & \cdots & \\ & & & 1 \\ -\alpha_k/\alpha_0 & -\alpha_{k-1}/\alpha_0 & \cdots & -\alpha_1/\alpha_0 \end{pmatrix}, \qquad \Phi^*(x, v, h) = \begin{pmatrix} 0 \\ \vdots \\ 0 \\ 1 \end{pmatrix} \psi^*(x, v, h)$$

where $\psi^* = \psi^*(x, v, h)$ (with $v = (v_0, \ldots, v_{k-1})^T$) is given by

$$-\alpha_0 \psi^* = \sum_{j=0}^{k-1} \beta_{k-j} f\left(x + (j - k + 1)h, v_j\right) + \beta_0 f\left(x + h, h\psi^* - \sum_{j=0}^{k-1} \frac{\alpha_{k-j}}{\alpha_0} v_j\right).$$

This shows that the adjoint method is again a linear multistep method. Its generating polynomials are

$$\varrho^*(\zeta) = -\zeta^k \varrho(\zeta^{-1}), \qquad \sigma^*(\zeta) = \zeta^k \sigma(\zeta^{-1}). \tag{9.26}$$

Our next aim is to prove that the adjoint method has exactly the same asymptotic expansion as the original method, with h replaced by $-h$. For this it is necessary that S^{-1} also be a stable matrix. Therefore all eigenvalues of S must lie on the unit circle.

Theorem 9.4. *Let the method (9.23) be stable, consistent of order p and assume that all eigenvalues of S satisfy $\zeta_j^q = 1$ for some positive integer q. Then the global error has an asymptotic expansion of the form* $(\omega = e^{2\pi i/q})$

$$u_h(x_n) - z(x_n, h) = \sum_{s=0}^{q-1} \omega^{ns} \left(e_{ps}(x_n)h^p + \ldots + e_{Ns}(x_n)h^N\right) + E(x_n, h)h^{N+1}, \tag{9.27}$$

valid for positive and negative h. The remainder $E(x, h)$ is uniformly bounded for $|h| \le h_0$ and $x_0 \le x \le \widehat{x}$.

Proof. As in the proof of Theorem 9.2 we consider q consecutive steps of method (9.23) as one new method. The assumption on the eigenvalues implies that $S^q = I =$ identity. Therefore the new method is essentially a one-step method. The only difference is that here the starting procedure and the correct value function may depend on h. A straightforward extension of Theorem 8.5 of Chapter II (Exercise 3) implies the existence of an expansion

$$u_h(x_{nq+i}) - z(x_{nq+i}, h) = \widetilde{e}_{pi}(x_{nq+i})h^p + \ldots + \widetilde{e}_{Ni}(x_{nq+i})h^N$$
$$+ E_i(x_{nq+i}, h)h^{N+1}.$$

This expansion is valid for positive and negative h; the remainder $E_i(x, h)$ is bounded for $|h| \leq h_0$ and $x_0 \leq x \leq \hat{x}$. The same argument as in the proof of Theorem 9.2 now leads to the desired expansion.   □

Symmetric Methods

The definition of symmetry for general linear methods is not as straightforward as for one-step methods. In Example 9.3 we saw that the components of the numerical solution of the adjoint method are in inverse order. Therefore, it is too restrictive to require that $\varphi(h) = \varphi(-h)$, $S = S^{-1}$ and $\Phi = \Phi^*$.

However, for many methods of practical interest the correct value function satisfies a *symmetry relation* of the form

$$z(x, h) = Qz(x + qh, -h) \tag{9.28}$$

where Q is a square matrix and q an integer. This is for instance the case for linear multistep methods, where the correct value function is given by

$$z(x, h) = \big(y(x + (k - 1)h), \ldots, y(x)\big)^T.$$

The relation (9.28) holds with

$$Q = \begin{pmatrix} & & 1 \\ & \cdot^{\cdot^{\cdot}} & \\ 1 & & \end{pmatrix} \qquad \text{and} \qquad q = k - 1. \tag{9.29}$$

Definition 9.5. Suppose that the correct value function satisfies (9.28). Method (9.23) is called *symmetric* (with respect to (9.28)), if the numerical solution satisfies its analogue

$$u_h(x) = Qu_{-h}(x + qh). \tag{9.30}$$

Example 9.6. Consider a linear multistep method and suppose that the generating polynomials of the adjoint method (9.26) satisfy

$$\varrho^*(\zeta) = \varrho(\zeta), \qquad \sigma^*(\zeta) = \sigma(\zeta). \tag{9.31}$$

This is equivalent to the requirement (cf. (3.24))

$$\alpha_{k-j} = -\alpha_j, \qquad \beta_{k-j} = \beta_j.$$

A straightforward calculation (using the formulas of Example 9.3) then shows that the symmetry relation (9.30) holds for all $x = x_0 + nh$ whenever it holds for $x = x_0$. This imposes an additional condition on the starting procedure $\varphi(h)$.

Let us finally demonstrate how Theorem 9.4 can be used to prove *asymptotic expansions in even powers* of h. Denote by $u_h^j(x)$ the jth component of $u_h(x)$. The symmetry relation (9.30) for multistep methods then implies

$$u_{-h}^k(x) = u_h^1(x - (k - 1)h)$$

Furthermore, for any multistep method we have

$$u_h^k(x) = u_h^1(x - (k-1)h)$$

so that

$$u_h^k(x) = u_{-h}^k(x)$$

for symmetric methods. As a consequence of Theorem 9.4 the asymptotic expansion of the global error is in even powers of h, whenever the multistep method is symmetric in the sense of Definition 9.5.

Exercises

1. Consider a strictly stable, pth order, linear multistep method written in the form (9.6) (see Example 9.3) and set

$$G(x) = \frac{\partial \Phi}{\partial u}(x, z(x, 0), 0).$$

   a) Prove that

   $$EG(x)\mathbb{1} = \mathbb{1}\frac{\partial f}{\partial y}(x, y(x))$$

   where E is the matrix given by (8.11) and $\mathbb{1} = (1, \ldots, 1)^T$.

   b) Show that the function $e_p(x)$ in the expansion (9.9) is given by $e_p(x) = \mathbb{1}\widehat{e}_p(x)$, where

   $$\widehat{e}'_p(x) = \frac{\partial f}{\partial y}(x, y(x))\widehat{e}_p(x) - Cy^{(p+1)}(x)$$

   and C is the error constant (cf. (2.13)). Compute also $\widehat{e}_p(x_0)$.

2. For the 3-step BDF-method, applied to $y' = -y$, $y(0) = 1$ with starting procedure (9.1c), compute the function $e_3(x)$ and the perturbations $\{\varepsilon_n^3\}_{n \geq 0}$ in the expansion (9.4). Compare your result with Fig. 9.1.

3. Consider the method

$$u_0 = \varphi(h), \qquad u_{n+1} = u_n + h\Phi(x_n, u_n, h) \qquad (9.32)$$

   with correct value function $z(x, h)$.

   a) Prove that the global error has an asymptotic expansion of the form

   $$u_n - z_n = e_p(x_n)h^p + \ldots + e_N(x_n)h^N + E(x_n, h)h^{N+1}$$

   where $E(x, h)$ is uniformly bounded for $0 \leq h \leq h_0$ and $x_0 \leq x \leq \widehat{x}$.

   b) Show that Theorem 8.5 of Chapter II remains valid for method (9.32).

III.10 Multistep Methods for Second Order Differential Equations

> En 1904 j'eus besoin d'une pareille méthode pour calculer les trajectoires des corpuscules électrisés dans un champ magnétique, et en essayant diverses méthodes déjà connues, mais sans les trouver assez commodes pour mon but, je fus conduit moi-même à élaborer une méthode assez simple, dont je me suis servi ensuite.
>
> (C. Störmer 1921)

Because of their importance, second order differential equations deserve some additional attention. We already saw in Section II.14 that for special second order differential equations certain direct one-step methods are more efficient than the classical Runge-Kutta methods. We now investigate whether a similar situation also holds for multistep methods.

Consider the second order differential equation

$$y'' = f(x, y, y') \tag{10.1}$$

where y is allowed to be a vector. We rewrite (10.1) in the usual way as a first order system and apply a multistep method

$$\sum_{i=0}^{k} \alpha_i y_{n+i} = h \sum_{i=0}^{k} \beta_i y'_{n+i}$$

$$\sum_{i=0}^{k} \alpha_i y'_{n+i} = h \sum_{i=0}^{k} \beta_i f(x_{n+i}, y_{n+i}, y'_{n+i}). \tag{10.2}$$

If the right hand side of the differential equation does not depend on y',

$$y'' = f(x, y), \tag{10.3}$$

it is natural to look for numerical methods which do not involve the first derivative. An elimination of $\{y'_n\}$ in the equations (10.2) results in

$$\sum_{i=0}^{2k} \widehat{\alpha}_i y_{n+i} = h^2 \sum_{i=0}^{2k} \widehat{\beta}_i f(x_{n+i}, y_{n+i}) \tag{10.4}$$

where the new coefficients $\widehat{\alpha}_i, \widehat{\beta}_i$ are given by

$$\sum_{i=0}^{2k} \widehat{\alpha}_i \zeta^i = \left(\sum_{i=0}^{k} \alpha_i \zeta^i \right)^2, \qquad \sum_{i=0}^{2k} \widehat{\beta}_i \zeta^i = \left(\sum_{i=0}^{k} \beta_i \zeta^i \right)^2. \tag{10.5}$$

In what follows we investigate (10.4) with coefficients that do not necessarily satisfy (10.5). It is hoped to achieve the same order with a smaller step number.

Explicit Störmer Methods

> Sein Vortrag ist übrigens ziemlich trocken und langweilig ...
> (B. Riemann's opinion about Encke, 1847)

> Had the Ast. Ges. Essay been entirely free from numerical blun-
> ders, ...   (P.H. Cowell & A.C.D. Crommelin 1910)

Since most differential equations of celestial mechanics are of the form (10.3) it is not surprising that the first attempts at developing special methods for these equations were made by astronomers.

For his extensive numerical calculations concerning the aurora borealis (see below), C. Störmer (1907) developed an accurate and simple method as follows: by adding the Taylor series for $y(x_n + h)$ and $y(x_n - h)$ we obtain

$$y(x_n+h) - 2y(x_n) + y(x_n-h) = h^2 y''(x_n) + \frac{h^4}{12} y^{(4)}(x_n) + \frac{h^6}{360} y^{(6)}(x_n) + \dots.$$

If we insert $y''(x_n)$ from the differential equation (10.3) and neglect higher terms, we get

$$y_{n+1} - 2y_n + y_{n-1} = h^2 f_n$$

as a first simple method, which is sometimes called Störmer's or Encke's method. For greater precision, we replace the higher derivatives of y by central differences of f

$$h^2 y^{(4)}(x_n) = \Delta^2 f_{n-1} - \frac{1}{12} \Delta^4 f_{n-2} + \dots$$

$$h^4 y^{(6)}(x_n) = \Delta^4 f_{n-2} + \dots$$

and obtain

$$y_{n+1} - 2y_n + y_{n-1} = h^2 \left(f_n + \frac{1}{12} \Delta^2 f_{n-1} - \frac{1}{240} \Delta^4 f_{n-2} + \dots \right). \qquad (10.6)$$

This formula is not yet very practical, since the differences of the right hand side contain the unknown expressions f_{n+1} and f_{n+2}. Neglecting fifth-order differences (i.e., putting $\Delta^4 f_{n-2} \approx \Delta^4 f_{n-4}$ and $\Delta^2 f_{n-1} = \Delta^2 f_{n-2} + \Delta^3 f_{n-3} + \Delta^4 f_{n-3} \approx \Delta^2 f_{n-2} + \Delta^3 f_{n-3} + \Delta^4 f_{n-4}$) one gets

$$y_{n+1} - 2y_n + y_{n-1} = h^2 f_n + \frac{h^2}{12} \left(\Delta^2 f_{n-2} + \Delta^3 f_{n-3} + \frac{19}{20} \Delta^4 f_{n-4} \right) \qquad (10.7)$$

("... formule qui est fondamentale dans notre méthode ...", C. Störmer 1907).

Some years later Cowell & Crommelin (1910) used the same ideas to investigate the motion of Halley's comet. They considered one additional term in the series (10.6), namely

$$\frac{31}{60480} \Delta^6 f_{n-3} \approx \frac{1}{1951} \Delta^6 f_{n-3}.$$

Arbitrary orders. Integrating equation (10.3) twice we obtain

$$y(x+h) = y(x) + hy'(x) + h^2 \int_0^1 (1-s) f(x+sh, y(x+sh)) \, ds. \qquad (10.8)$$

In order to eliminate the first derivative of $y(x)$ we write the same formula with h replaced by $-h$ and add the two expressions:

$$y(x+h) - 2y(x) + y(x-h) \qquad (10.9)$$

$$= h^2 \int_0^1 (1-s) \Big(f(x+sh, y(x+sh)) + f(x-sh, y(x-sh)) \Big) \, ds.$$

As in the derivation of the Adams formulas (Section III.1) we replace the unknown function $f(t, y(t))$ by the interpolation polynomial $p(t)$ of formula (1.4). This yields the *explicit* method

$$y_{n+1} - 2y_n + y_{n-1} = h^2 \sum_{j=0}^{k-1} \sigma_j \nabla^j f_n \qquad (10.10)$$

with coefficients σ_j given by

$$\sigma_j = (-1)^j \int_0^1 (1-s) \left(\binom{-s}{j} + \binom{s}{j} \right) ds. \qquad (10.11)$$

See Table 10.1 for their numerical values and Exercise 2 for their computation.

Table 10.1. Coefficients of the method (10.10)

j	0	1	2	3	4	5	6	7	8	9
σ_j	1	0	$\dfrac{1}{12}$	$\dfrac{1}{12}$	$\dfrac{19}{240}$	$\dfrac{3}{40}$	$\dfrac{863}{12096}$	$\dfrac{275}{4032}$	$\dfrac{33953}{518400}$	$\dfrac{8183}{129600}$

Special cases of (10.10) are

$$k=2: \quad y_{n+1} - 2y_n + y_{n-1} = h^2 f_n$$

$$k=3: \quad y_{n+1} - 2y_n + y_{n-1} = h^2 \left(\frac{13}{12} f_n - \frac{1}{6} f_{n-1} + \frac{1}{12} f_{n-2} \right) \qquad (10.10')$$

$$k=4: \quad y_{n+1} - 2y_n + y_{n-1} = h^2 \left(\frac{7}{6} f_n - \frac{5}{12} f_{n-1} + \frac{1}{3} f_{n-2} - \frac{1}{12} f_{n-3} \right).$$

Method (10.10) with $k=5$ is formula (10.7), the method used by Störmer (1907, 1921), and for $k=6$ one obtains the method used by Cowell & Crommelin (1910). The simplest of these methods ($k=1$ or $k=2$) has been successfully applied as the basis of an extrapolation method (Section II.14, formula (14.32)).

Implicit Störmer Methods

The first terms of (10.6)

$$y_{n+1} - 2y_n + y_{n-1} = h^2 \left(f_n + \frac{1}{12} \Delta^2 f_{n-1} \right)$$
$$= \frac{h^2}{12} \left(f_{n+1} + 10 f_n + f_{n-1} \right) \tag{10.12}$$

form an implicit equation for y_{n+1}. This can either be used in a predictor-corrector fashion, or, as advocated by B. Numerov (1924, 1927), by solving this implicit nonlinear equation directly for y_{n+1}.

To obtain more accurate formulas, analogous to the implicit Adams methods, we use the interpolation polynomial $p^*(t)$ of (1.8), which passes through the additional point (x_{n+1}, f_{n+1}). This yields the implicit method

$$y_{n+1} - 2y_n + y_{n-1} = h^2 \sum_{j=0}^{k} \sigma_j^* \nabla^j f_{n+1}, \tag{10.13}$$

where the coefficients σ_j^* are defined by

$$\sigma_j^* = (-1)^j \int_0^1 (1-s) \left(\binom{-s+1}{j} + \binom{s+1}{j} \right) ds \tag{10.14}$$

and are given in Table 10.2 for $j \leq 9$.

Table 10.2. Coefficients of the implicit method (10.13)

j	0	1	2	3	4	5	6	7	8	9
σ_j^*	1	-1	$\dfrac{1}{12}$	0	$\dfrac{-1}{240}$	$\dfrac{-1}{240}$	$\dfrac{-221}{60480}$	$\dfrac{-19}{6048}$	$\dfrac{-9829}{3628800}$	$\dfrac{-407}{172800}$

Further methods can be derived by using the ideas of Nyström and Milne for first order equations. With the substitutions $h \to 2h$, $2s \to s$ and $x \to x - h$ formula (10.9) becomes

$$y(x+h) - 2y(x-h) + y(x-3h) = h^2 \int_0^2 (2-s) \tag{10.15}$$
$$\cdot \Big(f\big(x+(s-1)h, y(x+(s-1)h)\big) + f\big(x-(s+1)h, y(x-(s+1)h)\big) \Big) \, ds.$$

If one replaces $f(t, y(t))$ by the polynomial $p(t)$ (respectively $p^*(t)$) one obtains the new classes of explicit (respectively implicit) methods.

Numerical Example

> Nous avons calculé plus de 120 trajectoires différentes, travail im-
> mense qui a exigé plus de 4500 heures ... Quand on est suffisam-
> ment exercé, on calcule environ trois points (R, z) par heure.
>
> (C. Störmer 1907)

We choose the historical problem treated by Störmer in 1907: Störmer's aim was to
confirm numerically the conjecture of Birkeland, who explained in 1896 the aurora
borealis as being produced by electrical particles emanating from the sun and danc-
ing in the earth's magnetic field. Suppose that an elementary magnet is situated at
the origin with its axis along to the z-axis. The trajectory $(x(s), y(s), z(s))$ of an
electrical particle in this magnetic field then satisfies

$$x'' = \frac{1}{r^5}\left(3yzz' - (3z^2 - r^2)y'\right)$$

$$y'' = \frac{1}{r^5}\left((3z^2 - r^2)x' - 3xzz'\right) \tag{10.16}$$

$$z'' = \frac{1}{r^5}\left(3xzy' - 3yzx'\right)$$

where $r^2 = x^2 + y^2 + z^2$. Introducing the polar coordinates

$$x = R\cos\varphi, \qquad y = R\sin\varphi \tag{10.17}$$

the system (10.16) becomes equivalent to

$$R'' = \left(\frac{2\gamma}{R} + \frac{R}{r^3}\right)\left(\frac{2\gamma}{R^2} + \frac{3R^2}{r^5} - \frac{1}{r^3}\right) \tag{10.18a}$$

$$z'' = \left(\frac{2\gamma}{R} + \frac{R}{r^3}\right)\frac{3Rz}{r^5} \tag{10.18b}$$

$$\varphi' = \left(\frac{2\gamma}{R} + \frac{R}{r^3}\right)\frac{1}{R} \tag{10.18c}$$

where now $r^2 = R^2 + z^2$ and γ is some constant arising from the integration of
φ''. The two equations (10.18a,b) constitute a second order differential equation
of type (10.3), which can be solved numerically by the methods of this section.
φ is then obtained by simple integration of (10.18c). Störmer found after long
calculations that the initial values

$$R_0 = 0.257453, \qquad z_0 = 0.314687, \qquad \gamma = -0.5,$$
$$R_0' = \sqrt{Q_0}\cos u, \qquad z_0' = \sqrt{Q_0}\sin u, \qquad u = 5\pi/4 \tag{10.18d}$$
$$r_0 = \sqrt{R_0^2 + z_0^2}, \qquad Q_0 = 1 - (2\gamma/R_0 + R_0/r_0^3)^2$$

produce a specially interesting solution curve approaching very closely the North
Pole. Fig. 10.1 shows 125 solution curves (in the x, y, z-space) with these and
neighbouring initial values to give an impression of how an aurora borealis comes
into being.

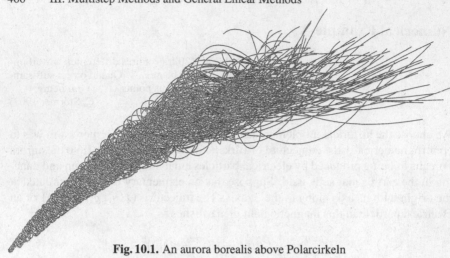

Fig. 10.1. An aurora borealis above Polarcirkeln

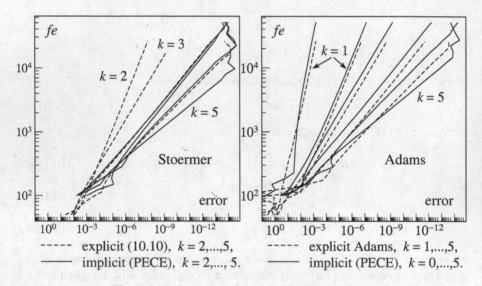

Fig. 10.2. Performance of Störmer and Adams methods

Fig. 10.2 compares the performance of the Störmer methods (10.10) and (10.13) (in PECE mode) with that of the Adams methods by integrating subsystem (10.18a,b) with initial values (10.18d) for $0 \leq s \leq 0.3$. The diagrams compare the Euclidean norm in $\mathbb{R}^2$ of the error of the final solution point (R, z) with the number of function evaluations fe. The step numbers used are $\{n = 50 \cdot 2^{0.3 \cdot i}\}_{i=0,1,\dots,30} = \{50, 61, 75, 93, 114, \dots, 25600\}$. The starting values were computed very precisely with an explicit Runge-Kutta method and step size $h_{RK} = h/10$. It can be observed that the Störmer methods are substantially more precise due to the smaller error constants (compare Tables 10.1 and 10.2 with Tables 1.1

and 1.2). In addition, they have lower overhead. However, they must be implemented carefully in order to avoid rounding errors (see below).

General Formulation

Our next aim is to study stability, consistency and convergence of general linear multistep methods for (10.3). We write them in the form

$$\sum_{i=0}^{k} \alpha_i y_{n+i} = h^2 \sum_{i=0}^{k} \beta_i f(x_{n+i}, y_{n+i}). \tag{10.19}$$

The generating polynomials of the coefficients α_i and β_i are again denoted by

$$\varrho(\zeta) = \sum_{i=0}^{k} \alpha_i \zeta^i, \qquad \sigma(\zeta) = \sum_{i=0}^{k} \beta_i \zeta^i. \tag{10.20}$$

If we apply method (10.19) to the initial value problem

$$y'' = f(x, y), \qquad y(x_0) = y_0, \qquad y'(x_0) = y_0' \tag{10.21}$$

it is natural to require that the starting values be consistent with both initial values, i.e., that

$$\frac{y_i - y_0 - ihy_0'}{h} \to 0 \qquad \text{for} \quad h \to 0, \qquad i = 0, 1, \ldots, k-1. \tag{10.22}$$

For the *stability condition* of method (10.19) we consider the simple problem

$$y'' = 0, \qquad y_0 = 0, \qquad y_0' = 0.$$

Its numerical solution satisfies a linear difference equation with $\varrho(\zeta)$ as characteristic polynomial. The same considerations as in the proof of Theorem 4.2 show that the following stability condition is necessary for convergence.

Definition 10.1. Method (10.19) is called *stable*, if the generating polynomial $\varrho(\zeta)$ satisfies:

 i) The roots of $\varrho(\zeta)$ lie on or within the unit circle;

 ii) The multiplicity of the roots on the unit circle is at most two.

For the *order conditions* we introduce, similarly to formula (2.3), the linear difference operator

$$L(y, x, h) = \varrho(E)y(x) - h^2 \sigma(E)y''(x)$$
$$= \sum_{i=0}^{k} \Big(\alpha_i y(x+ih) - h^2 \beta_i y''(x+ih) \Big), \tag{10.23}$$

where E is the shift operator. As in Definition 2.3 we now have:

Definition 10.2. Method (10.19) is *consistent of order* p if for all sufficiently smooth functions $y(x)$,

$$L(y, x, h) = \mathcal{O}(h^{p+2}).\tag{10.24}$$

The following theorem is then proved similarly to Theorem 2.4.

Theorem 10.3. *The multistep method (10.19) is of order p if and only if the following equivalent conditions hold:*

i) $\sum_{i=0}^{k} \alpha_i = 0, \qquad \sum_{i=0}^{k} i\alpha_i = 0$

  and $\sum_{i=0}^{k} \alpha_i i^q = q(q-1) \sum_{i=0}^{k} \beta_i i^{q-2}$ *for* $q = 2, \ldots, p+1$,

ii) $\varrho(e^h) - h^2 \sigma(e^h) = \mathcal{O}(h^{p+2})$ *for* $h \to 0$,

iii) $\dfrac{\varrho(\zeta)}{(\log \zeta)^2} - \sigma(\zeta) = \mathcal{O}\big((\zeta - 1)^p\big)$ *for* $\zeta \to 1$.

$\square$

As for Adams methods one easily verifies that the method (10.10) is of order k, and that (10.13) is of order $k+1$.

The following order barriers are similar to those of Theorems 3.5 and 3.9; their proofs are similar too (see, e.g., Dahlquist 1959, Henrici 1962):

Theorem 10.4. *The order p of a stable linear multistep method (10.19) satisfies*

$$p \le k + 2 \quad \textit{if } k \textit{ is even},$$
$$p \le k + 1 \quad \textit{if } k \textit{ is odd}.$$

$\square$

Theorem 10.5. *Stable multistep methods (10.19) of order $k+2$ are symmetric, i.e.,*

$$\alpha_j = \alpha_{k-j}, \qquad \beta_j = \beta_{k-j} \quad \textit{for all } j.$$

$\square$

Convergence

Theorem 10.6. *Suppose that method (10.19) is stable, of order p, and that the starting values satisfy*

$$y(x_j) - y_j = \mathcal{O}(h^{p+1}) \quad \textit{for } j = 0, 1, \ldots, k-1.\tag{10.25}$$

Then we have convergence of order p, i.e.,

$$\|y(x_n) - y_n\| \le Ch^p \quad \textit{for } 0 \le hn \le Const.$$

Proof. It is possible to develop a theory analogous to that of Sections III.2 - III.4. This is due to Dahlquist (1959) and can also be found in the book of Henrici (1962). We prefer to rewrite (10.19) *in a one-step formulation* of the form (8.4) and to apply directly the results of Section III.8 and III.9 (see Example 8.6). In order to achieve this goal, we could put $u_n = (y_{n+k-1}, \ldots, y_n)^T$, which seems to be a natural choice. But then the corresponding matrix S does not satisfy the stability condition of Definition 8.8 because of the double roots of modulus 1. To overcome this difficulty we separate these roots. We split the characteristic polynomial $\varrho(\zeta)$ into

$$\varrho(\zeta) = \varrho_1(\zeta) \cdot \varrho_2(\zeta) \tag{10.26}$$

such that each polynomial ($l + k = m$)

$$\varrho_1(\zeta) = \sum_{i=0}^{l} \gamma_i \zeta^i, \qquad \varrho_2(\zeta) = \sum_{i=0}^{m} \kappa_i \zeta^i \tag{10.27}$$

has only simple roots of modulus 1. Without loss of generality we assume in the sequel that $m \geq l$ and $\alpha_k = \gamma_l = \kappa_m = 1$. Using the shift operator E, method (10.19) can be written as

$$\varrho(E)y_n = h^2 \sigma(E) f_n.$$

The main idea is to introduce $\varrho_2(E)y_n$ as a new variable, say hv_n, so that the multistep formula becomes equivalent to the system

$$\varrho_1(E)v_n = h\sigma(E)f_n, \qquad \varrho_2(E)y_n = hv_n. \tag{10.28}$$

Introducing the vector

$$u_n = (v_{n+l-1}, \ldots, v_n, y_{n+m-1}, \ldots, y_n)^T$$

formula (10.28) can be written as

$$u_{n+1} = Su_n + h\Phi(x_n, u_n, h) \tag{10.29a}$$

where

$$S = \begin{pmatrix} G & 0 \\ 0 & K \end{pmatrix}, \qquad \Phi(x_n, u_n, h) = \begin{pmatrix} e_1 \psi(x_n, u_n, h) \\ e_1 v_n \end{pmatrix}. \tag{10.30}$$

The matrices G and K are the companion matrices

$$G = \begin{pmatrix} -\gamma_{l-1} & -\gamma_{l-2} & \cdots & & -\gamma_0 \\ 1 & 0 & \cdots & & 0 \\ & 1 & & & 0 \\ & & \vdots & \vdots & \\ & & & 1 & 0 \end{pmatrix}, \quad K = \begin{pmatrix} -\kappa_{m-1} & -\kappa_{m-2} & \cdots & & -\kappa_0 \\ 1 & 0 & \cdots & & 0 \\ & 1 & & & 0 \\ & & \vdots & \vdots & \\ & & & 1 & 0 \end{pmatrix},$$

$e_1 = (1, 0, \ldots, 0)^T$, and $\psi = \psi(x_n, u_n, h)$ is implicitly defined by

$$\psi = \sum_{j=0}^{k-1} \beta_j f(x_n + jh, y_{n+j}) + \beta_k f\left(x_n + kh, h^2\psi - \sum_{j=0}^{k-1} \alpha_j y_{n+j}\right). \tag{10.31}$$

In this formula ψ is written as a function of $x_n, (y_{n+k-1}, \ldots, y_n)$ and h. But the second relation of (10.28) shows that each value $y_{n+k-1}, \ldots, y_{n+m}$ can be expressed as a linear combination of the elements of u_n. Therefore ψ is in fact a function of (x_n, u_n, h).

Formula (10.29a) defines our forward step procedure. The corresponding starting procedure is

$$\varphi(h) = (v_{l-1}, \ldots, v_0, y_{m-1}, \ldots, y_0)^T \qquad (10.29b)$$

which, by (10.28), is uniquely determined by $(y_{k-1}, \ldots, y_0)^T$. As correct value function we have

$$z(x, h) = \left(\frac{1}{h} \varrho_2(E) y(x+(l-1)h), \ldots, \frac{1}{h} \varrho_2(E) y(x), \; y(x+(m-1)h, \ldots, y(x) \right)^T.$$
$$(10.29c)$$

By our choice of $\varrho_1(\zeta)$ and $\varrho_2(\zeta)$ (both have only simple roots of modulus 1) the matrices G and K are power bounded. Therefore S is also power bounded and method (10.29) *is stable* in the sense of Definition 8.8.

We now verify the conditions of Definition 8.10 and for this start with the error in the initial values

$$d_0 = z(x_0, h) - \varphi(h).$$

The first l components of this vector are

$$\frac{1}{h} \varrho_2(E) y(x_j) - v_j = \frac{1}{h} \sum_{i=0}^{m} \kappa_i \big(y(x_{i+j}) - y_{i+j} \big), \qquad j = 0, \ldots, l-1$$

and the last m components are just

$$y(x_j) - y_j, \qquad j = 0, \ldots, m-1.$$

Thus hypothesis (10.25) ensures that $d_0 = \mathcal{O}(h^p)$. Consider next the local error at x_n,

$$d_{n+1} = z(x_n + h, h) - Sz(x_n, h) - h\Phi\big(x_n, z(x_n, h), h\big).$$

All components of d_{n+1} vanish except the first, which equals

$$d_{n+1}^{(1)} = \frac{1}{h} \varrho(E) y(x_n) - h\psi(x_n, z(x_n, h), h).$$

Using formula (10.31), an application of the mean value theorem yields

$$d_{n+1}^{(1)} = \frac{1}{h} L(y, x_n, h) + h^2 \beta_k f'(x_{n+k}, \eta) \cdot d_{n+1}^{(1)} \qquad (10.32)$$

with η as in Lemma 2.2. We therefore have

$$d_{n+1} = \mathcal{O}(h^{p+1}) \qquad \text{since} \qquad L(y, x_n, h) = \mathcal{O}(h^{p+2}).$$

Finally Theorem 8.13 yields the stated convergence result. $\qquad\qquad\qquad\square$

Asymptotic Formula for the Global Error

Assume that the method (10.19) is stable and consistent of order p. The local truncation error of (10.29) is then given by

$$d_{n+1} = e_1 h^{p+1} C_{p+2} y^{(p+2)}(x_n) + \mathcal{O}(h^{p+2}) \qquad (10.33)$$

with

$$C_{p+2} = \frac{1}{(p+2)!} \sum_{i=0}^{k} \left(\alpha_i i^{p+2} - (p+2)(p+1)\beta_i i^p \right).$$

Formula (10.33) can be verified by developing $L(y, x_n, h)$ into a Taylor series in (10.32). An application of Theorem 9.1 (if 1 is the only root of modulus 1 of $\varrho(\zeta)$) or of Theorem 9.2 shows that the global error of method (10.29) is of the form

$$u_h(x) - z(x, h) = e(x)h^p + \mathcal{O}(h^{p+1})$$

where $e(x)$ is the solution of

$$e'(x) = E \frac{\partial \Phi}{\partial u}(x, z(x, 0), 0)e(x) - Ee_1 \cdot C_{p+2} y^{(p+2)}(x). \qquad (10.34)$$

Here E is the matrix defined in (8.12). Since no h^p-term is present in the local error (10.33), it follows from (9.16) that $e(x) = Ee(x)$. Therefore (see Exercise 4a) this function can be written as

$$e(x) = \begin{pmatrix} \gamma(x)\mathbb{1} \\ \kappa(x)\mathbb{1} \end{pmatrix}.$$

A straightforward calculation of $\frac{\partial \Phi}{\partial u}(x, z(x, 0), 0)$ and Ee_1 (for details see Exercise 4) shows that (10.34) becomes equivalent to the system

$$\gamma'(x) = \frac{\sigma(1)}{\varrho_1'(1)} \frac{\partial f}{\partial y}(x, y(x))\kappa(x) - \frac{C_{p+2}}{\varrho_1'(1)} y^{(p+2)}(x) \qquad (10.35a)$$

$$\kappa'(x) = \frac{1}{\varrho_2'(1)}\gamma(x). \qquad (10.35b)$$

Differentiating (10.35b) and inserting $\gamma'(x)$ from (10.35a), we finally obtain

$$\kappa''(x) = \frac{\partial f}{\partial y}(x, y(x))\kappa(x) - C y^{(p+2)}(x) \qquad (10.36)$$

with

$$C = \frac{C_{p+2}}{\sigma(1)}. \qquad (10.37)$$

Here we have used the relation $\sigma(1) = \varrho_1'(1) \cdot \varrho_2'(1)$, which is an immediate consequence of (10.26), and the assumption that the order of the method is at least 1. The constant C in (10.37) is called the *error constant* of method (10.19). It plays the same role as (2.13) for first order equations.

Since the last component of the vector u_n is y_n we have the desired result

$$y_n - y(x_n) = \kappa(x_n)h^p + \mathcal{O}(h^{p+1})$$

with $\kappa(x)$ satisfying (10.36). Further terms in the asymptotic expansion of the global error can also be obtained by specializing the results of III.9.

Rounding Errors

A *direct* implementation of Störmer's methods, for which (10.19) specializes to

$$y_{n+1} - 2y_n + y_{n-1} = h^2 \sum_{i=0}^{k} \beta_i f_{n+i-k+1}, \qquad (10.38)$$

by storing the y-values $y_0, y_1, \ldots, y_{k-1}$ and computing successively the values $y_k, y_{k+1}, \ldots$ with the help of (10.38) leads to numerical instabilities for small h. This instability is caused by the double root of $\varrho(\zeta)$ on the unit circle. It can be observed numerically in Fig. 10.3, where the left picture is a zoom of Fig. 10.2, while the right image contains the results of a code implementing (10.38) directly.

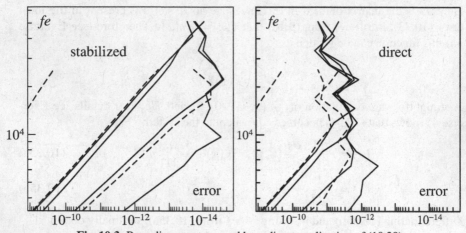

Fig. 10.3. Rounding errors caused by a direct application of (10.38)

In order to obtain the stabilized version of the algorithm, we apply the following two ideas:

a) Split, as in (10.26), the polynomial $\varrho(\zeta)$ as $(\zeta - 1)(\zeta - 1)$. Then (10.28) leads to $hv_n = y_{n+1} - y_n$ and (10.38) becomes the mathematically equivalent formulation

$$v_n - v_{n-1} = h \sum_{i=0}^{k} \beta_i f_{n+i-k+1}, \qquad y_{n+1} - y_n = hv_n. \qquad (10.38')$$

Here the corresponding matrix S of (10.30) is stable.

b) Avoid the use of $v_n = (y_{n+1} - y_n)/h$ for the computation of the starting values $v_0, v_1, \ldots, v_{k-2}$, since the difference is a numerically unstable operation. Instead, add up the increments of the Runge-Kutta method, which you use for the computation of the starting values, directly.

These two ideas together then produce the "stabilized" results in Fig. 10.3 and Fig. 10.2.

Exercises

1. Compute the solution of Störmer's problem (10.18) with one of the methods of this section.

2. a) Show that the generating functions of the coefficients σ_i and σ_j^* (defined in (10.11) and (10.14))

$$S(t) = \sum_{j=0}^{\infty} \sigma_j t^j, \qquad S^*(t) = \sum_{j=0}^{\infty} \sigma_j^* t^j$$

satisfy

$$S(t) = \left(\frac{t}{\log(1-t)}\right)^2 \frac{1}{1-t}, \qquad S^*(t) = \left(\frac{t}{\log(1-t)}\right)^2.$$

b) Compute the coefficients d_j of

$$\sum_{j=0}^{\infty} d_j t^j = \left(\frac{\log(1-t)}{t}\right)^2 = \left(1 + \frac{t}{2} + \frac{t^2}{3} + \frac{t^3}{4} + \ldots\right)^2$$

and derive a recurrence relation for the σ_j and σ_j^*.

c) Prove that $\sigma_j^* = \sigma_j - \sigma_{j-1}$.

3. Let $\varrho(\zeta)$ be a polynomial of degree k which has 1 as root of multiplicity 2. Then there exists a unique $\sigma(\zeta)$ such that the corresponding method is of order $k+1$.

4. Consider the method (10.29) and, for simplicity, assume the differential equation to be a scalar one.

a) For any vector w in $\mathbb{R}^k$ the image vector Ew, with E given by (8.12), satisfies

$$Ew = \begin{pmatrix} \gamma \mathbb{1} \\ \kappa \mathbb{1} \end{pmatrix}$$

where γ, κ are real numbers and $\mathbb{1}$ is the vector with all elements equal to 1. The dimensions of $\gamma \mathbb{1}$ and $\kappa \mathbb{1}$ are l and m, respectively.

b) Verify that for $e_1 = (1, 0, \ldots, 0)^T$,

$$E \begin{pmatrix} \alpha e_1 \\ \beta e_1 \end{pmatrix} = \begin{pmatrix} (\alpha/\varrho_1'(1))\mathbb{1} \\ (\beta/\varrho_2'(1))\mathbb{1} \end{pmatrix}.$$

c) Show that

$$E \frac{\partial \Phi}{\partial u}(x, z(x, 0), 0) \begin{pmatrix} \gamma\mathbb{1} \\ \kappa\mathbb{1} \end{pmatrix} = \begin{pmatrix} (\sigma(1)/\varrho_1'(1))(\partial f/\partial y)(x, y(x))\kappa\mathbb{1} \\ (1/\varrho_2'(1))\gamma\mathbb{1} \end{pmatrix}.$$

Hint. With $Y_n = (y_{n+k-1}, \ldots, y_n)^T$ the formula (10.31) expresses ψ as a function of (x_n, Y_n, h). The second formula of (10.28) relates Y_n and u_n as

$$KY_n = Lu_n + \mathcal{O}(h) \qquad \text{where} \qquad K\mathbb{1} = L\begin{pmatrix} 0 \\ \mathbb{1} \end{pmatrix}$$

and K is invertible. Use the chain rule for the computation of $\partial\psi/\partial u$. See also Exercise 2 of Section III.4 and Exercise 1 of Section III.9.

5. Compute the error constant (10.37) for the methods (10.10) and (10.13).

Result. σ_k and σ_{k+1}^*, respectively.

Appendix. Fortran Codes

> ... but the software is in various states of development from experimental (a euphemism for badly written) to what we might call
> ...   (C.W. Gear, in Aiken 1985)

Several Fortran codes have been developed for our numerical computations. Those of the first edition have been improved and several new options have been included, e.g., automatic choice of initial step size, stiffness detection, dense output. We have seen many of the ideas, which are incorporated in these codes, in the programs of P. Deuflhard, A.C. Hindmarsh and L.F. Shampine.

Experiences with all of our codes are welcome. The programs can be obtained from the authors' homepage (http://www.unige.ch/~hairer).

Address: Section de Mathématiques, Case postale 240,

CH-1211 Genève 24, Switzerland

E-mail: Ernst.Hairer@math.unige.ch   Gerhard.Wanner@math.unige.ch

Driver for the Code DOPRI5

The driver given here is for the differential equation (II.0.1) with initial values and x_{end} given in (II.0.2). This is the problem AREN of Section II.10. The subroutine FAREN ("F for AREN") computes the right-hand side of this differential equation. The subroutine SOLOUT ("Solution out"), which is called by DOPRI5 after every successful step, and the dense output routine CONTD5 are used to print the solution at equidistant points. The (optional) common block STATD5 gives statistical information after the call to DOPRI5. The common blocks COD5R and COD5I transfer the necessary information to CONTD5.

```
        IMPLICIT REAL*8 (A-H,O-Z)
        PARAMETER (NDGL=4,LWORK=8*NDGL+10,LIWORK=10)
        PARAMETER (NRDENS=2,LRCONT=5*NRDENS+2,LICONT=NRDENS+1)
        DIMENSION Y(NDGL),WORK(LWORK),IWORK(LIWORK)
        COMMON/STATD5/NFCN,NSTEP,NACCPT,NREJCT
        COMMON /COD5R/RCONT(LRCONT)
        COMMON /COD5I/ICONT(LICONT)
        EXTERNAL FAREN,SOLOUT
C --- DIMENSION OF THE SYSTEM
        N=NDGL
C --- OUTPUT ROUTINE (AND DENSE OUTPUT) IS USED DURING INTEGRATION
        IOUT=2
```

```
C --- INITIAL VALUES AND ENDPOINT OF INTEGRATION
        X=0.0D0
        Y(1)=0.994D0
        Y(2)=0.0D0
        Y(3)=0.0D0
        Y(4)=-2.00158510637908252240537862224D0
        XEND=17.0652165601579625588917206249D0
C --- REQUIRED (RELATIVE AND ABSOLUTE) TOLERANCE
        ITOL=0
        RTOL=1.0D-7
        ATOL=RTOL
C --- DEFAULT VALUES FOR PARAMETERS
        DO 10 I=1,10
        IWORK(I)=0
  10    WORK(I)=0.D0
C --- DENSE OUTPUT IS USED FOR THE TWO POSITION COORDINATES 1 AND 2
        IWORK(5)=NRDENS
        ICONT(2)=1
        ICONT(3)=2
C --- CALL OF THE SUBROUTINE DOPRI5
        CALL DOPRI5(N,FAREN,X,Y,XEND,
     +              RTOL,ATOL,ITOL,
     +              SOLOUT,IOUT,
     +              WORK,LWORK,IWORK,LIWORK,LRCONT,LICONT,IDID)
C --- PRINT FINAL SOLUTION
        WRITE (6,99) Y(1),Y(2)
  99    FORMAT(1X,'X = XEND    Y =',2E18.10)
C --- PRINT STATISTICS
        WRITE (6,91) RTOL,NFCN,NSTEP,NACCPT,NREJCT
  91    FORMAT('    tol=',D8.2,'   fcn=',I5,' step=',I4,
     +         '  accpt=',I4,' rejct=',I3)
        STOP
        END
C
        SUBROUTINE SOLOUT (NR,XOLD,X,Y,N,IRTRN)
C --- PRINTS SOLUTION AT EQUIDISTANT OUTPUT-POINTS BY USING "CONTD5"
        IMPLICIT REAL*8 (A-H,O-Z)
        DIMENSION Y(N)
        COMMON /INTERN/XOUT
        IF (NR.EQ.1) THEN
            WRITE (6,99) X,Y(1),Y(2),NR-1
            XOUT=X+2.0D0
        ELSE
  10        CONTINUE
            IF (X.GE.XOUT) THEN
                WRITE (6,99) XOUT,CONTD5(1,XOUT),CONTD5(2,XOUT),NR-1
                XOUT=XOUT+2.0D0
                GOTO 10
            END IF
        END IF
  99    FORMAT(1X,'X =',F6.2,'   Y =',2E18.10,'   NSTEP =',I4)
        RETURN
        END
C
        SUBROUTINE FAREN(N,X,Y,F)
C --- ARENSTORF ORBIT
        IMPLICIT REAL*8 (A-H,O-Z)
        DIMENSION Y(N),F(N)
        AMU=0.012277471D0
        AMUP=1.D0-AMU
```

```
F(1)=Y(3)
F(2)=Y(4)
R1=(Y(1)+AMU)**2+Y(2)**2
R1=R1*SQRT(R1)
R2=(Y(1)-AMUP)**2+Y(2)**2
R2=R2*SQRT(R2)
F(3)=Y(1)+2*Y(4)-AMUP*(Y(1)+AMU)/R1-AMU*(Y(1)-AMUP)/R2
F(4)=Y(2)-2*Y(3)-AMUP*Y(2)/R1-AMU*Y(2)/R2
RETURN
END
```

The result, obtained on an Apollo workstation, is the following:

```
X =   0.00    Y =  0.9940000000E+00   0.0000000000E+00    NSTEP =    0
X =   2.00    Y = -0.5798781411E+00   0.6090775251E+00    NSTEP =   60
X =   4.00    Y = -0.1983335270E+00   0.1137638086E+01    NSTEP =   73
X =   6.00    Y = -0.4735743943E+00   0.2239068118E+00    NSTEP =   91
X =   8.00    Y = -0.1174553350E+01  -0.2759466982E+00    NSTEP =  110
X =  10.00    Y = -0.8398073466E+00   0.4468302268E+00    NSTEP =  122
X =  12.00    Y =  0.1314712468E-01  -0.8385751499E+00    NSTEP =  145
X =  14.00    Y = -0.6031129504E+00  -0.9912598031E+00    NSTEP =  159
X =  16.00    Y =  0.2427110999E+00  -0.3899948833E+00    NSTEP =  177
X = XEND     Y =  0.9940021016E+00   0.8911185978E-05
     tol=0.10E-06    fcn= 1442 step= 240 accpt= 216 rejct= 22
```

Subroutine DOPRI5

Explicit Runge-Kutta code based on the method of Dormand & Prince (see Table 5.2 of Section II.5). It is provided with the step control algorithm of Section II.4 and the dense output of Section II.6.

```
      SUBROUTINE DOPRI5(N,FCN,X,Y,XEND,
     +                   RTOL,ATOL,ITOL,
     +                   SOLOUT,IOUT,
     +                   WORK,LWORK,IWORK,LIWORK,LRCONT,LICONT,IDID)
C -----------------------------------------------------------
C     NUMERICAL SOLUTION OF A SYSTEM OF FIRST ORDER
C     ORDINARY DIFFERENTIAL EQUATIONS  Y'=F(X,Y).
C     THIS IS AN EXPLICIT RUNGE-KUTTA METHOD OF ORDER (4)5
C     DUE TO DORMAND & PRINCE (WITH STEPSIZE CONTROL AND
C     DENSE OUTPUT).
C
C     AUTHORS: E. HAIRER AND G. WANNER
C              UNIVERSITE DE GENEVE, DEPT. DE MATHEMATIQUES
C              CH-1211 GENEVE 24, SWITZERLAND
C              E-MAIL:  HAIRER@ UNI2A.UNIGE.CH,  WANNER@ UNI2A.UNIGE.CH
C
C     THIS CODE IS DESCRIBED IN:
C         E. HAIRER, S.P. NORSETT AND G. WANNER, SOLVING ORDINARY
C         DIFFERENTIAL EQUATIONS I. NONSTIFF PROBLEMS. 2ND EDITION.
C         SPRINGER SERIES IN COMPUTATIONAL MATHEMATICS,
C         SPRINGER-VERLAG (1993)
C
```

```
C     VERSION OF OCTOBER 3, 1991
C
C     INPUT PARAMETERS
C     ----------------
C     N           DIMENSION OF THE SYSTEM
C
C     FCN         NAME (EXTERNAL) OF SUBROUTINE COMPUTING THE
C                 VALUE OF F(X,Y):
C                     SUBROUTINE FCN(N,X,Y,F)
C                     REAL*8 X,Y(N),F(N)
C                     F(1)=...     ETC.
C
C     X           INITIAL X-VALUE
C
C     Y(N)        INITIAL VALUES FOR Y
C
C     XEND        FINAL X-VALUE (XEND-X MAY BE POSITIVE OR NEGATIVE)
C
C     RTOL,ATOL   RELATIVE AND ABSOLUTE ERROR TOLERANCES. THEY
C                 CAN BE BOTH SCALARS OR ELSE BOTH VECTORS OF LENGTH N.
C
C     ITOL        SWITCH FOR RTOL AND ATOL:
C                     ITOL=0:  BOTH RTOL AND ATOL ARE SCALARS.
C                       THE CODE KEEPS, ROUGHLY, THE LOCAL ERROR OF
C                       Y(I) BELOW RTOL*ABS(Y(I))+ATOL
C                     ITOL=1:  BOTH RTOL AND ATOL ARE VECTORS.
C                       THE CODE KEEPS THE LOCAL ERROR OF Y(I) BELOW
C                       RTOL(I)*ABS(Y(I))+ATOL(I).
C
C     SOLOUT      NAME (EXTERNAL) OF SUBROUTINE PROVIDING THE
C                 NUMERICAL SOLUTION DURING INTEGRATION.
C                 IF IOUT.GE.1, IT IS CALLED AFTER EVERY SUCCESSFUL STEP.
C                 SUPPLY A DUMMY SUBROUTINE IF IOUT=0.
C                 IT MUST HAVE THE FORM
C                     SUBROUTINE SOLOUT (NR,XOLD,X,Y,N,IRTRN)
C                     REAL*8 X,Y(N)
C                     ....
C                 SOLOUT FURNISHES THE SOLUTION "Y" AT THE NR-TH
C                    GRID-POINT "X" (THEREBY THE INITIAL VALUE IS
C                    THE FIRST GRID-POINT).
C                 "XOLD" IS THE PRECEEDING GRID-POINT.
C                 "IRTRN" SERVES TO INTERRUPT THE INTEGRATION. IF IRTRN
C                    IS SET <0, DOPRI5 WILL RETURN TO THE CALLING PROGRAM.
C
C           ----- CONTINUOUS OUTPUT: -----
C                 DURING CALLS TO "SOLOUT", A CONTINUOUS SOLUTION
C                 FOR THE INTERVAL [XOLD,X] IS AVAILABLE THROUGH
C                 THE FUNCTION
C                       >>>   CONTD5(I,S)   <<<
C                 WHICH PROVIDES AN APPROXIMATION TO THE I-TH
C                 COMPONENT OF THE SOLUTION AT THE POINT S. THE VALUE
C                 S SHOULD LIE IN THE INTERVAL [XOLD,X].
C
C     IOUT        SWITCH FOR CALLING THE SUBROUTINE SOLOUT:
C                     IOUT=0:  SUBROUTINE IS NEVER CALLED
C                     IOUT=1:  SUBROUTINE IS USED FOR OUTPUT.
C                     IOUT=2:  DENSE OUTPUT IS PERFORMED IN SOLOUT
C                                (IN THIS CASE WORK(5) MUST BE SPECIFIED)
C
C     WORK        ARRAY OF WORKING SPACE OF LENGTH "LWORK".
```

```
C                       "LWORK" MUST BE AT LEAST  8*N+10
C
C      LWORK            DECLARED LENGHT OF ARRAY "WORK".
C
C      IWORK            INTEGER WORKING SPACE OF LENGHT "LIWORK".
C                       IWORK(1),...,IWORK(5) SERVE AS PARAMETERS
C                       FOR THE CODE. FOR STANDARD USE, SET THEM
C                       TO ZERO BEFORE CALLING.
C                       "LIWORK" MUST BE AT LEAST 10 .
C
C      LIWORK           DECLARED LENGHT OF ARRAY "IWORK".
C
C      LRCONT           DECLARED LENGTH OF COMMON BLOCK
C                       >>>   COMMON /COD5R/RCONT(LRCONT)   <<<
C                       WHICH MUST BE DECLARED IN THE CALLING PROGRAM.
C                       "LRCONT" MUST BE AT LEAST
C                                   5 * NRDENS + 2
C                       WHERE NRDENS=IWORK(5) (SEE BELOW).
C
C      LICONT           DECLARED LENGTH OF COMMON BLOCK
C                       >>>   COMMON /COD5I/ICONT(LICONT)   <<<
C                       WHICH MUST BE DECLARED IN THE CALLING PROGRAM.
C                       "LICONT" MUST BE AT LEAST
C                                   NRDENS + 1
C                       THESE COMMON BLOCKS ARE USED FOR STORING THE COEFFICIENTS
C                       OF THE CONTINUOUS SOLUTION AND MAKES THE CALLING LIST FOR
C                       THE FUNCTION "CONTD5" AS SIMPLE AS POSSIBLE.
C
C-------------------------------------------------------------------------
C
C      SOPHISTICATED SETTING OF PARAMETERS
C      -----------------------------------
C                       SEVERAL PARAMETERS (WORK(1),...,IWORK(1),...)   ALLOW
C                       TO ADAPT THE CODE TO THE PROBLEM AND TO THE NEEDS OF
C                       THE USER. FOR ZERO INPUT, THE CODE CHOOSES DEFAULT VALUES.
C
C      WORK(1)   UROUND, THE ROUNDING UNIT, DEFAULT 2.3D-16.
C
C      WORK(2)   THE SAFETY FACTOR IN STEP SIZE PREDICTION,
C                DEFAULT 0.9D0.
C
C      WORK(3), WORK(4)   PARAMETERS FOR STEP SIZE SELECTION
C                THE NEW STEP SIZE IS CHOSEN SUBJECT TO THE RESTRICTION
C                     WORK(3) <= HNEW/HOLD <= WORK(4)
C                DEFAULT VALUES: WORK(3)=0.2D0, WORK(4)=10.D0
C
C      WORK(5)   IS THE "BETA" FOR STABILIZED STEP SIZE CONTROL
C                (SEE SECTION IV.2).  LARGER VALUES OF BETA ( <= 0.1 )
C                MAKE THE STEP SIZE CONTROL MORE STABLE. DOPRI5 NEEDS
C                A LARGER BETA THAN HIGHAM & HALL. NEGATIVE WORK(5)
C                PROVOKE BETA=0.
C                DEFAULT 0.04D0.
C
C      WORK(6)   MAXIMAL STEP SIZE, DEFAULT XEND-X.
C
C      WORK(7)   INITIAL STEP SIZE, FOR WORK(7)=0.D0 AN INITIAL GUESS
C                IS COMPUTED WITH HELP OF THE FUNCTION HINIT
C
C      IWORK(1)  THIS IS THE MAXIMAL NUMBER OF ALLOWED STEPS.
C                THE DEFAULT VALUE (FOR IWORK(1)=0) IS 100000.
```

```
C
C     IWORK(2)   SWITCH FOR THE CHOICE OF THE COEFFICIENTS
C                IF IWORK(2).EQ.1  METHOD DOPRI5 OF DORMAND AND PRINCE
C                (TABLE 5.2 OF SECTION II.5).
C                AT THE MOMENT THIS IS THE ONLY POSSIBLE CHOICE.
C                THE DEFAULT VALUE (FOR IWORK(2)=0) IS IWORK(2)=1.
C
C     IWORK(3)   SWITCH FOR PRINTING ERROR MESSAGES
C                IF IWORK(3).LT.0 NO MESSAGES ARE BEING PRINTED
C                IF IWORK(3).GT.0 MESSAGES ARE PRINTED WITH
C                WRITE (IWORK(3),*) ...
C                DEFAULT VALUE (FOR IWORK(3)=0) IS IWORK(3)=6
C
C     IWORK(4)   TEST FOR STIFFNESS IS ACTIVATED AFTER STEP NUMBER
C                J*IWORK(4) (J INTEGER), PROVIDED IWORK(4).GT.0.
C                FOR NEGATIVE IWORK(4) THE STIFFNESS TEST IS
C                NEVER ACTIVATED; DEFAULT VALUE IS IWORK(4)=1000
C
C     IWORK(5)   = NRDENS = NUMBER OF COMPONENTS, FOR WHICH DENSE OUTPUT
C                IS REQUIRED; DEFAULT VALUE IS IWORK(5)=0;
C                FOR   0 < NRDENS < N   THE COMPONENTS (FOR WHICH DENSE
C                OUTPUT IS REQUIRED) HAVE TO BE SPECIFIED IN
C                ICONT(2),...,ICONT(NRDENS+1);
C                FOR   NRDENS=N  THIS IS DONE BY THE CODE.
C
C-----------------------------------------------------------------------
C
C     OUTPUT PARAMETERS
C     -----------------
C
C     X          X-VALUE FOR WHICH THE SOLUTION HAS BEEN COMPUTED
C                (AFTER SUCCESSFUL RETURN X=XEND).
C
C     Y(N)       NUMERICAL SOLUTION AT X
C
C     H          PREDICTED STEP SIZE OF THE LAST ACCEPTED STEP
C
C     IDID       REPORTS ON SUCCESSFULNESS UPON RETURN:
C                  IDID= 1  COMPUTATION SUCCESSFUL,
C                  IDID= 2  COMPUT. SUCCESSFUL (INTERRUPTED BY SOLOUT)
C                  IDID=-1  INPUT IS NOT CONSISTENT,
C                  IDID=-2  LARGER NMAX IS NEEDED,
C                  IDID=-3  STEP SIZE BECOMES TOO SMALL.
C                  IDID=-4  PROBLEM IS PROBABLY STIFF (INTERRUPTED).
C
C-----------------------------------------------------------------------
C *** *** *** *** *** *** *** *** *** *** *** *** *** ***
C          DECLARATIONS
C *** *** *** *** *** *** *** *** *** *** *** *** *** ***
      IMPLICIT REAL*8 (A-H,O-Z)
      DIMENSION Y(N),ATOL(1),RTOL(1),WORK(LWORK),IWORK(LIWORK)
      LOGICAL ARRET
      EXTERNAL FCN,SOLOUT
      COMMON/STATD5/NFCN,NSTEP,NACCPT,NREJCT
C --- COMMON STATD5 CAN BE INSPECTED FOR STATISTICAL PURPOSES:
C ---   NFCN       NUMBER OF FUNCTION EVALUATIONS
C ---   NSTEP      NUMBER OF COMPUTED STEPS
C ---   NACCPT     NUMBER OF ACCEPTED STEPS
C ---   NREJCT     NUMBER OF REJECTED STEPS (AFTER AT LEAST ONE STEP
C                  HAS BEEN ACCEPTED)
      .........
```

Subroutine DOP853

Explicit Runge-Kutta code of order 8 based on the method of Dormand & Prince, described in Section II.5. The local error estimation and the step size control is based on embedded formulas or orders 5 and 3 (see Section II.10). This method is provided with a dense output of order 7. In the following description we have omitted the parts which are identical to those for DOPRI5.

```
      SUBROUTINE DOP853(N,FCN,X,Y,XEND,
     +                  RTOL,ATOL,ITOL,
     +                  SOLOUT,IOUT,
     +                  WORK,LWORK,IWORK,LIWORK,LRCONT,LICONT,IDID)
C ------------------------------------------------------------
C     NUMERICAL SOLUTION OF A SYSTEM OF FIRST ORDER
C     ORDINARY DIFFERENTIAL EQUATIONS  Y'=F(X,Y).
C     THIS IS AN EXPLICIT RUNGE-KUTTA METHOD OF ORDER 8(5,3)
C     DUE TO DORMAND & PRINCE (WITH STEPSIZE CONTROL AND
C     DENSE OUTPUT)
C .........
C
C     VERSION OF NOVEMBER 29, 1992
C .........
C            ----- CONTINUOUS OUTPUT: -----
C                  DURING CALLS TO "SOLOUT", A CONTINUOUS SOLUTION
C                  FOR THE INTERVAL [XOLD,X] IS AVAILABLE THROUGH
C                  THE FUNCTION
C                      >>>   CONTD8(I,S)   <<<
C                  WHICH PROVIDES AN APPROXIMATION TO THE I-TH
C .........
C
C     WORK       ARRAY OF WORKING SPACE OF LENGTH "LWORK".
C                "LWORK" MUST BE AT LEAST   11*N+10
C .........
C
C     LRCONT     DECLARED LENGTH OF COMMON BLOCK
C                   >>>  COMMON /COD8R/RCONT(LRCONT)  <<<
C                WHICH MUST BE DECLARED IN THE CALLING PROGRAM.
C                "LRCONT" MUST BE AT LEAST
C                          8 * NRDENS + 2
C                WHERE NRDENS=IWORK(5) (SEE BELOW).
C
C     LICONT     DECLARED LENGTH OF COMMON BLOCK
C                   >>>  COMMON /COD8I/ICONT(LICONT)  <<<
C                WHICH MUST BE DECLARED IN THE CALLING PROGRAM.
C                "LICONT" MUST BE AT LEAST
C                          NRDENS + 1
C                THESE COMMON BLOCKS ARE USED FOR STORING THE COEFFICIENTS
C                OF THE CONTINUOUS SOLUTION AND MAKES THE CALLING LIST FOR
C                THE FUNCTION "CONTD8" AS SIMPLE AS POSSIBLE.
C .........
C
C     WORK(3), WORK(4)   PARAMETERS FOR STEP SIZE SELECTION
C                THE NEW STEP SIZE IS CHOSEN SUBJECT TO THE RESTRICTION
C                   WORK(3) <= HNEW/HOLD <= WORK(4)
C                DEFAULT VALUES: WORK(3)=0.333D0, WORK(4)=6.D0
C .........
```

Subroutine ODEX

Extrapolation code for $y' = f(x, y)$, based on the GBS algorithm (Section II.9). It uses variable order and variable step sizes and is provided with a high-order dense output. Again, the missing parts in the description are identical to those of DOPRI5.

```
          SUBROUTINE ODEX(N,FCN,X,Y,XEND,H,
         +                    RTOL,ATOL,ITOL,
         +                    SOLOUT,IOUT,
         +                    WORK,LWORK,IWORK,LIWORK,LRCONT,LICONT,IDID)
C     ------------------------------------------------------------
C         NUMERICAL SOLUTION OF A SYSTEM OF FIRST ORDER
C         ORDINARY DIFFERENTIAL EQUATIONS  Y'=F(X,Y).
C         THIS IS AN EXTRAPOLATION-ALGORITHM (GBS), BASED ON THE
C         EXPLICIT MIDPOINT RULE (WITH STEPSIZE CONTROL,
C       , ORDER SELECTION AND DENSE OUTPUT).
C
C         AUTHORS: E. HAIRER AND G. WANNER
C                  UNIVERSITE DE GENEVE, DEPT. DE MATHEMATIQUES
C                  CH-1211 GENEVE 24, SWITZERLAND
C                  E-MAIL:  HAIRER@ UNI2A.UNIGE.CH,   WANNER@ UNI2A.UNIGE.CH
C         DENSE OUTPUT WRITTEN BY E. HAIRER AND A. OSTERMANN
C     .........
C
C         VERSION DECEMBER 18, 1991
C     .........
C
C     H          INITIAL STEP SIZE GUESS;
C                H=1.D0/(NORM OF F'), USUALLY 1.D-1 OR 1.D-3, IS GOOD.
C                THIS CHOICE IS NOT VERY IMPORTANT, THE CODE QUICKLY
C                ADAPTS ITS STEP SIZE. WHEN YOU ARE NOT SURE, THEN
C                STUDY THE CHOSEN VALUES FOR A FEW
C                STEPS IN SUBROUTINE "SOLOUT".
C                (IF H=0.D0, THE CODE PUTS H=1.D-4).
C     .........
C
C              ----- CONTINUOUS OUTPUT (IF IOUT=2):  -----
C                DURING CALLS TO "SOLOUT", A CONTINUOUS SOLUTION
C                FOR THE INTERVAL [XOLD,X] IS AVAILABLE THROUGH
C                THE REAL*8 FUNCTION
C                    >>>    CONTEX(I,S)    <<<
C                WHICH PROVIDES AN APPROXIMATION TO THE I-TH
C                COMPONENT OF THE SOLUTION AT THE POINT S. THE VALUE
C                S SHOULD LIE IN THE INTERVAL [XOLD,X].
C     .........
C
C     WORK       ARRAY OF WORKING SPACE OF LENGTH "LWORK".
C                SERVES AS WORKING SPACE FOR ALL VECTORS.
C                "LWORK" MUST BE AT LEAST
C                   N*(KM+5)+5*KM+10+2*KM*(KM+1)*NRDENS
C                WHERE NRDENS=IWORK(8) (SEE BELOW) AND
C                      KM=9              IF IWORK(2)=0
C                      KM=IWORK(2)       IF IWORK(2).GT.0
C                WORK(1),...,WORK(10) SERVE AS PARAMETERS
C                FOR THE CODE. FOR STANDARD USE, SET THESE
C                PARAMETERS TO ZERO BEFORE CALLING.
C     .........
```

```
C
C      IWORK           INTEGER WORKING SPACE OF LENGTH "LIWORK".
C                      "LIWORK" MUST BE AT LEAST
C                               2*KM+10+NRDENS
C                      IWORK(1),...,IWORK(9) SERVE AS PARAMETERS
C                      FOR THE CODE. FOR STANDARD USE, SET THESE
C                      PARAMETERS TO ZERO BEFORE CALLING.
C.........
C
C      LRCONT          DECLARED LENGTH OF COMMON BLOCK
C                         >>> COMMON /CONTR/RCONT(LRCONT) <<<
C                      WHICH MUST BE DECLARED IN THE CALLING PROGRAM.
C                      "LRCONT" MUST BE AT LEAST
C                               ( 2 * KM + 5 ) * NRDENS + 2
C                      WHERE KM=IWORK(2) AND NRDENS=IWORK(8) (SEE BELOW).
C
C      LICONT          DECLARED LENGTH OF COMMON BLOCK
C                         >>> COMMON /CONTI/ICONT(LICONT) <<<
C                      WHICH MUST BE DECLARED IN THE CALLING PROGRAM.
C                      "LICONT" MUST BE AT LEAST
C                               NRDENS + 2
C                      THESE COMMON BLOCKS ARE USED FOR STORING THE COEFFICIENTS
C                      OF THE CONTINUOUS SOLUTION AND MAKES THE CALLING LIST FOR
C                      THE FUNCTION "CONTEX" AS SIMPLE AS POSSIBLE.
C.........
C
C      WORK(2)    MAXIMAL STEP SIZE, DEFAULT XEND-X.
C
C      WORK(3)    STEP SIZE IS REDUCED BY FACTOR WORK(3), IF THE
C                 STABILITY CHECK IS NEGATIVE, DEFAULT 0.5.
C
C      WORK(4), WORK(5)   PARAMETERS FOR STEP SIZE SELECTION
C                 THE NEW STEP SIZE FOR THE J-TH DIAGONAL ENTRY IS
C                 CHOSEN SUBJECT TO THE RESTRICTION
C                     FACMIN/WORK(5) <= HNEW(J)/HOLD <= 1/FACMIN
C                 WHERE FACMIN=WORK(4)**(1/(2*J-1))
C                 DEFAULT VALUES: WORK(4)=0.02D0, WORK(5)=4.D0
C
C      WORK(6), WORK(7)   PARAMETERS FOR THE ORDER SELECTION
C                 STEP SIZE IS DECREASED IF    W(K-1) <= W(K)*WORK(6)
C                 STEP SIZE IS INCREASED IF    W(K) <= W(K-1)*WORK(7)
C                 DEFAULT VALUES: WORK(6)=0.8D0, WORK(7)=0.9D0
C
C      WORK(8), WORK(9)   SAFETY FACTORS FOR STEP CONTROL ALGORITHM
C                 HNEW=H*WORK(9)*(WORK(8)*TOL/ERR)**(1/(J-1))
C                 DEFAULT VALUES: WORK(8)=0.65D0,
C                         WORK(9)=0.94D0  IF "HOPE FOR CONVERGENCE"
C                         WORK(9)=0.90D0  IF "NO HOPE FOR CONVERGENCE"
C.........
C
C      IWORK(2)   THE MAXIMUM NUMBER OF COLUMNS IN THE EXTRAPOLATION
C                 TABLE. THE DEFAULT VALUE (FOR IWORK(2)=0) IS 9.
C                 IF IWORK(2).NE.0 THEN IWORK(2) SHOULD BE .GE.3.
C
C      IWORK(3)   SWITCH FOR THE STEP SIZE SEQUENCE (EVEN NUMBERS ONLY)
C                 IF IWORK(3).EQ.1 THEN 2,4,6,8,10,12,14,16,...
C                 IF IWORK(3).EQ.2 THEN 2,4,8,12,16,20,24,28,...
C                 IF IWORK(3).EQ.3 THEN 2,4,6,8,12,16,24,32,...
C                 IF IWORK(3).EQ.4 THEN 2,6,10,14,18,22,26,30,...
C                 IF IWORK(3).EQ.5 THEN 4,8,12,16,20,24,28,32,...
```

```
C                  THE DEFAULT VALUE IS IWORK(3)=1 IF IOUT.LE.1;
C                  THE DEFAULT VALUE IS IWORK(3)=4 IF IOUT.GE.2.
C
C     IWORK(4)  STABILITY CHECK IS ACTIVATED AT MOST IWORK(4) TIMES IN
C                  ONE LINE OF THE EXTRAP. TABLE, DEFAULT IWORK(4)=1.
C
C     IWORK(5)  STABILITY CHECK IS ACTIVATED ONLY IN THE LINES
C                  1 TO IWORK(5) OF THE EXTRAP. TABLE, DEFAULT IWORK(5)=1.
C
C     IWORK(6)  IF  IWORK(6)=0 ERROR ESTIMATOR IN THE DENSE
C                  OUTPUT FORMULA IS ACTIVATED. IT CAN BE SUPPRESSED
C                  BY PUTTING IWORK(6)=1.
C                  DEFAULT IWORK(6)=0  (IF IOUT.GE.2).
C
C     IWORK(7)  DETERMINES THE DEGREE OF INTERPOLATION FORMULA
C                  MU = 2 * KAPPA - IWORK(7) + 1
C                  IWORK(7) SHOULD LIE BETWEEN 1 AND 6
C                  DEFAULT IWORK(7)=4  (IF IWORK(7)=0).
C
C     IWORK(8)  = NRDENS = NUMBER OF COMPONENTS, FOR WHICH DENSE OUTPUT
C                  IS REQUIRED
C
C     IWORK(10),...,IWORK(NRDENS+9) INDICATE THE COMPONENTS, FOR WHICH
C                  DENSE OUTPUT IS REQUIRED
.........
C
C     IDID       REPORTS ON SUCCESSFULNESS UPON RETURN:
C                  IDID=1  COMPUTATION SUCCESSFUL,
C                  IDID=-1 COMPUTATION UNSUCCESSFUL.
.........
```

Subroutine ODEX2

Extrapolation code for second order differential equations $y'' = f(x, y)$ (Section II.14). It uses variable order and variable step sizes and is provided with a high-order dense output. The missing parts of the description are identical to those of ODEX.

```
      SUBROUTINE ODEX2(N,FCN,X,Y,YP,XEND,H,
     +                 RTOL,ATOL,ITOL,
     +                 SOLOUT,IOUT,
     +                 WORK,LWORK,IWORK,LIWORK,LRCONT,LICONT,IDID)
C -----------------------------------------------------------
C     NUMERICAL SOLUTION OF A SYSTEM OF SECOND ORDER
C     ORDINARY DIFFERENTIAL EQUATIONS  Y''=F(X,Y).
C     THIS IS AN EXTRAPOLATION-ALGORITHM, BASED ON
C     THE STOERMER RULE (WITH STEPSIZE CONTROL
C     ORDER SELECTION AND DENSE OUTPUT).
.........
C
C     VERSION MARCH 30, 1992
.........
C
C     Y(N)       INITIAL VALUES FOR Y
C
```

```
C     YP(N)         INITIAL VALUES FOR Y'
.........
C
C     ITOL          SWITCH FOR RTOL AND ATOL:
C                     ITOL=0:  BOTH RTOL AND ATOL ARE SCALARS.
C                       THE CODE KEEPS, ROUGHLY, THE LOCAL ERROR OF
C                       Y(I)  BELOW  RTOL*ABS(Y(I))+ATOL
C                       YP(I) BELOW  RTOL*ABS(YP(I))+ATOL
C                     ITOL=1:  BOTH RTOL AND ATOL ARE VECTORS.
C                       THE CODE KEEPS THE LOCAL ERROR OF
C                       Y(I)  BELOW  RTOL(I)*ABS(Y(I))+ATOL(I).
C                       YP(I) BELOW  RTOL(I+N)*ABS(YP(I))+ATOL(I+N).
C
C     SOLOUT        NAME (EXTERNAL) OF SUBROUTINE PROVIDING THE
C                   NUMERICAL SOLUTION DURING INTEGRATION.
C                   IF IOUT>=1, IT IS CALLED AFTER EVERY SUCCESSFUL STEP.
C                   SUPPLY A DUMMY SUBROUTINE IF IOUT=0.
C                   IT MUST HAVE THE FORM
C                       SUBROUTINE SOLOUT (NR,XOLD,X,Y,YP,N,IRTRN)
C                       REAL*8 X,Y(N),YP(N)
C                       ....
C                   SOLOUT FURNISHES THE SOLUTIONS "Y, YP" AT THE NR-TH
C                       GRID-POINT "X" (THEREBY THE INITIAL VALUE IS
C                       THE FIRST GRID-POINT).
C                   "XOLD" IS THE PRECEEDING GRID-POINT.
C                   "IRTRN" SERVES TO INTERRUPT THE INTEGRATION. IF IRTRN
C                       IS SET <0, ODEX2 WILL RETURN TO THE CALLING PROGRAM.
C
C          -----    CONTINUOUS OUTPUT (IF IOUT=2):  -----
C                   DURING CALLS TO "SOLOUT", A CONTINUOUS SOLUTION
C                   FOR THE INTERVAL [XOLD,X] IS AVAILABLE THROUGH
C                   THE REAL*8 FUNCTION
C                          >>>   CONTX2(I,S)   <<<
C                   WHICH PROVIDES AN APPROXIMATION TO THE I-TH
C                   COMPONENT OF THE SOLUTION AT THE POINT S. THE VALUE
C                   S SHOULD LIE IN THE INTERVAL [XOLD,X].
.........
C
C     WORK          ARRAY OF WORKING SPACE OF LENGTH "LWORK".
C                   SERVES AS WORKING SPACE FOR ALL VECTORS.
C                   "LWORK" MUST BE AT LEAST
C                       N*(2*KM+6)+5*KM+10+KM*(2*KM+3)*NRDENS
C                   WHERE NRDENS=IWORK(8) (SEE BELOW) AND
C                       KM=9                IF IWORK(2)=0
C                       KM=IWORK(2)         IF IWORK(2).GT.0
C                   WORK(1),...,WORK(10) SERVE AS PARAMETERS
C                   FOR THE CODE. FOR STANDARD USE, SET THESE
C                   PARAMETERS TO ZERO BEFORE CALLING.
.........
C
C     IWORK         INTEGER WORKING SPACE OF LENGTH "LIWORK".
C                   "LIWORK" MUST BE AT LEAST
C                            KM+9+NRDENS
C                   IWORK(1),...,IWORK(9) SERVE AS PARAMETERS
C                   FOR THE CODE. FOR STANDARD USE, SET THESE
C                   PARAMETERS TO ZERO BEFORE CALLING.
.........
C
C     LRCONT        DECLARED LENGTH OF COMMON BLOCK
C                       >>>  COMMON /CONTR2/RCONT(LRCONT)  <<<
```

```
C                    WHICH MUST BE DECLARED IN THE CALLING PROGRAM.
C                    "LRCONT" MUST BE AT LEAST
C                               ( 2 * KM + 6 ) * NRDENS + 2
C                    WHERE KM=IWORK(2) AND NRDENS=IWORK(8) (SEE BELOW).
C
C    LICONT          DECLARED LENGTH OF COMMON BLOCK
C                       >>>  COMMON /CONTI2/ICONT(LICONT)  <<<
C                    WHICH MUST BE DECLARED IN THE CALLING PROGRAM.
C                    "LICONT" MUST BE AT LEAST
C                               NRDENS + 2
C                    THESE COMMON BLOCKS ARE USED FOR STORING THE COEFFICIENTS
C                    OF THE CONTINUOUS SOLUTION AND MAKES THE CALLING LIST FOR
C                    THE FUNCTION "CONTX2" AS SIMPLE AS POSSIBLE.
C .........
C
C    WORK(3)         STEP SIZE IS REDUCED BY FACTOR WORK(3), IF DURING THE
C                    COMPUTATION OF THE EXTRAPOLATION TABLEAU DIVERGENCE
C                    IS OBSERVED; DEFAULT 0.5.
C .........
C
C    IWORK(3)        SWITCH FOR THE STEP SIZE SEQUENCE (EVEN NUMBERS ONLY)
C                    IF IWORK(3).EQ.1 THEN 2,4,6,8,10,12,14,16,...
C                    IF IWORK(3).EQ.2 THEN 2,4,8,12,16,20,24,28,...
C                    IF IWORK(3).EQ.3 THEN 2,4,6,8,12,16,24,32,...
C                    IF IWORK(3).EQ.4 THEN 2,6,10,14,18,22,26,30,...
C                    THE DEFAULT VALUE IS IWORK(3)=1 IF IOUT.LE.1;
C                    THE DEFAULT VALUE IS IWORK(3)=4 IF IOUT.GE.2.
C .........
C
C    IWORK(7)        DETERMINES THE DEGREE OF INTERPOLATION FORMULA
C                    MU = 2 * KAPPA - IWORK(7) + 1
C                    IWORK(7) SHOULD LIE BETWEEN 1 AND 8
C                    DEFAULT IWORK(7)=6  (IF IWORK(7)=0).
C .........
```

Driver for the Code RETARD

We consider the delay equation (II.17.14) with initial values and initial functions given there. This is a 3-dimensional problem, but only the second component is used with retarded argument (hence NRDENS=1). We require that the points $1, 2, 3, \ldots, 9, 10, 20$ (points of discontinuity of the derivatives of the solution) are hitten exactly by the integration routine.

```
          IMPLICIT REAL*8 (A-H,O-Z)
          PARAMETER (NDGL=3,NGRID=11,LWORK=8*NDGL+11+NGRID,LIWORK=10)
          PARAMETER (NRDENS=1,LRCONT=500,LICONT=NRDENS+1)
          DIMENSION Y(NDGL),WORK(LWORK),IWORK(LIWORK)
          COMMON/STATRE/NFCN,NSTEP,NACCPT,NREJCT
          COMMON /CORER/RCONT(LRCONT)
          COMMON /COREI/ICONT(LICONT)
          EXTERNAL FCN,SOLOUT
C --- DIMENSION OF THE SYSTEM
          N=NDGL
C --- OUTPUT ROUTINE IS USED DURING INTEGRATION
```

```
          IOUT=1
C --- INITIAL VALUES AND ENDPOINT OF INTEGRATION
          X=0.0D0
          Y(1)=5.0D0
          Y(2)=0.1D0
          Y(3)=1.0D0
          XEND=40.D0
C --- REQUIRED (RELATIVE AND ABSOLUTE) TOLERANCE
          ITOL=0
          RTOL=1.0D-5
          ATOL=RTOL
C --- DEFAULT VALUES FOR PARAMETERS
          DO 10 I=1,10
          IWORK(I)=0
   10     WORK(I)=0.D0
C --- SECOND COMPONENT USES RETARDED ARGUMENT
          IWORK(5)=NRDENS
          ICONT(2)=2
C ---  USE AS GRID-POINTS
          IWORK(6)=NGRID
          DO 12 I=1,NGRID-1
   12     WORK(10+I)=I
          WORK(10+NGRID)=20.D0
C --- CALL OF THE SUBROUTINE RETARD
          CALL RETARD(N,FCN,X,Y,XEND,
     +                RTOL,ATOL,ITOL,
     +                SOLOUT,IOUT,
     +                WORK,LWORK,IWORK,LIWORK,LRCONT,LICONT,IDID)
C --- PRINT FINAL SOLUTION
          WRITE (6,99) Y(1),Y(2),Y(3)
   99     FORMAT(1X,'X = XEND      Y =',3E18.10)
C --- PRINT STATISTICS
          WRITE (6,91) RTOL,NFCN,NSTEP,NACCPT,NREJCT
   91     FORMAT('     tol=',D8.2,'    fcn=',I5,' step=',I4,
     +                 ' accpt=',I4,' rejct=',I3)
          STOP
          END
C
C
          SUBROUTINE SOLOUT (NR,XOLD,X,Y,N,IRTRN)
C --- PRINTS SOLUTION AT EQUIDISTANT OUTPUT-POINTS
          IMPLICIT REAL*8 (A-H,O-Z)
          DIMENSION Y(N)
          EXTERNAL PHI
          COMMON /INTERN/XOUT
          IF (NR.EQ.1) THEN
             WRITE (6,99) X,Y(1),NR-1
             XOUT=X+5.D0
          ELSE
   10        CONTINUE
             IF (X.GE.XOUT) THEN
                WRITE (6,99) X,Y(1),NR-1
                XOUT=XOUT+5.D0
                GOTO 10
             END IF
          END IF
   99     FORMAT(1X,'X =',F6.2,'    Y =',E18.10,'    NSTEP =',I4)
          RETURN
          END
C
```

```
      SUBROUTINE FCN(N,X,Y,F)
      IMPLICIT REAL*8 (A-H,O-Z)
      DIMENSION Y(N),F(N)
      EXTERNAL PHI
      Y2L1=YLAG(2,X-1.D0,PHI)
      Y2L10=YLAG(2,X-10.D0,PHI)
      F(1)=-Y(1)*Y2L1+Y2L10
      F(2)=Y(1)*Y2L1-Y(2)
      F(3)=Y(2)-Y2L10
      RETURN
      END
C
      FUNCTION PHI(I,X)
      IMPLICIT REAL*8 (A-H,O-Z)
      IF (I.EQ.2) PHI=0.1D0
      RETURN
      END
```

The result, obtained on an Apollo workstation, is the following:

```
X =  0.00    Y =  0.5000000000E+01    NSTEP =   0
X =  5.00    Y =  0.2533855892E+00    NSTEP =  18
X = 10.00    Y =  0.3328560326E+00    NSTEP =  32
X = 15.29    Y =  0.4539376456E+01    NSTEP =  40
X = 20.00    Y =  0.1706635702E+00    NSTEP =  52
X = 25.22    Y =  0.2524799457E+00    NSTEP =  62
X = 30.48    Y =  0.5134266860E+01    NSTEP =  68
X = 35.10    Y =  0.3610797907E+00    NSTEP =  78
X = 40.00    Y =  0.9125544555E-01    NSTEP =  89
X = XEND     Y =  0.9125544555E-01    0.2029882456E-01   0.5988445730E+01
     tol=0.10E-04    fcn=  586 step=  97 accpt=  89 rejct=  8
```

Subroutine RETARD

Modification of the code DOPRI5 for delay differential equations (see Section II.17). The missing parts of the description are identical to those of DOPRI5.

```
      SUBROUTINE RETARD(N,FCN,X,Y,XEND,
     +                  RTOL,ATOL,ITOL,
     +                  SOLOUT,IOUT,
     +                  WORK,LWORK,IWORK,LIWORK,LRCONT,LICONT,IDID)
C ------------------------------------------------------------
C     NUMERICAL SOLUTION OF A SYSTEM OF FIRST ORDER DELAY
C     ORDINARY DIFFERENTIAL EQUATIONS  Y'(X)=F(X,Y(X),Y(X-A),...).
C     THIS CODE IS BASED ON AN EXPLICIT RUNGE-KUTTA METHOD OF
C     ORDER (4)5 DUE TO DORMAND & PRINCE (WITH STEPSIZE CONTROL
C     AND DENSE OUTPUT).
C........
C
C     VERSION OF APRIL 24, 1992
C........
C
C     FCN        NAME (EXTERNAL) OF SUBROUTINE COMPUTING THE RIGHT-
C                HAND-SIDE OF THE DELAY EQUATION, E.G.,
```

```
C                           SUBROUTINE FCN(N,X,Y,F)
C                           REAL*8 X,Y(N),F(N)
C                           EXTERNAL PHI
C                           F(1)=(1.4D0-YLAG(1,X-1.D0,PHI))*Y(1)
C                           F(2)=...       ETC.
C                   FOR AN EXPLICATION OF YLAG SEE BELOW.
C                   DO NOT USE YLAG(I,X-0.D0,PHI) !
C                   THE INITIAL FUNCTION HAS TO BE SUPPLIED BY:
C                           FUNCTION PHI(I,X)
C                           REAL*8 PHI,X
C                   WHERE I IS THE COMPONENT AND X THE ARGUMENT
C .........
C
C     Y(N)          INITIAL VALUES FOR Y (MAY BE DIFFERENT FROM PHI (I,X),
C                   IN THIS CASE IT IS HIGHLY RECOMMENDED TO SET IWORK(6)
C                   AND WORK(11),..., SEE BELOW)
C .........
C
C           -----  CONTINUOUS OUTPUT: -----
C                   DURING CALLS TO "SOLOUT" AS WELL AS TO "FCN", A
C                   CONTINUOUS SOLUTION IS AVAILABLE THROUGH THE FUNCTION
C                           >>>   YLAG(I,S,PHI)   <<<
C                   WHICH PROVIDES AN APPROXIMATION TO THE I-TH
C                   COMPONENT OF THE SOLUTION AT THE POINT S. THE VALUE S
C                   HAS TO LIE IN AN INTERVAL WHERE THE NUMERICAL SOLUTION
C                   IS ALREADY COMPUTED. IT DEPENDS ON THE SIZE OF LRCONT
C                   (SEE BELOW) HOW FAR BACK THE SOLUTION IS AVAILABLE.
C
C     IOUT          SWITCH FOR CALLING THE SUBROUTINE SOLOUT:
C                     IOUT=0:  SUBROUTINE IS NEVER CALLED
C                     IOUT=1:  SUBROUTINE IS USED FOR OUTPUT.
C
C     WORK          ARRAY OF WORKING SPACE OF LENGTH "LWORK".
C                   "LWORK" MUST BE AT LEAST  8*N+11+NGRID
C                   WHERE NGRID=IWORK(6)
C .........
C
C     LRCONT        DECLARED LENGTH OF COMMON BLOCK
C                     >>>  COMMON /CORER/RCONT(LRCONT)  <<<
C                   WHICH MUST BE DECLARED IN THE CALLING PROGRAM.
C                   "LRCONT" MUST BE SUFFICIENTLY LARGE. IF THE DENSE
C                   OUTPUT OF MXST BACK STEPS HAS TO BE STORED, IT MUST
C                   BE AT LEAST
C                           MXST * ( 5 * NRDENS + 2 )
C                   WHERE NRDENS=IWORK(5) (SEE BELOW).
C
C     LICONT        DECLARED LENGTH OF COMMON BLOCK
C                     >>>  COMMON /COREI/ICONT(LICONT)  <<<
C                   WHICH MUST BE DECLARED IN THE CALLING PROGRAM.
C                   "LICONT" MUST BE AT LEAST
C                             NRDENS + 1
C                   THESE COMMON BLOCKS ARE USED FOR STORING THE COEFFICIENTS
C                   OF THE CONTINUOUS SOLUTION AND MAKES THE CALLING LIST FOR
C                   THE FUNCTION "CONTD5" AS SIMPLE AS POSSIBLE.
C .........
C
C   WORK(11),...,WORK(10+NGRID)  PRESCRIBED POINTS, WHICH THE
C             INTEGRATION METHOD HAS TO TAKE AS GRID-POINTS
C             X < WORK(11) < WORK(12) < ... < WORK(10+NGRID) <= XEND
C .........
```

```
C
C     IWORK(5)   = NRDENS = NUMBER OF COMPONENTS, FOR WHICH DENSE OUTPUT
C                  IS REQUIRED (EITHER BY "SOLOUT" OR BY "FCN");
C                  DEFAULT VALUE (FOR IWORK(5)=0) IS IWORK(5)=N;
C                  FOR   0 < NRDENS < N   THE COMPONENTS (FOR WHICH DENSE
C                  OUTPUT IS REQUIRED) HAVE TO BE SPECIFIED IN
C                  ICONT(2),...,ICONT(NRDENS+1);
C                  FOR  NRDENS=N  THIS IS DONE BY THE CODE.
C
C     IWORK(6)   = NGRID = NUMBER OF PRESCRIBED POINTS IN THE
C                  INTEGRATION INTERVAL WHICH HAVE TO BE GRID-POINTS
C                  IN THE INTEGRATION. USUALLY, AT THESE POINTS THE
C                  SOLUTION OR ONE OF ITS DERIVATIVE HAS A DISCONTINUITY.
C                  DEFINE THESE POINTS IN WORK(11),...,WORK(10+NGRID)
C                  DEFAULT VALUE:  IWORK(6)=0
C ........
C
C     IDID        REPORTS ON SUCCESSFULNESS UPON RETURN:
C                     IDID= 1   COMPUTATION SUCCESSFUL,
C                     IDID= 2   COMPUT. SUCCESSFUL (INTERRUPTED BY SOLOUT)
C                     IDID=-1   INPUT IS NOT CONSISTENT,
C                     IDID=-2   LARGER NMAX IS NEEDED,
C                     IDID=-3   STEP SIZE BECOMES TOO SMALL.
C                     IDID=-4   PROBLEM IS PROBABLY STIFF (INTERRUPTED).
C                     IDID=-5   COMPUT. INTERRUPTED BY YLAG
   ........
```

Bibliography

This bibliography includes the publications referred to in the text. Italic numbers in square brackets following a reference indicate the sections where the reference is cited.

N.H. Abel (1826): *Untersuchungen über die Reihe:*
$$1 + \frac{m}{1}x + \frac{m(m-1)}{1\cdot 2}x^2 + \frac{m(m-1)(m-2)}{1\cdot 2\cdot 3}x^3 + \ldots u.s.w.$$
Crelle J. f. d. r. u. angew. Math. (in zwanglosen Heften), Vol.1, p.311-339. *[III.8]*

N.H. Abel (1827): *Ueber einige bestimmte Integrale.* Crelle J. f. d. r. u. angew. Math., Vol.2, p.22-30. *[I.11]*

L. Abia & J.M. Sanz-Serna (1993): *Partitioned Runge-Kutta methods for separable Hamiltonian problems.* Math. Comp., Vol.60, p.617-634. *[II.16]*

L. Abia, see also J.M. Sanz-Serna & L. Abia.

M. Abramowitz & I.A. Stegun (1964): *Handbook of mathematical functions.* Dover, 1000 pages. *[II.7]*, *[II.8]*, *[II.9]*

J.C. Adams (1883): see F.Bashforth (1883).

R.C. Aiken ed. (1985): *Stiff computation.* Oxford, Univ. Press, 462pp. *[Appendix]*

A.C. Aitken (1932): *On interpolation by iteration of proportional parts, without the use of differences.* Proc. Edinburgh Math. Soc. Second ser., Vol.3, p.56-76. *[II.9]*

J. Albrecht (1955): *Beiträge zum Runge-Kutta-Verfahren.* ZAMM, Vol.35, p.100-110. *[II.13]*, *[II.14]*

P. Albrecht (1978): *Explicit, optimal stability functionals and their application to cyclic discretization methods.* Computing, Vol.19, p.233-249. *[III.8]*

P. Albrecht (1979): *Die numerische Behandlung gewöhnlicher Differentialgleichungen.* Akademie Verlag, Berlin; Hanser Verlag, München. *[III.8]*

P. Albrecht (1985): *Numerical treatment of O.D.E.s.: The theory of A-methods.* Numer. Math., Vol.47, p.59-87. *[III.8]*

V.M. Alekseev (1961): *An estimate for the perturbations of the solution of ordinary differential equations (Russian).* Vestn. Mosk. Univ., Ser.I, Math. Meh, 2, p.28-36. *[I.14]*

J.le Rond d'Alembert (1743): *Traité de dynamique, dans lequel les loix de l'équilibre & du mouvement des corps sont réduites au plus petit nombre possible, & démontrées d'une maniére nouvelle, & où l'on donne un principe général pour trouver le mouvement de plusieurs corps qui agissent les uns sur les autres, d'une maniére quelconque.* à Paris, MDCCXLIII, 186p., 70 figs. *[I.6]*

J.le Rond d'Alembert (1747): *Recherches sur la courbe que forme une corde tenduë mise en vibration.* Hist. de l'Acad. Royale de Berlin, Tom.3, Année MDCCXLVII, publ. 1749, p.214-219 et 220-249. *[I.6]*

J.le Rond d'Alembert (1748): *Suite des recherches sur le calcul intégral, quatrième partie: Méthodes pour intégrer quelques équations différentielles.* Hist. Acad. Berlin, Tom.IV, p.275-291. *[I.4]*

R.F. Arenstorf (1963): *Periodic solutions of the restricted three body problem representing analytic continuations of Keplerian elliptic motions.* Amer. J. Math., Vol.LXXXV, p.27-35. *[II.0]*

V.I. Arnol'd (1974): *Mathematical methods of classical mechanics.* Nauka, Moscow; French transl. Mir 1976; Engl. transl. Springer-Verlag 1978 (2nd edition 1989). *[I.14]*

C. Arzelà (1895): *Sulle funzioni di linee.* Memorie dell. R. Accad. delle Sc. di Bologna, 5e serie, Vol.V, p.225-244, see also: Vol.V, p.257-270, Vol.VI, (1896), p.131-140. *[I.7]*

U.M. Ascher, R.M.M. Mattheij & R.D. Russel (1988): *Numerical Solution of Boundary Value Problems for Ordinary Differential Equations.* Prentice Hall, Englewood Cliffs. *[I.15]*

L.S. Baca, see L.F. Shampine & L.S. Baca, L.F. Shampine, L.S. Baca & H.-J. Bauer.

H.F. Baker (1905): *Alternants and continuous groups.* Proc. London Math. Soc., Second Ser., Vol.3, p.24-47. *[II.16]*

N. Bakhvalov (1976): *Méthodes numériques.* Editions Mir, Moscou 600pp., russian edition 1973. *[I.9]*

F. Bashforth (1883): *An attempt to test the theories of capillary action by comparing the theoretical and measured forms of drops of fluid. With an explanation of the method of integration employed in constructing the tables which give the theoretical form of such drops, by J.C.Adams.* Cambridge Univ. Press. *[III.1]*

R.H. Battin (1976): *Resolution of Runge-Kutta-Nyström condition equations through eighth order.* AIAA J., Vol.14, p.1012-1021. *[II.14]*

F.L. Bauer, H. Rutishauser & E. Stiefel (1963): *New aspects in numerical quadrature.* Proc. of Symposia in Appl. Math., Vol.15, p.199-218, Am. Math. Soc. *[II.9]*

H.-J. Bauer, see L.F. Shampine, L.S. Baca & H.-J. Bauer.

P.A. Beentjes & W.J. Gerritsen (1976): *Higher order Runge-Kutta methods for the numerical solution of second order differential equations without first derivatives.* Report NW 34/76, Math. Centrum, Amsterdam. *[II.14]*

H. Behnke & F. Sommer (1962): *Theorie der analytischen Funktionen einer komplexen Veränderlichen.* Zweite Auflage. Springer Verlag, Berlin-Göttingen-Heidelberg. *[III.2]*

A. Bellen (1984): *One-step collocation for delay differential equations.* J. Comput. Appl. Math., Vol.10, p.275-283. *[II.17]*

A. Bellen & M. Zennaro (1985): *Numerical solution of delay differential equations by uniform corrections to an implicit Runge-Kutta method.* Numer. Math., Vol.47, p.301-316. *[II.17]*

R. Bellman & K.L. Cooke (1963): *Differential-Difference equations*. Academic Press, 462pp. *[II.17]*

I. Bendixson (1893): *Sur le calcul des intégrales d'un système d'équations différentielles par des approximations successives*. Stock. Akad. Öfversigt Förh., Vol.50, p.599-612. *[I.8]*

I. Bendixson (1901): *Sur les courbes définies par des équations différentielles*. Acta Mathematica, Vol.24, p.1-88. *[I.16]*

I.S. Berezin & N.P Zhidkov (1965): *Computing methods (Metody vychislenii)*. 2 Volumes, Fizmatgiz Moscow, Engl. transl.: Pergamon Press, 464 & 679pp. *[I.1]*

Dan. Bernoulli (1728): *Observationes de seriebus quae formantur ex additione vel substractione quacunque terminorum se mutuo consequentium, ubi praesertim earundem insignis usus pro inveniendis radicum omnium aequationum algebraicarum ostenditur.* Comm. Acad. Sci. Imperialis Petrop., Tom.III, 1728 (1732), p.85-100; Werke, Bd.2, p.49-70. *[III.3]*

Dan. Bernoulli (1732): *Theoremata de oscillationibus corporum filo flexili connexorum et catenae verticaliter suspensae*. Comm. Acad. Sci. Imperialis Petrop., Tom.VI, ad annus MDCCXXXII & MDCCXXXIII, p.108-122. *[I.6]*

Dan. Bernoulli (1760): *Essai d'une nouvelle analyse de la mortalité causée par la petite vérole, et des avantages de l'inoculation pour la prévenir.* Hist. et Mém. de l'Acad. Roy. Sciences Paris, 1760, p.1-45; Werke Bd. 2, p.235-267. *[II.17]*

Jac. Bernoulli (1690): *Analysis problematis ante hac propositi, de inventione lineae descensus a corpore gravi percurrendae uniformiter, sic ut temporibus aequalibus aequales altitudines emetiatur: & alterius cujusdam Problematis Propositio.* Acta Erudit. Lipsiae, Anno MDCLXXXX, p. 217-219. *[I.3]*

Jac. Bernoulli (1695): *Explicationes, Annotationes & Additiones ad ea, quae in Actis sup. anni de Curva Elastica, Isochrona Paracentrica, & Velaria, hinc inde memorata, & partim controversa legundur; ubi de Linea mediarum directionum, aliisque novis.* Acta Erudit. Lipsiae, Anno MDCXCV, p. 537-553. *[I.3]*

Jac. Bernoulli (1697): *Solutio Problematum Fraternorum, Peculiari Programmate Cal. Jan. 1697 Groningae, nec non Actorum Lips. mense Jun. & Dec. 1696, & Febr. 1697 propositorum: una cum Propositione reciproca aliorum.* Acta Erud. Lips. MDCXCVII, p.211-217. *[I.2]*

Joh. Bernoulli (1691): *Solutio problematis funicularii, exhibita à Johanne Bernoulli, Basil. Med. Cand..* Acta Erud. Lips. MDCXCI, p.274, Opera Omnia, Vol.I, p.48-51, Lausannae & Genevae 1742. *[I.3]*

Joh. Bernoulli (1696): *Problema novum Mathematicis propositorum.* Acta Erud. Lips. MDCXCVI, p.269, Opera Omnia, Vol.I, p.161 and 165, Lausannae & Genevae 1742. *[I.2]*

Joh. Bernoulli (1697): *De Conoidibus et Sphaeroidibus quaedam. Solutio analytica Aequationis in Actis A. 1695, pag. 553 propositae.* Acta Erud. Lips., MDCXCVII, p.113-118. Opera Omnia, Vol.I, p.174-179. *[I.3]*

Joh. Bernoulli (1697b): *Solutioque Problematis a se in Actis 1696, p.269, proposit, de invenienda Linea Brachystochrona.* Acta Erud.Lips. MDCXCVII, p.206, Opera Omnia, Vol.I, p.187-193. *[I.2]*

Joh. Bernoulli (1727): *Meditationes de chordis vibrantibus* .... Comm. Acad. Sci. Imperialis Petrop., Tom.III, p.13; Opera, Vol.III, p.198-210. *[I.6]*

J. Berntsen & T.O. Espelid (1991): *Error estimation in automatic quadrature routines.* ACM Trans. on Math. Software, Vol.17, p.233-255. *[II.10]*

D.G. Bettis (1973): *A Runge-Kutta Nyström algorithm.* Celestial Mechanics, Vol.8, p.229-233. *[II.14]*

L. Bieberbach (1923): *Theorie der Differentialgleichungen.* Grundlehren Bd.VI, Springer Verlag. *[II.3]*

L. Bieberbach (1951): *On the remainder of the Runge-Kutta formula in the theory of ordinary differential equations.* ZAMP, Vol.2, p.233-248. *[II.3]*

J. Binney (1981): *Resonant excitation of motion perpendicular to galactic planes.* Mon. Not. R. astr. Soc., Vol.196, p.455-467. *[II.16]*

J. Binney & S. Tremaine (1987): *Galactic dynamics.* Princeton Univ. Press, 733pp. *[II.16]*

J.B. Biot (1804): *Mémoire sur la propagation de la chaleur, lu à la classe des sciences math. et phys. de l'Institut national.* Bibl. britann. Sept 1804, 27, p.310. *[I.6]*

G. Birkhoff & R.S. Varga (1965): *Discretization errors for well-set Cauchy problems I.* Journal of Math. and Physics, Vol.XLIV, p.1-23. *[I.13]*

H.G. Bock (1981): *Numerical treatment of inverse problems in chemical reaction kinetics.* In: Modelling of chemical reaction systems, ed. by K.H. Ebert, P. Deuflhard & W. Jäger, Springer Series in Chem. Phys., Vol.18, p.102-125. *[II.6]*

H.G. Bock & J. Schlöder (1981): *Numerical solution of retarded differential equations with statedependent time lages.* ZAMM, Vol.61, p.269-271. *[II.17]*

P. Bogacki & L.F. Shampine (1989): *An efficient Runge-Kutta (4,5) pair.* SMU Math Rept 89-20. *[II.6]*

N. Bogoliuboff, see N. Kryloff & N. Bogoliuboff.

R.W. Brankin, I. Gladwell, J.R. Dormand, P.J. Prince & W.L. Seward (1989): *Algorithm 670. A Runge-Kutta-Nyström code.* ACM Trans. Math. Softw., Vol.15, p.31-40. *[II.14]*

R.W. Brankin, see also I. Gladwell, L.F. Shampine & R.W. Brankin.

P.N. Brown, G.D. Byrne & A.C. Hindmarsh (1989): *VODE: a variable-coefficient ODE solver.* SIAM J. Sci. Stat. Comput., Vol.10, p.1038-1051. *[III.7]*

H. Brunner & P.J. van der Houwen (1986): *The numerical solution of Volterra equations.* North-Holland, Amsterdam, 588pp. *[II.17]*

R. Bulirsch & J. Stoer (1964): *Fehlerabschätzungen und Extrapolation mit rationalen Funktionen bei Verfahren vom Richardson-Typus.* Num. Math., Vol.6, p.413-427. *[II.9]*

R. Bulirsch & J. Stoer (1966): *Numerical treatment of ordinary differential equations by extrapolation methods.* Num. Math., Vol.8, p.1-13. *[II.9]*

K. Burrage (1985): *Order and stability properties of explicit multivalue methods.* Appl. Numer. Anal., Vol.1, p.363-379. *[III.8]*

K. Burrage & J.C. Butcher (1980): *Non-linear stability of a general class of differential equation methods.* BIT, Vol.20, p.185-203. *[III.8]*

K. Burrage & P. Moss (1980): *Simplifying assumptions for the order of partitioned multivalue methods.* BIT, Vol.20, p.452-465. *[III.8]*

J.C. Butcher (1963): *Coefficients for the study of Runge-Kutta integration processes.* J. Austral. Math. Soc., Vol.3, p.185-201. *[II.2]*

J.C. Butcher (1963a): *On the integration process of A. Huťa.* J. Austral. Math. Soc., Vol.3, p.202-206. *[II.2]*

J.C. Butcher (1964a): *Implicit Runge-Kutta Processes.* Math. Comput., Vol.18, p.50-64. *[II.7]*, *[II.16]*

J.C. Butcher (1964b): *On Runge-Kutta processes of high order.* J.Austral. Math. Soc., Vol.IV, Part2, p.179-194. *[II.1]*, *[II.5]*

J.C. Butcher (1964c): *Integration processes based on Radau quadrature formulas.* Math. Comput., Vol.18, p.233-244. *[II.7]*

J.C. Butcher (1965a): *A modified multistep method for the numerical integration of ordinary differential equations.* J. ACM, Vol.12, p.124-135. *[III.8]*

J.C. Butcher (1965b): *On the attainable order of Runge-Kutta methods.* Math. of Comp., Vol.19, p.408-417. *[II.5]*

J.C. Butcher (1966): *On the convergence of numerical solutions to ordinary differential equations.* Math. Comput., Vol.20, p.1-10. *[III.4]*, *[III.8]*

J.C. Butcher (1969): *The effective order of Runge-Kutta methods.* in: Conference on the numerical solution of differential equations, Lecture Notes in Math., Vol.109, p.133-139. *[II.12]*

J.C. Butcher (1981): *A generalization of singly-implicit methods.* BIT, Vol.21, p.175-189. *[III.8]*

J.C. Butcher (1984): *An application of the Runge-Kutta space.* BIT, Vol.24, p.425-440. *[II.12]*, *[III.8]*

J.C. Butcher (1985a): *General linear method: a survey.* Appl. Num. Math., Vol.1, p.273-284. *[III.8]*

J.C. Butcher (1985b): *The non-existence of ten stage eighth order explicit Runge-Kutta methods.* BIT, Vol.25, p.521-540. *[II.5]*

J.C. Butcher (1987): *The numerical analysis of ordinary differential equations. Runge-Kutta and general linear methods.* John Wiley & Sons, Chichester, 512pp. *[II.16]*

J.C. Butcher, see also K. Burrage & J.C. Butcher.

G.D. Byrne & A.C. Hindmarsh (1975): *A polyalgorithm for the numerical solution of ordinary differential equations.* ACM Trans. on Math. Software, Vol.1, No.1, p.71-96. *[III.6]*, *[III.7]*

G.D. Byrne & R.J. Lambert (1966): *Pseudo-Runge-Kutta methods involving two points.* J. Assoc. Comput. Mach., Vol.13, p.114-123. *[III.8]*

G.D. Byrne, see also P.N. Brown, G.D. Byrne & A.C. Hindmarsh.

R. Caira, C. Costabile & F. Costabile (1990): *A class of pseudo Runge-Kutta methods.* BIT, Vol.30, p.642-649. *[III.8]*

M. Calvé & R. Vaillancourt (1990): *Interpolants for Runge-Kutta pairs of order four and five.* Computing, Vol.45, p.383-388. *[II.6]*

M. Calvo, J.I. Montijano & L. Rández (1990): *A new embedded pair of Runge-Kutta formulas of orders 5 and 6.* Computers Math. Applic., Vol.20, p.15-24. *[II.6]*

M. Calvo, J.I. Montijano & L. Rández (1992): *New continuous extensions for the Dormand and Prince RK method.* In: Computational ordinary differential equations, ed. by J.R. Cash & I. Gladwell, Clarendon Press, Oxford, p.135-164. *[II.6]*

M.P. Calvo & J.M. Sanz-Serna (1992): *Order conditions for canonical Runge-Kutta-Nyström methods.* BIT, Vol.32, p.131-142. *[II.16]*

M.P. Calvo & J.M. Sanz-Serna (1992b): *High order symplectic Runge-Kutta-Nyström methods.* SIAM J. Sci. Stat. Comput., Vol.14 (1993), p.1237-1252. *[II.16]*

M.P. Calvo & J.M. Sanz-Serna (1992c): *Reasons for a failure. The integration of the two-body problem with a symplectic Runge-Kutta method with step changing facilities.* Intern. Conf. on Differential Equations, Vol. 1, 2 (Barcelona, 1991), 93-102, World Sci. Publ., River Edge, NJ, 1993. *[II.16]*

J.M. Carnicer (1991): *A lower bound for the number of stages of an explicit continuous Runge-Kutta method to obtain convergence of given order.* BIT, Vol.31, p.364-368. *[II.6]*

E. Cartan (1899): *Sur certaines expressions différentielles et le problème de Pfaff.* Ann. Ecol. Normale, Vol.16, p.239-332, Oeuvres partie II, p.303-396. *[I.14]*

A.L. Cauchy (1824): *Résumé des Leçons données à l'Ecole Royale Polytechnique. Suite du Calcul Infinitésimal;* published: Equations différentielles ordinaires, ed. Chr. Gilain, Johnson 1981. *[I.2], [I.7], [I.9], [II.3], [II.7]*

A.L. Cauchy (1831): *Sur la mecanique celeste et sur un nouveau calcul appelé calcul des limites.* lu à l'acad. de Turin le 11 oct 1831; also: exerc. d'anal. et de pysique math, 2, Paris 1841; oeuvres (2), 12. *[III.3]*

A.L. Cauchy (1839-42): *Several articles in Comptes Rendus de l'Acad. des Sciences de Paris.* (Aug. 5, Nov. 21, 1839, June 29, Oct. 26, 1840, etc). *[I.8]*

A. Cayley (1857): *On the theory of the analytic forms called trees.* Phil. Magazine, Vol.XIII, p.172-176, Mathematical Papers, Vol.3, Nr.203, p.242-246. *[II.2]*

A. Cayley (1858): *A memoir on the theory of matrices.* Phil. Trans. of Royal Soc. of London, Vol.CXLVIII, p.17-37, Mathematical Papers, Vol.2, Nr.152, p.475.

F. Ceschino (1961): *Modification de la longueur du pas dans l'intégration numérique par les méthodes à pas liés.* Chiffres, Vol.2, p.101-106. *[II.4], [III.5]*

F. Ceschino (1962): *Evaluation de l'erreur par pas dans les problèmes différentiels.* Chiffres, Vol.5, p.223-229. *[II.4]*

F. Ceschino & J. Kuntzmann (1963): *Problèmes différentiels de conditions initiales (méthodes numériques).* Dunod Paris, 372pp.; english translation: Numerical solutions of initial value problems, Prentice Hall 1966 *[II.5], [II.7]*

P.J. Channell & C. Scovel (1990): *Symplectic integration of Hamiltonian systems.* Nonlinearity, Vol.3, p.231-259. *[II.16]*

A.C. Clairaut (1734): *Solution de plusieurs problèmes où il s'agit de trouver des courbes dont la propriété consiste dans une certaine relation entre leurs branches, exprimée par une Equation donnée.* Mémoires de Math. et de Phys. de l'Acad. Royale des Sciences, Paris, Année MDCCXXXIV, p.196-215. *[I.2]*

L. Collatz (1951): *Numerische Behandlung von Differentialgleichungen.* Grundlehren Band LX, Springer Verlag, 458pp; second edition 1955; third edition and english translation 1960. *[II.7]*

L. Collatz (1967): *Differentialgleichungen. Eine Einführung unter besonderer Berücksichtigung der Anwendungen.* Leitfäden der angewandten Mathematik, Teubner 226pp. English translation: *Differential equations. An introduction with applications,* Wiley, 372pp., (1986). *[I.15]*

P. Collet & J.P. Eckmann (1980): *Iterated maps on the interval as dynamical systems.* Birkhäuser, 248pp. *[I.16]*

K.L. Cooke, see R. Bellman & K.L. Cooke.

G.J. Cooper (1978): *The order of convergence of general linear methods for ordinary differential equations.* SIAM, J. Numer. Anal., Vol.15, p.643-661. *[III.8]*

G.J. Cooper (1987): *Stability of Runge-Kutta methods for trajectory problems.* IMA J. Numer. Anal., Vol.7, p.1-13. *[II.16]*

G.J. Cooper & J.H. Verner (1972): *Some explicit Runge-Kutta methods of high order.* SIAM J.Numer. Anal., Vol.9, p.389-405. *[II.5]*

S.A. Corey (1906): *A method of approximation.* Amer. Math. Monthly, Vol.13, p.137-140. *[II.9]*

C. Costabile, see R. Caira, C. Costabile & F. Costabile.

F. Costabile, see R. Caira, C. Costabile & F. Costabile.

C.A. de Coulomb (1785): *Théorie des machines simples, en ayant égard au frottement de leurs parties, et a la roideur des cordages. Pièce qui a remporté le Prix double de l'Académie des Sciences pour l'année 1781.* Mémoires des Savans Etrangers, tome X, p. 163-332; réimprimé 1809 chez Bachelier, Paris. *[II.6]*

P.H. Cowell & A.C.D. Crommelin (1910): *Investigation of the motion of Halley's comet from 1759 to 1910.* Appendix to Greenwich Observations for 1909, Edinburgh, p.1-84. *[III.10]*

J.W. Craggs, see A.R. Mitchell & J.W. Craggs.

D.M. Creedon & J.J.H. Miller (1975): *The stability properties of q-step backward-difference schemes.* BIT, Vol.15, p.244-249. *[III.3]*

A.C.D. Crommelin, see P.H. Cowell & A.C.D. Crommelin.

M. Crouzeix (1975): *Sur l'approximation des équations différentielles operationnelles linéaires par des méthodes de Runge-Kutta.* Thèse d'état, Univ. Paris 6, 192pp. *[II.2], [II.7]*

M. Crouzeix & F.J. Lisbona (1984): *The convergence of variable-stepsize, variable formula, multistep methods.* SIAM J. Num. Anal., Vol.21, p.512-534. *[III.5]*

C.W. Cryer (1971): *A proof of the instability of backward-difference multistep methods for the numerical integration of ordinary differential equations.* Tech. Rep. No.117, Comp. Sci. Dept., Univ. of Wisconsin, p.1-52. *[III.3]*

C.W. Cryer (1972): *On the instability of high order backward-difference multistep methods.* BIT, Vol.12, p.17-25. *[III.3]*

W.J. Cunningham (1954): *A nonlinear differential-difference equation of growth.* Proc. Mat. Acad. Sci., USA, Vol.40, p.708-713. *[II.17]*

A.R. Curtis (1970): *An eighth order Runge-Kutta process with eleven function evaluations per step.* Numer. Math., Vol.16, p.268-277. *[II.5]*

A.R. Curtis (1975): *High-order explicit Runge-Kutta formulae, their uses, and limitations.* J.Inst. Maths Applics, Vol.16, p.35-55. *[II.5]*

C.F. Curtiss & J.O. Hirschfelder (1952): *Integration of stiff equations.* Proc. of the National Academy of Sciences of U.S., Vol.38, p.235-243. *[III.1]*

G. Dahlquist (1956): *Convergence and stability in the numerical integration of ordinary differential equations.* Math. Scand., Vol.4, p.33-53. *[III.2]*, *[III.3]*, *[III.4]*

G. Dahlquist (1959): *Stability and error bounds in the numerical integration of ordinary differential equations.* Trans. of the Royal Inst. of Techn., Stockholm, Sweden, Nr.130, 87pp. *[I.10]*, *[III.2]*, *[III.10]*

G. Dahlquist (1985): *33 years of numerical instability, Part I.* BIT, Vol.25, p.188-204. *[III.3]*

G. Dahlquist & R. Jeltsch (1979): *Generalized disks of contractivity for explicit and implicit Runge-Kutta methods.* Report TRITA-NA-7906, NADA, Roy. Inst. Techn. Stockholm. *[II.12]*

G. Darboux (1876): *Sur les développements en série des fonctions d'une seule variable.* J. des Mathématiques pures et appl., 3ème série, t. II, p.291-312. *[II.13]*

G. H. Darwin (Sir George) (1898): *Periodic orbits.* Acta Mathematica, Vol.21, p.99-242, plates I-IV. *[II.0]*

S.M. Davenport, see L.F. Shampine, H.A. Watts & S.M. Davenport.

F. Debaune (1638): *Letter to Descartes.* lost; answer of Descartes: Feb 20, 1639. *[I.2]*

J.P. Den Hartog (1930): *Forced vibrations with combined viscous and Coulomb damping.* Phil. Mag. Ser.7, Vol.9, p.801-817. *[II.6]*

J. Descloux (1963): *A note on a paper by A. Nordsieck.* Report No.131, Dept. of Comp. Sci., Univ. of Illinois at Urbana-Champaign. *[III.6]*

P. Deuflhard (1980): *Recent advances in multiple shooting techniques.* In: Computational techniques for ordinary differential equations (Gladwell-Sayers, ed.), Section 10, p.217-272, Academic Press. *[I.15]*

P. Deuflhard (1983): *Order and stepsize control in extrapolation methods.* Num. Math., Vol.41, p.399-422. *[II.9]*, *[II.10]*

P. Deuflhard (1985): *Recent progress in extrapolation methods for ordinary differential equations.* SIAM Rev., Vol.27, p.505-535. *[II.14]*

P. Deuflhard & U. Nowak (1987): *Extrapolation integrators for quasilinear implicit ODEs.* In: P. Deuflhard, B. Engquist (eds.), Large-scale scientific computing, Birkhäuser, Boston. *[II.9]*

E. de Doncker-Kapenga, see R. Piessens, E. de Doncker-Kapenga, C.W. Überhuber & D.K. Kahaner.

J. Donelson & E. Hansen (1971): *Cyclic composite multistep predictor-corrector methods.* SIAM, J. Numer. Anal., Vol.8, p.137-157. *[III.8]*

J.R. Dormand, M.E.A. El-Mikkawy & P.J. Prince (1987): *High-order embedded Runge-Kutta-Nystrom formulae.* IMA J. Numer. Anal., Vol.7, p.423-430. *[II.14]*

J.R. Dormand & P.J. Prince (1978): *New Runge-Kutta algorithms for numerical simulation in dynamical astronomy.* Celestial Mechanics, Vol.18, p.223-232. *[II.14]*

J.R. Dormand & P.J. Prince (1980): *A family of embedded Runge-Kutta formulae.* J.Comp. Appl. Math., Vol.6, p.19-26. *[II.5]*

J.R. Dormand & P.J. Prince (1986): *Runge-Kutta triples.* Comp. & Maths. with Applc., Vol.12A, p.1007-1017. *[II.6]*

J.R. Dormand & P.J. Prince (1987): *Runge-Kutta-Nystrom triples.* Comput. Math. Applic., Vol.13(12), p.937-949. *[II.14]*

J.R. Dormand & P.J. Prince (1989): *Practical Runge-Kutta processes.* SIAM J. Sci. Stat. Comput., Vol.10, p.977-989. *[II.5]*

J.R. Dormand, see also P.J. Prince & J.R. Dormand, R.W. Brankin, I. Gladwell, J.R. Dormand, P.J. Prince & W.L. Seward.

R.D. Driver (1977): *Ordinary and delay differential equations.* Applied Math. Sciences 20, Springer Verlag, 501pp. *[II.17]*

J.P. Eckmann, see P. Collet & J.P. Eckmann.

B.L. Ehle (1968): *High order A-stable methods for the numerical solution of systems of D.E.'s.* BIT, Vol.8, p.276-278. *[II.7]*

E. Eich (1992): *Projizierende Mehrschrittverfahren zur numerischen Lösung von Bewegungsgleichungen technischer Mehrkörpersysteme mit Zwangsbedingungen unmd Unstetigkeiten.* Fortschritt-Ber. VDI, Reihe 18, Nr.109, VDI-Verlag Düsseldorf, 188pp. *[II.6]*

N.F. Eispack (1974): *B.T.Smith, J.M. Boyle, B.S.Garbow, Y.Jkebe, V.C.Klema, C.B.Moler: Matrix Eigensystem Routines.* (Fortran-translations of algorithms published in Reinsch & Wilkinson), Lecture Notes in Computer Science, Vol.6, Springer Verlag. *[I.12], [I.13]*

M.E.A. El-Mikkawy, see J.R. Dormand, M.E.A. El-Mikkawy & P.J. Prince.

H. Eltermann (1955): *Fehlerabschätzung bei näherungsweiser Lösung von Systemen von Differentialgleichungen erster Ordnung.* Math. Zeitschr., Vol.62, p.469-501. *[I.10]*

R. England (1969): *Error estimates for Runge-Kutta type solutions to systems of ordinary differential equations.* The Computer J. Vol.12, p.166-170. *[II.4]*

W.H. Enright & D.J. Higham (1991): *Parallel defect control.* BIT, Vol.31, p.647-663. *[II.11]*

W.H. Enright, K.R. Jackson, S.P. Nørsett & P.G. Thomson (1986): *Interpolants for Runge-Kutta formulas.* ACM Trans. Math. Softw., Vol.12, p.193-218. *[II.6] [II.6]*

W.H. Enright, K.R. Jackson, S.P. Nørsett & P.G. Thomson (1988): *Effective solution of discontinuous IVPs using a Runge-Kutta formula pair with interpolants.* Appl. Math. and Comput., Vol.27, p.313-335. *[II.6]*

T.O. Espelid, see J. Berntsen & T.O. Espelid.

L. Euler (1728): *Nova methodus innumerabiles aequationes differentiales secundi gradus reducendi ad aequationes differentiales primi gradus.* Comm. acad. scient. Petrop., Vol.3, p.124-137; Opera Omnia, Vol.XXII, p.1-14. *[I.3]*

L. Euler (1743): *De integratione aequationum differentialium altiorum graduum.* Miscellanea Berolinensia, Vol.7, p.193-242; Opera Omnia, Vol.XXII, p.108-149. See also: Letter from Euler to Joh. Bernoulli, 15.Sept.1739. *[I.4]*

L. Euler (1744): *Methodus inveniendi lineas curvas maximi minimive proprietate gaudentes* ... Lausannae & Genevae, Opera Omnia (intr. by Caratheodory) Vol.XXIV, p.1-308. *[I.2]*

L. Euler (1747): *Recherches sur le mouvement des corps celestes en général.* Hist. de l'Acad. Royale de Berlin, Tom.3, Année MDCCXLVII, publ. 1749, p.93-143. *[I.6]*

L. Euler (1750): *Methodus aequationes differentiales altiorum graduum integrandi ulterius promota.* Novi Comment. acad. scient. Petrop., Vol.3, p.3-35; Opera Omnia, Vol.XXII, p.181-213. *[I.4]*

L. Euler (1755): *Institutiones calculi differentialis cum eius vsu in analysi finitorum ac doctrina serierum.* Imp. Acad. Imper. Scient. Petropolitanae, Opera Omnia, Vol.X, *[I.6]*

L. Euler (1756): *Elementa calculi variationum.* presented September 16, 1756 at the Acad. of Science, Berlin; printed 1766, Opera Omnia, Vol.XXV,p.141-176. *[I.2]*

L. Euler (1758): *Du mouvement de rotation des corps solides autour d'un axe variable.* Hist. de l'Acad. Royale de Berlin, Tom.14, Année MDCCLVIII, pp.154-193. Opera Omnia Ser.II, Vol.8, p.200-235. *[II.10]*

L. Euler (1768): *Institutionum Calculi Integralis.* Volumen Primum, Opera Omnia, Vol.XI. *[I.7], [I.8], [II.1]*

L. Euler (1769): *Institutionum Calculi Integralis.* Volumen Secundum, Opera Omnia, Vol.XII. *[I.3], [I.5]*

L. Euler (1769b): *De formulis integralibus duplicatis.* Novi Comment. acad. scient. Petrop., Vol.14, I, 1770, p.72-103; Opera Omnia, Vol.XVII, p.289-315. *[I.14]*

L. Euler (1778): *Specimen transformationis singularis serienum.* Nova acta. acad. Petrop., Vol.12 (1794), p.58-70, Opera Omnia, Vol.XVI, Sectio Altera, p.41-55. *[I.5]*

E. Fehlberg (1958): *Eine Methode zur Fehlerverkleinerung beim Runge-Kutta-Verfahren.* ZAMM, Vol.38, p.421-426. *[II.13]*

E. Fehlberg (1964): *New high-order Runge-Kutta formulas with step size control for systems of first and second order differential equations.* ZAMM, Vol.44, Sonderheft T17-T19. *[II.4], [II.13]*

E. Fehlberg (1968): *Classical fifth-, sixth-, seventh-, and eighth order Runge-Kutta formulas with step size control.* NASA Technical Report 287 (1968); extract published in Computing, Vol.4, p.93-106 (1969). *[II.4], [II.5]*

E. Fehlberg (1969): *Low-order classical Runge-Kutta formulas with step size control and their application to some heat transfer problems.* NASA Technical Report 315 (1969), extract published in Computing, Vol.6, p.61-71 (1970). *[II.4], [II.5]*

E. Fehlberg (1972): *Classical eighth- and lower-order Runge-Kutta-Nyström formulas with stepsize control for special second-order differential equations.* NASA Technical Report R-381. *[II.14]*

M. Feigenbaum (1978): *Quantitative universality for a class of nonlinear transformations.* J.Stat. Phys., Vol.19, p.25-52, Vol.21 (1979), p.669-706. *[I.16]*

Feng Kang (冯康) (1985): *On difference schemes and symplectic geometry.* Proceedings of the 5-th Intern. Symposium on differential geometry & differential equations, August 1984, Beijing, p.42-58. *[II.16]*

Feng Kang (1986): *Difference schemes for Hamiltonian formalism and symplectic geometry.* J. Comp. Math., Vol.4, p.279-289. *[II.16]*

Feng Kang (1991): *How to compute properly Newton's equation of motion?* Proceedings of the second conference on numerical methods for partial differential equations, Nankai Univ., Tianjin, China, Eds. Ying Lungan & Guo Benyu, World Scientific, p.15-22. *[II.16]*

Feng Kang (1991b): *Formal power series and numerical algorithms for dynamical systems.* Proceedings of international conference on scientific computation, Hangzhou, China, Eds. Tony Chan & Zhong-Ci Shi, Series on Appl. Math., Vol.1, pp.28-35. *[II.16]*

Feng Kang, Wu Hua-mo, Qin Meng-zhao & Wang Dao-liu (1989): *Construction of canonical difference schemes for Hamiltonian formalism via generating functions.* J. Comp. Math., Vol.11, p.71-96. *[II.16]*

J.R. Field & R.M. Noyes (1974): *Oscillations in chemical systems. IV. Limit cycle behavior in a model of a real chemical reaction.* J. Chem. Physics, Vol.60, p.1877-1884. *[I.16]*

S. Filippi & J. Gräf (1986): *New Runge-Kutta-Nyström formula-pairs of order 8(7), 9(8), 10(9) and 11(10) for differential equations of the form* $y'' = f(x,y)$. J. Comput. and Applied Math., Vol.14, p.361-370. *[II.14]*

A.F. Filippov (1960): *Differential equations with discontinuous right-hand side.* Mat. Sbornik (N.S.) Vol.51(93), p.99-128; Amer. Math. Soc. Transl. Ser.2, Vol.42, p.199-231. *[II.6]*

J.M. Fine (1987): *Interpolants for Runge-Kutta-Nyström methods.* Computing, Vol.39, p.27-42. *[II.14]*

R. Fletcher & D.C. Sorensen (1983): *An algorithmic derivation of the Jordan canonical form.* Amer. Math. Monthly, Vol.90, No.1, p.12-16. *[I.12]*

C.V.D. Forrington (1961-62): *Extensions of the predictor-corrector method for the solution of systems of ordinary differential equations.* Comput. J. 4, p.80-84. *[III.5]*

J.B.J. Fourier (1807): *Sur la propagation de la chaleur.* Unpublished manuscript; published: La théorie analytique de la chaleur, Paris 1822. *[I.6]*

R.A. Frazer, W.P. Jones & S.W. Skan (1937): *Approximations to functions and to the solutions of differential equations.* Reports and Memoranda Nr.1799 (2913), Aeronautical Research Committee. 33pp. *[II.7]*

A. Fricke (1949): *Ueber die Fehlerabschätzung des Adamsschen Verfahrens zur Integration gewöhnlicher Differentialgleichungen erster Ordnung.* ZAMM, Vol.29, p.165-178. *[III.4]*

G. Frobenius (1873): *Ueber die Integration der linearen Differentialgleichungen durch Reihen.* Journal für Math. LXXVI, p.214-235 *[I.5]*

M. Frommer (1934): *Ueber das Auftreten von Wirbeln und Strudeln (geschlossener und spiraliger Integralkurven) inder Umgebung rationaler Unbestimmtheitsstellen.* Math. Ann., Vol.109, p.395-424. *[I.16]*

L. Fuchs (1866, 68): *Zur Theorie der linearen Differentialgleichungen mit veränderlichen Coefficienten.* Crelle J. f. d. r. u. angew. Math., Vol.66, p.121-160 (prepublished in "Programm der städtischen Gewerbeschule zu Berlin, Ostern 1865"). Ergänzung: J. f. Math. LXVIII, p. 354-385. *[I.5], [I.11]*

C.F. Gauss (1812): *Disquisitiones generales circa seriem infinitam*
$$1 + \frac{\alpha\beta}{1\cdot\gamma}x + \frac{\alpha(\alpha+1)\beta(\beta+1)}{1\cdot2\cdot\gamma(\gamma+1)}xx + \frac{\alpha(\alpha+1)(\alpha+2)\beta(\beta+1)(\beta+2)}{1\cdot2\cdot3\cdot\gamma(\gamma+1)(\gamma+2)}x^3 + etc,$$
Werke, Vol.3, p.123-162. *[I.5]*

W. Gautschi (1962): *On inverses of Vandermonde and confluent Vandermonde matrices.* Numer. Math., Vol.4, p.117-123. *[II.13]*

C.W. Gear (1965): *Hybrid methods for initial value problems in ordinary differential equations.* SIAM J. Numer. Anal., ser.B, Vol.2, p.69-86. *[III.8]*

C.W. Gear (1971): *Numerical initial value problems in ordinary differential equations.* Prentice-Hall, 253pp. *[II.2], [III.1], [III.7]*

C.W. Gear (1987): *The potential for parallelism in ordinary differential equations.* In: Computational mathematics II, Proc. 2nd Int. Conf. Numer. Anal. Appl., Benin City/Niger. 1986, Conf. Ser. Boole Press 11, p. 33-48. *[II.11]*

C.W. Gear (1988): *Parallel methods for ordinary differential equations.* Calcolo, Vol.25, No.1/2, p. 1-20. *[II.11]*

C.W. Gear & O. Østerby (1984): *Solving ordinary differential equations with discontinuities.* ACM Trans. Math. Softw., Vol.10, p.23-44. *[II.6]*

C.W. Gear & K.W. Tu (1974): *The effect of variable mesh size on the stability of multistep methods.* SIAM J. Num. Anal., Vol.11, p.1025-1043. *[III.5]*

C.W. Gear & D.S. Watanabe (1974): *Stability and convergence of variable order multistep methods.* SIAM J. Num. Anal., Vol.11, p.1044-1058. *[III.3]*

W.J. Gerritsen, see P.A. Beentjes & W.J. Gerritsen.

A. Gibbons (1960): *A program for the automatic integration of differential equations using the method of Taylor series.* Computer J., Vol.3, p.108-111. *[I.8]*

S. Gill (1951): *A process for the step-by-step integration of differential equations in an automatic digital computing machine.* Proc. Cambridge Philos. Soc., Vol.47, p.95-108. *[II.1], [II.2]*

S. Gill (1956): Discussion in Merson (1957). *[II.2]*

B. Giovannini, L. Weiss-Parmeggiani & B.T. Ulrich (1978): *Phase locking in coupled Josephson weak links.* Helvet. Physica Acta, Vol.51, p.69-74. *[I.16]*

I. Gladwell, L.F. Shampine & R.W. Brankin (1987): *Automatic selection of the initial step size for an ODE solver.* J. Comp. Appl. Math., Vol.18, p.175-192. *[II.4]*

I. Gladwell, see also R.W. Brankin, I. Gladwell, J.R. Dormand, P.J. Prince & W.L. Seward.

G.H. Golub & J.H. Wilkinson (1976): *Ill-conditioned eigensystems and the computation of the Jordan canonical form.* SIAM Review, Vol.18, p.578-619. *[I.12]*

M.K. Gordon, see L.F. Shampine & M.K. Gordon.

J. Gräf, see S. Filippi & J. Gräf.

W.B. Gragg (1964): *Repeated extrapolation to the limit in the numerical solution of ordinary differential equations.* Thesis, Univ. of California; see also SIAM J. Numer. Anal., Vol.2, p.384-403 (1965). *[II.8], [II.9]*

W.B. Gragg (1965): *On extrapolation algorithms for ordinary initial value problems.* SIAM J. Num. Anal., ser.B, Vol.2, p.384-403. *[II.14]*

W.B. Gragg & H.J. Stetter (1964): *Generalized multistep predictor-corrector methods.* J. ACM, Vol.11, p.188-209. *[III.8]*

E. Griepentrog (1978): *Gemischte Runge-Kutta-Verfahren für steife Systeme.* In: Seminar-bericht Nr. 11, Sekt. Math., Humboldt-Univ. Berlin, p.19-29. *[II.15]*

R.D. Grigorieff (1977): *Numerik gewöhnlicher Differentialgleichungen 2.* Teubner Studien-bücher, Stuttgart. *[III.3], [III.4]*

R.D. Grigorieff (1983): *Stability of multistep-methods on variable grids.* Numer. Math. 42, p.359-377. *[III.5]*

W. Gröbner (1960): *Die Liereihen und ihre Anwendungen.* VEB Deutscher Verlag der Wiss., Berlin 1960, 2nd ed. 1967. *[I.14], [II.16]*

T.H. Gronwall (1919): *Note on the derivatives with respect to a parameter of the solutions of a system of differential equations.* Ann. Math., Vol.20, p.292-296. *[I.10], [I.14]*

A. Guillou & J.L. Soulé (1969): *La résolution numérique des problèmes différentiels aux conditions initiales par des méthodes de collocation.* R.I.R.O, No R-3, p.17-44. *[II.7]*

P. Habets, see N. Rouche, P. Habets & M. Laloy.

H. Hahn (1921): *Theorie der reellen Funktionen.* Springer Verlag Berlin, 600pp. *[I.7]*

W. Hahn (1967): *Stability of motion.* Springer Verlag, 446pp. *[I.13]*

E. Hairer (1977): *Méthodes de Nyström pour l'équation différentielle $y'' = f(x,y)$.* Numer. Math., Vol.27, p.283-300. *[II.14]*

E. Hairer (1978): *A Runge-Kutta method of order 10.* J.Inst. Maths Applics, Vol.21, p.47-59. *[II.5]*

E. Hairer (1981): *Order conditions for numerical methods for partitioned ordinary differential equations.* Numer. Math., Vol.36, p.431-445. *[II.15]*

E. Hairer (1982): *A one-step method of order 10 for $y'' = f(x,y)$.* IMA J. Num. Anal., Vol.2, p.83-94. *[II.14]*

E. Hairer & Ch. Lubich (1984): *Asymptotic expansions of the global error of fixed-stepsize methods.* Numer. Math., Vol.45, p.345-360. *[II.8], [III.9]*

E. Hairer & A. Ostermann (1990): *Dense output for extrapolation methods.* Numer. Math., Vol.58, p.419-439. *[III.9]*

E. Hairer & A. Ostermann (1992): *Dense output for the GBS extrapolation method.* In: Computational ordinary differential equations, ed. by J.R. Cash & I. Gladwell, Clarendon Press, Oxford, p.107-114. *[II.9]*

E. Hairer & G. Wanner (1973): *Multistep-multistage-multiderivative methods for ordinary differential equations.* Computing, Vol.11, p.287-303. *[II.13], [III.8]*

E. Hairer & G. Wanner (1974): *On the Butcher group and general multi-value methods.* Computing, Vol.13, p.1-15. *[II.2], [II.12], [II.13]*

E. Hairer & G. Wanner (1976): *A theory for Nyström methods.* Numer. Math., Vol.25, p.383-400. *[II.14]*

E. Hairer & G. Wanner (1983): *On the instability of the BDF formulas.* SIAM J. Numer. Anal., Vol.20, No.6, p.1206-1209. *[III.3]*

Sir W. R. Hamilton (1833): *On a general method of expressing the paths of light, and of the planets, by the coefficients of a characteristic function.* Dublin Univ. Review, p.795-826; Math. Papers, Vol.I, p.311-332. *[I.6]*

504   Bibliography

Sir W. R. Hamilton (1834): *On a general method in dynamics; by which the study of the motions of all free systems of attracting or repelling points is reduced to the search and differentiation of one central relation, or characteristic function.* Phil. Trans. Roy. Soc. Part II for 1834, p.247-308; Math. Papers, Vol.II, p.103-161. *[I.6]*

Sir W. R. Hamilton (1835): *Second essay on a general method in dynamics.* Phil. Trans. Roy. Soc. Part I for 1835, p.95-144; Math. Papers, Vol.II, p.162-211. *[I.6]*

P.C. Hammer & J.W. Hollingsworth (1955): *Trapezoidal methods of approximating solutions of differential equations.* MTAC, Vol.9, p.92-96. *[II.7]*

E. Hansen, see J. Donelson & E. Hansen.

F. Hausdorff (1906): *Die symbolische Exponentialformel in der Gruppentheorie.* Berichte ü. d. Verh. Königl. Sächs. Ges. d. Wiss. Leipzig, Math.-Phys. Klasse, Vol.58, p.19-48. *[II.16]*

K. Hayashi, see M. Okamoto & K. Hayashi.

N.D. Hayes (1950): *Roots of the transzendental equation associated with a certain difference-differential equation.* J. of London Math. Soc., Vol.25, p.226-232. *[II.17]*

H.M. Hebsacker (1982): *Conditions for the coefficients of Runge-Kutta methods for systems of n -th order differential equations.* J. Comput. Appl. Math., Vol.8, p.3-14. *[II.14]*

P. Henrici (1962): *Discrete variable methods in ordinary differential equations.* John Wiley & Sons, Inc., New-York-London-Sydney. *[II.2]*, *[II.8]*, *[III.1]*, *[III.2]*, *[III.10]*

P. Henrici (1974): *Applied and computational complex analysis.* Volume 1, John Wiley & Sons, New York, 682pp. *[I.13]*, *[III.3]*

Ch. Hermite (1878): *Extrait d'une lettre de M. Ch. Hermite à M. Borchardt sur la formule d'interpolation de Lagrange.* J. de Crelle, Vol.84, p.70; Oeuvres, tome III, p.432-443. *[II.13]*

K. Heun (1900): *Neue Methode zur approximativen Integration der Differentialgleichungen einer unabhängigen Veränderlichen.* Zeitschr. für Math. u. Phys., Vol.45, p.23-38. *[I.5]*, *[II.1]*

D.J. Higham, see W.H. Enright & D.J. Higham.

A.C. Hindmarsh (1972): *GEAR: ordinary differential equation system solver.* UCID-30001, Rev.2, LLL, Livermore, Calif. *[III.7]*

A.C. Hindmarsh (1980): *LSODE and LSODI, two new initial value ordinary differential equation solvers.* ACM Signum Newsletter 15,4. *[II.4]*

A.C. Hindmarsh, see also P.N. Brown, G.D. Byrne & A.C. Hindmarsh, G.D. Byrne & A.C. Hindmarsh.

J.O. Hirschfelder, see C.F. Curtiss & J.O. Hirschfelder.

E.W. Hobson (1921): *The theory of functions of a real variable.* Vol.I, Cambridge, 670pp. *[I.10]*

E. Hofer (1976): *A partially implicit method for large stiff systems of ODEs with only few equations introducing small time-constants.* SIAM J. Numer. Anal., vol 13, No.5, p.645-663. *[II.15]*

J.W. Hollingsworth, see P.C. Hammer & J.W. Hollingsworth.

G.'t Hooft (1974): *Magnetic monopoles in unified gauge theories.* Nucl. Phys., Vol.B79, p.276-284. *[I.6]*

E. Hopf (1942): *Abzweigung einer periodischen Lösung von einer stationären Lösung eines Differentialsystems.* Ber. math. physik. Kl. Akad. d. Wiss. Leipzig, Bd.XCIV, p.3-22. *[I.16]*

M.K. Horn (1983): *Fourth and fifth-order scaled Runge-Kutta algorithms for treating dense output.* SIAM J.Numer. Anal., Vol.20, p.558-568. *[II.6]*

P.J. van der Houwen (1977): *Construction of integration formulas for initial value problems.* North-Holland Amsterdam, 269pp. *[II.1]*

P.J. van der Houwen & B.P. Sommeijer (1990): *Parallel iteration of high-order Runge-Kutta methods with step size control.* J. Comput. Appl. Math., Vol.29, p.111-127. *[II.11]*

P.J. van der Houwen, see also H. Brunner & P.J. van der Houwen.

T.E. Hull (1967): A search for optimum methods for the numerical integration of ordinary differential equations. SIAM Rev., Vol.9, p.647-654. *[II.1]*, *[II.3]*

T.E. Hull & R.L. Johnston (1964): *Optimum Runge-Kutta methods.* Math. Comput., Vol.18, p.306-310. *[II.3]*

B.L. Hulme (1972): *One-step piecewise polynomial Galerkin methods for initial value problems.* Math. of Comput., Vol.26, p.415-426. *[II.7]*

W.H. Hundsdorfer & M.N. Spijker (1981): *A note on B-stability of Runge-Kutta methods.* Num. Math., Vol.36, p.319-331. *[II.12]*

A. Hurwitz (1895): *Ueber die Bedingungen, unter welchen eine Gleichung nur Wurzeln mit negativen reellen Theilen besitzt.* Math. Ann., Vol.46, p.273-284; Werke, Vol.2, p.533ff. *[I.13]*

A. Huťa (1956): *Une amélioration de la méthode de Runge-Kutta-Nyström pour la résolution numérique des équations différentielles du premier ordre.* Acta Fac. Rerum Natur. Univ. Comenianae (Bratislava) Math., Vol.1, p.201-224. *[II.5]*

B.J. Hyett, see R.F. Warming & B.J. Hyett.

E.L. Ince (1944): *Ordinary differential equations.* Dover Publications, New York, 558pp. *[I.2]*, *[I.3]*

A. Iserles & S.P. Nørsett (1990): On the theory of parallel Runge-Kutta methods. IMA J. Numer. Anal., Vol.10, p.463-488. *[II.11]*

Z. Jackiewicz & M. Zennaro (1992): *Variable stepsize explicit two-step Runge-Kutta methods.* Math. Comput., Vol.59, p.421-438. *[III.8]*

K.R. Jackson (1991): *A survey of parallel numerical methods for initial value problems for ordinary differential equations.* IEEE Trans. on Magnetics, Vol.27, p.3792-3797. *[II.11]*

K.R. Jackson & S.P. Nørsett (1992): *The potential of parallelism in Runge-Kutta methods. Part 1: RK formulas in standard form.* Report. *[II.11]*

K.R. Jackson, see also W.H. Enright, K.R. Jackson, S.P. Nørsett & P.G. Thomson.

C.G.J. Jacobi (1841): *De determantibus functionalibus.* Crelle J. f. d. r. u. angew. Math, Vol.22, p.319-359, Werke, Vol.III, p.393-438. *[I.14]*

C.G.J. Jacobi (1842/43): *Vorlesungen über Dynamik,* gehalten an der Universität zu Königsberg im Wintersemester 1842–1843 und nach einem von C.W. Borchardt ausgearbeiteten Hefte, edited 1866 by A. Clebsch, Werke, Vol. VIII. *[I.6],[I.14]*

C.G.J. Jacobi (1845): *Theoria novi multiplicatoris systemati aequationum differentialum vulgarium applicandi.* Crelle J. f. d. r. u. angew. Math, Vol.29, p.213-279, 333-376. Werke, Vol.IV, p.395-509. *[I.11]*

R. Jeltsch, see G. Dahlquist & R. Jeltsch.

R.L. Johnston, see T.E. Hull & R.L. Johnston.

W.P. Jones, see R.A. Frazer, W.P. Jones & S.W. Skan.

C. Jordan (1870): *Traité des Substitutions et des équations algébriques.* Paris 667pp. *[I.12]*

C. Jordan (1928): *Sur une formule d'interpolation.* Atti Congresso Bologna, vol 6, p.157-177 *[II.9]*

B. Kågström & A. Ruhe (1980): *An algorithm for numerical computation of the Jordan normal form of a complex matrix.* ACM Trans. Math. Software, Vol.6, p.398-419. (Received May 1975, revised Aug. 1977, accepted May 1979). *[I.12]*

D.K. Kahaner, see R. Piessens, E. de Doncker-Kapenga, C.W. Überhuber & D.K. Kahaner.

S. Kakutani & L. Marcus (1958): *On the non-linear difference-differential equation* $y'(t) = [A - By(t - \tau)]y(t)$. In: Contributions to the theory of nonlinear oscillations, ed. by S.Lefschetz, Princeton, Vol.IV, p.1-18. *[II.17]*

E. Kamke (1930): *Ueber die eindeutige Bestimmtheit der Integrale von Differentialgleichungen II.* Sitz. Ber. Heidelberg Akad. Wiss. Math. Naturw. Kl., 17. Abhandl., see also Math. Zeitschr., Vol.32, p.101-107. *[I.10]*

E. Kamke (1942): *Differentialgleichungen, Lösungsmethoden und Lösungen.* Becker & Erler, Leipzig, 642pp. *[I.3]*

K.H. Kastlunger & G. Wanner (1972): *Runge Kutta processes with multiple nodes.* Computing, Vol.9, p.9-24. *[II.13]*

K.H. Kastlunger & G. Wanner (1972b): *On Turan type implicit Runge-Kutta methods.* Computing, Vol.9, p.317-325. *[II.13]*

A.G.Mc. Kendrick, see W.O. Kermack & A.G.Mc. Kendrick.

W.O. Kermack & A.G.Mc. Kendrick (1927): *Contributions to the mathematical theory of epidemics (Part I).* Proc. Roy. Soc., A, Vol.115, p.700-721. *[II.17]*

H. Knapp & G. Wanner (1969): *LIESE II, A program for ordinary differential equations using Lie-series.* MRC Report No.1008, Math. Research Center, Univ. Wisconsin, Madison, Wisc. 53706. *[I.8]*

H. König, see C. Runge & H. König.

F.T. Krogh (1969): *A variable step variable order multistep method for the numerical solution of ordinary differential equations.* Information Processing 68, North-Holland, Amsterdam, p.194-199. *[III.5]*

F.T. Krogh (1973): *Algorithms for changing the step size.* SIAM J. Num. Anal. 10, p.949-965. *[III.5]*

F.T. Krogh (1974): *Changing step size in the integration of differential equations using modified devided differences.* Proceedings of the Conference on the Num. Sol. of ODE, Lecture Notes in Math. No.362, Springer Verlag New York, p.22-71. *[III.5]*

N. Kryloff & N. Bogoliuboff (1947): *Introduction to non-linear Mechanics.* Free translation by S. Lefschetz, Princeton Univ. Press, 105pp. *[I.16]*

E.E. Kummer (1839): *Note sur l'intégration de l'équation $d^n y / dx^n = x^m y$ par des intégrales définies.* Crelle J. f. d. r. u. angew. Math., Vol.19, p.286-288. *[I.11]*

J. Kuntzmann (1961): *Neuere Entwickelungen der Methode von Runge-Kutta.* ZAMM, Vol.41, p.28-31. *[II.7]*

J. Kuntzmann, see also F. Ceschino & J. Kuntzmann.

W. Kutta (1901): *Beitrag zur näherungsweisen Integration totaler Differentialgleichungen.* Zeitschr. für Math. u. Phys., Vol.46, p.435-453. *[II.1], [II.2], [II.3], [II.5]*

J.L.de Lagrange (1759): *Recherches sur la nature et la propagation du son.* Miscell. Taurinensia t.I, Oeuvres t.1, p.39-148. *[I.6]*

J.L.de Lagrange (1762): *Solution de différents problèmes de Calcul Intégral.* Miscell. Taurinensa, t.III, Oeuvres t.1, p.471-668. *[I.6]*

J.L.de Lagrange (1774): *Sur les Intégrales particulières des Equations différentielles.* Oeuvres, tom.4, p.5-108. *[I.2]*

J.L.de Lagrange (1775): *Recherche sur les Suites Récurrentes.* Nouveaux Mém. de l'Acad. royale des Sciences et Belles-Lettres, Berlin. Oeuvres, Vol.4, p.159. *[I.4], [III.3]*

J.L.de Lagrange (1788): *Mécanique analytique.* Paris, Oeuvres t.11 et 12. *[I.4], [I.6], [I.12]*

J.L.de Lagrange (1792): *Mémoire sur l'expression du terme géneral des séries récurrentes, lorsque l'équation génératrice a des racines égales.* Nouv. Mém. de l'Acad. royale des Sciences de Berlin, Oeuvres t.5, p.627-641. *[III.3]*

J.L.de Lagrange (1797): *Théorie des fonctions analytiques, contenant les principes du calcul différentiel, dégagés de toute considération d'infiniment petits, d'évanouissants, de limites et de fluxions, et réduits à l'analyse algébrique des quantités finies.* Paris, 1797, nouv. ed. 1813, Oeuvres Tome 9. *[II.3]*

E. Laguerre (1883): *Mémoire sur la théorie des équations numériques.* J. Math. pures appl. (3e série), Vol.9, p.99-146 (also in *Oeuvres* I, p.3-47). *[II.9]*

J.D. Lambert (1987): *Developments in stability theory for ordinary differential equations.* In: The state of the art in numerical analysis, ed. by A. Iserles & M.J.D. Powell, Clarendon Press, Oxford, p.409-431. *[I.13]*

M. Laloy, see N. Rouche, P. Habets & M. Laloy.

R.J. Lambert, see G.D. Byrne & R.J. Lambert.

P.S. Laplace (An XIII = 1805): *Supplément au dixìeme livre du Traité de mécanique céleste sur l'action capillaire.* Paris chez Courcier, 65+78pp. *[II.1]*

F.M. Lasagni (1988): *Canonical Runge-Kutta methods.* ZAMP Vol.39, p.952-953. *[II.16]*

F.M. Lasagni (1990): *Integration methods for Hamiltonian differential equations.* Unpublished manuscript. *[II.16]*

508   Bibliography

P.D. Lax & R.D. Richtmyer (1956): *Survey of the stability of linear limite difference equations.* Comm. Pure Appl. Math., Vol.9, p.267-293. *[III.4]*

R. Lefever & G. Nicolis (1971): *Chemical Instabilities and sustained oscillations.* J. theor. Biol., Vol.30, p.267-284. *[I.16]*

A.M. Legendre (1787): *Mémoire sur l'intégration de quelques équations aux différences partielles.* Histoire Acad. R. Sciences, Paris, Année MDCCLXXXVII, à Paris MDC-CLXXXIX, p.309-351. *[I.6]*

G.W. Leibniz (1684): *Nova methodus pro maximis et minimis, itemque tangentibus, quae nec fractas, nec irrationales quantitates moratur, & singulare pro illis calculi genus.* Acta Eruditorum, Lipsiae, MDCLXXXIV, p.467-473. *[I.2]*

G.W. Leibniz (1691): *Methodus, qua innummerarum linearum construction ex data proprietate tangentium seu aequatio inter abscissam et ordinatam ex dato valore subtangentialis, exhibetur.* Letter to Huygens, in: C.I. Gerhardt, Leibnizens math. Schriften, 1850, Band II, p.116-121. *[I.3]*

G.W. Leibniz (1693) (Gothofredi Guilielmi Leibnitzii): *Supplementum Geometriae Dimensoriae seugeneralissima omnium tetra gonismorum effectio per motum: Similiterque multiplex constructio linea ex data tangentium conditione.* Acta Eruditorum, Lipsiae, p.385-392; german translation: G. Kowalewski, Leibniz über die Analysis des Unendlichen, Ostwalds Klassiker Nr.162 (1908), p.24-34. *[I.2]*

A.M. Liapunov (1892): *Problème général de la stabilité du mouvement.* Russ., trad. en français 1907 (Annales de la Faculté des Sciences de Toulouse), reprinted 1947 Princeton Univ. Press, 474pp. *[I.13]*

A.M. Liénard (1928): *Etude des oscillations entretenues.* Revue générale de l'Electricité, tome XXIII, p. 901-912 et 946-954. *[I.16]*

B. Lindberg (1972): *A simple interpolation algorithm for improvement of the numerical solution of a differential equation.* SIAM J. Numer. Anal., Vol.9, p.662-668. *[II.9]*

E. Lindelöf (1894): *Sur l'application des méthodes d'approximation successives à l'étude des intégrales réelles des équations différentielles ordinaires.* J. de Math., 4e série, Vol.10, p.117-128. *[I.8]*

W. Liniger, see W.L. Miranker & W. Liniger.

J. Liouville (1836): *Sur le développement des fonctions ou parties de fonctions en séries dont les divers termes sont assujétis à satisfaire à une même équation différentielle du second ordre, contenant un paramètre variable.* Journ. de Math. pures et appl., Vol.1, p.253-265. *[I.8], [I.15]*

J. Liouville (1838): *Sur la Théorie de la variation des constantes arbitraires.* Liouville J. de Math., Vol.3, p.342-349. *[I.8], [I.11]*

J. Liouville (1841): *Remarques nouvelles sur l'équation de Riccati.* J. des Math. pures et appl., Vol.6, p.1-13. *[I.3]*

R. Lipschitz (1876): *Sur la possibilité d'intégrer complètement un système donné d'équations différentielles.* Bulletin des Sciences Math. et Astr., Paris, Vol.10, p.149-159. *[I.7]*

F.J. Lisbona, see M. Crouzeix & F.J. Lisbona.

R. Lobatto (1852): *Lessen over Differentiaal- en Integraal-Rekening.* 2 Vol., La Haye 1851-52. *[II.7]*

E.N. Lorenz (1979): *On the prevalence of aperiodicity in simple systems.* Global Analysis, Calgary 1978, ed. by M.Grmela and J.E.Marsden, Lecture Notes in Mathematics, Vol.755, p.53-75. *[I.16]*

F.R. Loscalzo (1969): *An introduction to the application of spline functions to initial value problems.* In: Theory and Applications of spline functions, ed. T.N.E. Greville, Acad. Press 1969, p.37-64. *[II.13]*

F.R. Loscalzo & I.J. Schoenberg (1967): *On the use of spline functions for the approximation of solutions of ordinary differential equations.* Tech. Summ. Rep. # 723, Math. Res. Center, Univ. Wisconsin, Madison. *[II.13]*

M. Lotkin (1951): *On the accuracy of Runge-Kutta methods.* MTAC Vol.5, p.128-132. *[II.3]*

Ch. Lubich (1989): *Linearly implicit extrapolation methods for differential-algebraic systems.* Numer. Math., Vol.55, p.197-211. *[II.9]*

Ch. Lubich, see also E. Hairer & Ch. Lubich.

G.I. Marchuk (1975): *Prostejshaya matematicheskaya model virusnogo zabolevaniya.* Novosibirsk, VTS SO AN SSSR. Preprint. *[II.17]*

G.I. Marchuk (1983): *Mathematical models in immunology.* Translation series, Optimization Software, New York, Springer Verlag, 351pp. *[II.17]*

L. Marcus, see S. Kakutani & L. Marcus.

M. Marden (1966): *Geometry of polynomials.* American Mathematical Society, Providence, Rhode Island, 2nd edition. *[III.3]*

R.M. May (1976): *Simple mathematical models with very complicated dynamics.* Nature, Vol.261, p.459-467 *[I.16]*

R.M.M. Mattheij, see U.M. Ascher, R.M.M. Mattheij & R.D. Russel.

R.H. Merson (1957): *An operational method for the study of integration processes.* Proc. Symp. Data Processing, Weapons Research Establishment, Salisbury, Australia, p.110-1 to 110-25. *[II.2]*, *[II.4]*, *[II.14]*

S. Miesbach & H.J. Pesch (1992): *Symplectic phase flow approximation for the numerical integration of canonical systems.* Numer.Math., Vol.61, p.501-521. *[II.16]*

J.J.H. Miller, see D.M. Creedon & J.J.H. Miller.

W.E. Milne (1926): *Numerical integration of ordinary differential equations.* Amer. Math. Monthly, Vol.33, p.455-460. *[III.1]*

W.E. Milne (1970): *Numerical solution of differential equations.* Dover Publications, Inc., New York, second edition. *[III.1]*

W.L. Miranker (1971): *A survey of parallelism in numerical analysis.* SIAM Review, Vol.13, p.524-547. *[II.11]*

W.L. Miranker & W. Liniger (1967): *Parallel methods for the numerical integration of ordinary differential equations.* Math. Comput., Vol.21, p. 303-320. *[II.11]*

R. von Mises (1930): *Zur numerischen Integration von Differentialgleichungen.* ZAMM, Vol.10, p.81-92. *[III.4]*

A.R. Mitchell & J.W. Craggs (1953): *Stability of difference relations in the solution of ordinary differential equations*. Math. Tables Aids Comput., Vol.7, p.127-129. *[III.1], [III.3]*

C. Moler & C. Van Loan (1978): *Nineteen dubious ways to compute the exponential of a matrix;* SIAM Review, Vol.20, p.801-836. *[I.12]*

J.I. Montijano, see M. Calvo, J.I. Montijano & L. Rández.

R.E. Moore (1966): *Interval Analysis*. Prentice-Hall, Inc, 145pp. *[I.8]*

R.E. Moore (1979): *Methods and applications of interval analysis*. SIAM studies in Appl. Math., 190pp. *[I.8]*

P. Moss, see K. Burrage & P. Moss.

F.R. Moulton (1926): *New methods in exterior ballistics*. Univ. Chicago Press. *[III.1]*

M. Müller (1926): *Ueber das Fundamentaltheorem in der Theorie der gewöhnlichen Differentialgleichungen*. Math. Zeitschr., Vol.26, p.619-645. (Kap.III). *[I.10]*

F.D. Murnaghan, see A. Wintner & F.D. Murnaghan.

O. Nevanlinna (1989): *Remarks on Picard-Lindelöf iteration*. BIT, Vol.29, p.328-346 and 535-562. *[I.8]*

E.H. Neville (1934): *Iterative interpolation*. Ind. Math. Soc. J. Vol.20, p.87-120. *[II.9]*

I. Newton (1671): *Methodus Fluxionum et Serierum Infinitarum*. edita Londini 1736, Opuscula mathematica, Vol.I, Traduit en français par M.de Buffon, Paris MDCCXL. *[I.2]*

I. Newton (1687): *Philosophiae naturalis principia mathematica*. Imprimatur S. Pepys, Reg. Soc. Praeses, julii 5, 1686, Londini anno MDCLXXXVII. *[I.6], [II.14]*

I. Newton (1711): *Methodus differentialis (Analysis per quantitatum, series, fluxiones, ac differentias: cum enumeratione linearum tertii ordinis)*. London 1711. *[III.1]*

G. Nicolis, see R. Lefever & G. Nicolis.

J. Nievergelt (1964): *Parallel methods for integrating ordinary differential equations*. Comm. ACM, Vol.7, p.731-733. *[II.11]*

S.P. Nørsett (1974a): *One-step methods of Hermite type for numerical integration of stiff systems*. BIT, Vol.14, p.63-77. *[II.13]*

S.P. Nørsett (1974b): *Semi explicit Runge-Kutta methods*. Report No.6/74, ISBN 82-7151-009-6, Dept. Math. Univ. Trondheim, Norway, 68+7pp. *[II.7]*

S.P. Nørsett & G. Wanner (1981): *Perturbed collocation and Runge-Kutta methods*. Numer. Math., Vol.38, p.193-208. *[II.7]*

S.P. Nørsett, see also A. Iserles & S.P. Nørsett, K.R. Jackson & S.P. Nørsett, W.H. Enright, K.R. Jackson, S.P. Nørsett & P.G. Thomson.

A. Nordsieck (1962): *On numerical integration of ordinary differential equations*. Math. Comp., Vol.16, p.22-49. *[III.6]*

U. Nowak, see P. Deuflhard & U. Nowak.

R.M. Noyes, see J.R. Field & R.M. Noyes.

B. Numerov (B.V.Noumerov) (1924): *A method of extrapolation of perturbations*. Monthly notices of the Royal Astronomical Society, Vol.84, p.592-601. *[III.10]*

B. Numerov (1927): *Note on the numerical integration of $d^2x/dt^2 = f(x,t)$.* Astron. Nachrichten, Vol.230, p.359-364. *[III.10]*

E.J. Nyström (1925): *Ueber die numerische Integration von Differentialgleichungen.* Acta Soc. Sci. Fenn., Vol.50, No.13, p.1-54. *[II.2], [II.14], [III.1]*

H.J. Oberle & H.J. Pesch (1981): *Numerical treatment of delay differential equations by Hermite interpolation.* Numer. Math., Vol.37, p.235-255. *[II.17]*

N. Obreschkoff (1940): *Neue Quadraturformeln.* Abh. der Preuss. Akad. der Wiss., Math.-naturwiss. Klasse, Nr.4, Berlin. *[II.13]*

M. Okamoto & K. Hayashi (1984): *Frequency conversion mechanism in enzymatic feedback systems.* J. Theor. Biol., Vol.108, p.529-537. *[II.17]*

D. Okunbor & R.D. Skeel (1992): *An explicit Runge-Kutta-Nyström method is canonical if and only if its adjoint is explicit.* SIAM J. Numer. Anal., Vol.29, p. 521-527. *[II.16]*

D. Okunbor & R.D. Skeel (1992b): *Explicit canonical methods for Hamiltonian systems.* Math. Comput., Vol.59, p.439-455. *[II.16]*

D. Okunbor & R.D. Skeel (1992c): *Canonical Runge-Kutta-Nyström methods of orders 5 and 6,* Working Document 92-1, Dep. Computer Science, Univ. Illinois. *[II.16]*

J. Oliver (1975): *A curiosity of low-order explicit Runge-Kutta methods.* Math. Comp., Vol.29, p.1032-1036. *[II.1]*

J. Oppelstrup (1976): *The RKFHB4 method for delay-differential equations.* Lect. Notes Math., Nr. 631, p.133-146. *[II.17]*

M.R. Osborne (1966): *On Nordsieck's method for the numerical solution of ordinary differential equations.* BIT, Vol.6, p.51-57. *[III.6]*

O. Østerby, see C.W. Gear & O. Østerby.

A. Ostermann, see E. Hairer & A. Ostermann.

B. Owren & M. Zennaro (1991): *Order barriers for continuous explicit Runge-Kutta methods.* Math. Comput., Vol.56, p.645-661. *[II.6]*

B. Owren & M. Zennaro (1992): *Derivation of efficient, continuous, explicit Runge-Kutta methods.* SIAM J. Sci. Stat. Comput., Vol.13, p.1488-1501. *[II.6]*

B.N. Parlett (1976): *A recurrence among the elements of functions of triangular matrices.* Linear Algebra Appl., Vol.14, p.117-121. *[I.12]*

G. Peano (1888): *Intégration par séries des équations différentielles linéaires.* Math. Annalen, Vol.32, p.450-456. *[I.8], [I.9]*

G. Peano (1890): *Démonstration de l'intégràbilité des équations différentielles ordinaires, Math.* Annalen, Vol.37, p.182-228; see also the german translation and commentation: G. Mie, Math. Annalen, Vol.43 (1893), p.553-568. *[I.1], [I.7], [I.9], [I.10]*

G. Peano (1913): *Resto nelle formule di quadratura, espresso con un integrale definito.* Atti Della Reale Accad. Dei Lincei, Rendiconti, Vol.22, N.9, p.562-569, Roma. *[III.2]*

R. Pearl & L.J. Reed (1922): *A further note on the mathematical theory of population growth.* Proceedings of the National Acad. of Sciences, Vol.8, No.12, p.365-368. *[II.17]*

L.M. Perko (1984): *Limit cycles of quadratic systems in the plane.* Rocky Mountain J. of Math., Vol.14, p.619-645. *[I.16]*

O. Perron (1915): *Ein neuer Existenzbeweis für die Integrale der Differentialgleichung* $y' = f(x, y)$. Math.Annalen, Vol.76, p.471-484. *[I.10]*

O. Perron (1918, zur Zeit im Felde): *Ein neuer Existenzbeweis für die Integrale eines Systems gewöhnlicher Differentialgleichungen.* Math. Annalen, Vol.78, p.378-384. *[I.7]*

O. Perron (1930): *Ueber ein vermeintliches Stabilitätskriterium.* Nachrichten Göttingen, (1930) p.28-29 (see also Fort.d.Math. 1930 I, p.380.) *[I.13]*

H.J. Pesch, see H.J. Oberle & H.J. Pesch, S. Miesbach & H.J. Pesch.

D. Pfenniger (1990): *Stability of the Lagrangian points in stellar bars.* Astron. Astrophys., Vol.230, p.55-66. *[II.16]*

D. Pfenniger, see also T. de Zeeuw & D. Pfenniger.

E. Picard (1890): *Mémoire sur la théorie des équations aux dérivées partielles et la méthode des approximations successives.* J. de Math. pures et appl., 4e série, Vol.6, p.145-210. *[I.8]*

E. Picard (1891-96): *Traité d'Analyse.* 3 vols. Paris. *[I.7]*, *[I.8]*

R. Piessens, E. de Doncker-Kapenga, C.W. Überhuber & D.K. Kahaner (1983): *QUAD-PACK. A subroutine package for automatic integration.* Springer Series in Comput. Math., Vol.1, 301pp. *[II.10]*

P. Piotrowsky (1969): *Stability, consistency and convergence of variable k -step methods for numerical integration of large systems of ordinary differential equations.* Lecture Notes in Math., 109, Dundee 1969, p.221-227. *[III.5]*

H. Poincaré (1881,82,85): *Mémoire sur les courbes définies par les équations différentielles.* J. de Math., 3e série, t.7, p.375-422, 3e série, t.8, p.251-296, 4e série, t.1, p.167-244, Oeuvres t.1, p.3-84, 90-161. *[I.12]*, *[I.16]*

H. Poincaré (1892,1893,1899): *Les méthodes nouvelles de la mécanique céleste.* Tome I 385pp., Tome II 480pp., Tome III 414pp., Gauthier-Villars Paris. *[I.6]*, *[I.16]*, *[I.14]*, *[II.8]*

S.D. Poisson (1835): *Théorie mathématique de la chaleur.* Paris, Bachelier, 532pp., Supplément 1837, 72pp. *[I.15]*

B. Van der Pol (1926): *On "Relaxation Oscillations".* Phil. Mag., Vol.2, p.978-992; reproduced in: B. van der Pol, Selected Scientific Papers, Vol.I, North. Holland Publ. Comp. Amsterdam (1960). *[I.16]*

G. Pólya & G. Szegö (1925): *Aufgaben und Lehrsätze aus der Analysis.* Two volumes, Springer Verlag; many later editions and translations. *[II.9]*

P. Pouzet (1963): *Etude en vue de leur traitement numérique des équations intégrales de type Volterra.* Rev. Français Traitement Information (Chiffres), Vol.6, p.79-112. *[II.17]*

P.J. Prince & J.R. Dormand (1981): *High order embedded Runge-Kutta formulae.* J. Comp. Appl. Math., Vol.7, p.67-75. *[II.5]*

P.J. Prince, see also J.R. Dormand, M.E.A. El-Mikkawy & P.J. Prince, J.R. Dormand & P.J. Prince, R.W. Brankin, I. Gladwell, J.R. Dormand, P.J. Prince & W.L. Seward.

H. Prüfer (1926): *Neue Herleitung der Sturm-Liouvillschen Reihenentwicklung stetiger Funktionen.* Math. Annalen, Vol.95, p.499-518. *[I.15]*

D.I. Pullin & P.G. Saffman (1991): *Long-time symplectic integration: the example of four-vortex motion.* Proc. R. Soc. London, A, Vol.432, p.481-494. *[II.16]*

Qin Meng-Zhao & Zhu Wen-Jie (1991): *Canonical Runge-Kutta-Nyström (RKN) methods for second order ordinary differential equations.* Computers Math. Applic., Vol.22, p.85-95. *[II.16]*

Qin Meng-Zhao & Zhu Wen-Jie (1992): *Construction of higher order symplectic schemes by composition.* Computing, Vol.47, p.309-321. *[II.16]*

Qin Meng-zhao, see also Feng Kang, Wu Hua-mo, Qin Meng-zhao & Wang Dao-liu.

R. Radau (1880): *Étude sur les formes d'approximation qui servent à calculer la valeur numérique d'une intégrale définie.* Liouville J. de Mathém. pures et appl., 3eser., tome VI, p.283-336. (Voir p.307). *[II.7]*

A. Ralston (1962): *Runge-Kutta methods with minimum error bounds.* Math. Comput., Vol.16, p.431-437, corr., Vol.17, p.488. *[II.1], [II.3], [III.7]*

L. Rández, see M. Calvo, J.I. Montijano & L. Rández.

Lord Rayleigh (1883): *On maintained vibrations.* Phil. Mag. Ser.5, Vol.15, p.229-235. *[I.16]*

L.J. Reed, see R. Pearl & L.J. Reed.

W.T. Reid (1980): *Sturmian theory for ordinary differential equations.* Springer Verlag, Appl. Math., Serie31, 559pp. *[I.15]*

C. Reinsch, see J.H. Wilkinson & C. Reinsch.

R. Reissig (1954): *Erzwungene Schwingungen mit zäher und trockener Reibung.* Math. Nach-richten, Vol.11, p.345-384; see also p.231. *[II.6]*

P. Rentrop (1985): *Partitioned Runge-Kutta methods with stiffness detection and stepsize control.* Numer. Math., Vol.47, p.545-564. *II.15*

J. Riccati (1712): *Soluzione generale del Problema inverso intorno à raggi osculatori,.., determinar la curva, a cui convenga una tal'espressione.* Giornale de'Letterati d'Italia, Vol.11, p.204-220. *[I.3]*

J. Riccati (1723): *Animadversiones in aequationes differentiales secundi gradus.* Acta Erud. Lips., anno MDCCXXIII, p.502-510. *[I.3]*

L.F. Richardson (1910): *The approximate arithmetical solution by finite differences of phys-ical problems including differential equations, with an application to the stresses in a masonry dam.* Phil. Trans., A, Vol.210, p.307-357. *[II.4]*

L.F. Richardson (1927): *The deferred approach to the limit.* Phil. Trans., A, Vol.226, p.299-349. *[II.4], [II.9]*

R.D. Richtmyer, see P.D. Lax & R.D. Richtmyer.

B. Riemann (1854): *Ueber die Darstellbarkeit einer Function durch eine trigonometrische Reihe.* Von dem Verfasser behufs seiner Habilitation an der Universität zu Göttingen der philosophischen Facultät eingereicht; collected works p. 227-265. *[I.6]*

W. Romberg (1955): *Vereinfachte numerische Integration.* Norske Vid. Selsk. Forhdl, Vol.28, p.30-36. *[II.8], [II.9]*

E. Rothe (1930): *Zweidimensionale parabolische Randwertaufgaben als Grenzfall eindi-mensionaler Randwertaufgaben.* Math. Annalen, Vol.102, p. 650-670. *[I.1]*

514   Bibliography

N. Rouche, P. Habets & M. Laloy (1977): *Stability theory by Liapunov's direct method.* Appl. Math. Sci. 22, Springer Verlag, 396pp. *[I.13]*

E.J. Routh (1877): *A Treatise on the stability of a given state of motions.* Being the essay to which the Adams prize was adjudged in 1877, in the University of Cambridge. London 108pp. *[I.13]*

E.J. Routh (1884): *A Treatise on the dynamics of a system of rigid bodies, part I and II.* 4th edition (1st ed. 1860, 6th ed. 1897, german translation with remarks of F.Klein 1898). *[I.12]*

D. Ruelle & F. Takens (1971): *On the nature of turbulence.* Commun. Math. Physics, Vol.20, p.167-192. *[I.16]*

A. Ruhe, see B. Kågström & A. Ruhe.

C. Runge (1895): *Ueber die numerische Auflösung von Differentialgleichungen.* Math. Ann., Vol.46, p.167-178. *[II.1], [II.4]*

C. Runge (1905): *Ueber die numerische Auflösung totaler Differentialgleichungen.* Göttinger Nachr., p.252-257. *[II.1], [II.3]*

C. Runge & H. König (1924): *Vorlesungen über numerisches Rechnen.* Grundlehren XI, Springer Verlag, 372pp. *[I.8], [II.1]*

R.D. Russel, see U.M. Ascher, R.M.M. Mattheij & R.D. Russel.

R.D. Ruth (1983): *A canonical integration technique.* IEEE Trans. Nuclear Scince, Vol.NS-30, p.2669-2671. *[II.16]*

H. Rutishauser (1952): *Ueber die Instabilität von Methoden zur Integration gewöhnlicher Differentialgleichungen.* ZAMP, Vol.3, p.65-74. *[III.3]*

H. Rutishauser, see also F.L. Bauer, H. Rutishauser & E. Stiefel.

P.G. Saffman, see D.I. Pullin & P.G. Saffman.

J.M. Sanz-Serna (1988): *Runge-Kutta schemes for Hamiltonian systems.* BIT Vol.28, p.877-883. *[II.16]*

J.M. Sanz-Serna (1992): *Symplectic integrators for Hamiltonian problems: an overview.* Acta Numerica, Vol.1, p.243-286. *[II.16]*

J.M. Sanz-Serna (1992b): *The numerical integration of Hamiltonian systems.* In: Computational ordinary differential equations, ed. by J.R. Cash & I. Gladwell, Clarendon Press, Oxford, p.437-449. *[II.16]*

J.M. Sanz-Serna & L. Abia (1991): *Order conditions for canonical Runge-Kutta schemes.* SIAM J. Numer. Anal., Vol.28, p. 1081-1096. *[II.16]*

J.M. Sanz-Serna, see also M.P. Calvo & J.M. Sanz-Serna, L. Abia & J.M. Sanz-Serna.

D. Sarafyan (1966): *Error estimation for Runge-Kutta methods through pseudo-iterative formulas.* Techn. Rep. No 14, Lousiana State Univ., New Orleans, May 1966. *[II.4]*

L. Scheeffer (1884): *Zur Theorie der stetigen Funktionen einer reellen Veränderlichen.* Acta Mathematica, Vol.5, p.183-194. *[I.10]*

J. Schlöder, see H.G. Bock & J. Schlöder.

I.J. Schoenberg, see F.R. Loscalzo & I.J. Schoenberg.

I. Schur (1909): *Ueber die charakteristischen Wurzeln einer linearen Substitution mit einer Anwendung auf die Theorie der Integralgleichungen.* Math. Ann., Vol.66, p.488-510. *[I.12]*

C. Scovel, see P.J. Channell & C. Scovel.

W.L. Seward, see R.W. Brankin, I. Gladwell, J.R. Dormand, P.J. Prince & W.L. Seward.

L.F. Shampine (1979): *Storage reduction for Runge-Kutta codes.* ACM Trans. Math. Software, Vol.5, p.245-250. *[II.5]*

L.F. Shampine (1985): *Interpolation for Runge-Kutta methods.* SIAM J. Numer. Anal., Vol.22, p.1014-1027. *[II.6]*

L.F. Shampine (1986): *Some practical Runge-Kutta formulas.* Math. Comp., Vol.46, p.135-150. *[II.5], [II.6]*

L.F. Shampine & L.S. Baca (1983): *Smoothing the extrapolated midpoint rule.* Numer. Math., Vol.41, p.165-175. *[II.9]*

L.F. Shampine & L.S. Baca (1986): *Fixed versus variable order Runge-Kutta.* ACM Trans. Math. Softw., Vol.12, p.1-23. *[II.9]*

L.F. Shampine, L.S. Baca & H.-J. Bauer (1983): *Output in extrapolation codes.* Comp. & Maths. with Appls., Vol.9, p.245-255. *[II.9]*

L.F. Shampine & M.K. Gordon (1975): *Computer Solution of Ordinary Differential Equations, The Initial Value Problem.* Freeman and Company, San Francisco, 318pp. *[III.7]*

L.F. Shampine & H.A. Watts (1979): *The art of writing a Runge-Kutta code. II.* Appl. Math. Comput., Vol.5, p.93-121. *[II.4], [III.7]*

L.F. Shampine, H.A. Watts & S.M. Davenport (1976): *Solving nonstiff ordinary differential equations - The state of the art.* SIAM Rev., Vol.18, p.376-410. *[II.6]*

L.F. Shampine, see also I. Gladwell, L.F. Shampine & R.W. Brankin, P. Bogacki & L.F. Shampine.

E.B. Shanks (1966): *Solutions of differential equations by evaluations of functions.* Math. of Comp., Vol.20, p.21-38. *[II.5]*

Shi Songling (1980): *A concrete example of the existence of four limit cycles for plane quadratic systems.* Sci. Sinica, Vol.23, p.153-158. *[I.16]*

G.F. Simmons (1972): *Differential equations with applications and historical notes.* MC Graw-Hill, 465pp. *[I.16]*

H.H. Simonsen (1990): *Extrapolation methods for ODE's: continuous approximations, a parallel approach.* Dr.Ing. Thesis, Norwegian Inst. Tech., Div. of Math. Sciences. *[II.9]*

S.W. Skan, see R.A. Frazer, W.P. Jones & S.W. Skan.

R. Skeel (1976): *Analysis of fixed-stepsize methods.* SIAM J. Numer. Anal., Vol.13, p.664-685. *[III.4], [III.8], [III.9]*

R.D. Skeel (1979): *Equivalent forms of multistep formulas.* Math. Comput., Vol.33, p.1229-1250. *[III.6]*

R.D. Skeel, see also D. Okunbor & R.D. Skeel.

B.P. Sommeijer, see P.J. van der Houwen & B.P. Sommeijer.

516   Bibliography

D. Sommer (1965): *Numerische Anwendung impliziter Runge-Kutta-Formeln*. ZAMM, Vol. 45, Sonderheft, p. T77-T79. *[II.7]*

F. Sommer, see H. Behnke & F. Sommer.

A. Sommerfeld (1942): *Vorlesungen über theoretische Physik*. Bd.1., Mechanik; translated from the 4th german ed.: Acad. Press. *[II.10]*, *[II.14]*

D.C. Sorensen, see R. Fletcher & D.C. Sorensen.

J.L. Soulé, see A. Guillou & J.L. Soulé.

M.N. Spijker (1971): *On the structure of error estimates for finite difference methods*. Numer. Math., Vol.18, pp.73-100. *[III.8]*

M.N. Spijker, see also W.H. Hundsdorfer & M.N. Spijker.

D.D. Stancu, see A.H. Stroud & D.D. Stancu.

J.F. Steffensen (1956): *On the restricted problem of three bodies*. K. danske Vidensk. Selsk., Mat-fys. Medd. 30 Nr.18. *[I.8]*

I.A. Stegun, see M. Abramowitz & I.A. Stegun.

H.J. Stetter (1970): *Symmetric two-step algorithms for ordinary differential equations*. Computing, Vol.5, p.267-280. *[II.9]*

H.J. Stetter (1971): *Local estimation of the global discretization error*. SIAM J. Numer. Anal., Vol.8, p.512-523. *[II.12]*

H.J. Stetter (1973): *Analysis of discretization methods for ordinary differential equations*. Springer Verlag, Berlin-Heidelberg-New York. *[II.8]*, *[II.12]*, *[III.2]*, *[III.8]*, *[III.9]*

H.J. Stetter, see also W.B. Gragg & H.J. Stetter.

D. Stewart (1990): *A high accuracy method for solving ODEs with discontinuous right-hand side*. Numer. Math., Vol.58, p.299-328. *[II.6]*

E. Stiefel, see F.L. Bauer, H. Rutishauser & E. Stiefel.

J. Stoer, see R. Bulirsch & J. Stoer.

C. Störmer (1907): *Sur les trajectoires des corpuscules électrisés*. Arch. sci. phys. nat., Genève, Vol.24, p.5-18, 113-158, 221-247. *[III.10]*

C. Störmer (1921): *Méthodes d'intégration numérique des équations différentielles ordinaires*. C.R. congr. intern. math., Strasbourg, p.243-257. *[II.14]*, *[III.10]*

A.H. Stroud & D.D. Stancu (1965): *Quadrature formulas with multiple Gaussian nodes*. SIAM J. Numer. Anal., ser.B., Vol.2, p.129-143. *[II.13]*

Ch. Sturm (1829): *Bulletin des Sciences de Férussac*. Tome XI, p.419, see also: Algèbre de Choquet et Mayer (1832). *[I.13]*

Ch. Sturm (1836): *Sur les équations différentielles linéaires du second ordre*. Journal de Math. pures et appl. (Liouville), Vol.1, p.106-186 (see also p.253, p.269, p.373 of this volume). *[I.15]*

Sun Geng (孙 耿) (1992): *Construction of high order symplectic Runge-Kutta Methods*. Comput. Math., Vol.11 (1993), p.250-260. *[II.16]*

Y.B. Suris (1989): *The canonicity of mappings generated by Runge-Kutta type methods when integrating the systems* $\ddot{x} = -\partial U/\partial x$. Zh. Vychisl. Mat. i Mat. Fiz., vol 29, p.202-211 (in Russian); same as U.S.S.R. Comput. Maths. Phys., vol 29., p.138-144. *[II.16]*

Y.B. Suris (1990): *Hamiltonian Runge-Kutta type methods and their variational formulation.* Mathematical Simulation, Vol.2, p.78-87 (Russian). *[II.16]*

V. Szebehely (1967): *Theory of orbits. The restricted problem of three bodies.* Acad. Press, New York, 668pp. *[II.0]*

G. Szegö, see G. Pólya & G. Szegö.

P.G.Tait, see W. Thomson (Lord Kelvin) & P.G.Tait.

F. Takens, see D. Ruelle & F. Takens.

K. Taubert (1976): *Differenzenverfahren für Schwingungen mit trockener und zäher Reibung und für Regelungssysteme.* Numer. Math., Vol.26, p.379-395. *[II.6]*

K. Taubert (1976): *Eine Erweiterung der Theorie von G. Dahlquist.* Computing, Vol.17, p.177-185. *[III.4]*

B. Taylor (1715): *Methodus incrementorum directa et inversa.* Londini 1715. *[I.6]*

W. Thomson (Lord Kelvin) & P.G.Tait (1879): *Treatise on natural philosophy (Vol.I., Part I).* Cambridge; New edition 1890, 508pp. *[I.12]*

P.G. Thomson, see W.H. Enright, K.R. Jackson, S.P. Nørsett & P.G. Thomson.

J. Todd (1950): *Notes on modern numerical analysis, I.* Math. Tables Aids Comput., Vol.4, p.39-44. *[III.3]*

W. Tollmien (1938): *Ueber die Fehlerabschätzung beim Adamsschen Verfahren zur Integration gewöhnlicher Differentialgleichungen.* ZAMM, Vol.18, p.83-90. *[III.4]*

S. Tremaine, see J. Binney & S. Tremaine.

K.W. Tu, see C.W. Gear & K.W. Tu.

C.W. Überhuber, see R. Piessens, E. de Doncker-Kapenga, C.W. Überhuber & D.K. Kahaner.

W. Uhlmann (1957): *Fehlerabschätzungen bei Anfangswertaufgaben gewöhnlicher Differentialgleichungssysteme 1. Ordnung.* ZAMM, Vol.37, p.88-99. *[I.10]*

B.T. Ulrich, see B. Giovannini, L. Weiss-Parmeggiani & B.T. Ulrich.

R. Vaillancourt, see M. Calvé & R. Vaillancourt.

C. Van Loan, see C. Moler & C. Van Loan.

R.S. Varga, see G. Birkhoff & R.S. Varga,

P.F. Verhulst (1845): *Recherches mathématiques sur la loi d'accroissement de la population.* Nuov. Mem. Acad. Roy. Bruxelles, Vol.18, p.3-38. *[II.17]*

J.H. Verner (1971): *On deriving explicit Runge-Kutta methods.* Proc. Conf. on Appl. Numer. Analysis, Lecture Notes in Mathematics 228, Springer Verlag, p.340-347. *[II.5]*

J.H. Verner (1978): *Explicit Runge-Kutta methods with estimates of the local truncation error.* SIAM J.Numer. Anal., Vol.15, p.772-790. *[II.5]*

J.H. Verner, see also G.J. Cooper & J.H. Verner.

L. Vietoris (1953): *Der Richtungsfehler einer durch das Adamssche Interpolationsverfahren gewonnenen Näherungslösung einer Gleichung* $y' = f(x, y)$. Oesterr. Akad. Wiss., Math.-naturw. Kl., Abt. IIa, Vol.162, p.157-167 and p.293-299. *[III.4]*

R. de Vogelaere (1956): *Methods of integration which preserve the contact transformation property of the Hamiltonian equations.* Report No. 4, Dept. Mathem., Unive. of Notre Dame, Notre Dame, Ind. *[II.16]*

V. Volterra (1934): *Remarques sur la Note de M. Régnier et Mlle Lambin.* C.R.Acad. Sc. t. CXCIX, p.1682. See also: V.Volterra - U.d'Ancona , Les associations biologiques au point de vue mathématique, Paris 1935. *[II.17]*

W. Walter (1970): *Differential and integral inequalities.* Springer Verlag 352pp., german edition 1964. *[I.10]*

W. Walter (1971): *There is an elementary proof of Peano's existence theorem.* Amer. Math. Monthly, Vol.78, p.170-173. *[I.7]*

Wang Dao-liu, see Feng Kang, Wu Hua-mo, Qin Meng-zhao & Wang Dao-liu.

G. Wanner (1969): *Integration gewöhnlicher Differentialgleichungen, Lie Reihen, Runge-Kutta-Methoden.* BI Mannheim Htb. 831/831a, 182pp. *[I.8]*

G. Wanner (1973): *Runge-Kutta methods with expansions in even powers of* h. Computing, Vol.11, p.81-85. *[II.8]*

G. Wanner (1983): *On Shi's counter example for the 16th Hilbert problem.* Internal Rep. Sect. de Math., Univ. Genève 1982; in german in: Jahrbuch Ueberblicke Mathematik 1983, ed. Chatterji, Fenyö, Kulisch, Laugwitz, Liedl, BI Mannheim, p.9-24. *[I.13]*, *[I.16]*

G. Wanner, see also K.H. Kastlunger & G. Wanner, S.P. Nørsett & G. Wanner, E. Hairer & G. Wanner, H. Knapp & G. Wanner.

R.F. Warming & B.J. Hyett (1974): *The modified equation approach to the stability and accuracy analysis of finite-difference methods.* J. Comp. Phys., Vol.14, p.159-179. *[II.16]*

D.S. Watanabe, see C.W. Gear & D.S. Watanabe.

H.A. Watts (1983): *Starting stepsize for an ODE solver.* J. Comp. Appl. Math., Vol.9, p.177-191. *[II.4]*

H.A. Watts, see also L.F. Shampine & H.A. Watts, L.F. Shampine, H.A. Watts & S.M. Davenport.

K. Weierstrass (1858): *Ueber ein die homogenen Functionen zweiten Grades betreffendes Theorem, nebst Anwendung desselben auf die Theorie der kleinen Schwingungen.* Monatsber. der Königl. Akad. der Wiss., 4. März 1858, Werke Bd.I, p.233-246. *[I.6]*

L. Weiss-Parmeggiani, see B. Giovannini, L. Weiss-Parmeggiani & B.T. Ulrich.

J. Weissinger (1950): *Eine verschärfte Fehlerabschätzung zum Extrapolationsverfahren von Adams.* ZAMM, Vol.30, p.356-363. *[III.4]*

H. Weyl (1939): *The classical groups.* Princeton, 302pp. *[I.14]*

O. Wilde (1892): *Lady Windermere's Fan, Comedy in four acts. [I.7]*

J.H. Wilkinson (1965): *The algebraic eigenvalue problem, Monographs on numerical analysis.* Oxford, 662pp. *[I.9]*

J.H. Wilkinson & C. Reinsch (1970): *Linear Algebra.* Grundlehren Band 186, Springer Verlag, 439pp. *[I.12]*

J.H. Wilkinson, see also G.H. Golub & J.H. Wilkinson.

A. Wintner & F.D. Murnaghan (1931): *A canonical form for real matrices under orthogonal transformations.* Proc. Nat. Acad. Sci. U.S.A., Vol.17, p.417-420. *[I.12]*

E.M. Wright (1945): *On a sequence defined by a non-linear recurrence formula.* J. of London Math. Soc., Vol.20, p.68-73. *[II.17]*

E.M. Wright (1946): *The non-linear difference-differential equation.* Quart. J. of Math., Vol.17, p.245-252. *[II.17]*

E.M. Wright (1955): *A non-linear difference-differential equation.* J.f.d.r.u. angew. Math., Vol.194, p.66-87. *[II.17]*

K. Wright (1970): *Some relationships between implicit Runge-Kutta collocation and Lanczos τ methods, and their stability properties.* BIT Vol.10, p.217-227. *[II.7]*

H. Wronski (1810): *Premier principe des méthodes algorithmiques comme base de la technique algorithmique.* Publication refused by the Acad. de Paris (for more details see: S.Dickstein, Int. Math. Congress 1904, p.515). *[I.11]*

Wu Hua-mo, see Feng Kang, Wu Hua-mo, Qin Meng-zhao & Wang Dao-liu.

H. Yoshida (1990): *Construction of higher order symplectic integrators.* Phys. Lett. A, Vol.150, p.262-268. *[II.16]*

H. Yoshida (1993): *Recent progress in the theory and application of symplectic integrators.* Celestial Mechanics Dynam. Astr., Vol.56, p.27-43. *[II.16]*

T. de Zeeuw & D. Pfenniger (1988): *Potential-density pairs for galaxies.* Mon. Not. R. astr. Soc., Vol.235, p.949-995. *[II.16]*

M. Zennaro (1986): *Natural continuous extensions of Runge-Kutta methods.* Math. Comput., Vol.46, p.119-133. *[II.17]*

M. Zennaro, see also A. Bellen & M. Zennaro, B. Owren & M. Zennaro, Z. Jackiewicz & M. Zennaro.

N.P Zhidkov, see I.S. Berezin & N.P Zhidkov.

Zhu Wen-Jie, see Qin Meng-Zhao & Zhu Wen-Jie.

J.A. Zonneveld (1963): *Automatic integration of ordinary differential equations.* Report R743, Mathematisch Centrum, Postbus 4079, 1009AB Amsterdam. Appeared in book form 1964. *[II.4]*

R. Zurmühl (1948): *Runge-Kutta-Verfahren zur numerischen Integration von Differentialgleichungen n-ter Ordnung.* ZAMM, Vol.28, p.173-182. *[II.14]*

R. Zurmühl (1952): *Runge-Kutta Verfahren unter Verwendung höherer Ableitungen.* ZAMM, Vol.32, p.153-154. *[II.13]*

Symbol Index

Subject Index